BERRU Predictive Modeling

Dan Gabriel Cacuci

BERRU Predictive Modeling

Best Estimate Results with Reduced Uncertainties

Dan Gabriel Cacuci
University of South Carolina
Columbia, USA

Additional material to this book can be downloaded from http://extras.springer.com.

ISBN 978-3-662-58393-7 ISBN 978-3-662-58395-1 (eBook)
https://doi.org/10.1007/978-3-662-58395-1

Library of Congress Control Number: 2018962394

This Springer imprint is published by the registered company Springer-Verlag GmbH, DE part of Springer Nature
The registered company address is: Heidelberger Platz 3, 14197 Berlin, Germany

Preface

The results of measurements and computations are never perfectly accurate. On the one hand, results of measurements inevitably reflect the influence of experimental errors, imperfect instruments, or imperfectly known calibration standards. Around any reported experimental value, therefore, there always exists a range of values that may also be plausibly representative of the true but unknown value of the measured quantity. On the other hand, computations are afflicted by errors stemming from numerical procedures, uncertain model parameters, boundary and initial conditions, and/or imperfectly known physical processes or problem geometry. Therefore, knowing just the nominal values of experimentally measured or computed quantities is insufficient for applications. The quantitative uncertainties accompanying measurements and computations are also needed, along with the respective nominal values. Extracting "best estimate" values for model parameters and predicted results, together with "best estimate" uncertainties for these parameters and results requires the combination of experimental and computational data, including their accompanying uncertainties (standard deviations and correlations). The goal of predictive modeling is to perform such a combination, which requires reasoning from incomplete, error-afflicted, and occasionally discrepant information, to predict future outcomes based on all recognized errors and uncertainties. In contradistinction to the customary methods used currently for data assimilation, the BERRU predictive modeling methodology presented in this book uses the maximum entropy principle to avoid the need for minimizing an arbitrarily user-chosen "cost functional" (usually a quadratic functional that represents the weighted errors between measured and computed responses), thus generalizing and significantly extending the customary "data adjustment" and/or 4D-VAR data assimilation procedures. The acronym BERRU stands for "Best-Estimate Results with Reduced Uncertainties," because the application of the BERRU predictive modeling methodology reduces the predicted standard deviations of both the best-estimate predicted responses and parameters. The BERRU predictive modeling methodology also provides a quantitative indicator, constructed from response sensitivities and response and parameter covariance matrices, for determining the consistency (agreement or disagreement) among the a priori computational and experimental information available for parameters and responses. Furthermore, the maximum entropy principles ensures that the more information is assimilated, the more the predicted standard deviations of the predicted responses and parameters are reduced, since the introduction of additional knowledge reduces the state of ignorance (as long as the additional information is consistent with the physical underlying system), as would also be expected based on principles of information theory.

In general terms, the modeling of a physical system and/or the result of an indirect experimental measurement requires consideration of the following modeling components:

(a) a mathematical model comprising linear and/or nonlinear equations (algebraic, differential, and integral) that relate the system's independent variables and parameters to the system's state (i.e., dependent) variables;
(b) inequality and/or equality constraints that delimit the ranges of the system's parameters;
(c) one or several computational results, customarily referred to as system responses (or objective functions, or indices of performance), which are computed using the mathematical model; and
(d) experimentally measured responses, with their respective nominal (mean) values and uncertainties (variances, covariances, skewness, kurtosis, etc.).

Predictive modeling comprises three key elements, namely model calibration, quantification of the validation domain, and model extrapolation. Model calibration addresses the integration of experimental data for the purpose of updating the data of the computer/numerical simulation model. Important components underlying model calibration include quantification of uncertainties in the data and the model, quantification of the biases between model predictions and experimental data, and the computation of the sensitivities of the model responses to the model's parameters. For large-scale models, the current model calibration methods are hampered by the significant computational effort required for computing exhaustively and exactly the requisite response sensitivities. Reducing this computational effort is paramount, and methods based on adjoint sensitivity models show great promise in this regard, as will be demonstrated in this book. The quantification of the validation domain underlying the model under investigation requires estimation of contours of constant uncertainty in the high-dimensional space that characterizes the application of interest. In practice, this involves the identification of areas where the predictive estimation of uncertainty meets specified requirements for the performance, reliability, or safety of the system of interest. The conceptual and mathematical development of methods for quantifying the validation domain is in a relatively incipient stage. Model extrapolation aims at quantifying the uncertainties in predictions under new environments or conditions, including both untested regions of the parameter space and higher levels of system complexity in the validation hierarchy. Extrapolation of models and the resulting increase of uncertainty are poorly understood, particularly the estimation of uncertainty that results from nonlinear coupling of two or more physical phenomena that were not coupled in the existing validation database.

The material presented in this book is structured as follows: Chapter 1 presents the predictive modeling methodology developed by Cacuci and Ionescu-Bujor (2010a), which will be called the "BERRU-SMS Predictive Modeling" methodology, since it is conceived for a *single multiphysics system*, for which it combines uncertain computational and experimental data in order to predict best-estimate values for model responses and parameters, along with reduced predicted uncertainties for these best-estimate values. In contradistinction to the customary methods used currently for data assimilation, the BERRU-SMS predictive modeling methodology developed by Cacuci and Ionescu-Bujor (2010a) uses the maximum entropy principle to avoid the need for minimizing an arbitrarily user-chosen "cost functional" (usually a quadratic functional that represents the weighted errors between measured and computed responses). The BERRU-SMS predictive modeling methodology also provides a quantitative indicator, constructed from sensitivity and covariance matrices, for determining the consistency (agreement or disagreement) among the a priori computational and experimental information available for parameters and responses. This consistency indicator measures, in a chi-square-like metric, the deviations between the experimental and nominally computed responses and can be evaluated directly from the originally given data (i.e., given parameters and responses, together with their original uncertainties), once the response sensitivities have become available. Chapter 1 also presents applications of the BERRU-SMS predictive modeling methodology to perform predictive modeling of several illustrative systems involving thermal-hydraulics and coupled neutron kinetics/thermal-hydraulics numerical simulations characterized by many (thousands of) imprecisely known model parameters, yielding significant reductions in the predicted standard deviation of the optimally predicted model responses and parameters.

Chapter 2 presents the application of the BERRU-SMS methodology to perform forward and inverse predictive modeling of a paradigm dissolver model that is representative for material separations and international safeguards activities. This dissolver model comprises eight active compartments in which the time-dependent nonlinear differential equations modeling the physical and chemical processes include sixteen time-dependent spatially dependent state functions and 635 model parameters related to the model's equation of state and inflow conditions. The most important response for the dissolver model is the computed nitric acid in the compartment furthest away from the inlet, where measurements are available for comparisons. The sensitivities to all model parameters of the acid concentrations at each of these instances in time are computed exactly and efficiently using the adjoint sensitivity analysis methodology for nonlinear systems (ASAM). When using the traditional "ASAM for functional-type responses," 307 adjoint computations are needed for computing *exactly* the sensitivities of the time-dependent nitric concentration response in compartment #1, at the measured time instances, to all model parameters. Alternatively, the "ASAM for operator-type responses" developed by Cacuci (1981b) reduces the number of required adjoint computations from 307 to 17, with an overall loss of accuracy of less than 0.05% over the entire time-span of 10.5

h (307 time steps). Similarly significant reductions in the number of computations when using the "ASAM for operator-type responses" were also obtained when computing the sensitivities of the time-dependent responses (nitric acid concentrations) in the other dissolver compartments, e.g., compartments #4 and #7. These sensitivities are subsequently used for uncertainty analysis and predictive modeling, combining the computational results with experiments that were performed solely in the compartment furthest from the inlet. The results of applying the BERRU-SMS predictive modeling methodology yields optimally calibrated values for all 635 model parameters, with reduced predicted uncertainties, as well as optimal ("best-estimate") predicted values for the acid concentrations, also with reduced predicted uncertainties. Notably, even though the experimental data pertains solely to the compartment furthest from the inlet (where the data was measured), the BERRU-SMS methodology actually improves the predictions and reduces the predicted uncertainties not only in the compartment in which the data was actually measured, but throughout the entire dissolver, including the compartment furthest from the measurements (i.e., at the inlet). This is because the BERRU-SMS methodology combines and transmits information simultaneously over the entire phase-space, comprising all time steps and spatial locations. Chapter 2 also presents an application of the BERRU-SMS predictive modeling methodology in the *inverse* mode to determine, within a tight a priori specified convergence criterion and overall accuracy, an unknown time-dependent boundary condition for the dissolver model by using measurements of the state function at a specified location. The maximum entropy principle which underlies the BERRU-SMS predictive modeling methodology enables the construction of an intrinsic regularizing metric for solving any inverse problem. For the dissolver model, the unknown time-dependent boundary condition is predicted by the BERRU-SMS methodology within an a priori selected convergence criterion, without user intervention and/or introduction of arbitrary "regularization parameters," as the currently popular procedures for analyzing inverse problems need to do.

The BERRU-SMS methodology was extended by Cacuci (2014b) from a single multiphysics system to *coupled multiphysics systems* (CMS), which will be called the "BERRU-CMS" methodology. The BERRU-CMS methodology, which is presented in Chapter 3, is built on the same principles as the BERRU-CMS methodology presented in Chapter 1. Thus, the maximum entropy principle is used within the BERRU-CMS methodology to construct an optimal approximation of the unknown a priori distribution based on a priori known mean values and covariance matrices that characterize the parameters and responses for both multiphysics systems. This "maximum entropy"-approximate a priori distribution is combined, using Bayes' theorem, with the "likelihood" provided by the multi-physics simulation models. Subsequently, the posterior distribution thus obtained is evaluated using the saddle-point method to obtain analytical expressions for the optimally predicted values for the multi-physics models parameters and responses along with corresponding reduced uncertainties. Noteworthy, the predictive modeling methodology for the coupled systems is constructed such that the systems can be considered sequentially rather than simultaneously, while preserving exactly the same results as if the systems were treated simultaneously. Consequently, very large coupled systems, which could perhaps exceed available computational resources if treated simultaneously, can be treated sequentially using the BERRU-CMS methodology, without any loss of generality or information, while requiring only the computational resources that would be needed for treating each system separately.

Chapter 4 illustrates applications of the BERRU-CMS methodology to paradigm reactor physics systems, starting with the predictive modeling of a paradigm neutron diffusion model of a one-dimensional pool of water containing distributed neutron sources (e.g., spent fuel rods) and continuing with the predictive modeling of the OECD/NEA ICSBEP (2010) benchmarks Godiva, Jezebel-239 and Jezebel-240. The paradigm neutron diffusion model characterizes the main physical processes that would occur, for example, in a spent-fuel storage pool. Thus problem was chosen because it is amenable to closed-form analytical solutions which highlight the salient features of the BERRU-CMS methodology. In addition, this example will also illustrate the basic concepts of computing response sensitivities using the ASAM (Cacuci 1981a, b, 2003). In contradistinction to this didactical example involving the simple neutron diffusion equation, the application of the BERRU-CMS methodology to the benchmarks Godiva, Jezebel-239, and Jezebel-240 involves large-scale neutron transport computations characterized by many imprecisely-known model parameters, as follows: 2241 imprecisely known model parameters for Jezebel-239, 1458 imprecisely known parameters for Jezebel-240, and 2916 imprecisely known parameters for Godiva. Eight responses (for which experimental measurements are available) have been considered for Jezebel-239, namely: the effective multiplication factor, the center core fission rates for ^{233}U, ^{238}U, ^{237}Np, and ^{239}Pu, and the center core radiative capture rates for ^{55}Mn, ^{93}Nb and ^{63}Cu. Three responses (the effective multiplication factor, the center core fission rates for ^{233}U and ^{237}Np) were selected for Jezebel-240, and eleven responses were selected for Godiva (the reaction rate types listed for Jezebel-239, along with the radiative capture rates for ^{107}Ag, ^{127}I and ^{81}Br). The advantages of applying the BERRU-CMS-methodology in the response-space (as opposed to the parameter-space) for these benchmark systems will become evident, as this methodology will be shown to reduce significantly the computational memory requirements for predictive modeling of systems involving many model parameters. The illustrative examples presented in Chapter 4 also underscore the fact that the BERRU-CMS methodology ensures that increasing the amount of consistent information will increase the accuracy of the predicted results, while reducing their accompanying predicted standard deviations.

Chapter 5 presents a paradigm problem of inverse prediction of the thickness of a homogeneous slab of material containing uniformly distributed sources emitting gamma-rays, using detector responses located outside the slab and affected by counting uncertainties. This inverse problem is of interest in proliferation detection. It is shown that for *optically thin* slabs, both the traditional inverse-problem methods (which are based on minimizing chi-square-like functionals) and the BERRU methodology predict the slab's thickness accurately. However, the BERRU methodology is considerably more efficient computationally, and a single application of the BERRU methodology predicts the thin slab's thickness at least as precisely as the converged iterations of the traditional chi-square-minimization methods, even though the measurements used in the BERRU methodology were ten times less accurate than the ones used for the traditional chi-square-minimization method. For *optically thick* slabs, however, the results presented in this Chapter show that the traditional inverse-problem methods based on the minimization of chi-square-type functionals fail to predict the slab's thickness, while the BERRU methodology correctly predicts the slab's actual physical thickness when precise experimental results are assimilated, while also predicting the physically correct response within the selected precision criterion. Also, the BERRU methodology is vastly more efficient computationally while yielding significantly more accurate results than the traditional chi-square-minimization methodologies.

In Chapter 6, the BERRU methodology is applied to the model of a mechanical draft cooling tower located at Savannah River National Laboratory (SRNL/USA) in order to improve the predictions of this model by combining computational information with measurements of outlet air humidity, outlet air and outlet water temperatures, which are considered to be the model's responses. At the outlet of this cooling tower, where measurements of the responses are available, the BERRU methodology reduces the predicted standard deviations for these responses to values that are smaller than the corresponding standard deviations of both the computed and the measured responses. The BERRU methodology has also been applied to locations inside the tower's fill section, where no direct measurements are available. The results obtained for these locations demonstrate that the predicted standard deviations of the predicted responses are reduced by the BERRU methodology to values that are smaller than the computed response standard deviations that arise from the imprecisely known model parameters.

The numerical results presented in this book were obtained by the author in collaboration with his former doctoral students, Drs. Madalina C. Badea, Erkan Arslan, Christine Latten, James J. Peltz and Federico Di Rocco, to whom the author wishes to express his deep gratitude for their contributions. The author is also grateful to Drs. Aurelian F. Badea, Ruxian Fang, and Jeffrey A. Favorite for their collaboration and contributions. Special thanks are due to the author's long-time collaborator, Dr. Mihaela Ionescu-Bujor, for her very significant contributions to the BERRU predictive modeling methodology, and for constructively reviewing this book. Last, but not least, the author is grateful to the Springer Editorial team, especially to Dr. Christoph Baumann, Springer Nature, for their guidance throughout the publication process.

Dan Gabriel Cacuci

Contents

1 BERRU PREDICTIVE MODELING FOR SINGLE MULTIPHYSICS SYSTEMS (BERRU-SMS)

Abstract: This Chapter presents the predictive modeling methodology developed by Cacuci and Ionescu-Bujor (2010a), which will be called the "BERRU-SMS Predictive Modeling" methodology, since it is conceived for a single, albeit large-scale, nonlinear multi-physics system, for which it combines uncertain computational and experimental data in order to predict best-estimate values for model responses and parameters, along with reduced predicted uncertainties for these best-estimate values. In contradistinction to the customary methods used currently for data assimilation, the BERRU-SMS predictive modeling methodology developed by Cacuci and Ionescu-Bujor (2010a) uses the maximum entropy principle to avoid the need for minimizing an arbitrarily user-chosen "cost functional" (usually a quadratic functional that represents the weighted errors between measured and computed responses), thus generalizing and significantly extending the customary "data adjustment" and/or 4D-VAR data assimilation procedures. The BERRU-SMS predictive modeling methodology also provides a quantitative indicator, constructed from sensitivity and covariance matrices, for determining the consistency (agreement or disagreement) among the a priori computational and experimental information available for parameters and responses. This consistency indicator measures, in the corresponding metric, the deviations between the experimental and nominally computed responses. This consistency indicator can be evaluated directly from the originally given data (i.e., given parameters and responses, together with their original uncertainties), once the response sensitivities have become available. Section 1.2 presents the mathematical framework underlying the BERRU-SMS methodology of Cacuci and Ionescu-Bujor (2010a) for both time-independent and time-dependent physical systems. Sections 1.3 and 1.4 present the application of this methodology to perform predictive modeling of two-phase water-flow and, respectively, liquid sodium flow, illustrating the reduction of predicted uncertainties for both the optimally-predicted model responses and calibrated model parameters. Section 1.5 presents the application of the BERRU-SMS predictive modeling methodology to a multi-physics benchmark (BWR-TT2) which involves coupled neutron kinetics/thermal-hydraulics numerical simulations characterized by 6 660 imprecisely known model parameters (6090 macroscopic neutron cross sections, and 570 thermal-hydraulics parameters involved in modeling the phase-slip correlation, transient outlet pressure, and total mass flow), showing reductions of up to 50% in the predicted standard deviation of the optimally predicted benchmark's power response.

D. G. Cacuci, *BERRU Predictive Modeling*,
https://doi.org/10.1007/978-3-662-58395-1_1

1.1 INTRODUCTION

The results of measurements and computations are never perfectly accurate. On the one hand, results of measurements inevitably reflect the influence of experimental errors, imperfect instruments, or imperfectly known calibration standards. Around any reported experimental value, therefore, there always exists a range of values that may also be plausibly representative of the true but unknown value of the measured quantity. On the other hand, computations are afflicted by errors stemming from numerical procedures, uncertain model parameters, boundary and initial conditions, and/or imperfectly known physical processes or problem geometry. Therefore, knowing just the nominal values experimentally measured or computed quantities is insufficient for applications. The quantitative uncertainties accompanying measurements and computations are also needed, along with the respective nominal values. Extracting "best estimate" values for model parameters and predicted results, together with "best estimate" uncertainties for these parameters and results requires the combination of experimental and computational data and their uncertainties. Such a combination process requires reasoning from incomplete, error-afflicted, and occasionally discrepant information.

The discrepancies between experimental and computational results provide the basic motivation for performing quantitative model verification, validation, qualification and predictive estimation. Loosely speaking, "code or model verification" means "are you solving the mathematical model correctly?" "Code or model qualification" means certifying that a proposed simulation and/or design methodology satisfies all performance and safety specifications regarding the system under consideration. "Code or model validation" means "does the model represent reality?" Model validation aims at assessing model accuracy by comparing computational results that are characterized by probability distributions with corresponding experimental results. Model validation can be done only by selected benchmarking, taking into account systematically (i.e., using sensitivities) all relevant uncertainties (computational, experimental, etc.).

Predictive modeling commences by identifying and characterizing the uncertainties involved in every step in the sequence of the numerical simulation processes that ultimately lead to a prediction. This includes characterizing: (a) errors and uncertainties in the data used in the simulation (e.g., input data, model parameters, initial conditions, boundary conditions, sources and forcing functions), (b) numerical discretization errors, and (c) uncertainties in (e.g., lack of knowledge of) the processes being modeled. Under ideal circumstances, the result of predictive modeling is a probabilistic description of possible future outcomes based on all recognized errors and uncertainties.

Predictive modeling comprises three key elements, namely model calibration, model extrapolation, and estimation of the validation domain. Model calibration addresses the integration of experimental data for the purpose of updating the data of the computer model. Important components include the estimation of discrepancies in the data, and of the biases between model predictions and experimental data.

The state-of-the-art of model calibration is fairly well developed, but current methods are still hampered in practice by the significant computational effort required for performing exhaustive (i.e., including all of the model's parameters) model calibration for large-scale models. Reducing the computational effort is paramount, and methods based on adjoint models show great promise in his regard, as will be demonstrated in this book. Model extrapolation addresses, in principle, the uncertainties in predictions under new environments or conditions, including both untested regions of the parameter space and higher levels of system complexity in the validation hierarchy. Extrapolation of models and the resulting increase of uncertainty are poorly understood, particularly the estimation of uncertainty that results from nonlinear coupling of two or more physical phenomena that were not coupled in the existing validation database. The quantification of the validation domain underlying the models of interest requires estimation of contours of constant uncertainty in the high-dimensional space that characterizes the application of interest. In practice, this involves the identification of areas where the predictive estimation of uncertainty meets specified requirements for the performance, reliability, or safety of the system of interest. The conceptual and mathematical development of methods for quantifying the validation domain is in an incipient stage.

In general terms, the modeling of a physical system and/or the result of an indirect experimental measurement requires consideration of the following modeling components:

(a) a mathematical model comprising linear and/or nonlinear equations that relate the system's independent variables and parameters to the system's state (i.e., dependent) variables;

(b) inequality and/or equality constraints that delimit the ranges of the system's parameters;

(c) one or several computational results, customarily referred to as system responses (or objective functions, or indices of performance), which are computed using the mathematical model; and

(d) experimentally measured responses, with their respective nominal (mean) values and uncertainties (variances, covariances, skewness, kurtosis, etc.).

The earliest systematic activities aimed at extracting best-estimate values for model parameters appear to have been initiated in the 1960s (see, e.g., Humi et al, 1964; Cecchini et al, 1964; Usachev, 1964), in the course of evaluating neutron cross sections by using time-independent reactor physics experiments for measuring "integral quantities" (also called "system responses") such as reaction rates and multiplication factors. A decade later, these activities had reached conceptual maturity under the name of "cross section adjustment" (see, e.g., Rowlands et al, 1973; Gandini and Petilli, 1973), a methodology which essentially amounted to using a weighted least-square procedure (with response sensitivities used as weighting functions) for combining uncertainties in the model parameters with uncertainties in the experimental data, subject to the constraint imposed by the linearized model. The resulting "adjusted" neutron cross sections (model parameters) and their "adjusted" uncertainties were subsequently employed in the respective reactor core model to predict improved model responses (e.g., improved reaction rates and reaction rate ratios, reactor multiplication factors, Doppler coefficients) in an extended application domain (e.g., a new or improved reactor core design). By the late 1970s, adjoint neutron fluxes were introduced and used (see, e.g., Kuroi and Mitani, 1975; Dragt et al, 1977; Weisbin et al, 1978) to compute efficiently the first-order response sensitivities, which appeared as weighting functions in the least squares adjustment procedure. It is important to note that all of these works dealt with the time-independent linear neutron transport or diffusion equation, as encountered in reactor physics and shielding, for which the corresponding adjoint equations were already known and readily available. For *nonlinear* systems, the adjoint method for computing efficiently response sensitivities to model parameters was generally formulated by Cacuci (1981a, 1981b), who pioneered its introduction into the earth sciences (see, e.g., Práger and Kelemen, 2013) and applied it to many problems, in various fields, as will be seen from the examples presented in this book.

A pioneering formalism for performing data adjustment in the context of time-dependent nonlinear system was presented by Cacuci's group at Oak Ridge National Laboratory (Barhen, Cacuci et al, 1982). However, this advanced (for its time) methodology stagnated in the field of nuclear engineering although it may have inspired later developments in the geophysical sciences.

In the late 1980s and during the 1990s, the fundamental concepts underlying "data adjustment" seem to have been rediscovered in the geophysical sciences while developing the so-called "data assimilation" procedure, in that the concepts underlying data assimilation are the same as those underlying the previously developed "data adjustment" procedure. Since then, numerous works have been published on data assimilation; the most representative can be found cited in the books by Lewis et al (2006), Lahoz et al (2010), and Cacuci et al (2013).

In contradistinction to the customary methods used currently for data assimilation, the BERRU-SMS predictive modeling methodology developed by Cacuci and Ionescu-Bujor (2010a) dispenses with the need to minimize an a priori chosen "cost functional" (usually a quadratic functional that represents the weighted errors between measured and computed responses), thus generalizing and significantly extending the "data adjustment" methods customarily used in nuclear engineering, as well as those underlying the so-called 4D-VAR data assimilation procedures in the geophysical sciences (see, e.g., Lahoz et al, 2010, and Cacuci et al., 2013). The BERRU-SMS predictive modeling methodology also provides a quantitative indicator, constructed from sensitivity and covariance matrices, for determining the consistency (agreement or disagreement) among the a priori computational and experimental data for parameters and responses. This consistency indicator measures, in the corresponding metric, the deviations between the experimental and nominally computed responses. Note that this consistency indicator can be evaluated directly from the originally given data (i.e., given parameters and responses, together with their original uncertainties), once the response sensitivities have become available. When the numerical value of this consistency indicator is close to unity (per degrees of freedom), the respective data is considered to be consistent "within the respective error norms" (usually under quadratic loss). However, when the numerical value of this consistency indicator differs considerably from unity, which usually occurs when the distance between the mean values of two (sets of) measurements or two (sets of) computations of the same quantity are larger than the sum of the two accompanying standard deviations, the respective (measured or computed) data points are usually considered to be inconsistent or discrepant. Note, though, that a nonzero probability still exists for two non-discrepant (i.e. belonging to the same distribution) measurements that are separated by more than 2 standard deviations (thus giving the appearance of being discrepant!) to occur. For example, for a Gaussian sampling distribution, the probability that two equally precise measurements would be separated by more than 2 standard deviations is 15.7%. However, this probability is rather small; therefore it is much more likely that apparently discrepant data actually indicate the presence of unrecognized errors. Methods for treating unrecognized errors have been developed by Cacuci and Ionescu-Bujor (2010b), by applying the maximum entropy principle under quadratic loss to the discrepant data. Once the inconsistent data, if any, is discarded, the BERRU-SMS predictive modeling methodology predicts best-estimate values for parameters and predicted responses, as well as best-estimate reduced uncertainties (i.e., "smaller" values for the variance-covariance matrices) for the predicted best-estimate parameters and responses.

This Chapter is structured as follows: Section 1.2 presents the BERRU-SMS methodology of Cacuci and Ionescu-Bujor (2010a) for both time-independent and time-dependent physical systems. Sections 1.3 and 1.4 present the application of the BERRU-SMS methodology to perform predictive modeling of two-phase water-flow and, respectively, liquid sodium flow. Section 1.5 presents the application of the BERRU-SMS predictive modeling methodology to a multi-physics benchmark (BWR-TT2) which involves coupled neutron kinetics/thermal-hydraulics numerical simulations characterized by 6 660 imprecisely known model parameters (6090 macroscopic neutron cross sections, and 570 thermal-hydraulics parameters involved in modeling the phase-slip correlation, transient outlet pressure, and total mass flow), showing reductions of up to 50% in the predicted standard deviation of the optimally predicted benchmark's power response.

1.2 BERRU-SMS PREDICTIVE MODELING METHODOLOGY

1.2.1 *Time-Independent "Perfect" Model*

Consider a mathematical model which comprises N parameters (e. g., material properties, correlations, etc.), denoted by α_n, which will be considered for subsequent mathematical derivations to be components of a column vector denoted as $\boldsymbol{\alpha} \triangleq (\alpha_1,\ldots,\alpha_N)$, where the symbol "$\triangleq$" signifies "by definition". The exact values of these model parameters are unknown in practice, so they will be are considered to be variates with known nominal (mean) values, $\alpha_n^0 \triangleq \langle \alpha_n \rangle$ and known covariance (uncertainty) matrix, $\mathbf{C}_\alpha$, defined as

$$\mathbf{C}_\alpha \triangleq \langle \delta\boldsymbol{\alpha}\delta\boldsymbol{\alpha}^\dagger \rangle \triangleq \left\langle (\boldsymbol{\alpha} - \boldsymbol{\alpha}^0)(\boldsymbol{\alpha} - \boldsymbol{\alpha}^0)^\dagger \right\rangle, \tag{1.1}$$

where the components of the N-dimensional vector $\delta\boldsymbol{\alpha}$ are defined as $\delta\alpha_n \triangleq \alpha_n - \alpha_n^0$, where the dagger $(\dagger)$ signifies transposition, and where $\langle , \rangle$ denotes integration over the unknown probability distribution of the imprecisely known model parameters and responses (as detailed further in the following). Note that "transposition" will be indicated only when necessary to avoid misinterpretation. All vectors are considered to be *column vectors*, even though they will be usually written in row-format, without always indicating "transposition" explicitly, in order to keep the notation as simple as possible. The diagonal elements of $\mathbf{C}_\alpha$ are the variances, $(C_\alpha)_{nn} \triangleq var(\alpha_n) \triangleq \left\langle (\delta\alpha_n)^2 \right\rangle$, of the N parameters, while the off-diagonal components are the covariances, $(C_\alpha)_{mn} \triangleq cov(\alpha_m, \alpha_n) \triangleq \langle \delta\alpha_m\, \delta\alpha_n \rangle$, of the corresponding pairs of model parameters.

The results computed using the mathematical model are customarily called model (or system) *responses*. The ith-response will be denoted as r_i; we consider that the model produces I responses of interest. Therefore, the corresponding model responses will be considered to constitute the components of the I-dimensional column vector $\mathbf{r} \triangleq (r_1,\ldots,r_I)$. Note that the actual values of these responses are not known exactly, since $\mathbf{r}$ depends on the variates $\boldsymbol{\alpha} \triangleq (\alpha_1,\ldots,\alpha_N)$; hence, $\mathbf{r}$ is also considered to be a variate.

Consider, furthermore, that experimental measurements corresponding to the computed responses are also available. The nominal values of the ith experimentally measured response will be denoted as r_m^i; these measured responses will be considered to be the components of the I-dimensional vector, denoted as $\mathbf{r}_m \triangleq \left(r_m^1,\ldots,r_m^I\right)$, of nominal values of measured responses. The associated uncertainty matrix for the experimentally measured responses will be denoted by $\mathbf{C}_m$, and is defined as

$$\mathbf{C}_m \triangleq \left\langle \delta\mathbf{r}_m \delta\mathbf{r}_m^\dagger \right\rangle \triangleq \left\langle (\mathbf{r}-\mathbf{r}_m)(\mathbf{r}-\mathbf{r}_m)^\dagger \right\rangle, \; \delta\mathbf{r}_m \triangleq \mathbf{r}-\mathbf{r}_m \,. \tag{1.2}$$

In the most general case, the measured responses may be correlated to the parameters $\boldsymbol{\alpha} \triangleq (\alpha_1,\ldots,\alpha_N)$ through a response-parameter correlation matrix of order $I \times N$, of the form

$$\mathbf{C}_{r\alpha} \triangleq \left\langle \delta\mathbf{r}\delta\boldsymbol{\alpha}^\dagger \right\rangle \triangleq \left\langle (\mathbf{r}-\mathbf{r}_m)(\boldsymbol{\alpha}-\boldsymbol{\alpha}^0)^\dagger \right\rangle = \left\langle (\boldsymbol{\alpha}-\boldsymbol{\alpha}^0)(\mathbf{r}-\mathbf{r}_m)^\dagger \right\rangle, \; \mathbf{C}_{r\alpha}^\dagger = \mathbf{C}_{\alpha r}. \tag{1.3}$$

Applying now the maximum entropy algorithm described in *Appendix 1* (located at the end of Section 1.2) to the computational and experimental information described in Eqs. (1.1) through (1.3) indicates that the most objective probability distribution for the information at hand is a multivariate Gaussian of the form:

$$p(\mathbf{z}|\mathbf{C})d(\mathbf{z})=\frac{exp\left[-\frac{1}{2}Q(\mathbf{z})\right]}{det(2\pi\mathbf{C})^{1/2}}d(\mathbf{z}),\ Q(\mathbf{z})\triangleq\mathbf{z}^{\dagger}\mathbf{C}^{-1}\mathbf{z},\ -\infty<z_j<\infty,\tag{1.4}$$

where the $(N+I)\times(N+I)$ partitioned matrix $\mathbf{C}$ represents the joint covariance matrix of the parameters and responses

$$\mathbf{C}\triangleq\begin{pmatrix}\mathbf{C}_{\alpha} & \mathbf{C}_{\alpha r}\\ \mathbf{C}_{r\alpha} & \mathbf{C}_{m}\end{pmatrix},\tag{1.5}$$

while the vector $\mathbf{z}$ denotes the joint $(I+N)$-dimensional partitioned vector

$$\mathbf{z}\triangleq\begin{pmatrix}\boldsymbol{\alpha}-\boldsymbol{\alpha}^{0}\\ \mathbf{r}-\mathbf{r}_{m}\end{pmatrix}.\tag{1.6}$$

As was already mentioned, the responses $\mathbf{r}\triangleq(r_{1},\dots,r_{I})$ are functions of the parameters $\boldsymbol{\alpha}\triangleq(\alpha_{1},\dots,\alpha_{N})$. In this Section, the dependence on $\boldsymbol{\alpha}$ of each response r_i is modeled generically through the function $r_i=R_i(\boldsymbol{\alpha})$, which leads to the vector-form representation $\mathbf{r}=\mathbf{R}(\boldsymbol{\alpha})$. The simplest and most commonly used model to represent explicitly the dependence of the computed response on the model parameters is a linearized model, in which the computed response is linearized via a functional Taylor-series expansion around the nominal values, $\boldsymbol{\alpha}^{0}$, of the parameters $\boldsymbol{\alpha}$, as follows:

$$\mathbf{r}=\mathbf{R}(\boldsymbol{\alpha}^{0}+\delta\boldsymbol{\alpha})=\mathbf{R}(\boldsymbol{\alpha}^{0})+\mathbf{S}(\boldsymbol{\alpha}^{0})(\boldsymbol{\alpha}-\boldsymbol{\alpha}^{0})+\textit{higher order terms},\tag{1.7}$$

where $\mathbf{R}(\boldsymbol{\alpha}^{0})$ denotes the *computed response* at the nominal parameter values $\boldsymbol{\alpha}^{0}$, while the "response sensitivity" matrix $\mathbf{S}(\boldsymbol{\alpha})$ represents the $(I\times N)$-dimensional matrix containing the first Gateaux-derivatives of the computed responses with respect to the parameters, defined as

$$\mathbf{S}(\boldsymbol{\alpha})\triangleq\begin{pmatrix}s_{11} & \cdots & s_{1N}\\ \vdots & \ddots & \vdots\\ s_{I1} & \cdots & s_{IN}\end{pmatrix}\triangleq\begin{pmatrix}\frac{\partial R_1(\boldsymbol{\alpha})}{\partial\alpha_1} & \cdots & \frac{\partial R_1(\boldsymbol{\alpha})}{\partial\alpha_N}\\ \vdots & \ddots & \vdots\\ \frac{\partial R_I(\boldsymbol{\alpha})}{\partial\alpha_1} & \cdots & \frac{\partial R_I(\boldsymbol{\alpha})}{\partial\alpha_N}\end{pmatrix}.\tag{1.8}$$

In Eq. (1.7), the notation $\mathbf{S}\left(\boldsymbol{\alpha}^0\right)$ indicates that the sensitivity matrix is evaluated at the nominal parameter values $\boldsymbol{\alpha}^0$. The expectation value, $\langle \mathbf{r} \rangle$, of the response $\mathbf{r}$, and the corresponding covariance matrix, $\mathbf{C}_{rc}\left(\boldsymbol{\alpha}^0\right)$, of the computed responses can be obtained by formally integrating Eq. (1.7) over the unknown joint distributions of the parameters $\boldsymbol{\alpha}$, to obtain,

$$\langle \mathbf{r} \rangle = \mathbf{R}\left(\boldsymbol{\alpha}^0\right), \tag{1.9}$$

and

$$\mathbf{C}_{rc}\left(\boldsymbol{\alpha}^0\right) \triangleq \left\langle \delta\mathbf{r}\delta\mathbf{r}^{\dagger} \right\rangle = \left[\mathbf{S}\left(\boldsymbol{\alpha}^0\right)\right]\left\langle \delta\boldsymbol{\alpha}\delta\boldsymbol{\alpha}^{\dagger} \right\rangle\left[\mathbf{S}\left(\boldsymbol{\alpha}^0\right)\right]^{\dagger} = \left[\mathbf{S}\left(\boldsymbol{\alpha}^0\right)\right]\mathbf{C}_{\alpha}\left[\mathbf{S}\left(\boldsymbol{\alpha}^0\right)\right]^{\dagger}. \tag{1.10}$$

As indicated by Eq. (1.9), the expectation value of the computed responses for linearized models in which the numerical errors are neglected is given by the value of the response computed at the nominal parameter-values.

The next task is to condense the posterior information contained in Eqs. (1.4) and (1.7) into a recommended best-estimate value $\mathbf{z}^{be}$ for the parameters $\boldsymbol{\alpha} = \left(\alpha_1, \ldots, \alpha_N\right)$ and responses $\mathbf{r} = \left(r_1, \ldots, r_I\right)$, together with corresponding best-estimate recommended uncertainties for these quantities. If a loss function is given, decision theory indicates how these best-estimate quantities are to be computed. If no specific loss function is provided, the recommended best-estimate updated posterior mean vector $\mathbf{z}^{be}$ and its respective best-estimate posterior covariance matrix will be evaluated by using the "saddle-point method," which extracts the bulk of the contribution to the distribution $p\left(\mathbf{z}|\mathbf{C}\right)$ in Eq. (1.4) at the point in phase space where the respective exponent attains its minimum, subject to the relation provided by Eq. (1.7). When the numerical errors are also neglected in Eq. (1.7), in addition to neglecting the higher-order terms, then Eq. (1.7) is imposed as a hard constraint. In this case, the resulting constrained minimization problem is solved by introducing an I-dimensional vector of Lagrange multipliers, $\boldsymbol{\lambda}$, to obtain the following unconstrained minimization problem

$$\begin{aligned} &P\left(\mathbf{z}, \boldsymbol{\lambda}\right) \triangleq Q(\mathbf{z}) + 2\boldsymbol{\lambda}^{\dagger}\left[-\mathbf{r} + \mathbf{R}\left(\boldsymbol{\alpha}^0\right) + \mathbf{S}\left(\boldsymbol{\alpha}^0\right)\left(\boldsymbol{\alpha} - \boldsymbol{\alpha}^0\right)\right] = min, \\ &at\ \mathbf{z} = \mathbf{z}^{be} \triangleq \begin{pmatrix} \boldsymbol{\alpha}^{be} - \boldsymbol{\alpha}^0 \\ \mathbf{r}^{be} - \mathbf{r}_m \end{pmatrix}. \end{aligned} \tag{1.11}$$

In the above expression, the superscript "be" denotes "best estimated values", and the factor "2" was introduced for convenience in front of $\boldsymbol{\lambda}$ in order to simplify the subsequent algebraic derivations. The point $\mathbf{z}^{be}$ where the functional $P\left(\mathbf{z}, \boldsymbol{\lambda}\right)$ attains its extremum (minimum) is defined as the point where its derivative with respect to $\mathbf{z}$ vanishes. This point can be conveniently determined by rewriting $P(\mathbf{z},\boldsymbol{\lambda})$ in the form

$$P(\mathbf{z},\boldsymbol{\lambda}) \triangleq Q(\mathbf{z}) + 2\boldsymbol{\lambda}^{\dagger}\left\{\left[\mathbf{S}(\boldsymbol{\alpha}),-\mathbf{I}_I\right]\mathbf{z}+\mathbf{d}\right\} = min, \;\; at\; \mathbf{z}=\mathbf{z}^{be}, \tag{1.12}$$

where

$$\mathbf{d} \triangleq \mathbf{R}\left(\boldsymbol{\alpha}^0\right) - \mathbf{r}_m, \tag{1.13}$$

is an I -dimensional vector of "deviations" reflecting the discrepancies between the nominal computations and the nominally measured responses.
The functional $P(\mathbf{z})$ becomes stationary at the point $\mathbf{z}=\mathbf{z}^{be}$, which is defined implicitly through the conditions

$$\nabla_{\mathbf{z}} P(\mathbf{z},\boldsymbol{\lambda}) = \mathbf{0}, \quad \nabla_{\boldsymbol{\lambda}} P(\mathbf{z},\boldsymbol{\lambda}) = \mathbf{0}, \quad at\; \mathbf{z}=\mathbf{z}^{be}. \tag{1.14}$$

The condition $\nabla_{\boldsymbol{\lambda}} P(\mathbf{z},\boldsymbol{\lambda}) = \mathbf{0}$ ensures that the constraint represented by Eq. (1.7) is fulfilled at $\mathbf{z}=\mathbf{z}^{be}$, while the condition $\nabla_{\mathbf{z}} P(\mathbf{z},\boldsymbol{\lambda}) = \mathbf{0}$ yields

$$\begin{aligned}\nabla_{\mathbf{z}} P(\mathbf{z},\boldsymbol{\lambda}) &= \nabla_{\mathbf{z}}\left\{\mathbf{z}^{\dagger}\mathbf{C}^{-1}\mathbf{z} + 2\left[\boldsymbol{\lambda}^{\dagger}\mathbf{S}(\boldsymbol{\alpha}),-\boldsymbol{\lambda}^{\dagger}\right]\mathbf{z} + 2\boldsymbol{\lambda}^{\dagger}\mathbf{d}\right\} \\ &= 2\mathbf{C}^{-1}\mathbf{z} + 2\begin{pmatrix}\mathbf{S}^{\dagger}(\boldsymbol{\alpha})\boldsymbol{\lambda} \\ -\boldsymbol{\lambda}\end{pmatrix} = \mathbf{0}, \quad at\; \mathbf{z}=\mathbf{z}^{be}.\end{aligned} \tag{1.15}$$

Multiplying the last line of the above equation on the left by $\mathbf{C}$ and solving it for $\mathbf{z}^{be}$ gives:

$$\mathbf{z}^{be} = \mathbf{C}\begin{pmatrix}\mathbf{S}^{\dagger}(\boldsymbol{\alpha})\boldsymbol{\lambda} \\ -\boldsymbol{\lambda}\end{pmatrix} = \begin{pmatrix}\mathbf{C}_{\alpha} & \mathbf{C}_{\alpha r} \\ \mathbf{C}_{r\alpha} & \mathbf{C}_m\end{pmatrix}\begin{pmatrix}\mathbf{S}^{\dagger}(\boldsymbol{\alpha})\boldsymbol{\lambda} \\ -\boldsymbol{\lambda}\end{pmatrix}. \tag{1.16}$$

Writing the above expression in component form gives the following results for the calibrated best-estimate parameters and responses, respectively:

$$\boldsymbol{\alpha}^{be} = \boldsymbol{\alpha}^0 + \left(\mathbf{C}_{\alpha r} - \mathbf{C}_{\alpha}\left[\mathbf{S}\left(\boldsymbol{\alpha}^0\right)\right]^{\dagger}\right)\boldsymbol{\lambda}, \tag{1.17}$$

$$\mathbf{r}_c\left(\boldsymbol{\alpha}^{be}\right) = \mathbf{r}_m + \left(\mathbf{C}_m - \mathbf{C}_{r\alpha}\left[\mathbf{S}\left(\boldsymbol{\alpha}^0\right)\right]^{\dagger}\right)\boldsymbol{\lambda}. \tag{1.18}$$

Evaluating Eq. (1.7) at $\mathbf{z}^{be}$ while using Eqs. (1.17) and (1.18) yields the following expression:

$$\left[\mathbf{C}_{rc}\left(\boldsymbol{\alpha}^0\right)-\mathbf{C}_{r\alpha}\left[\mathbf{S}\left(\boldsymbol{\alpha}^0\right)\right]^{\dagger}-\left[\mathbf{S}\left(\boldsymbol{\alpha}^0\right)\right]\mathbf{C}_{\alpha r}+\mathbf{C}_m\right]\boldsymbol{\lambda}=\mathbf{d}. \tag{1.19}$$

In Eq. (1.19), the matrix-valued expression that multiplies $\boldsymbol{\lambda}$ is actually the covariance-matrix, $\mathbf{C}_d\left(\boldsymbol{\alpha}^0\right)$, of the vector of response-deviations, $\mathbf{d}$, as shown below:

$$\begin{aligned}\mathbf{C}_d\left(\boldsymbol{\alpha}^0\right)&\triangleq\left\langle\mathbf{d}\mathbf{d}^{\dagger}\right\rangle=\left\langle\left(\delta\mathbf{r}-\mathbf{S}\left(\boldsymbol{\alpha}^0\right)\delta\boldsymbol{\alpha}\right)\left(\delta\mathbf{r}^{\dagger}-\delta\boldsymbol{\alpha}^{\dagger}\left[\mathbf{S}\left(\boldsymbol{\alpha}^0\right)\right]^{\dagger}\right)\right\rangle\\&=\mathbf{C}_{rc}\left(\boldsymbol{\alpha}^0\right)-\mathbf{C}_{r\alpha}\left[\mathbf{S}\left(\boldsymbol{\alpha}^0\right)\right]^{\dagger}-\left[\mathbf{S}\left(\boldsymbol{\alpha}^0\right)\right]\mathbf{C}_{\alpha r}+\mathbf{C}_m.\end{aligned} \tag{1.20}$$

Introducing Eq. (1.20) into Eq. (1.19) and solving the resulting equation yields the following expression for the Lagrange multiplier $\boldsymbol{\lambda}$ at $\mathbf{z}^{be}$:

$$\boldsymbol{\lambda}=\left[\mathbf{C}_d\left(\boldsymbol{\alpha}^0\right)\right]^{-1}\mathbf{d}. \tag{1.21}$$

Note that the second and third terms in the last equality in Eq. (1.20) are square matrices of order I resulting from the multiplication of two rectangular matrices. Consequently, the matrix $\mathbf{C}_d\left(\boldsymbol{\alpha}^0\right)$ is a symmetric matrix of order I, which is important when computing its inverse since, in practical problems, the number of computed or measured responses is typically much less than the number N of model parameters.

Replacing now Eq. (1.21) in Eqs. (1.17) and (1.18), respectively, yields the following expressions for the nominal values of the calibrated (adjusted) best-estimate responses and parameters:

$$\boldsymbol{\alpha}^{be}=\boldsymbol{\alpha}^0+\left(\mathbf{C}_{\alpha r}-\mathbf{C}_{\alpha}\left[\mathbf{S}\left(\boldsymbol{\alpha}^0\right)\right]^{\dagger}\right)\left[\mathbf{C}_d\left(\boldsymbol{\alpha}^0\right)\right]^{-1}\mathbf{d} \tag{1.22}$$

$$\mathbf{r}\left(\boldsymbol{\alpha}^{be}\right)=\mathbf{r}_m+\left(\mathbf{C}_m-\mathbf{C}_{r\alpha}\left[\mathbf{S}\left(\boldsymbol{\alpha}^0\right)\right]^{\dagger}\right)\left[\mathbf{C}_d\left(\boldsymbol{\alpha}^0\right)\right]^{-1}\mathbf{d}. \tag{1.23}$$

Using Eqs. (1.21), (1.16) and (1.17) in Eq. (1.11) yields the following expression for the minimum of $Q(\mathbf{z})$:

$$\begin{aligned}
Q_{min} &\triangleq Q\left(\mathbf{z}^{be}\right) = \left[-\boldsymbol{\lambda}^{\dagger}\mathbf{Z}\left(\boldsymbol{\alpha}^{0}\right)\mathbf{C}\right]\mathbf{C}^{-1}\mathbf{z}^{be} \\
&= -\boldsymbol{\lambda}^{\dagger}\mathbf{Z}\left(\boldsymbol{\alpha}^{0}\right)\mathbf{z}^{be} = \boldsymbol{\lambda}^{\dagger}\mathbf{d} = \mathbf{d}^{\dagger}\left[\mathbf{C}_{d}\left(\boldsymbol{\alpha}^{0}\right)\right]^{-1}\mathbf{d} \\
&= \chi^{2}.
\end{aligned} \tag{1.24}$$

The best-estimate covariances, $\mathbf{C}_{\alpha}^{be}$ and $\mathbf{C}_{r}^{be}$, corresponding to the best-estimate parameters $\boldsymbol{\alpha}^{be}$ and responses $\mathbf{r}\left(\boldsymbol{\alpha}^{be}\right)$, together with the best-estimate parameter-response covariance matrix $\mathbf{C}_{\alpha r}^{be}$ are defined as follows:

$$\mathbf{C}_{\alpha}^{be} \triangleq \left\langle \left(\boldsymbol{\alpha}-\boldsymbol{\alpha}^{be}\right)\left(\boldsymbol{\alpha}-\boldsymbol{\alpha}^{be}\right)^{\dagger} \right\rangle, \tag{1.25}$$

$$\mathbf{C}_{r}^{be} \triangleq \left\langle \left(\mathbf{r}-\mathbf{r}\left(\boldsymbol{\alpha}^{be}\right)\right)\left(\mathbf{r}-\mathbf{r}\left(\boldsymbol{\alpha}^{be}\right)\right)^{\dagger} \right\rangle, \tag{1.26}$$

$$\mathbf{C}_{r\alpha}^{be} = \mathbf{C}_{\alpha r}^{be} \triangleq \left\langle \left(\boldsymbol{\alpha}-\boldsymbol{\alpha}^{be}\right)\left(\mathbf{r}-\mathbf{r}\left(\boldsymbol{\alpha}^{be}\right)\right)^{\dagger} \right\rangle. \tag{1.27}$$

Recalling from Eqs. (1.7) and (1.13) that

$$\mathbf{d} \triangleq \mathbf{R}\left(\boldsymbol{\alpha}^{0}\right) - \mathbf{r}_{m} = \mathbf{r} - \mathbf{r}_{m} - \mathbf{S}\left(\boldsymbol{\alpha}^{0}\right)\left(\boldsymbol{\alpha}-\boldsymbol{\alpha}^{0}\right) \tag{1.28}$$

and replacing Eqs. (1.28) and (1.22) in Eq. (1.25) yields the following expression:

$$\begin{aligned}
\mathbf{C}_{\alpha}^{be} &\triangleq \left\langle \left(\boldsymbol{\alpha}-\boldsymbol{\alpha}^{0}\right)\left(\boldsymbol{\alpha}-\boldsymbol{\alpha}^{0}\right)^{\dagger} \right\rangle - \left\langle \left(\boldsymbol{\alpha}-\boldsymbol{\alpha}^{0}\right)\mathbf{d}^{\dagger} \right\rangle \left[\mathbf{C}_{d}\left(\boldsymbol{\alpha}^{0}\right)\right]^{-1}\left(\mathbf{C}_{\alpha r} - \left[\mathbf{S}\left(\boldsymbol{\alpha}^{0}\right)\right]\mathbf{C}_{\alpha}\right) \\
&- \left(\mathbf{C}_{\alpha r} - \mathbf{C}_{\alpha}\left[\mathbf{S}\left(\boldsymbol{\alpha}^{0}\right)\right]^{\dagger}\right)\left[\mathbf{C}_{d}\left(\boldsymbol{\alpha}^{0}\right)\right]^{-1}\left\langle \mathbf{d}\left(\boldsymbol{\alpha}-\boldsymbol{\alpha}^{0}\right)^{\dagger} \right\rangle \\
&+ \left(\mathbf{C}_{\alpha r} - \mathbf{C}_{\alpha}\left[\mathbf{S}\left(\boldsymbol{\alpha}^{0}\right)\right]^{\dagger}\right)\left[\mathbf{C}_{d}\left(\boldsymbol{\alpha}^{0}\right)\right]^{-1}\left\langle \mathbf{d}\mathbf{d}^{\dagger} \right\rangle \left[\mathbf{C}_{d}\left(\boldsymbol{\alpha}^{0}\right)\right]^{-1}\left(\mathbf{C}_{\alpha r} - \left[\mathbf{S}\left(\boldsymbol{\alpha}^{0}\right)\right]\mathbf{C}_{\alpha}\right).
\end{aligned} \tag{1.29}$$

The above expression can be simplified by using Eq. (1.20), and by noting that

$$\mathbf{C}_{\alpha d}\left(\boldsymbol{\alpha}^{0}\right) \triangleq \left\langle \left(\boldsymbol{\alpha}-\boldsymbol{\alpha}^{0}\right)\mathbf{d}^{\dagger} \right\rangle = \mathbf{C}_{\alpha r} - \mathbf{C}_{\alpha}\left[\mathbf{S}\left(\boldsymbol{\alpha}^{0}\right)\right]^{\dagger}, \tag{1.30}$$

$$\mathbf{C}_{d\alpha}\left(\boldsymbol{\alpha}^{0}\right) \triangleq \left\langle \mathbf{d}\left(\boldsymbol{\alpha}-\boldsymbol{\alpha}^{0}\right)^{\dagger} \right\rangle = \mathbf{C}_{\alpha r} - \left[\mathbf{S}\left(\boldsymbol{\alpha}^{0}\right)\right]\mathbf{C}_{\alpha} = \left[\mathbf{C}_{\alpha d}\left(\boldsymbol{\alpha}^{0}\right)\right]^{\dagger}. \tag{1.31}$$

Replacing Eqs. (1.29) through (1.31) in the expression of $\mathbf{C}_{\alpha}^{be}$ yields

$$\mathbf{C}_{\alpha}^{be} = \mathbf{C}_{\alpha} - \left(\mathbf{C}_{\alpha r} - \mathbf{C}_{\alpha}\left[\mathbf{S}\left(\boldsymbol{\alpha}^{0}\right)\right]^{\dagger}\right)\left[\mathbf{C}_{d}\left(\boldsymbol{\alpha}^{0}\right)\right]^{-1}\left(\mathbf{C}_{\alpha r} - \left[\mathbf{S}\left(\boldsymbol{\alpha}^{0}\right)\right]\mathbf{C}_{\alpha}\right)$$
$$= \mathbf{C}_{\alpha} - \left[\mathbf{C}_{\alpha d}\left(\boldsymbol{\alpha}^{0}\right)\right]\left[\mathbf{C}_{d}\left(\boldsymbol{\alpha}^{0}\right)\right]^{-1}\left[\mathbf{C}_{\alpha d}\left(\boldsymbol{\alpha}^{0}\right)\right]^{\dagger}. \tag{1.32}$$

Next, it is convenient to define the following matrices:

$$\mathbf{C}_{rd}\left(\boldsymbol{\alpha}^{0}\right) \triangleq \left\langle \left(\mathbf{r} - \mathbf{r}_{m}\right)\mathbf{d}^{\dagger}\right\rangle = \mathbf{C}_{m} - \mathbf{C}_{r\alpha}\left[\mathbf{S}\left(\boldsymbol{\alpha}^{0}\right)\right]^{\dagger},$$
$$\mathbf{C}_{dr}\left(\boldsymbol{\alpha}^{0}\right) \triangleq \left\langle \mathbf{d}\left(\mathbf{r} - \mathbf{r}_{m}\right)^{\dagger}\right\rangle = \mathbf{C}_{m} - \left[\mathbf{S}\left(\boldsymbol{\alpha}^{0}\right)\right]\mathbf{C}_{r\alpha} = \left[\mathbf{C}_{rd}\left(\boldsymbol{\alpha}^{0}\right)\right]^{\dagger}, \tag{1.33}$$

and to replace the above expressions in Eq. (1.26) to obtain the following expression for the best-estimate response covariance matrix $\mathbf{C}_{r}^{be}$:

$$\mathbf{C}_{r}^{be} = \mathbf{C}_{m} - \left(\mathbf{C}_{m} - \mathbf{C}_{r\alpha}\left[\mathbf{S}\left(\boldsymbol{\alpha}^{0}\right)\right]^{\dagger}\right)\left[\mathbf{C}_{d}\left(\boldsymbol{\alpha}^{0}\right)\right]^{-1}\left(\mathbf{C}_{m} - \left[\mathbf{S}\left(\boldsymbol{\alpha}^{0}\right)\right]\mathbf{C}_{\alpha r}\right)$$
$$= \mathbf{C}_{m} - \left[\mathbf{C}_{rd}\left(\boldsymbol{\alpha}^{0}\right)\right]\left[\mathbf{C}_{d}\left(\boldsymbol{\alpha}^{0}\right)\right]^{-1}\left[\mathbf{C}_{rd}\left(\boldsymbol{\alpha}^{0}\right)\right]^{\dagger}. \tag{1.34}$$

A similar sequence of computations leads to the following expression for the best-estimate response-parameter covariance matrix:

$$\mathbf{C}_{r\alpha}^{be} = \mathbf{C}_{\alpha r}^{be}$$
$$= \mathbf{C}_{r\alpha} - \left(\mathbf{C}_{m} - \mathbf{C}_{r\alpha}\left[\mathbf{S}\left(\boldsymbol{\alpha}^{0}\right)\right]^{\dagger}\right)\left[\mathbf{C}_{d}\left(\boldsymbol{\alpha}^{0}\right)\right]^{-1}\left(\mathbf{C}_{\alpha r} - \left[\mathbf{S}\left(\boldsymbol{\alpha}^{0}\right)\right]\mathbf{C}_{\alpha}\right) \tag{1.35}$$
$$= \mathbf{C}_{r\alpha} - \left[\mathbf{C}_{rd}\left(\boldsymbol{\alpha}^{0}\right)\right]\left[\mathbf{C}_{d}\left(\boldsymbol{\alpha}^{0}\right)\right]^{-1}\left[\mathbf{C}_{\alpha d}\left(\boldsymbol{\alpha}^{0}\right)\right]^{\dagger}.$$

As indicated in Eq. (1.32), a symmetric positive matrix is subtracted from the initial parameter covariance matrix $\mathbf{C}_{\alpha}$, which implies that the best-estimated parameter variances, which are the components of the main diagonal of $\mathbf{C}_{\alpha}^{be}$ must have smaller values than the initial parameter variances, which are the corresponding elements of the main diagonal of $\mathbf{C}_{\alpha}$. In this sense, the best-estimate parameter uncertainty matrix $\mathbf{C}_{\alpha}^{be}$ has been "reduced" by the "calibration (adjustment) procedure," namely by the introduction of new information from experiments. Similarly in Eq. (1.34), a symmetric positive matrix is subtracted from the initial covariance matrix $\mathbf{C}_{m}$ of the experimentally-measured responses. Hence, the best-estimate response covariance matrix $\mathbf{C}_{r}^{be}$ has been improved (reduced) through the addition of new experimental information. Furthermore, Eq. (1.35) indicates that the calibration (adjustment) procedure will introduce correlations between the calibrated (adjusted) parameters and responses even if the parameters and response were initially uncorrelated, since $\mathbf{C}_{r\alpha}^{be} \neq 0$ even if $\mathbf{C}_{r\alpha} = 0$, i.e.,

$$\mathbf{C}_{r\alpha}^{be} = \mathbf{C}_m \left[\mathbf{C}_{rc}\left(\boldsymbol{\alpha}^0\right) + \mathbf{C}_m \right]^{-1} \left[\mathbf{S}\left(\boldsymbol{\alpha}^0\right) \right] \mathbf{C}_\alpha \text{, when } \mathbf{C}_{r\alpha} = 0 . \qquad (1.36)$$

As the above expression indicates, the adjustment (calibration) modifies the correlations among the parameters through couplings introduced by the sensitivities of the participating responses; these sensitivities relate the initial parameter-covariances and experimental-response covariances. In summary, the incorporation of additional (experimental) information in the calibration (adjustment) process reduces the variances of the adjusted parameters and responses while also modifying the correlations between parameters and responses.
Note that Eq. (1.34) expresses the best-estimate response covariance matrix $\mathbf{C}_r^{be}$ in terms of the initial covariance matrix $\mathbf{C}_m$ of the experimentally-measured responses. Alternatively, it is of interest to derive the expression of the computed best-estimate response covariance matrix, $\mathbf{C}_{rc}^{be}$, directly from the model (the subscript "rc", denotes "computed response", to distinguish it from the covariance $\mathbf{C}_r^{be}$, which is obtained directly from the calibration/adjustment process). The starting point for computing $\mathbf{C}_{rc}^{be}$ is the linearization of the model, similar to that shown in Eq. (1.37), but around $\boldsymbol{\alpha}^{be}$ instead of $\boldsymbol{\alpha}^0$, i.e.

$$\mathbf{r} = \mathbf{R}\left(\boldsymbol{\alpha}^{be}\right) + \mathbf{S}\left(\boldsymbol{\alpha}^{be}\right)\left(\boldsymbol{\alpha} - \boldsymbol{\alpha}^{be}\right) + \textit{higher order terms} . \qquad (1.37)$$

It follows from Eqs. (1.37) that

$$\begin{aligned}
\mathbf{C}_{rc}^{be} &= \left\langle \left(\mathbf{r} - \mathbf{R}\left(\boldsymbol{\alpha}^{be}\right)\right)\left(\mathbf{r} - \mathbf{R}\left(\boldsymbol{\alpha}^{be}\right)\right)^\dagger \right\rangle = \left[\mathbf{S}\left(\boldsymbol{\alpha}^0\right)\right] \left\langle \left(\boldsymbol{\alpha} - \boldsymbol{\alpha}^{be}\right)\left(\boldsymbol{\alpha} - \boldsymbol{\alpha}^{be}\right)^\dagger \right\rangle \left[\mathbf{S}\left(\boldsymbol{\alpha}^0\right)\right]^\dagger \\
&= \left[\mathbf{S}\left(\boldsymbol{\alpha}^{be}\right)\right] \mathbf{C}_\alpha^{be} \left[\mathbf{S}\left(\boldsymbol{\alpha}^{be}\right)\right]^\dagger \\
&= \left[\mathbf{S}\left(\boldsymbol{\alpha}^{be}\right)\right] \left[\mathbf{C}_\alpha - \left(\mathbf{C}_{\alpha r} - \mathbf{C}_\alpha \left[\mathbf{S}\left(\boldsymbol{\alpha}^0\right)\right]^\dagger \right) \left[\mathbf{C}_d\left(\boldsymbol{\alpha}^0\right)\right]^{-1} \left(\mathbf{C}_{\alpha r} - \left[\mathbf{S}\left(\boldsymbol{\alpha}^0\right)\right] \mathbf{C}_\alpha \right) \right] \left[\mathbf{S}\left(\boldsymbol{\alpha}^{be}\right)\right]^\dagger .
\end{aligned} \qquad (1.38)$$

Comparing Eq. (1.38) to Eq. (1.34) reveals that $\mathbf{C}_{rc}^{be} \neq \mathbf{C}_r^{be}$, in general, since $\mathbf{S}\left(\boldsymbol{\alpha}^{be}\right) \neq \mathbf{S}\left(\boldsymbol{\alpha}^0\right)$. However, when the model is exactly linear, then the sensitivity matrix $\mathbf{S}$ is independent of the parameter values $\boldsymbol{\alpha}$, i.e.,

$$\mathbf{S}\left(\boldsymbol{\alpha}^{be}\right) = \mathbf{S}\left(\boldsymbol{\alpha}^0\right) = \mathbf{S} \text{, for linear models.} \qquad (1.39)$$

Replacing Eq. (1.39) into Eq. (1.38) yields the following result:

$$\begin{aligned}\mathbf{C}_{rc}^{be} &= \mathbf{S}\left[\mathbf{C}_{\alpha} - \left(\mathbf{C}_{\alpha r} - \mathbf{C}_{\alpha}\mathbf{S}^{\dagger}\right)\mathbf{C}_{d}^{-1}\left(\mathbf{C}_{\alpha r} - \mathbf{S}\mathbf{C}_{\alpha}\right)\right]\mathbf{S}^{\dagger} \\ &= \mathbf{C}_{rc} - \left(\mathbf{C}_{rc} - \mathbf{S}\mathbf{C}_{\alpha r}\right)\left[\mathbf{C}_{rc} + \mathbf{C}_{e} - \mathbf{C}_{r\alpha}\mathbf{S}^{\dagger} - \mathbf{S}\mathbf{C}_{\alpha r}\right]^{-1}\left(\mathbf{C}_{rc} - \mathbf{C}_{\alpha r}\mathbf{S}^{\dagger}\right) \\ &= \mathbf{C}_{r}^{be}, \quad \textbf{for linear models.}\end{aligned} \tag{1.40}$$

The equality shown in Eq. (1.40) can be demonstrated by using the following identity which holds for non-singular square matrices $\mathbf{A}$, $\mathbf{B}$ and $\mathbf{C}$,

$$\begin{aligned}&\mathbf{A} - \left(\mathbf{A} - \mathbf{C}^{\dagger}\right)\left(\mathbf{A} + \mathbf{B} - \mathbf{C} - \mathbf{C}^{\dagger}\right)^{-1}\left(\mathbf{A} - \mathbf{C}\right) \\ &\equiv \mathbf{B} - \left(\mathbf{B} - \mathbf{C}\right)\left(\mathbf{A} + \mathbf{B} - \mathbf{C} - \mathbf{C}^{\dagger}\right)^{-1}\left(\mathbf{B} - \mathbf{C}^{\dagger}\right).\end{aligned} \tag{1.41}$$

Effecting the replacements $\mathbf{A} \to \mathbf{C}_{rc}$, $\mathbf{B} \to \mathbf{C}_{e}$ $\mathbf{C} \to \mathbf{C}_{\alpha r}\mathbf{S}^{\dagger}$ in the identity expressed by Eq. (1.41) yields the result shown in Eq. (1.40). The relation expressed by Eq. (1.41) can also be obtained by starting from the identity

$$\mathbf{I} \equiv \left(\mathbf{A} - \mathbf{C}^{\dagger}\right)\left(\mathbf{A} + \mathbf{B} - \mathbf{C} - \mathbf{C}^{\dagger}\right)^{-1} + \left(\mathbf{B} - \mathbf{C}\right)\left(\mathbf{A} + \mathbf{B} - \mathbf{C} - \mathbf{C}^{\dagger}\right)^{-1}, \tag{1.42}$$

and by multiplying Eq. (1.42) on the right by $\left(\mathbf{A} - \mathbf{C}\right)$ to obtain

$$\begin{aligned}\mathbf{A} - \mathbf{C} &= \left(\mathbf{A} - \mathbf{C}^{\dagger}\right)\left(\mathbf{A} + \mathbf{B} - \mathbf{C} - \mathbf{C}^{\dagger}\right)^{-1}\left(\mathbf{A} - \mathbf{C}\right) \\ &+ \left(\mathbf{B} - \mathbf{C}\right)\left(\mathbf{A} + \mathbf{B} - \mathbf{C} - \mathbf{C}^{\dagger}\right)^{-1}\left(\mathbf{A} - \mathbf{C} + \mathbf{B} - \mathbf{B} + \mathbf{C}^{\dagger} - \mathbf{C}^{\dagger}\right) \\ &= \left(\mathbf{A} - \mathbf{C}^{\dagger}\right)\left(\mathbf{A} + \mathbf{B} - \mathbf{C} - \mathbf{C}^{\dagger}\right)^{-1}\left(\mathbf{A} - \mathbf{C}\right) \\ &+ \left(\mathbf{B} - \mathbf{C}\right)\left(\mathbf{A} + \mathbf{B} - \mathbf{C} - \mathbf{C}^{\dagger}\right)^{-1}\left(\mathbf{A} + \mathbf{B} - \mathbf{C} - \mathbf{C}^{\dagger}\right) \\ &- \left(\mathbf{B} - \mathbf{C}\right)\left(\mathbf{A} + \mathbf{B} - \mathbf{C} - \mathbf{C}^{\dagger}\right)^{-1}\left(\mathbf{B} - \mathbf{C}^{\dagger}\right).\end{aligned} \tag{1.43}$$

The result obtained in Eq. (1.43) reduces to Eq. (1.41).

It is important to note that the computation of the best estimate parameter and response values, together with their corresponding best-estimate covariance matrices, requires the inversion of a single matrix, namely the matrix $\mathbf{C}_{d}\left(\boldsymbol{\alpha}^{0}\right)$ defined in Eq. (1.20). Note also that $\mathbf{C}_{d}\left(\boldsymbol{\alpha}^{0}\right)$ is matrix of order I, which is computationally advantageous to invert in practice, since the number of measured (or computed responses) is most often considerably smaller that the number of model parameters N.

On the other hand, for the relatively rarely encountered practical instances when the number of model parameters significantly exceeds the number of model responses, i.e., when $I \gg N$, it is also possible to derive alternative expressions for the best-estimate calibrated parameters and their corresponding best-estimate covariances, by using the linearized model, namely Eq. (1.7) to eliminate at the outset the response $\mathbf{r}$, and carry out the minimization procedure solely for the parameters $\boldsymbol{\alpha}$, thus performing all derivations in the N-dimensional "parameter space" rather than in the I-dimensional "response space". These derivations are quite tedious to perform, and a considerable shortcut can be achieved by rewriting the matrix $\left[\mathbf{C}_d\left(\boldsymbol{\alpha}^0\right)\right]^{-1}$ in an alternative way, by employing the Sherman-Morrison-Woodbury extension, namely:

$$\left(\mathbf{A}+\mathbf{CBD}^{\dagger}\right)^{-1}=\mathbf{A}^{-1}-\mathbf{A}^{-1}\mathbf{C}\left(\mathbf{B}^{-1}+\mathbf{D}^{\dagger}\mathbf{A}^{-1}\mathbf{C}\right)^{-1}\mathbf{D}^{\dagger}\mathbf{A}^{-1}, \tag{1.44}$$

with $\mathbf{A}$ and $\mathbf{B}$ are invertible, and $\mathbf{D}=\mathbf{C}$. Thus, applying Eq. (1.44) to Eq. (1.20) leads to

$$\begin{aligned}
\mathbf{C}_d^{-1} &\triangleq \left(\mathbf{C}_{rc}-\mathbf{C}_{r\alpha}\mathbf{S}^{\dagger}-\mathbf{S}\mathbf{C}_{\alpha r}+\mathbf{C}_e\right)^{-1}\\
&=\mathbf{A}^{-1}-\mathbf{A}^{-1}\mathbf{S}\left(\mathbf{C}_{\alpha}^{-1}+\mathbf{S}^{\dagger}\mathbf{A}^{-1}\mathbf{S}\right)^{-1}\mathbf{S}^{\dagger}\mathbf{A}^{-1},\\
\mathbf{A} &\triangleq \mathbf{C}_e-\mathbf{C}_{r\alpha}\mathbf{S}^{\dagger}-\mathbf{S}\mathbf{C}_{\alpha r}.
\end{aligned} \tag{1.45}$$

The above expression provides the bridge between the "response-space" and "parameter-space" formulations, highlighting the fact that the response-space formulation requires a single inversion of an I-dimensional square symmetric matrix, while the "parameter space" formulation require the inversion of three symmetric matrices, two of which are N-dimensional and one is I-dimensional.

When the parameters and responses are initially uncorrelated, i.e., if $\mathbf{C}_{r\alpha}=\mathbf{0}$, then the expressions in parameter space of the best-estimate calibrated (adjusted) quantities can be simplified somewhat by using the following special form of Eq. (1.45)

$$\mathbf{BC}^{\dagger}\left(\mathbf{A}+\mathbf{CBC}^{\dagger}\right)^{-1}=\left(\mathbf{B}^{-1}+\mathbf{C}^{\dagger}\mathbf{A}^{-1}\mathbf{C}\right)^{-1}\mathbf{C}^{\dagger}\mathbf{A}^{-1}, \tag{1.46}$$

in which case Eq. (1.45) can be rewritten in the form

$$\mathbf{C}_{\alpha}\mathbf{S}^{\dagger}\mathbf{C}_d^{-1}=\left(\mathbf{C}_{\alpha}^{-1}+\mathbf{S}^{\dagger}\mathbf{C}_x^{-1}\mathbf{S}\right)^{-1}\mathbf{S}^{\dagger}\mathbf{C}_x^{-1},\ \text{when } \mathbf{C}_{r\alpha}=\mathbf{0}. \tag{1.47}$$

Consequently, when $\mathbf{C}_{r\alpha}=\mathbf{0}$, the expressions for $\boldsymbol{\alpha}^{be}$ and $\mathbf{C}_{\alpha}^{be}$ in the "parameter-space" become

$$\boldsymbol{\alpha}^{be}=\boldsymbol{\alpha}^0-\left(\mathbf{C}_{\alpha}^{-1}+\mathbf{S}^{\dagger}\mathbf{C}_m^{-1}\mathbf{S}\right)^{-1}\mathbf{S}^{\dagger}\mathbf{C}_m^{-1}\mathbf{d},\quad \text{when } \mathbf{C}_{r\alpha}=\mathbf{0}, \tag{1.48}$$

and

$$\mathbf{C}_{\alpha}^{be}=\left(\mathbf{C}_{\alpha}^{-1}+\mathbf{S}^{\dagger}\mathbf{C}_{m}^{-1}\mathbf{S}\right)^{-1}, \quad \text{when } \mathbf{C}_{r\alpha}=\mathbf{0}. \tag{1.49}$$

The computational evaluation of the above expressions still requires the inversion of two N-dimensional one I-dimensional symmetric matrices. From a computational standpoint, therefore, the "parameter-space" formulas should be avoided whenever possible, using the "response-space" formulas instead.

1.2.2 Time-Dependent "Perfect" Model

Consider that the physical problem under consideration is time-dependent, and consider that the time span of interest is partitioned into $(N_t - 1)$ intervals. For notational simplicity, the quantities J_t, J_{α}^{ν}, and J_r^{ν} will be used in the sequel to denote the ordered set of integers $J_t \triangleq \{1,...,N_t\}$, $J_{\alpha}^{\nu} \triangleq \{1,...,N_{\alpha}^{\nu}\}$, and $J_r^{\nu} \triangleq \{1,...,N_r^{\nu}\}$, respectively, where N_{α}^{ν} and N_r^{ν} denote the numbers of distinct system parameters and distinct responses, respectively, at every time instance ν. Furthermore, the notation $i \in J$, where i is an index and J is an ordered set of integers, will be used to signify that the index i takes on (i.e., runs through) all the integer values contained in J, i.e., $\nu \in J_t$ will signify that $\nu = 1,2,...,N_t$. Hence, at every time instance ν, the (column) vector $\boldsymbol{\alpha}^{\nu}$ of J_{α}^{ν} system parameters, and the (column) vector $\mathbf{r}^{\nu}$ of J_r^{ν} *measured responses* can be represented, in component form as

$$\boldsymbol{\alpha}^{\nu}=\left\{\alpha_n^{\nu}\,|\,n\in J_{\alpha}^{\nu}\right\}, \quad \mathbf{r}^{\nu}=\left\{r_i^{\nu}\,|\,i\in J_r^{\nu}\right\}, \quad \nu\in J_t. \tag{1.50}$$

The imprecisely known system parameters are considered to be variates having mean values $\left(\boldsymbol{\alpha}^0\right)^{\nu}$, at every time instance ν. The correlations between two parameters α_i^{ν} and α_j^{μ}, at two time instances μ and ν, with $\left(\nu,\mu\in J_t\right)$, are defined as follows:

$$c_{\alpha,ij}^{\nu\mu} \triangleq \left\langle\left[\alpha_i^{\nu}-\left(\alpha_i^{\nu}\right)^0\right]\left[\alpha_j^{\mu}-\left(\alpha_j^{\mu}\right)^0\right]\right\rangle, \tag{1.51}$$

and are considered to constitute the elements of symmetric covariance matrices of the form

$$\mathbf{C}_{\alpha}^{\mu\nu} \triangleq \left\langle\left(\boldsymbol{\alpha}-\boldsymbol{\alpha}^0\right)^{\mu}\left[\left(\boldsymbol{\alpha}-\boldsymbol{\alpha}^0\right)^{\nu}\right]^{\dagger}\right\rangle=\left(\mathbf{C}_{\alpha}^{\mu\nu}\right)^{\dagger}=\mathbf{C}_{\alpha}^{\nu\mu}=\left(\mathbf{C}_{\alpha}^{\nu\mu}\right)^{\dagger}. \tag{1.52}$$

Similarly, the imprecisely known measured responses are characterized by mean values $(\mathbf{r}_e)^\nu$ at a time instance ν, and covariance matrices, $\mathbf{C}_m^{\mu\nu}$, between two time instances μ and ν defined as follows:

$$\mathbf{C}_m^{\mu\nu} \triangleq \left\langle (\mathbf{r}-\mathbf{r}_m)^\mu \left[(\mathbf{r}-\mathbf{r}_m)^\nu \right]^\dagger \right\rangle = \left(\mathbf{C}_m^{\mu\nu}\right)^\dagger = \mathbf{C}_m^{\nu\mu} = \left(\mathbf{C}_m^{\nu\mu}\right)^\dagger . \tag{1.53}$$

In the most general case, the measured responses may be correlated to the parameters through a response-parameter covariance matrix, $\mathbf{C}_{r\alpha}^{\mu\nu}$, of order $I \times N$, of the form

$$\mathbf{C}_{r\alpha}^{\mu\nu} \triangleq \left\langle (\mathbf{r}-\mathbf{r}_m)^\mu \left[\left(\boldsymbol{\alpha}-\boldsymbol{\alpha}^0\right)^\nu \right]^\dagger \right\rangle = \left(\mathbf{C}_{r\alpha}^{\mu\nu}\right)^\dagger = \mathbf{C}_{r\alpha}^{\nu\mu} = \left(\mathbf{C}_{r\alpha}^{\nu\mu}\right)^\dagger . \tag{1.54}$$

At any given time instance ν, a response r_i^ν can be a function of not only the system parameters at time instance ν, but also of the system parameters at all previous time instances μ, $1 \le \mu \le \nu$; this means that $\mathbf{r}^\nu = \mathbf{R}^\nu(\mathbf{p}^\nu)$, where $\mathbf{p}^\nu \triangleq \left(\boldsymbol{\alpha}^1,\dots,\boldsymbol{\alpha}^\mu,\dots,\boldsymbol{\alpha}^\nu\right)$. As in the previous Section, the *computed response* is considered to depend *linearly on the model parameters*, i.e., the computed response is linearized via a functional Taylor-series expansion around the nominal values, $\mathbf{p}_0^\nu \triangleq \left(\left(\boldsymbol{\alpha}^0\right)^1,\dots,\left(\boldsymbol{\alpha}^0\right)^\mu,\dots,\left(\boldsymbol{\alpha}^0\right)^\nu\right)$, of the parameters $\mathbf{p}^\nu$, as follows:

$$\mathbf{r}^\nu \triangleq \mathbf{R}^\nu(\mathbf{p}^\nu) = \mathbf{R}^\nu(\mathbf{p}_0^\nu) + \sum_{\mu=1}^{\nu} \mathbf{S}^{\nu\mu}(\mathbf{p}_0^\mu)\left[\boldsymbol{\alpha}^\mu - \left(\boldsymbol{\alpha}^0\right)^\mu\right] + \textit{higher order terms},\ \nu \in J_t, \tag{1.55}$$

where $\mathbf{R}^\nu(\mathbf{p}_0^\nu)$ denotes the vector of computed responses at a time instance ν, at the nominal parameter values $\mathbf{p}_0^\nu$, while $\mathbf{S}^{\nu\mu}(\mathbf{p}_0^\mu)$, $1 \le \mu \le \nu$, represents the $\left(J_r^\nu \times J_\alpha^\mu\right)$-dimensional matrix containing the first Gateaux-derivatives of the computed responses with respect to the parameters, defined as follows:

$$\mathbf{S}^{\nu\mu}\left(\mathbf{p}_0^{\mu}\right) \triangleq \begin{pmatrix} s_{11}^{\nu\mu} & \cdots & s_{1N}^{\nu\mu} \\ \vdots & s_{in}^{\nu\mu} & \vdots \\ s_{I1}^{\nu\mu} & \cdots & s_{IN}^{\nu\mu} \end{pmatrix} \triangleq \begin{pmatrix} \dfrac{\partial R_1^{n}\left(\mathbf{p}_0^{\mu}\right)}{\partial \alpha_1^{\mu}} & \cdots & \dfrac{\partial R_1^{n}\left(\mathbf{p}_0^{\mu}\right)}{\partial \alpha_N^{\mu}} \\ \vdots & \dfrac{\partial R_i^{\nu}}{\partial \alpha_n^{\mu}} & \vdots \\ \dfrac{\partial R_I^{\nu}\left(\mathbf{p}_0^{\mu}\right)}{\partial \alpha_1^{\mu}} & \cdots & \dfrac{\partial R_I^{\nu}\left(\mathbf{p}_0^{\mu}\right)}{\partial \alpha_N^{\mu}} \end{pmatrix}, \; 1 \le \mu \le \nu . \tag{1.56}$$

Since the response $\mathbf{R}^{\nu}\left(\mathbf{p}_0^{\nu}\right)$ at time instance ν can depend only on parameters $\left(\boldsymbol{\alpha}^{0}\right)^{\mu}$ which appear up to the current time instance ν, it follows that $\mathbf{S}^{\nu\mu} = \mathbf{0}$ when $\mu > \nu$, implying that non-zero terms in the expansion shown in Eq. (1.55) can only occur in the range $1 \le \mu \le \nu$. By introducing the block matrix

$$\mathbf{S} \triangleq \begin{pmatrix} \mathbf{S}^{11} & \cdots & \mathbf{0} \\ \vdots & \ddots & \vdots \\ \mathbf{S}^{N_t 1} & \cdots & \mathbf{S}^{N_t N_t} \end{pmatrix}, \tag{1.57}$$

and the (block) column vectors

$$\begin{aligned} \boldsymbol{\alpha} &\triangleq \left(\boldsymbol{\alpha}^{1},\ldots,\boldsymbol{\alpha}^{\mu},\ldots,\boldsymbol{\alpha}^{N_t}\right) \\ \mathbf{r} &\triangleq \left(\mathbf{r}^{1},\ldots,\mathbf{r}^{\mu},\ldots,\mathbf{r}^{N_t}\right) \\ \mathbf{R}\left(\boldsymbol{\alpha}^{0}\right) &\triangleq \left(\mathbf{R}^{1},\ldots,\mathbf{R}^{\mu},\ldots,\mathbf{R}^{N_t}\right) \end{aligned} , \tag{1.58}$$

the system shown in Eq. (1.55) can be written in the same form as Eq. (1.7), namely

$$\mathbf{r} = \mathbf{R}\left(\boldsymbol{\alpha}^{0}\right) + \mathbf{S}\left(\boldsymbol{\alpha} - \boldsymbol{\alpha}^{0}\right) + \textit{higher order terms} . \tag{1.59}$$

The covariance matrix of the computed responses, $\mathbf{C}_{rc}$, defined as

$$\mathbf{C}_{rc} \triangleq \begin{pmatrix} \mathbf{C}_{rc}^{11} & \cdots & \mathbf{C}_{rc}^{1N_t} \\ \vdots & \ddots & \vdots \\ \mathbf{C}_{rc}^{N_t 1} & \cdots & \mathbf{C}_{rc}^{N_t N_t} \end{pmatrix} = \mathbf{S}\,\mathbf{C}_{\alpha}\mathbf{S}^{\dagger}, \tag{1.60}$$

is obtained by multiplying each side of Eq. (1.59) by its own transpose and subsequently computing the expectation value of the resulting expression. The matrix

$\mathbf{C}_{rc}$ is a symmetric positive-definite matrix, and is hence a proper covariance matrix. The components of $\mathbf{C}_{rc}$ are the covariance matrices $\mathbf{C}_{rc}^{\nu\mu}$ of responses calculated at time instances ν and μ, having the following expressions:

$$\mathbf{C}_{rc}^{\nu\mu} = \sum_{\eta=1}^{\nu}\sum_{\rho=1}^{\mu} \mathbf{S}^{\nu\eta}\mathbf{C}_{\alpha}^{\eta\rho}\left(\mathbf{S}^{\mu\rho}\right)^{\dagger} = \left(\mathbf{C}_{rc}^{\mu\nu}\right)^{\dagger}; \quad \nu,\mu \in J_t . \tag{1.61}$$

1.2.2.1 BERRU-SMS Methodology which Assimilates all Available Information Simultaneously

Applying now the maximum entropy algorithm (which is detailed in Appendix 1, at the end of Section 1.2) to the computational and experimental information described in Eqs. (1.50) through (1.61) indicates that the most objective probability distribution for this information is a multivariate Gaussian of the same form as Eq. (1.4), namely,

$$p(\mathbf{z}|\mathbf{C})d(\mathbf{z}) = \frac{exp\left[-\frac{1}{2}Q(\mathbf{z})\right]}{det(2\pi\mathbf{C})^{1/2}}d(\mathbf{z}),\ Q(\mathbf{z}) \triangleq \mathbf{z}^{\dagger}\mathbf{C}^{-1}\mathbf{z},\ -\infty < z_j < \infty, \tag{1.62}$$

where:

$$\mathbf{z} \triangleq \begin{pmatrix} \boldsymbol{\alpha}-\boldsymbol{\alpha}^0 \\ \mathbf{r}-\mathbf{r}_x \end{pmatrix}, \qquad \boldsymbol{\alpha}^0 \triangleq \left[\left(\boldsymbol{\alpha}^0\right)^1, ..., \left(\boldsymbol{\alpha}^0\right)^{\mu}, ..., \left(\boldsymbol{\alpha}^0\right)^{N_t}\right]^{\dagger}, \tag{1.63}$$

$$\mathbf{C} \triangleq \begin{pmatrix} \mathbf{C}_{\alpha} & \mathbf{C}_{\alpha r} \\ \mathbf{C}_{r\alpha} & \mathbf{C}_{m} \end{pmatrix},\ \mathbf{C}_{\alpha} \triangleq \begin{pmatrix} \mathbf{C}_{\alpha}^{11} & \mathbf{C}_{\alpha}^{12} & \cdots & \cdots \\ \mathbf{C}_{\alpha}^{21} & \mathbf{C}_{\alpha}^{22} & \cdots & \cdots \\ \cdots & \cdots & \cdots & \cdots \\ \cdots & \cdots & \cdots & \mathbf{C}_{\alpha}^{N_tN_t} \end{pmatrix},\ \mathbf{C}_{\alpha r} \triangleq \begin{pmatrix} \mathbf{C}_{\alpha r}^{11} & \mathbf{C}_{\alpha r}^{12} & \cdots & \cdots \\ \mathbf{C}_{\alpha r}^{21} & \mathbf{C}_{\alpha r}^{22} & \cdots & \cdots \\ \cdots & \cdots & \cdots & \cdots \\ \cdots & \cdots & \cdots & \mathbf{C}_{\alpha r}^{N_tN_t} \end{pmatrix},$$

$$\mathbf{C}_{m} \triangleq \begin{pmatrix} \mathbf{C}_{m}^{11} & \mathbf{C}_{m}^{12} & \cdots & \cdots \\ \mathbf{C}_{m}^{21} & \mathbf{C}_{m}^{22} & \cdots & \cdots \\ \cdots & \cdots & \cdots & \cdots \\ \cdots & \cdots & \cdots & \mathbf{C}_{m}^{N_tN_t} \end{pmatrix}. \tag{1.64}$$

The posterior information contained in Eqs. (1.62) and (1.59) can now be condensed into a recommended best-estimate value $\left(\mathbf{z}^{be}\right)^{\nu}$ at a time node ν for the parameters $\boldsymbol{\alpha}^{\nu}$ and responses $\mathbf{r}^{\nu}$, together with corresponding best-estimate uncertainties for these quantities, following the procedure used in Subsection 1.2.1. Thus, the bulk of the contribution to the distribution $p(\mathbf{z}|\mathbf{C})$ in Eq. (1.62) is computed using the "saddle-point method" at the point in phase space where the exponent in the expression of $p(\mathbf{z}|\mathbf{C})$ attains its minimum, subject to the relation provided by Eq. (1.59). When the numerical errors are also neglected in Eq. (1.59), in addition to neglecting the higher order terms, then Eq. (1.59) is imposed as a hard constraint, which can be conveniently written in the form

$$\mathbf{Z}\left(\boldsymbol{\alpha}^0\right)\mathbf{z}+\mathbf{d}=\mathbf{0}, \quad \mathbf{d}\triangleq\mathbf{R}\left(\boldsymbol{\alpha}^0\right)-\mathbf{r}_m, \tag{1.65}$$

where $\mathbf{r}_m \triangleq \left(\mathbf{r}_m^1,\dots,\mathbf{r}_m^\mu,\dots,\mathbf{r}_m^{N_t}\right)$, $\mathbf{Z}$ denotes the partitioned matrix

$$\mathbf{Z}\triangleq\begin{pmatrix}\mathbf{S} & \mathbf{U}\end{pmatrix}; \quad \mathbf{U}\triangleq\begin{pmatrix} -\mathbf{I}^{11} & \cdots & \mathbf{0} \\ \vdots & \ddots & \vdots \\ \mathbf{0} & \cdots & -\mathbf{I}^{N_tN_t}\end{pmatrix}, \tag{1.66}$$

and the quantities $\mathbf{I}^{jj}, j=1,\dots,N_t$ denote the identity matrices of corresponding dimensions. Thus, the augmented Lagrangian functional $P(\mathbf{z})$ to be minimized becomes formally identical to the expression given in Eq. (1.11), namely

$$P(\mathbf{z})\triangleq Q(\mathbf{z})+2\boldsymbol{\lambda}^\dagger\left[\mathbf{Z}\left(\boldsymbol{\alpha}^0\right)\mathbf{z}+\mathbf{d}\right]=min,\ at\ \mathbf{z}=\mathbf{z}^{be}, \tag{1.67}$$

where $\boldsymbol{\lambda}=\left(\boldsymbol{\lambda}^1,\dots,\boldsymbol{\lambda}^\nu,\dots,\boldsymbol{\lambda}^{N_t}\right)$ denotes the corresponding vector of Lagrange multipliers. Minimizing the augmented Lagrangian in Eq. (1.67) leads to expressions that are formally identical to Eqs. (1.22) and (1.23) for the best-estimate calibrated parameters and responses, respectively, and to expressions that are formally identical to Eqs. (1.32) through (1.35) for their corresponding calibrated best-estimate covariance matrices. Recall that the inversion of the symmetric matrix $\mathbf{C}_d\triangleq\mathbf{C}_{rc}-\mathbf{C}_{r\alpha}\mathbf{S}^\dagger-\mathbf{S}\mathbf{C}_{\alpha r}+\mathbf{C}_m$ plays a central role in the computation of the calibrated best estimate results. In the present (time-dependent) setting, the matrix $\mathbf{C}_d$ has the following form:

$$\mathbf{C}_d \triangleq \begin{pmatrix} \mathbf{C}_d^{11} & \cdots & \mathbf{C}_d^{1N_t} \\ \vdots & \ddots & \vdots \\ \mathbf{C}_d^{N_t 1} & \cdots & \mathbf{C}_d^{N_t N_t} \end{pmatrix} = \begin{pmatrix} \mathbf{C}_{rc}^{11} + \mathbf{C}_m^{11} & \cdots & \mathbf{C}_{rc}^{1N_t} + \mathbf{C}_m^{1N_t} \\ \vdots & \ddots & \vdots \\ \mathbf{C}_{rc}^{N_t 1} + \mathbf{C}_m^{N_t 1} & \cdots & \mathbf{C}_{rc}^{N_t N_t} + \mathbf{C}_m^{N_t N_t} \end{pmatrix}$$
$$- \begin{pmatrix} \mathbf{C}_{r\alpha}^{11}\left(\mathbf{S}^{\dagger}\right)^{11} + \mathbf{S}^{11}\mathbf{C}_{\alpha r}^{11} & \cdots & \mathbf{S}^{11}\mathbf{C}_{\alpha r}^{1N_t} + \sum_{\rho=1}^{N_t}\mathbf{C}_{r\alpha}^{1\rho}\left(\mathbf{S}^{\dagger}\right)^{N_t\rho} \\ \vdots & \ddots & \vdots \\ \mathbf{C}_{r\alpha}^{N_t 1}\left(\mathbf{S}^{\dagger}\right)^{11} + \sum_{\rho=1}^{N_t}\mathbf{S}^{N_t\rho}\mathbf{C}_{\alpha r}^{\rho 1} & \cdots & \sum_{\rho=1}^{N_t}\left[\mathbf{C}_{r\alpha}^{N_t\rho}\left(\mathbf{S}^{\dagger}\right)^{N_t\rho} + \mathbf{S}^{N_t\rho}\mathbf{C}_{\alpha r}^{\rho N_t}\right] \end{pmatrix}. \quad (1.68)$$

In component form, the resulting expression of the calibrated best-estimate parameter values has the same formal structure as shown in Eq. (1.22), namely:

$$\begin{pmatrix} \left(\boldsymbol{\alpha}^{be}\right)^{1} \\ \vdots \\ \left(\boldsymbol{\alpha}^{be}\right)^{N_t} \end{pmatrix} = \begin{pmatrix} \left(\boldsymbol{\alpha}^{0}\right)^{1} \\ \vdots \\ \left(\boldsymbol{\alpha}^{0}\right)^{N_t} \end{pmatrix} +$$
$$\begin{pmatrix} \mathbf{C}_{\alpha r}^{11} - \mathbf{C}_{\alpha}^{11}\left(\mathbf{S}^{\dagger}\right)^{11} & \cdots & \mathbf{C}_{\alpha r}^{1N_t} - \sum_{\rho=1}^{N_t}\mathbf{C}_{\alpha}^{1\rho}\left(\mathbf{S}^{\dagger}\right)^{N_t\rho} \\ \vdots & \ddots & \vdots \\ \mathbf{C}_{\alpha r}^{N_t 1} - \mathbf{C}_{\alpha}^{N_t 1}\left(\mathbf{S}^{\dagger}\right)^{11} & \cdots & \mathbf{C}_{\alpha r}^{N_t N_t} - \sum_{\rho=1}^{N_t}\mathbf{C}_{\alpha}^{N_t\rho}\left(\mathbf{S}^{\dagger}\right)^{N_t\rho} \end{pmatrix} \times \begin{pmatrix} \sum_{\eta=1}^{N_t}\mathbf{K}_d^{1\eta}\mathbf{d}^{\eta} \\ \vdots \\ \sum_{\eta=1}^{N_t}\mathbf{K}_d^{N_t\eta}\mathbf{d}^{\eta} \end{pmatrix}, \quad (1.69)$$

where $\mathbf{K}_d^{\nu\eta}$ denotes the corresponding (ν,η)-element of the block-matrix $\mathbf{C}_d^{-1}$. Written in component form, Eq. (1.69) indicates that the vector $\left(\boldsymbol{\alpha}^{be}\right)^{\nu}$, representing the calibrated best-estimates for the system parameters at a specific time instance ν, takes on the expression

$$\left(\boldsymbol{\alpha}^{be}\right)^{\nu} = \left(\boldsymbol{\alpha}^{0}\right)^{\nu} + \sum_{\mu=1}^{N_t}\left\{\left[\mathbf{C}_{\alpha r}^{\nu\mu} - \sum_{\rho=1}^{\mu}\mathbf{C}_{\alpha}^{\nu\rho}\left(\mathbf{S}^{\dagger}\right)^{\mu\rho}\right]\left[\sum_{\eta=1}^{N_t}\mathbf{K}_d^{\mu\eta}\mathbf{d}^{\eta}\right]\right\}, \quad \nu \in J_t. \quad (1.70)$$

The calibrated best-estimate covariance matrix, $\mathbf{C}_{\alpha}^{be}$, corresponding to the calibrated best-estimates system parameters is derived as in Subsection 1.2.1, ultimately obtaining an expression that has formally the same structure as Eq. (1.32), namely

$$\mathbf{C}_{\alpha}^{be} \triangleq \begin{pmatrix} \left(\mathbf{C}_{\alpha}^{be}\right)^{11} & \cdots & \left(\mathbf{C}_{\alpha}^{be}\right)^{1N_t} \\ \vdots & \ddots & \vdots \\ \left(\mathbf{C}_{\alpha}^{be}\right)^{N_t 1} & \cdots & \left(\mathbf{C}_{\alpha}^{be}\right)^{N_t N_t} \end{pmatrix} = \begin{pmatrix} \mathbf{C}_{\alpha}^{11} & \cdots & \mathbf{C}_{\alpha}^{1N_t} \\ \vdots & \ddots & \vdots \\ \mathbf{C}_{\alpha}^{N_t 1} & \cdots & \mathbf{C}_{\alpha}^{N_t N_t} \end{pmatrix}$$
$$- \begin{pmatrix} \mathbf{C}_{\alpha d}^{11} & \cdots & \mathbf{C}_{\alpha d}^{1N_t} \\ \vdots & \ddots & \vdots \\ \mathbf{C}_{\alpha d}^{N_t 1} & \cdots & \mathbf{C}_{\alpha d}^{N_t N_t} \end{pmatrix} \begin{pmatrix} \mathbf{K}_{d}^{11} & \cdots & \mathbf{K}_{d}^{1N_t} \\ \vdots & \ddots & \vdots \\ \mathbf{K}_{d}^{N_t 1} & \cdots & \mathbf{K}_{d}^{N_t N_t} \end{pmatrix} \begin{pmatrix} \mathbf{C}_{\alpha d}^{11} & \cdots & \mathbf{C}_{\alpha d}^{1N_t} \\ \vdots & \ddots & \vdots \\ \mathbf{C}_{\alpha d}^{N_t 1} & \cdots & \mathbf{C}_{\alpha d}^{N_t N_t} \end{pmatrix}^{\dagger} \tag{1.71}$$

where

$$\begin{pmatrix} \mathbf{C}_{\alpha d}^{11} & \cdots & \mathbf{C}_{\alpha d}^{1N_t} \\ \vdots & \ddots & \vdots \\ \mathbf{C}_{\alpha d}^{N_t 1} & \cdots & \mathbf{C}_{\alpha d}^{N_t N_t} \end{pmatrix} \triangleq \begin{pmatrix} \mathbf{C}_{\alpha r}^{11} - \mathbf{C}_{\alpha}^{11}\left(\mathbf{S}^{\dagger}\right)^{11} & \cdots & \mathbf{C}_{\alpha r}^{1N_t} - \sum_{\rho=1}^{N_t} \mathbf{C}_{\alpha}^{1\rho}\left(\mathbf{S}^{\dagger}\right)^{N_t\rho} \\ \vdots & \ddots & \vdots \\ \mathbf{C}_{\alpha r}^{N_t 1} - \mathbf{C}_{\alpha}^{N_t 1}\left(\mathbf{S}^{\dagger}\right)^{11} & \cdots & \mathbf{C}_{\alpha r}^{N_t N_t} - \sum_{\rho=1}^{N_t} \mathbf{C}_{\alpha}^{N_t\rho}\left(\mathbf{S}^{\dagger}\right)^{N_t\rho} \end{pmatrix}. \tag{1.72}$$

The block-matrix expression in Eq. (1.71) can be written in component form, for the calibrated best-estimate parameter covariance matrix $\left(\mathbf{C}_{\alpha}^{be}\right)^{\nu\mu}$ between two (distinct or not) time instances $\nu, \mu \in J_t$, as follows

$$\left(\mathbf{C}_{\alpha}^{be}\right)^{\nu\mu} = \mathbf{C}_{\alpha}^{\nu\mu} - \sum_{\eta=1}^{N_t}\sum_{\rho=1}^{N_t}\left[\mathbf{C}_{\alpha r}^{\nu\rho} - \sum_{\pi=1}^{\rho} \mathbf{C}_{\alpha}^{\nu\pi}\left(\mathbf{S}^{\dagger}\right)^{\rho\pi}\right]\mathbf{K}_{d}^{\rho\eta}\left[\mathbf{C}_{r\alpha}^{\eta\mu} - \sum_{\pi=1}^{\eta} \mathbf{S}^{\eta\pi}\mathbf{C}_{\alpha}^{\pi\mu}\right]. \tag{1.73}$$

The vector $\mathbf{R}\left(\boldsymbol{\alpha}^{be}\right) \triangleq \mathbf{r}^{be}$, representing the calibrated best-estimate system responses at all time instances $\nu \in J_t$, is also derived by following the procedure outlined in Subsection 1.2.1, ultimately obtaining the following expression [which formally has the same structure as Eq. (1.23)]:

$$\begin{pmatrix} \left(\mathbf{r}^{be}\right)^{1} \\ \vdots \\ \left(\mathbf{r}^{be}\right)^{N_t} \end{pmatrix} = \begin{pmatrix} \left(\mathbf{r}_{m}\right)^{1} \\ \vdots \\ \left(\mathbf{r}_{m}\right)^{N_t} \end{pmatrix} +$$
$$\begin{pmatrix} \mathbf{C}_{m}^{11} - \mathbf{C}_{r\alpha}^{11}\left(\mathbf{S}^{\dagger}\right)^{11} & \cdots & \mathbf{C}_{m}^{1N_t} - \sum_{\rho=1}^{N_t} \mathbf{C}_{r\alpha}^{1\rho}\left(\mathbf{S}^{\dagger}\right)^{N_t\rho} \\ \vdots & \ddots & \vdots \\ \mathbf{C}_{m}^{N_t 1} - \mathbf{C}_{r\alpha}^{N_t 1}\left(\mathbf{S}^{\dagger}\right)^{11} & \cdots & \mathbf{C}_{m}^{N_t N_t} - \sum_{\rho=1}^{N_t} \mathbf{C}_{r\alpha}^{N_t\rho}\left(\mathbf{S}^{\dagger}\right)^{N_t\rho} \end{pmatrix} \times \begin{pmatrix} \sum_{\eta=1}^{N_t} \mathbf{K}_{d}^{1\eta}\mathbf{d}^{\eta} \\ \vdots \\ \sum_{\eta=1}^{N_t} \mathbf{K}_{d}^{N_t\eta}\mathbf{d}^{\eta} \end{pmatrix}. \tag{1.74}$$

Written in component form, Eq. (1.74) gives the following expression for the vector $\left(\mathbf{r}^{be}\right)^{\nu}$, of calibrated best-estimates for the responses at a specific time instance ν:

$$\left(\mathbf{r}^{be}\right)^{\nu}=\left(\mathbf{r}_{m}\right)^{\nu}+\sum_{\mu=1}^{N_t}\left\{\left[\mathbf{C}_{x}^{\nu\mu}-\sum_{\rho=1}^{\mu}\mathbf{C}_{r\alpha}^{\nu\rho}\left(\mathbf{S}^{\dagger}\right)^{\mu\rho}\right]\left[\sum_{\eta=1}^{N_t}\mathbf{K}_{d}^{\mu\eta}\mathbf{d}^{\eta}\right]\right\},\quad \nu\in J_t\,. \tag{1.75}$$

The expression of the calibrated best-estimate covariance block-matrix, $\mathbf{C}_{r}^{be}$, for the best-estimate responses is also obtained by following the procedure used in Subsection 1.2.1, ultimately arriving at an expression that has the same formal structure as Eq. (1.34), namely:

$$\mathbf{C}_{r}^{be}\triangleq\begin{pmatrix}\left(\mathbf{C}_{r}^{be}\right)^{11} & \cdots & \left(\mathbf{C}_{r}^{be}\right)^{1N_t}\\ \vdots & \ddots & \vdots\\ \left(\mathbf{C}_{r}^{be}\right)^{N_t1} & \cdots & \left(\mathbf{C}_{r}^{be}\right)^{N_tN_t}\end{pmatrix}=\begin{pmatrix}\mathbf{C}_{m}^{11} & \cdots & \mathbf{C}_{m}^{1N_t}\\ \vdots & \ddots & \vdots\\ \mathbf{C}_{m}^{N_t1} & \cdots & \mathbf{C}_{m}^{N_tN_t}\end{pmatrix}$$
$$-\begin{pmatrix}\mathbf{C}_{rd}^{11} & \cdots & \mathbf{C}_{rd}^{1N_t}\\ \vdots & \ddots & \vdots\\ \mathbf{C}_{rd}^{N_t1} & \cdots & \mathbf{C}_{rd}^{N_tN_t}\end{pmatrix}\begin{pmatrix}\mathbf{K}_{d}^{11} & \cdots & \mathbf{K}_{d}^{1N_t}\\ \vdots & \ddots & \vdots\\ \mathbf{K}_{d}^{N_t1} & \cdots & \mathbf{K}_{d}^{N_tN_t}\end{pmatrix}\begin{pmatrix}\mathbf{C}_{rd}^{11} & \cdots & \mathbf{C}_{rd}^{1N_t}\\ \vdots & \ddots & \vdots\\ \mathbf{C}_{rd}^{N_t1} & \cdots & \mathbf{C}_{rd}^{N_tN_t}\end{pmatrix}^{\dagger}, \tag{1.76}$$

where

$$\begin{pmatrix}\mathbf{C}_{rd}^{11} & \cdots & \mathbf{C}_{rd}^{1N_t}\\ \vdots & \ddots & \vdots\\ \mathbf{C}_{rd}^{N_t1} & \cdots & \mathbf{C}_{rd}^{N_tN_t}\end{pmatrix}=\begin{pmatrix}\mathbf{C}_{m}^{11}-\mathbf{C}_{r\alpha}^{11}\left(\mathbf{S}^{\dagger}\right)^{11} & \cdots & \mathbf{C}_{m}^{1N_t}-\sum_{\rho=1}^{N_t}\mathbf{C}_{r\alpha}^{1\rho}\left(\mathbf{S}^{\dagger}\right)^{N_t\rho}\\ \vdots & \ddots & \vdots\\ \mathbf{C}_{m}^{N_t1}-\mathbf{C}_{r\alpha}^{N_t1}\left(\mathbf{S}^{\dagger}\right)^{11} & \cdots & \mathbf{C}_{m}^{N_tN_t}-\sum_{\rho=1}^{N_t}\mathbf{C}_{r\alpha}^{N_t\rho}\left(\mathbf{S}^{\dagger}\right)^{N_t\rho}\end{pmatrix}. \tag{1.77}$$

The block-matrix expression given in Eq. (1.76) can be written in component form, where each of the calibrated best-estimate parameter covariance matrix $\left(\mathbf{C}_{r}^{be}\right)^{\nu\mu}$ between two (distinct or not) time instances $\nu,\mu\in J_t$, has the following form:

$$\left(\mathbf{C}_{r}^{be}\right)^{\nu\mu}=\mathbf{C}_{m}^{\nu\mu}-\sum_{\eta=1}^{N_t}\sum_{\rho=1}^{N_t}\left[\mathbf{C}_{m}^{\nu\rho}-\sum_{\pi=1}^{\rho}\mathbf{C}_{r\alpha}^{\nu\pi}\left(\mathbf{S}^{\dagger}\right)^{\rho\pi}\right]\mathbf{K}_{d}^{\rho\eta}\left[\mathbf{C}_{m}^{\eta\mu}-\sum_{\pi=1}^{\eta}\mathbf{S}^{\eta\pi}\mathbf{C}_{\alpha r}^{\pi\mu}\right]. \tag{1.78}$$

A similar sequence of computations leads to the following expression, having the same formal structure as Eq. (1.35), for the best-estimate response-parameter covariance block-matrix $\mathbf{C}_{\alpha r}^{be}$:

$$\mathbf{C}_{r\alpha}^{be} \triangleq \begin{pmatrix} \left(\mathbf{C}_{r\alpha}^{be}\right)^{11} & \cdots & \left(\mathbf{C}_{r\alpha}^{be}\right)^{1N_t} \\ \vdots & \ddots & \vdots \\ \left(\mathbf{C}_{r\alpha}^{be}\right)^{N_t 1} & \cdots & \left(\mathbf{C}_{r\alpha}^{be}\right)^{N_t N_t} \end{pmatrix} = \begin{pmatrix} \mathbf{C}_{r\alpha}^{11} & \cdots & \mathbf{C}_{r\alpha}^{1N_t} \\ \vdots & \ddots & \vdots \\ \mathbf{C}_{r\alpha}^{N_t 1} & \cdots & \mathbf{C}_{r\alpha}^{N_t N_t} \end{pmatrix} - \begin{pmatrix} \mathbf{C}_{rd}^{11} & \cdots & \mathbf{C}_{rd}^{1N_t} \\ \vdots & \ddots & \vdots \\ \mathbf{C}_{rd}^{N_t 1} & \cdots & \mathbf{C}_{rd}^{N_t N_t} \end{pmatrix} \begin{pmatrix} \mathbf{K}_{d}^{11} & \cdots & \mathbf{K}_{d}^{1N_t} \\ \vdots & \ddots & \vdots \\ \mathbf{K}_{d}^{N_t 1} & \cdots & \mathbf{K}_{d}^{N_t N_t} \end{pmatrix} \begin{pmatrix} \mathbf{C}_{\alpha d}^{11} & \cdots & \mathbf{C}_{\alpha d}^{1N_t} \\ \vdots & \ddots & \vdots \\ \mathbf{C}_{\alpha d}^{N_t 1} & \cdots & \mathbf{C}_{\alpha d}^{N_t N_t} \end{pmatrix}^{\dagger} . \quad (1.79)$$

Each of the calibrated best-estimate parameter-response covariance matrices $\left(\mathbf{C}_{\alpha r}^{be}\right)^{\nu\mu}$, $\nu,\mu \in J_t$, which appear as block-matrix components of the matrix $\mathbf{C}_{r\alpha}^{be}$ in Eq. (1.79), has the following expression:

$$\left(\mathbf{C}_{r\alpha}^{be}\right)^{\nu\mu} = \mathbf{C}_{r\alpha}^{\nu\mu} - \sum_{\eta=1}^{N_t}\sum_{\rho=1}^{N_t}\left[\mathbf{C}_{x}^{\nu\rho} - \sum_{\pi=1}^{\rho}\mathbf{C}_{r\alpha}^{\nu\pi}\left(\mathbf{S}^{\dagger}\right)^{\rho\pi}\right]\mathbf{K}_{d}^{\rho\eta}\left[\mathbf{C}_{\alpha r}^{\eta\mu} - \sum_{\pi=1}^{\eta}\mathbf{S}^{\eta\pi}\mathbf{C}_{\alpha}^{\pi\mu}\right]. \quad (1.80)$$

Computing the calibrated best-estimate quantities by using Eqs. (1.70), (1.73), (1.75), (1.78), and (1.80) is definitely more advantageous in terms of storage requirements than the direct computations of the corresponding full block-matrices. The largest requirement of computational resources arises for inverting the matrix $\mathbf{C}_d$. In view of Eq. (1.68), it is important to note that *the inverse matrix,* $\mathbf{C}_d^{-1}$, *incorporates simultaneously all of the available information about the system parameters and responses at all time instances* (i.e., $\nu = 1,2,...,N_t$). In other words, at any time instance ν, $\mathbf{C}_d^{-1}$ incorporates information not only from time instances prior to, and at, ν (i.e., information regarding the "past" and "present" states of the system) but also from time instances posterior to ν (i.e., information about the "future" states of the system). Therefore, at any specified time instance ν, the calibrated best-estimates parameters $\left(\boldsymbol{\alpha}^{be}\right)^{\nu}$ and responses $\mathbf{r}\left(\boldsymbol{\alpha}^{be}\right) \triangleq \mathbf{r}^{be}$ together with the corresponding calibrated best-estimate covariance matrices $\left(\mathbf{C}_{\alpha}^{be}\right)^{\nu\mu}$, $\left(\mathbf{C}_{r}^{be}\right)^{\nu\mu}$, and $\left(\mathbf{C}_{\alpha r}^{be}\right)^{\nu\mu}$ will also incorporate automatically, through the matrix $\mathbf{C}_d^{-1}$, *all of the available information about the system parameters and responses at all time instances*, i.e., $\left(\nu = 1,2,...,N_t\right)$. In this respect, the BERRU-SMS predictive modeling methodology presented in this Section is conceptually related to the "foresight" aspects encountered in decision analysis.

It is also important to note that, in practice, the application of the BERRU-SMS predictive modeling methodology involves two distinct computational stages. The first computational stage involves the generation of the requisite a priori information, including the generation of a complete sensitivity data base (comprising the sensitivities $s_{ni}^{\nu\mu}$ at all times instances $\nu,\mu \in J_t$). The second stage involves the simultaneous execution of "data assimilation" and "model calibration" (or data adjustment) together with the computation of the calibrated best-estimate covariance matrices (the "uncertainty analysis" stage), which involves combining the sensitivities with covariance matrices.

Because of the "foresight" and "off-line" characteristics, the methodology presented in this Subsection can be called the *"off-line with foresight" BERRU-SMS methodology*, to serve as a reminder that all sensitivities are generated separately, "off-line," prior to performing the uncertainty analysis, and that foresight characteristics are included automatically in the procedure. Since the incorporation of foresight effects involves the inversion of the matrix $\mathbf{C}_d$, this methodology is best suited for problems involving relatively few time steps. For large-scale problems that involve many time steps, the matrix $\mathbf{C}_d$ becomes very large, so its inversion may become prohibitively expensive. These difficulties can be reduced at the expense of using less than the complete information available at any specific time instance. For example, in time-dependent problems in which the entire time history is known (e.g., transient behavior of reactor systems), one may nevertheless choose to use only information up to the current time index, and disregard the information about "future" system states. On the other hand, in dynamical problems such as climate or weather prediction, in which the time variable advances continuously and states beyond the current time are not known, information about future states cannot be reliably accounted for anyway. Thus, the most common way of reducing the dimensionality of the data assimilation and model calibration problem is to disregard information about future states and limit the amount of information assimilated about "past states". Data assimilation and model calibration procedure using such a limited amount of information can be performed either off-line or on-line, assimilating the new data as the time index advances. The simplest such case occurs when the data assimilation and model calibration is carried out by using information from only two successive time-steps, on-line; this particular case will be presented in the next Subsection.

1.2.2.2 BERRU-SMS "Two-Time-Steps" Sequential Predictive Modeling Methodology

When only the information from two consecutive time instances, $\nu = k-1,k;\quad k=(1,2,...,N_t)$, is considered, Eq. (1.55) becomes

$$\begin{pmatrix} \mathbf{r}^{k-1} \\ \mathbf{r}^{k} \end{pmatrix} = \begin{pmatrix} \mathbf{R}^{k-1} \\ \mathbf{R}^{k} \end{pmatrix} + \begin{pmatrix} \mathbf{S}^{k-1,k-1} & \mathbf{0} \\ \mathbf{S}^{k,k-1} & \mathbf{S}^{k,k} \end{pmatrix} \begin{pmatrix} \boldsymbol{\alpha}^{k-1} - \left(\boldsymbol{\alpha}^{0}\right)^{k-1} \\ \boldsymbol{\alpha}^{k} - \left(\boldsymbol{\alpha}^{0}\right)^{k} \end{pmatrix}. \tag{1.81}$$

Corresponding to the relation above, Eq. (1.61) reduces to the following form:

$$\begin{pmatrix} \mathbf{C}_{rc}^{k-1,k-1} & \mathbf{C}_{rc}^{k-1,k} \\ \mathbf{C}_{rc}^{k,k-1} & \mathbf{C}_{rc}^{k,k} \end{pmatrix} = \begin{pmatrix} \mathbf{S}^{k-1,k-1} & \mathbf{0} \\ \mathbf{S}^{k,k-1} & \mathbf{S}^{k,k} \end{pmatrix} \begin{pmatrix} \mathbf{C}_{\alpha}^{k-1,k-1} & \mathbf{C}_{\alpha}^{k-1,k} \\ \mathbf{C}_{\alpha}^{k,k-1} & \mathbf{C}_{\alpha}^{k,k} \end{pmatrix} \begin{pmatrix} \left(\mathbf{S}^{k-1,k-1}\right)^{\dagger} & \left(\mathbf{S}^{k,k-1}\right)^{\dagger} \\ \mathbf{0} & \left(\mathbf{S}^{k,k}\right)^{\dagger} \end{pmatrix}. \tag{1.82}$$

Hence, the explicit expressions of the components of the covariance matrix of the computed responses, $\mathbf{C}_r$, are as follows:

$$\mathbf{C}_{rc}^{k-1,k-1} = \mathbf{S}^{k-1,k-1}\mathbf{C}_{\alpha}^{k-1,k-1}\left(\mathbf{S}^{k-1,k-1}\right)^{\dagger} = \left(\mathbf{C}_{rc}^{k-1,k-1}\right)^{\dagger}, \tag{1.83}$$

$$\mathbf{C}_{rc}^{k-1,k} = \mathbf{S}^{k-1,k-1}\left[\mathbf{C}_{\alpha}^{k-1,k-1}\left(\mathbf{S}^{k,k-1}\right)^{\dagger} + \mathbf{C}_{\alpha}^{k-1,k}\left(\mathbf{S}^{k,k}\right)^{\dagger}\right] = \left(\mathbf{C}_{rc}^{k,k-1}\right)^{\dagger}, \tag{1.84}$$

$$\mathbf{C}_{rc}^{k,k-1} = \mathbf{S}^{k,k-1}\mathbf{C}_{\alpha}^{k-1,k-1}\left(\mathbf{S}^{k-1,k-1}\right)^{\dagger} + \mathbf{S}^{k,k}\mathbf{C}_{\alpha}^{k,k-1}\left(\mathbf{S}^{k-1,k-1}\right)^{\dagger} = \left(\mathbf{C}_{rc}^{k-1,k}\right)^{\dagger}, \tag{1.85}$$

$$\begin{aligned} \mathbf{C}_{rc}^{k,k} &= \mathbf{S}^{k,k-1}\left[\mathbf{C}_{\alpha}^{k-1,k-1}\left(\mathbf{S}^{k,k-1}\right)^{\dagger} + \mathbf{C}_{\alpha}^{k-1,k}\left(\mathbf{S}^{k,k}\right)^{\dagger}\right] \\ &+ \mathbf{S}^{k,k}\left[\mathbf{C}_{\alpha}^{k,k-1}\left(\mathbf{S}^{k,k-1}\right)^{\dagger} + \mathbf{C}_{\alpha}^{k,k}\left(\mathbf{S}^{k,k}\right)^{\dagger}\right] = \left(\mathbf{C}_{rc}^{k,k}\right)^{\dagger}. \end{aligned} \tag{1.86}$$

Recall that the inversion of the symmetric matrix $\mathbf{C}_d \triangleq \mathbf{C}_{rc} - \mathbf{C}_{r\alpha}\mathbf{S}^{\dagger} - \mathbf{S}\mathbf{C}_{\alpha r} + \mathbf{C}_m$ plays a central role in the computation of the calibrated best estimate results. When only information from two consecutive time instances, $\nu = k-1,k;\quad k=(1,2,...,N_t)$, is considered, the matrix $\mathbf{C}_d$ reduces to the following expression:

$$\mathbf{C}_d \triangleq \begin{pmatrix} \mathbf{C}_d^{k-1,k-1} & \mathbf{C}_d^{k-1,k} \\ \mathbf{C}_d^{k,k-1} & \mathbf{C}_d^{k,k} \end{pmatrix} = \begin{pmatrix} \mathbf{C}_{rc}^{k-1,k-1} + \mathbf{C}_m^{k-1,k-1} & \mathbf{C}_{rc}^{k-1,k} + \mathbf{C}_m^{k-1,k} \\ \mathbf{C}_{rc}^{k,k-1} + \mathbf{C}_m^{k,k-1} & \mathbf{C}_{rc}^{k,k} + \mathbf{C}_m^{k,k} \end{pmatrix}$$
$$- \begin{pmatrix} \mathbf{C}_{r\alpha}^{k-1,k-1}\left(\mathbf{S}^{\dagger}\right)^{k-1,k-1} + \mathbf{S}^{k-1,k-1}\mathbf{C}_{\alpha r}^{k-1,k-1} & \mathbf{S}^{k-1,k-1}\mathbf{C}_{\alpha r}^{k-1,k} + \sum_{\rho=k-1}^{k} \mathbf{C}_{r\alpha}^{k-1,\rho}\left(\mathbf{S}^{\dagger}\right)^{k\rho} \\ \mathbf{C}_{r\alpha}^{k,k-1}\left(\mathbf{S}^{\dagger}\right)^{k-1,k-1} + \sum_{\rho=k-1}^{k} \mathbf{S}^{k,\rho}\mathbf{C}_{\alpha r}^{\rho k-1} & \sum_{\rho=k-1}^{k}\left[\mathbf{C}_{r\alpha}^{k,\rho}\left(\mathbf{S}^{\dagger}\right)^{k\rho} + \mathbf{S}^{k\rho}\mathbf{C}_{\alpha r}^{\rho k}\right] \end{pmatrix}. \tag{1.87}$$

Since both $\mathbf{C}_d^{k-1,k-1}$ and $\mathbf{C}_d^{k,k}$ are nonsingular sub-matrices, the matrix $\mathbf{C}_d$ can be inverted directly "by partitioning". The expressions of the components $\mathbf{K}_d^{\nu\eta}$ of the inverse matrix

$$\mathbf{C}_d^{-1} \triangleq \begin{pmatrix} \mathbf{K}_d^{k-1,k-1} & \mathbf{K}_d^{k-1,k} \\ \mathbf{K}_d^{k,k-1} & \mathbf{K}_d^{k,k} \end{pmatrix} \tag{1.88}$$

are obtained as follows:

$$\begin{aligned} \mathbf{K}_d^{k-1,k-1} &= \left[\mathbf{C}_d^{k-1,k-1} - \mathbf{C}_d^{k-1,k}\left(\mathbf{C}_d^{k,k}\right)^{-1}\mathbf{C}_d^{k,k-1}\right]^{-1} \\ &= \left(\mathbf{C}_d^{k-1,k-1}\right)^{-1} + \left(\mathbf{C}_d^{k-1,k-1}\right)^{-1}\mathbf{C}_d^{k-1,k}\mathbf{K}_d^{k,k}\mathbf{C}_d^{k,k-1}\left(\mathbf{C}_d^{k-1,k-1}\right)^{-1} \end{aligned} \tag{1.89}$$

$$\begin{aligned} \mathbf{K}_d^{k-1,k} &= -\left(\mathbf{C}_d^{k-1,k-1}\right)^{-1}\mathbf{C}_d^{k-1,k}\left[\mathbf{C}_d^{k,k} - \mathbf{C}_d^{k,k-1}\left(\mathbf{C}_d^{k-1,k-1}\right)^{-1}\mathbf{C}_d^{k-1,k}\right]^{-1} \\ &= -\left(\mathbf{C}_d^{k-1,k-1}\right)^{-1}\mathbf{C}_d^{k-1,k}\mathbf{K}_d^{k,k} \end{aligned} \tag{1.90}$$

$$\begin{aligned} \mathbf{K}_d^{k,k} &= \left[\mathbf{C}_d^{k,k} - \mathbf{C}_d^{k,k-1}\left(\mathbf{C}_d^{k-1,k-1}\right)^{-1}\mathbf{C}_d^{k-1,k}\right]^{-1} \\ &= \left(\mathbf{C}_d^{k,k}\right)^{-1} + \left(\mathbf{C}_d^{k,k}\right)^{-1}\mathbf{C}_d^{k,k-1}\mathbf{K}_d^{k-1,k-1}\mathbf{C}_d^{k-1,k}\left(\mathbf{C}_d^{k,k}\right)^{-1} \end{aligned} \tag{1.91}$$

$$\begin{aligned} \mathbf{K}_d^{k,k-1} &= -\left(\mathbf{C}_d^{k,k}\right)^{-1}\mathbf{C}_d^{k,k-1}\left[\mathbf{C}_d^{k-1,k-1} - \mathbf{C}_d^{k-1,k}\left(\mathbf{C}_d^{k,k}\right)^{-1}\mathbf{C}_d^{k,k-1}\right]^{-1}. \\ &= -\left(\mathbf{C}_d^{k,k}\right)^{-1}\mathbf{C}_d^{k,k-1}\mathbf{K}_d^{k-1,k-1} \end{aligned} \tag{1.92}$$

Consequently, the expressions for the calibrated best-estimate parameter values, cf. Eq. (1.69), become

$$\begin{pmatrix} \left(\boldsymbol{\alpha}^{be}\right)^{k-1} \\ \left(\boldsymbol{\alpha}^{be}\right)^{k} \end{pmatrix} = \begin{pmatrix} \left(\boldsymbol{\alpha}^{0}\right)^{k-1} \\ \left(\boldsymbol{\alpha}^{0}\right)^{k} \end{pmatrix} + \begin{pmatrix} \sum_{\eta=k-1}^{k} \mathbf{K}_{d}^{k-1,\eta}\mathbf{d}^{\eta} \\ \sum_{\eta=k-1}^{k} \mathbf{K}_{d}^{k,\eta}\mathbf{d}^{\eta} \end{pmatrix}$$
$$\times \begin{pmatrix} \mathbf{C}_{\alpha r}^{k-1,k-1} - \mathbf{C}_{\alpha}^{k-1,k-1}\left(\mathbf{S}^{\dagger}\right)^{k-1,k-1} & \mathbf{C}_{\alpha r}^{k-1,k} - \sum_{\rho=k-1}^{k} \mathbf{C}_{\alpha}^{1,\rho}\left(\mathbf{S}^{\dagger}\right)^{k,\rho} \\ \mathbf{C}_{\alpha r}^{k,k-1} - \mathbf{C}_{\alpha}^{k,k-1}\left(\mathbf{S}^{\dagger}\right)^{k-1,k-1} & \mathbf{C}_{\alpha r}^{k,k} - \sum_{\rho=k-1}^{k} \mathbf{C}_{\alpha}^{k,\rho}\left(\mathbf{S}^{\dagger}\right)^{k,\rho} \end{pmatrix}. \tag{1.93}$$

Written in component form, Eq. (1.93) takes on the following particular form of Eq. (1.70) at time node k:

$$\left(\boldsymbol{\alpha}^{be}\right)^{k} = \left(\boldsymbol{\alpha}^{0}\right)^{k} + \sum_{\mu=k-1}^{k} \left\{ \left[\mathbf{C}_{\alpha r}^{k\mu} - \sum_{\rho=1}^{\mu} \mathbf{C}_{\alpha}^{k\rho}\left(\mathbf{S}^{\dagger}\right)^{\mu\rho} \right] \left[\sum_{\eta=k-1}^{k} \mathbf{K}_{d}^{\mu\eta}\mathbf{d}^{\eta} \right] \right\}. \tag{1.94}$$

The calibrated best-estimate covariance matrix, $\mathbf{C}_{\alpha}^{be}$, for the above calibrated best-estimates system parameters is obtained by particularizing Eq. (1.71) to two consecutive time instances $(k-1,k)$, which leads to the following expression:

$$\mathbf{C}_{\alpha}^{be} \triangleq \begin{pmatrix} \left(\mathbf{C}_{\alpha}^{be}\right)^{k-1,k-1} & \left(\mathbf{C}_{\alpha}^{be}\right)^{k-1,k} \\ \left(\mathbf{C}_{\alpha}^{be}\right)^{k,k-1} & \left(\mathbf{C}_{\alpha}^{be}\right)^{k,k} \end{pmatrix} = \begin{pmatrix} \mathbf{C}_{\alpha}^{k-1,k-1} & \mathbf{C}_{\alpha}^{k-1,k} \\ \mathbf{C}_{\alpha}^{k,k-1} & \mathbf{C}_{\alpha}^{k,k} \end{pmatrix}$$
$$- \begin{pmatrix} \mathbf{C}_{\alpha d}^{k-1,k-1} & \mathbf{C}_{\alpha d}^{k-1,k} \\ \mathbf{C}_{\alpha d}^{k,k-1} & \mathbf{C}_{\alpha d}^{k,k} \end{pmatrix} \begin{pmatrix} \mathbf{K}_{d}^{k-1,k-1} & \mathbf{K}_{d}^{k-1,k} \\ \mathbf{K}_{d}^{k,k-1} & \mathbf{K}_{d}^{k,k} \end{pmatrix} \begin{pmatrix} \mathbf{C}_{\alpha d}^{k-1,k-1} & \mathbf{C}_{\alpha d}^{k-1,k} \\ \mathbf{C}_{\alpha d}^{k,k-1} & \mathbf{C}_{\alpha d}^{k,k} \end{pmatrix}^{\dagger}, \tag{1.95}$$

where

$$\begin{pmatrix} \mathbf{C}_{\alpha d}^{k-1,k-1} & \mathbf{C}_{\alpha d}^{k-1,k} \\ \mathbf{C}_{\alpha d}^{k,k-1} & \mathbf{C}_{\alpha d}^{k,k} \end{pmatrix} = \begin{bmatrix} \mathbf{C}_{\alpha r}^{k-1,k-1} - \mathbf{C}_{\alpha}^{k-1,k-1}\left(\mathbf{S}^{\dagger}\right)^{k-1,k-1} & \mathbf{C}_{\alpha r}^{k-1,k} - \sum_{\rho=k-1}^{k} \mathbf{C}_{\alpha}^{k-1,\rho}\left(\mathbf{S}^{\dagger}\right)^{k,\rho} \\ \mathbf{C}_{\alpha r}^{k,k-1} - \mathbf{C}_{\alpha}^{k,k-1}\left(\mathbf{S}^{\dagger}\right)^{k-1,k-1} & \mathbf{C}_{\alpha r}^{k,k} - \sum_{\rho=k-1}^{k} \mathbf{C}_{\alpha}^{k,\rho}\left(\mathbf{S}^{\dagger}\right)^{k,\rho} \end{bmatrix}. \tag{1.96}$$

The components $\left(\mathbf{C}_{\alpha}^{be}\right)^{\nu\mu}$, $\nu,\mu = k-1,k$, of the (block) covariance matrix $\mathbf{C}_{\alpha}^{be}$ shown in Eq. (1.95) can be written in the following particular form of Eq. (1.73):

$$\left(\mathbf{C}_{\alpha}^{be}\right)^{\nu\mu}=\mathbf{C}_{\alpha}^{\nu\mu}-\sum_{\eta=k-1}^{k}\sum_{\rho=k-1}^{k}\left[\mathbf{C}_{\alpha r}^{\nu\rho}-\sum_{\pi=1}^{\rho}\mathbf{C}_{\alpha}^{\nu\pi}\left(\mathbf{S}^{\dagger}\right)^{\rho\pi}\right]\mathbf{K}_{d}^{\rho\eta}\left[\mathbf{C}_{r\alpha}^{\eta\mu}-\sum_{\pi=1}^{\eta}\mathbf{S}^{\eta\pi}\mathbf{C}_{\alpha}^{\pi\mu}\right]. \tag{1.97}$$
$$for \quad \nu=k-1,k; \quad and \quad \mu=k-1,k;$$

The vector $\left(\mathbf{r}^{be}\right)^{k}$, representing the calibrated best-estimates for the system parameters at a time instance k, is a particular form of Eq. (1.75), as follows:

$$\left(\mathbf{r}^{be}\right)^{k}=\left(\mathbf{r}_{m}\right)^{k}+\sum_{\mu=k-1}^{k}\left\{\left[\mathbf{C}_{m}^{k\mu}-\sum_{\rho=1}^{\mu}\mathbf{C}_{r\alpha}^{k\rho}\left(\mathbf{S}^{\dagger}\right)^{\mu\rho}\right]\left[\sum_{\eta=k-1}^{k}\mathbf{K}_{d}^{\mu\eta}\mathbf{d}^{\eta}\right]\right\}. \tag{1.98}$$

Similarly, the calibrated best-estimate covariance block-matrix $\mathbf{C}_{r}^{be}$ for the best-estimate responses takes on a particular form of Eq. (1.76), having in this case four components $\left(\mathbf{C}_{r}^{be}\right)^{\nu\mu}$, $\nu,\mu=k-1,k$, expressed as follows:

$$\left(\mathbf{C}_{r}^{be}\right)^{\nu\mu}=\mathbf{C}_{m}^{\nu\mu}-\sum_{\eta=k-1}^{k}\sum_{\rho=k-1}^{k}\left[\mathbf{C}_{x}^{\nu\rho}-\sum_{\pi=k-1}^{\rho}\mathbf{C}_{r\alpha}^{\nu\pi}\left(\mathbf{S}^{\dagger}\right)^{\rho\pi}\right]\mathbf{K}_{d}^{\rho\eta}\left[\mathbf{C}_{m}^{\eta\mu}-\sum_{\pi=k-1}^{\eta}\mathbf{S}^{\eta\pi}\mathbf{C}_{\alpha r}^{\pi\mu}\right]. \tag{1.99}$$

A similar sequence of computations leads to the following particular form of Eq. (1.80), for the four matrix-valued components $\left(\mathbf{C}_{\alpha r}^{be}\right)^{\nu\mu}$, $\nu,\mu=k-1,k$, of the best-estimate response-parameter covariance matrix $\mathbf{C}_{\alpha r}^{be}$:

$$\left(\mathbf{C}_{r\alpha}^{be}\right)^{\nu\mu}=\mathbf{C}_{r\alpha}^{\nu\mu}-\sum_{\eta=k-1}^{k}\sum_{\rho=k-1}^{k}\left[\mathbf{C}_{m}^{\nu\rho}-\sum_{\pi=k-1}^{\rho}\mathbf{C}_{r\alpha}^{\nu\pi}\left(\mathbf{S}^{\dagger}\right)^{\rho\pi}\right]\mathbf{K}_{d}^{\rho\eta}\left[\mathbf{C}_{r\alpha}^{\eta\mu}-\sum_{\pi=k-1}^{\eta}\mathbf{S}^{\eta\pi}\mathbf{C}_{\alpha}^{\pi\mu}\right]. \tag{1.100}$$

1.2.3 Data Consistency and Rejection Criteria

The actual application of the predictive modeling results presented in Eqs. (1.70), (1.73), (1.75), (1.78) and (1.80) to a physical system characterized by nominal values and uncertainties for model parameters together with the computed and measured responses is straightforward, in principle, although it can become computationally very demanding regarding both data handling and computational speed. Care must be exercised, however, since the indiscriminate incorporation of all seemingly relevant experimental-response data could produce a set of calibrated (adjusted) parameter values that might differ unreasonably much from the corresponding original nominal values, and might even fail to improve the agreement between the calculated and measured values of some of the very responses that were used to calibrate the model parameters. When calibrating (adjusting) a library of model pa-

rameters, it is tacitly assumed that the given parameters are basically "correct," except that they are not sufficiently accurate for the objective at hand. The calibration procedure uses additional data (e.g., experimentally measured responses) for improving the parameter values while reducing their uncertainties. Although such additional information induces modifications of the original parameter values, the adjusted parameters are still generally expected to remain consistent with their original nominal values, within the range of their original uncertainties. However, calibration of a library of model parameters by experimental responses which significantly deviate from their respective computed values would significantly modify the resulting adjusted parameters, perhaps even violating the restriction of linearity expressed by Eq. (1.55). Such unlikely adjustments would most probably lead to failure of even reproducing the original experimental responses.
On the other hand, calibrating model parameter by using measured responses that are very close to their respective computed values would cause minimal parameter modifications and a nearly perfect reproduction of the given responses by the calibrated parameters, as would be expected. In such a case, the given responses would be considered as being consistent with the parameter library, in contradistinction to adjustment by inconsistent experimental information, in which case the adjustment (calibration) could fail because of inconsistencies. These considerations clearly underscore the need for using a quantitative indicator to measure the mutual and joint consistency of the information available for model calibration.
The value of χ^2 [i.e., the minimum of the functional $Q(\mathbf{z})$] for a time-independent system is provided in Eq. (1.24). Using Eq. (1.62) leads to the same formal expression for Q_{min} as shown in Eq. (1.24), namely

$$
\begin{aligned}
Q_{min} &\triangleq Q\left(\mathbf{z}^{be}\right) = \left[-\boldsymbol{\lambda}^{\dagger}\mathbf{Z}\left(\boldsymbol{\alpha}^{0}\right)\mathbf{C}\right]\mathbf{C}^{-1}\mathbf{z}^{be} \\
&= -\boldsymbol{\lambda}^{\dagger}\mathbf{Z}\left(\boldsymbol{\alpha}^{0}\right)\mathbf{z}^{be} = \boldsymbol{\lambda}^{\dagger}\mathbf{d} = \mathbf{d}^{\dagger}\left[\mathbf{C}_{d}\left(\boldsymbol{\alpha}^{0}\right)\right]^{-1}\mathbf{d} = \chi^{2}.
\end{aligned}
\tag{1.101}
$$

As the above expression indicates, the minimal value, $Q_{min} \triangleq Q\left(\mathbf{z}^{be}\right)$, represents the square of the length of the vector $\mathbf{d}$, measuring (in the corresponding metric) the deviations between the experimental and nominally computed responses. The quantity $Q_{min} \triangleq Q\left(\mathbf{z}^{be}\right)$ can be evaluated directly from the given data (i.e., given parameters and responses, together with their original uncertainties) after having inverted the deviation-vector covariance matrix $\mathbf{C}_{d}\left(\boldsymbol{\alpha}^{0}\right)$. It is also very important to note that $Q_{min} \triangleq Q\left(\mathbf{z}^{be}\right)$ is independent of calibrating (or adjusting) the original data. As the dimension of $\mathbf{d}$ indicates, the number of degrees of freedom characteristic of the calibration under consideration is equal to the number of experimental responses. In the extreme case of absence of experimental responses, no actual calibration takes place since $\mathbf{d} = \mathbf{R}\left(\boldsymbol{\alpha}^{0}\right)$, so that the best-estimate parameter values are just the original nominal values, i.e., $(\boldsymbol{\alpha}^{be})^k = (\boldsymbol{\alpha}^0)^k$. An actual calibration (adjustment) occurs only when including at least one experimental response.

Replacing Eq. (1.101) in Eq. (1.62) shows that the bulk of the contribution to the joint posterior probability distribution, which comes from the point $\mathbf{z} = \mathbf{z}^{be}$, takes on the form of the following multivariate Gaussian:

$$
\begin{aligned}
&p(\mathbf{z}^{be} \,|\mathbf{C}) \sim exp\left[-\frac{1}{2}Q(\mathbf{z}^{be})\right] = \\
&= exp\left\{-\frac{1}{2}\left[\mathbf{r}_m - \mathbf{R}(\boldsymbol{\alpha}^0)\right]^{\dagger}\left[\mathbf{C}_d(\boldsymbol{\alpha}^0)\right]^{-1}\left[\mathbf{r}_m - \mathbf{R}(\boldsymbol{\alpha}^0)\right]\right\}.
\end{aligned}
\tag{1.102}
$$

The above relation indicates that imprecisely known experimental responses can be considered as variates approximately described by a multivariate Gaussian distribution with means located at the nominal values of the computed responses, and with a covariance matrix $\mathbf{C}_d(\boldsymbol{\alpha}^0)$. In turn, the variate Q_{min} follows a χ^2-distribution with n degrees of freedom, where n denotes the total number of experimental responses considered in the calibration (adjustment) procedure. The quantity Q_{min} is the "χ^2 of the calibration (adjustment) at hand", and can be used as an indicator of the agreement between the computed and measured responses, measuring essentially the consistency of the measured responses with the model parameters. Recall that the χ^2 (chi-square) distribution with n degrees of freedom of the continuous variable ($0 \le x < \infty$) is defined as

$$
\begin{aligned}
&P(x < \chi^2 < x + dx) \triangleq k_n(x)dx \\
&= \frac{1}{2^{n/2}\Gamma(n/2)} x^{n/2-1} e^{-x/2} dx, \; x > 0, \; (n = 1, 2, \ldots).
\end{aligned}
\tag{1.103}
$$

The χ^2- distribution is a measure of the deviation of a "true distribution" (in this case –the distribution of experimental responses) from the hypothetic one (in this case –a Gaussian). The mean and variance of x, $0 \le x < \infty$, are $\langle x \rangle = n$ and $var(x) = 2n$; additional practically useful asymptotic properties of the χ^2- distribution for $n \to \infty$ are: (i) x is asymptotically normal with mean n and variance $2n$; (ii) x / n is asymptotically normal with mean 1 and variance $2 / n$; (iii) $\sqrt{2x}$ is asymptotically normal with mean $\sqrt{2n-1}$ and variance 1. Although the χ^2-distribution is extensively tabulated, the notation is not uniform in the literature for the various derived quantities (in particular, for the corresponding cumulative distribution functions and fractiles). The cumulative distributions, denoted here by $P_n\left(\chi^2\right)$ and $Q_n\left(\chi^2\right)$, are defined as

$$\begin{aligned} P_n\left(\chi_0^2\right) &\triangleq P\left(\chi^2 \le \chi_0^2\right) \triangleq \int_0^{\chi_0^2} k_n(t)\,dt; \\ Q_n\left(\chi_0^2\right) &\triangleq P\left(\chi^2 \ge \chi_0^2\right) \triangleq \int_{\chi_0^2}^{\infty} k_n(t)\,dt = 1 - P_n\left(\chi_0^2\right). \end{aligned} \tag{1.104}$$

In practice, one rejects a hypothesis using the χ^2-distribution when, for a given significance level α and number of degrees of freedom n, the value of $Q_{min} \equiv \chi^2$ exceeds a chosen critical fractile value $\chi_\alpha^2(n)$. Published tables often show $\chi_{1-\alpha}^2(n)$ versus α. When the number of degrees of freedom is large ($n > 30$), a useful asymptotic approximation is $\chi_\alpha^2(n) \approx 1/2\left(\sqrt{2n-1} + z_{2\alpha}\right)^2$, with $z_{2\alpha}$ denoting the corresponding fractile of the standard normal distribution $\Phi_0(z)$, computed by solving $2\Phi_0\left(z_{2\alpha}\right) = 1 - 2\alpha$ using the tabulated tables for $\Phi_0(z)$. For large or small values of α, a more accurate approximation is $\chi_\alpha^2(n) \approx n\left(1 - \frac{2}{9m} + z_{2\alpha}\sqrt{\frac{2}{9m}}\right)^3$. Often, it is convenient to transform χ^2 to the variate $t = \chi^2/n$ (i.e., "χ^2 per degree of freedom"), in which case the transformed distribution, $g_n(t)$, becomes $g_n(t) = nk_n(nt)$, with mean value $\langle t \rangle = 1$ and variance $2/n$.

For predictive modeling it is important to assess if: (i) the response and data measurements are free of gross errors (blunders such as wrong settings, mistaken readings, etc.), and (ii) the measurements are consistent with the assumptions regarding the respective means, variances, and covariances. For example, the measurements are very likely to be both free of gross errors and consistent with the assumptions if $\chi^2/n \simeq 1$. However, if $\chi^2/n \gg 1$ or $\chi^2/n \ll 1$, the measurements (or at least some measurements), the assumptions, or both are suspect. In particular, unusually large values $\chi^2/n \gg 1$ could be obtained when the original variances are underestimated; increasing them beyond their assumed nominal values would cause the adjusted values of χ^2/n and $P_n\left(\chi^2\right)$ to decrease accordingly. The reverse argument would apply if the a priori values of χ^2/n and $P_n\left(\chi^2\right)$ were unusually small (e.g., $\chi^2/n \ll 1$, $P_n\left(\chi^2\right) \sim 10^{-4}$), which could be the consequence of a priori overestimated variances. A practical quantitative criterion for the "acceptance" or "rejection" of experimental results in conjunction with a given "theoretical" model (i.e., in conjunction with the assumptions regarding the imprecisely known responses and parameters underlying the model) is to accept the value of χ^2/n whenever $0.15 < P_n(\chi^2) < 0.85$, in analogy to the $"1\sigma"$ range of normal distributions. Note that, when setting an acceptance criterion for χ^2/n of the general form

$$\alpha < P_n(\chi^2) < 1-\alpha\,, \tag{1.105}$$

the exact value of α is not essential and is subject to personal judgment. This is because the probability $P_n(\chi^2)$ is still sensitive to the value of χ^2/n due to the fact that $\chi^2/n \simeq 1 \pm \sqrt{(2/n)}$, as long as $n \geq 5$, so the acceptable range of χ^2/n narrows as $1/\sqrt{n}$ (see also the previously noted asymptotic forms for χ^2_α/n). In other words, moderate changes in χ^2/n lead to significant relative changes in $P_n(\chi^2)$. For example, the central 50%-range of $\chi^2/20$ is (0.77, 1.19), and the corresponding 90%-range is (0.54, 1.57), implying that values of $\chi^2/20$ below $\simeq 0.4$ or above $\simeq 2.0$ would be unacceptable.

In addition to measuring the overall consistency of a given set of model parameters and responses, the quantity χ^2/n also measures the consistency among the measured responses. Hence, an entire data set (model parameters and/or experimental responses) should not be indiscriminately disqualified because of a “too high” or “too low” value of χ^2/n, since even a single “outlying” response could significantly degrade the set’s overall consistency. Note that a simple-minded assessment and ranking of “questionable responses” according to the values of the “individual consistencies” (i.e., the values of χ^2 obtained for each response, as if it were the only response available for calibrating the entire set of parameters), would probably lead to false conclusions. This is because the sum of the respective “individual consistencies” [which would numerically be obtained by dividing the squares of the deviations, d_i^2, through the sum of the respective variances of the computed and measured responses $var(r_i^{comp}) + var(r_i^{exp})$], would not be equal to the “joint consistency” (i.e., the joint χ^2) of the entire set of experimental responses. This is because the “deviation-vector covariance matrix” $\mathbf{C}_d(\boldsymbol{\alpha}^0) \triangleq \mathbf{C}_{rc}(\boldsymbol{\alpha}^0) - \mathbf{C}_{r\alpha}\left[\mathbf{S}(\boldsymbol{\alpha}^0)\right]^\dagger - \left[\mathbf{S}(\boldsymbol{\alpha}^0)\right]\mathbf{C}_{\alpha r} + \mathbf{C}_m$ is generally non-diagonal, unless the responses and parameters are uncorrelated and both $\mathbf{C}_{rc}(\boldsymbol{\alpha}^0)$ and $\mathbf{C}_m$ are diagonal. On the other hand, verifying the consistency of all partial sets of the array of n responses with respect to their consistency with the given library is usually impractical, since the number of partial sets of an array of n responses is $2^n - 1$;

hence, such a verification would be practically feasible only when the number of measured responses is very small.
A procedure that has been successfully used to identify successively the responses which are least consistent with a given library is based on leaving out one response at a time and evaluating $\chi^2_{n-1}(1)$ for the remaining $n-1$ responses. The response left out is subsequently returned to the response set, another (response "two") response is eliminated, and the corresponding $\chi^2_{I-1}(2)$ is evaluated. This procedure is continued until all remaining $\chi^2_{I-1}(i), i=3,\ldots,n,$ are successively evaluated. The response that yields the lowest χ^2_{I-1} when eliminated is considered to be "the least consistent", and is thus ranked "last" in the consistency sequence, and eliminated from further consideration. The evaluation procedure is then repeated for the remaining $n-1$ ("more consistent") responses, to identify the "second least consistent response", which is then ranked next-to-last. The procedure is then repeatedly applied to the successive, fewer and fewer, partial response sets until establishing the complete consistency sequence. Establishing such a consistency sequence requires only $n(n+1)/2$ computations of χ^2, as compared to (2^n-1) calculations needed to assign χ^2 values to all possible partial sets of n responses.
The quantity χ^2/n measures the consistency of any set of n experimentally measured responses with a given library of model parameters, in the sense that if χ^2_1 refers to a specific set of n experimental responses and χ^2_2 to another set of n responses, then $\chi^2_1 < \chi^2_2$ means that the first set is more consistent with the library of model parameters than the second. On the other hand, when varying the number of responses, it is not a priori obvious whether the set yielding a smaller χ^2/n is also necessarily the most consistent with the given model parameters. As an example, consider the value $P_n(\chi^2)=0.85$, which can correspond to both $\chi^2/5=1.623$ and also to $\chi^2/10=1.453$. If, for example, one set of 5 responses would give a computed value $\chi^2/5=1.6$, and second set of 10 responses would give $\chi^2/10=1.5$, the first set would be considered to be the "more consistent", for it falls within the "central 70% range," whereas the second does not. In such situations, it is preferable to use the quantity $Q_n(\chi^2)=1-P_n(\chi^2)$, as an additional measure of consistency.
Quite generally, therefore, the calibration (adjustment) of a set of model parameters and experimental responses must include the verification of their mutual consistency, which is performed by first generating the consistency sequence, and then determining the probabilities $Q_i(\chi^2)$, when $i=1,2,\ldots,n$, while generating the sequence. The less consistent responses will show up at the end of the sequence, and the probabilities $Q_i(\chi^2)$ will generally decrease as i approaches the total number of responses. Such an analysis would identify the significantly less-consistent responses, and would also indicate the level of consistency of all response subsets

along the consistency sequence. In parallel, the irregular model parameters, if any, must also be identified; this can be done by computing not only χ^2 for any response subset, but also computing from Eq. (1.70) the corresponding best-estimate (correspondingly adjusted) parameters $\boldsymbol{\alpha}^{be} = \boldsymbol{\alpha}^0 + \left(\mathbf{C}_{\alpha r} - \mathbf{C}_{\alpha}\left[\mathbf{S}\left(\boldsymbol{\alpha}^0\right)\right]^{\dagger}\right)\left[\mathbf{C}_d\left(\boldsymbol{\alpha}^0\right)\right]^{-1}\mathbf{d}$. This way, the actual individual parameter calibrations (adjustments) induced by the respective response subset are also examined while proceeding step-by-step along the consistency sequence, noting which parameters vary more than others, and by how much. Usually, the parameter-adjustments induced by the more consistent subsets of responses tend to be marginal. The less-consistent responses and the questionable parameters would tend to exhibit more substantial adjustments, likely requiring additional specific further examinations.

APPENDIX 1: Construction of A Priori Probability Distribution Using the Maximum entropy Principle

While establishing information theory, Shannon (1948) proved that the lack of information implied by a discrete probability distribution, p_n, with mutually exclusive alternatives can be expressed quantitatively (up to a constant) by its information entropy, S, defined as

$$S = -\sum_{n=1}^{N} p_n \ln p_n \,. \tag{1.106}$$

Shannon proved that S is the only measure of indeterminacy that satisfies the following three requirements:

(i) S is a smooth function of the p_i;

(ii) If there are N alternatives, all equally probable, then the indeterminacy and hence S must grow monotonically as N increases; and

(iii) Grouping of alternatives leaves S unchanged (i.e., adding the entropy quantifying ignorance about the true group, and the suitably weighted entropies quantifying ignorance about the true member within each group, must yield the same overall entropy S as for ungrouped alternatives).

For continuous distributions with probability density $p(x)$, the expression for its information entropy becomes

$$S = -\int dx\, p(x) \ln \frac{p(x)}{m(x)}, \tag{1.107}$$

where $m(x)$ is a prior density that ensures form invariance under change of variable. When only certain information about the underlying distribution $p(x)$ is available but $p(x)$ itself is unknown and needs to be determined, the principle of maximum entropy provides the optimal compatibility with the available information, while simultaneously ensuring minimal spurious information content.

As an application of the principle of maximal entropy, suppose that an unknown distribution $p(x)$ needs to be determined when the only available information comprises the (possibly non-informative) prior $m(x)$ and "integral data" in the form of moments of several known functions $F_k(x)$ over the unknown distribution $p(x)$, namely

$$\langle F_k \rangle = \int dx\, p(x) F_k(x)\,, \qquad k = 1,2,\ldots,K\,. \tag{1.108}$$

According to the principle of maximum entropy, the probability density $p(x)$ would satisfy the "available information," i.e., would comply with the constraints expressed in Eq. (1.108) without implying any spurious information or hidden assumptions if $p(x)$ would maximize the information entropy defined by Eq. (1.107) subject to the known constraints given in Eq. (1.108). This is a variational problem that can be solved by using Lagrange multipliers, λ_k, to obtain the following expression:

$$p(x) = \frac{1}{Z} m(x) exp\left[-\sum_{k=1}^{K} \lambda_k F_k(x) \right]. \tag{1.109}$$

The normalization constant Z in Eq. (1.109) is defined as

$$Z \triangleq \int dx\, m(x) exp\left[-\sum_{k=1}^{K} \lambda_k F_k(x) \right]. \tag{1.110}$$

In statistical mechanics, the normalization constant Z is called the partition function (or sum over states), and carries all of the information available about the possible states of the system.

The expected integral data is obtained by differentiating Z with respect to the respective Lagrange multiplier, i.e.,

$$\langle F_k \rangle = -\frac{\partial}{\partial \lambda_k} ln Z, \quad k = 1,2,\ldots,K\,. \tag{1.111}$$

In the case of discrete distributions, when the integral data $\langle F_k \rangle$ are not yet known, then $m(x)=1$, and the maximum entropy algorithm described above yields the uniform distribution, as would be required by the principle of insufficient reason. Thus, the principle of maximum entropy provides a generalization of the principle of insufficient reason, and can be applied to discrete and to continuous distributions, ranging from problems in which only information about discrete alternatives is available, to problems in which global or macroscopic information is available.

As already discussed in the foregoing, data reported in the form $\langle x \rangle \pm \Delta x$, $\Delta x \triangleq \sqrt{var(x)}$, customarily implies "best estimates under quadratic loss," which, in turn, implies the availability of the first and second moments, $\langle x \rangle$ and $\langle x^2 \rangle = \langle x \rangle^2 + (\Delta x)^2$, respectively. In such a case, the maximum entropy algorithm above can be used with $K=2$, in which case Eq. (1.109) yields $p(x) \sim exp(-\lambda_1 x - \lambda_2 x^2)$ as the most objective probability density for further inference. In terms of the known moments $\langle x \rangle$ and $\langle x^2 \rangle$, $p(x)$ would therefore be a Gaussian having the following specific form:

$$p(x \mid \langle x \rangle, \Delta x)\, dx = \frac{exp\left[-\frac{1}{2}\left(\frac{x-\langle x \rangle}{\Delta x}\right)^2\right]}{\left[2\pi(\Delta x)^2\right]^{1/2}} dx, \quad -\infty < x < \infty . \tag{1.112}$$

When several observables x_i, $i=1,2,\ldots,n,$ are simultaneously measured, the respective results are customarily reported in the form of "best values" $\langle x_i \rangle$, together with covariances matrix elements $c_{ij} \triangleq \langle \varepsilon_i \varepsilon_j \rangle \triangleq cov(x_j, x_k) = c_{ji}$, $c_{jj} \triangleq var\, x_j$, where the errors are defined through the relation $\varepsilon_j \triangleq (x_j - \langle x_j \rangle)$. In this case, Eq. (1.109) takes on the form

$$p(\varepsilon \mid C)\, d^n\varepsilon = \frac{1}{Z} exp\left(-\sum_{i,j} \varepsilon_i \lambda_{ij} \varepsilon_j\right) d^n\varepsilon = \frac{1}{Z} exp\left(-\frac{1}{2}\boldsymbol{\varepsilon}^\dagger \boldsymbol{\Lambda} \boldsymbol{\varepsilon}\right) d^n\varepsilon , \tag{1.113}$$

where: the Lagrange multiplier λ_{ij} corresponds to c_{ij}; the normalization constant is defined as $Z \equiv \int d^n\varepsilon\, exp\left(-\frac{1}{2}\boldsymbol{\varepsilon}^\dagger \boldsymbol{\Lambda} \boldsymbol{\varepsilon}\right)$; $\boldsymbol{\varepsilon} = (\varepsilon_1, \ldots, \varepsilon_n)$ denotes the n-component the vector of errors; $\mathbf{C} \triangleq (c_{ij})_{n \times n}$ is the $n \times n$ covariance matrix for the observables x_i, $i=1,2,\ldots,n$; and $\boldsymbol{\Lambda} = (\lambda_{ij})_{n \times n}$ denotes the $n \times n$ matrix of Lagrange multipliers.

The explicit form of the normalization constant Z in Eq. (1.113) is obtained by performing the respective integrations to obtain

$$Z = \sqrt{\frac{\pi^n}{det(\mathbf{\Lambda})}} . \tag{1.114}$$

The relationship between the Lagrange multiplier λ_{ij} and the covariance c_{ij} is obtained by differentiating $ln Z$ with respect to the respective Lagrange multiplier, cf. Eq. (1.111), to obtain

$$c_{ij} = -\frac{\partial}{\partial \lambda_{ij}} ln Z = \frac{1}{2}\left(\mathbf{\Lambda}^{-1}\right)_{ij}, \tag{1.115}$$

since differentiation of the determinant $det(\mathbf{\Lambda})$ with respect to an element of $\mathbf{\Lambda}$ yields the cofactor for this element (which, for a nonsingular matrix, is equal to the corresponding element of the inverse matrix times the determinant).
Replacing Eqs. (1.114) and (1.115) in Eq. (1.113) transforms the latter into the following form:

$$p(\boldsymbol{\varepsilon} | \mathbf{C}) d^n \varepsilon = \frac{exp\left(-\frac{1}{2}\boldsymbol{\varepsilon}^{\dagger}\mathbf{C}^{-1}\boldsymbol{\varepsilon}\right)}{\sqrt{det(2\pi\mathbf{C})}} d^n \varepsilon, \; -\infty < \varepsilon_i < \infty , \tag{1.116}$$

with $\langle \boldsymbol{\varepsilon} \rangle = \mathbf{0}$, $\langle \boldsymbol{\varepsilon}\boldsymbol{\varepsilon}^{\dagger} \rangle = \mathbf{C}$, which is an n-variate Gaussian centered at the origin. In terms of the expected values $\langle x_i \rangle$, the expression in Eq. (1.116) becomes:

$$p(\mathbf{x} | \langle \mathbf{x} \rangle, \mathbf{C}) d\mathbf{x} = \frac{exp\left[-\frac{1}{2}(\mathbf{x} - \langle \mathbf{x} \rangle)^{\dagger} \mathbf{C}^{-1} (\mathbf{x} - \langle \mathbf{x} \rangle)\right] d\mathbf{x},}{\sqrt{det(2\pi\mathbf{C})}}, \; -\infty < x_j < \infty. \tag{1.117}$$

Thus, the considerations leading to Eq. (1.117) demonstrate that, when only means and covariances are known, the maximum entropy algorithm yields the Gaussian probability distribution shown in Eq. (1.117) as the most objective probability distribution, where: $\mathbf{x}$ is the data vector with coordinates x_j, $\mathbf{C}$ is the covariance matrix with elements C_{jk}, and $dx \triangleq \prod_j dx_j$ is the volume element in the data space.
It often occurs in practice that the variances c_{ii} are known but the covariances c_{ij} are not, in which case the covariance matrix $\mathbf{C}$ would *a priori* be diagonal. In such a case, only the Lagrange parameters λ_{ii} would appear in Eq. (1.113), so that the

matrix $\mathbf{\Lambda}$ would also be *a priori* diagonal. In other words, in the absence of information about correlations, the maximum entropy algorithm indicates that unknown covariances/correlations can be taken to be zero. This is another example of a general property of maximum entropy distributions: all expectation values vanish unless the constraints demand otherwise.

Gaussian distributions are often considered appropriate only if many independent random deviations act in concert such that the central limit theorem is applicable. Nevertheless, if only "best values" and their (co)variances are available, the maximum entropy principle indicates that the corresponding Gaussian is the best choice for all further inference, regardless of the actual form of the unknown true distribution. Furthermore, in contrast to the central limit theorem, the maximum entropy principle is also valid for correlated data. The maximum entropy principle can also be employed to address systematic errors when their possible magnitudes can be (at least vaguely) inferred, but their signs are not known. The maximum entropy principle indicates that such errors should be described by a Gaussian distribution with zero mean and a width corresponding to the (vaguely) known magnitude, rather than by a rectangular distribution.

The maximum entropy principle is a powerful tool for the assignment of prior (or any other) probabilities, in the presence of incomplete information. Although the above results have been derived for observables x_j that vary in the interval $-\infty < x_j < \infty$, these results can also be used for positive observable x_j, $0 < x_j < \infty$, by considering a logarithmic scale (or lognormal distributions on the original scale).

1.3 BERRU-SMS APPLICATIONS TO TWO-PHASE FLOW COMPUTATIONS AND EXPERIMENTS

This Section presents an application of the BERRU-SMS predictive modeling methodology to the calibration of the three-dimensional reactor thermal-hydraulics simulation and design tool FLICA4 (Fillion et al., 2007) using experimental information from the international OECD/NRC BWR Full Size Fine-Mesh Bundle Tests (BFBT) benchmarks, which were designed by the Nuclear Power Engineering Corporation (NUPEC) of Japan. The results presented in this Section are based on the work by Badea, Cacuci, and Badea (2012). The following specific BFBT experiments have been used for calibrating parameters and boundary conditions for FLICA4: (i) one-dimensional pressure drops; (ii) axial void fraction distributions; and (iii) three-dimensional transverse void fraction distributions. The resulting uncertainties (standard deviations) for the best-estimate parameters and distributions of pressure drops and void fractions are shown to be smaller than the a priori experimental and computed uncertainties, thus demonstrating the successful calibration of a large-scale reactor core thermal-hydraulics code using the BFBT benchmark-grade experiments.

1.3.1 Description of the BFBT Experiments

From 1987 to 1995, the Nuclear Power Engineering Corporation of Japan (NUPEC) performed measurements (Inoue et al, 1995) of void fraction distribution in full-size mock-up fuel bundles for both boiling water reactors (BWRs) and pressurized water reactors (PWRs). The void fraction distributions were visualized using computer tomography (CT) technology under actual plant conditions for mesh sizes smaller than a sub-channel. In addition to measuring void fraction distributions, NUPEC also performed steady state and transient measurement of critical power in equivalent full-size mock-ups. The NUPEC measurements are internationally considered to be highly reliable because of the high reliability of the experimental facility, including control of the system pressure, inlet sub-cooling, and rod surface temperature. Thus, these measurements provide internationally a comprehensive database for the development of consistent mechanistic models for predicting void fraction distributions and boiling transition in sub-channels.

Gaining accurate knowledge about boiling transition and void fraction distribution is essential for the quantification of nuclear reactor safety margins. However, the theoretical principles underlying the numerical modeling of sub-channel void distribution are incompletely known, and the correlations replacing first-principles are not generally applicable to the wide range of geometrical arrangements and operating conditions characterizing the various types of LWRs. The international OECD/NRC BWR Full-Size Fine-Mesh Bundle Tests (BFBT) benchmarks (Neykov et al, 2006) were established based on the NUPEC database to motivate research on insufficiently known two-phase flow regimes by facilitating a systematic comparison of full-scale experimental data to predictions of numerical simulation models. These benchmarks are particularly well suited for quantifying uncertainties in the prediction of detailed distributions of sub-channel void fractions and critical powers.

The design and data acquisition systems of NUPEC's facility enable both macroscopic and microscopic measurements. In this context, measurements of sub-channel void fractions are considered as macroscopic data, while the digitized computer graphic images are considered as microscopic data. Thus, the BFBT measurements of void fraction distributions and critical powers in a multi-rod assembly under typical reactor power and fluid conditions enable comparisons with predictions of computational models and encourage the development of more accurate theoretical models for describing microscopic processes.

Figure 1.1 shows the diagram of the BFBT. By using an electrically heated rod assembly that simulates a full scale BWR fuel assembly, NUPEC's BWR Full-size Fine-mesh Bundle Test (BFBT) facility is able to simulate the full range of BWR steady-state operating conditions, as well as time-dependent BWR operational transients. The limiting operating conditions for the facility are as follows: pressure up to 10.3 MPa, temperature up to 315 °C, power up to 12 MW, and flow rate up to 75 t/h.

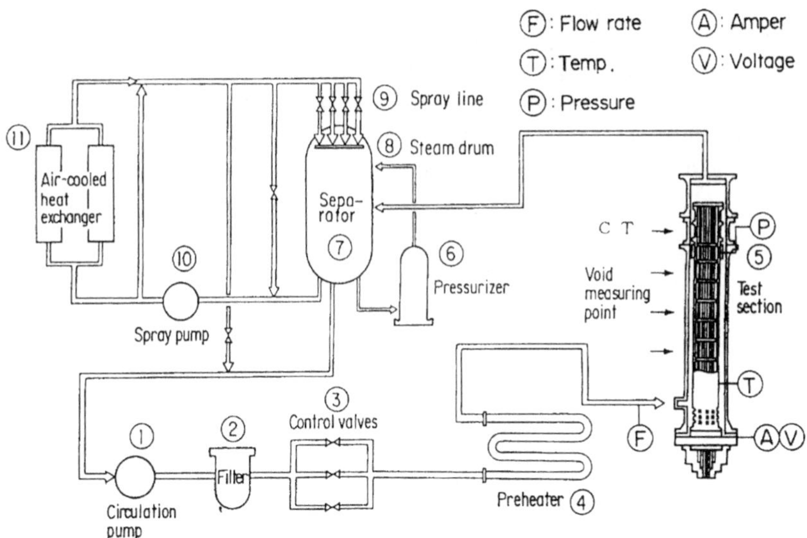

Figure 1.1: NUPEC Rod Bundle Test Facility

The main structural components are made of stainless steel (SUS304). Demineralized water is used as a cooling fluid. As depicted in Figure 1.1, water is circulated by the circulation pump (1); the three valves (3) of different sizes control the coolant flow rate. A direct-heating tubular pre-heater (4) controls the inlet fluid temperature for the test section (5). Sub-cooled coolant flows upward into the test bundle (5), in which it is heated to become a two-phase mixture. The steam is separated from the steam-water mixture in the separator (7) and is condensed using a spray of sub-cooled water in the steam drum (8). The condensed water is then returned to the circulation pump (1). The spray lines (9), which have four different-sized valves, control the system pressure in both steady and transient state. The pressurizer (6) controls the system pressure when the power in the test section is low. After the water is cooled with two air-cooled heat exchangers (11), the spray pump (10) forces a spray into the steam-drum.

Figure 1.2 depicts the test section, which comprises a pressure vessel, a mock-up flow channel, and a full-scale BWR mock-up fuel assembly installed within the vessel. Although two types of BWR bundles (a current 8x8 type and a high burn-up 8x8 type) were simulated in the BFBT experiments, the illustrative application of the BERRU-SMS methodology presented in this Section assimilates just the experimental data involving the "high burn-up" 8×8 mock-up. The geometry and characteristic dimensions of the high burn-up 8x8 mock-up are listed in Table 1.1. Each rod consists of an electrical heater made of nichrome (outer diameter 7.3 mm), surrounded by an insulator made of boron nitride (outer diameter 9.7 mm), which is in turn wrapped in a cladding (1.3 mm thick) made of inconel 600. The thermo-mechanical properties of these materials are based on the MATPRO model used in the TRAC code (TRAC-PF1/MOD2 Theory Manual20) and are listed in Table 1.2. The heated rod is single-ended and electrically grounded. The surface temperature of the rod is measured by 0.5 mm-diameter chromel-alumel thermocouples. Additional thermocouples are embedded in the cladding surface, which are positioned axially just upstream of the spacers. Each heated rod is joined to an X-ray transmission section, which is of the same diameter as the heated rod but with cladding made of beryllium in order to facilitate the transmission of X-ray.

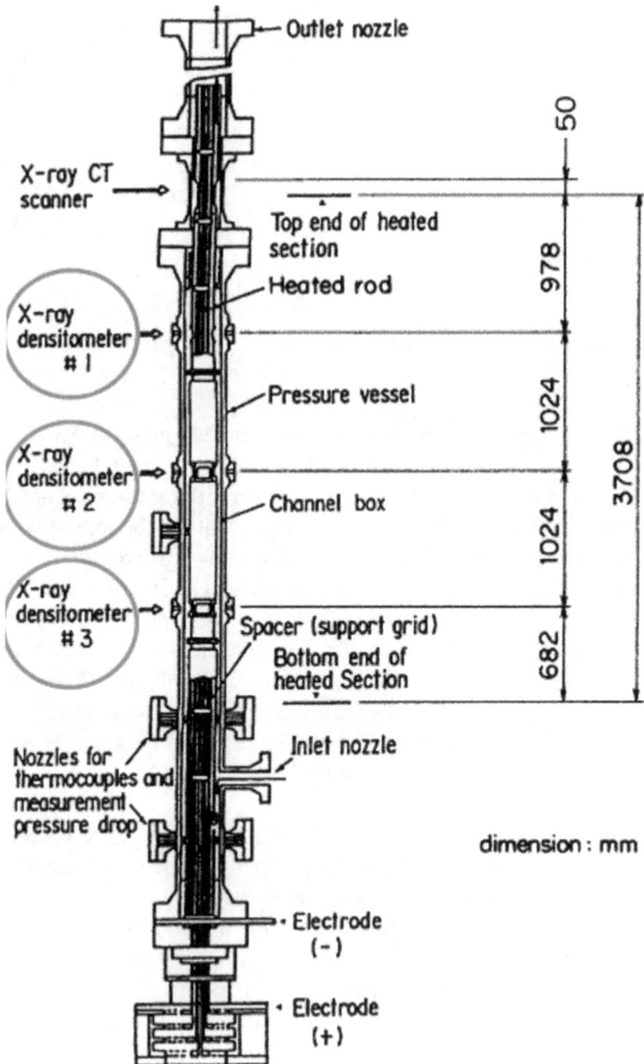

Figure 1.2: Cross sectional view of the test section with void fraction measurement system.

Table 1.1: Geometry and Characteristics of BWR Test Bundles

Item	Data
Test assembly 4	
Simulated fuel assembly type	High burn-up 8×8
Number of heated rods	60
Heated rods outer diameter (mm)	12.3
Heated rods pitch (mm)	16.2
Axial heated length (mm)	3708
Number of water rods	1
Water rods outer diameter (mm)	34.0
Channel box inner width (mm)	132.5
Channel box corner radius (mm)	8.0
In channel flow area (mm^2)	9463
Spacer type	Ferrule
Number of spacers	7
Spacer pressure loss coefficients	1.2
Spacer location (mm)	455, 967, 1479, 1991, 2503, 3015, 3527 mm (Distance from bottom of heated length to spacer bottom face)
Transversal power shape	A
Axial power shape	Uniform

Table 1.2: Thermo-mechanical properties of a heated rod

	Density ρ (kg/m³)
Nichrome	8393.4
Boron Nitride	2002
Inconel 600	$16.01846\times(5.261008\times10^{2}-1.345453\times10^{-2}T_f$ $-1.194357x10^{-7}T_f^{\,2})$
	Specific heat c_p (J/kg.K)
Nichrome	$110T_f^{\,0.2075}$
Boron Nitride	$760.59+1.7955T_f-8.6704x10^{-4}T_f^{\,2}$ $+1.7955x10^{-7}T_f^{\,3}$
Inconel 600	$4186.8\times(0.1014+4.378952\times10^{-5}T_f$ $-2.046138\times10^{-8}T_f^{\,2}+1.7955\times10^{-7}T_f^{\,3}$ $-2.060318\times10^{-13}T_f^{\,4}+3.682836\times10^{-16}T_f^{\,5}$ $-2.458648\times10^{-19}T_f^{\,6}+5.597571\times10^{-23}T_f^{\,7})$
	Thermal Conductivity k(W/m.K)
Nichrome	$29.18+2.683x10^{-3}\left(T_f-100\right)$
Boron Nitride	$25.27-1.365\times10.3T_f$
Inconel 600	$1.729577\times(8.011332+4.643719\times10^{-3}T_f$ $+1.872857\times10^{-6}T_f^{\,2}-3.914512\times10^{-9}T_f^{\,3}$ $+3.475513\times10^{-12}T_f^{\,4}-9.936696\times10^{-16}T_f^{\,5})$

The X-ray computerized tomography (CT) scanner and the X-ray densitometer shown in Figure 1.2 were used for measuring void fraction distributions. The X-ray CT measurement system comprises an X-ray tube and 512 detectors, attaining a spatial resolution of 0.3 mm × 0.3 mm. Fine-mesh void distributions were measured under steady-state conditions using the X-ray CT scanner at a point 50 mm above the heated length. The pressure vessel is made of titanium, while the channel wall and the cladding of the heater rods at this location are made of beryllium in order to minimize X-ray attenuation in the structure.

The steady state measurements were performed using thermal-hydraulic conditions that would envelope the steady-state parameters characterizing the actual operation of a BWR in terms of the bundle's geometrical configuration, power shape and two-phase flow. The range of test conditions included pressures ranging from 1MPa to 8.6 MPa, flow rates ranging from 284 kg/m^2/s to 1988 kg/m^2/s, and exit qualities ranging from 1 to 25%.

The X-ray densitometer was employed for performing transient measurements of the cross-sectional averaged transient void fraction distribution resulting from the combined effects of pressure, flow rate, and power for the following four operational transients: (a) turbine trip without bypass; (b) one pump trip; (c) re-circulation pump tripped; and (d) malfunction of pressure control system (pressure increase). The cross-sectional average transient void fraction distributions were measured by the X-ray densitometer at 3 elevations (at elevations 682, 1706, 2730 mm, from bottom to top, in Figure 1.2). As previously mentioned, to enable these measurements, the channel sections at these elevations were made of beryllium, and the heater rods were clad with beryllium having the same diameter as the inconel portion of the heater rod.

Pressure drop measurements were performed while the loop facility was operated under normal BWR operational conditions, including typical transient conditions. The transversal power profiles for the beginning of operation (type A) are listed in Table 1.3. The steady-state pressure drops were measured in both single-phase flow and two-phase flow conditions covering the normal BWR operational behavior.

Table 1.3: Transversal Power Shape of Type A for the Test Assembly Type 4.

1.15	1.30	1.15	1.30	1.30	1.15	1.30	1.15
1.30	0.45	0.89	0.89	0.89	0.45	1.15	1.30
1.15	0.89	0.89	0.89	0.89	0.89	0.45	1.15
1.30	0.89	0.89			0.89	0.89	1.15
1.30	0.89	0.89			0.89	0.89	1.15
1.15	0.45	0.89	0.89	0.89	0.89	0.45	1.15
1.30	1.15	0.45	0.89	0.89	0.45	1.15	1.30
1.15	1.30	1.15	1.15	1.15	1.15	1.30	1.15

1.3.2 BERRU-SMS Predictive Modeling of Void Fraction Distribution

Three sets of measurements were used to calibrate the high fidelity simulations performed using FLICA4, as follows: (i) pressure drops (steady-state one-dimensional simulations); (ii) axial void fractions distributions (transient one-dimensional simulations); and (iii) transversal void fraction distributions (steady-state three-dimensional simulations at sub-channel level with cross-flow). The steady-state and transient conditions, respectively, for the benchmark measurements have been modeled in FLICA4 by designing axial meshes with surfaces that match perfectly the measurement coordinates on the vertical-axis (see Figure 1.2) .The specific features of the spatial meshes used in the FLICA4 simulations are as follows:

(A) For the *one-dimensional* computations of *pressure drops* and *axial void fraction distributions*: axial meshes, from bottom to top, as follows: 10 meshes of 64 mm each, followed by one mesh of 42 mm, followed by 46 meshes of 64 mm each, and finally followed by one mesh of 82 mm.

(B) For the *three-dimensional* computations of *transversal void fraction distributions*:

(i) axial meshes, from bottom to top, as follows: 50 meshes of 74.16 mm each for modeling the heated length of 3.708 m, followed by 2 meshes of 25 mm each for modeling the X-ray CT-scanner surface (i.e., the unheated length of 50 mm in Figure 6); and

(ii) 64 cross-sectional meshes, each modeling one of the sub-channels depicted in the figure inserted in Table 1.1, as follows: (a) four central sub-channels for modeling the center of the mock-up assembly; and (b) 60 sub-channels containing the heated fuel rods. The four sub-channels located in the assembly's four corners (having one rounded corner and two surfaces facing the exterior) and the 24 lateral sub-channels (having only one surface facing the exterior) are modeled using mesh-shapes that differ from those for the interior sub-channels.

The thermo-mechanical properties of the heater rods (see Table 1.2) have also been implemented in FLICA4. These properties play a particularly important role for modeling the transient benchmark, where the thermal coupling between heater rods and fluid is essential for describing correctly the transient behavior. The following correlations have been used in FLICA for simulating the benchmark measurements:

- friction model type F3 (Fillion et al, 2007), corrected by the default model for the wall heating and the biphasic multiplying model of Friedel (Friedel, 1979);
- turbulent mixing and diffusivity coefficients computed with the F3 model (Fillion et al, 2007);
- recondensation model of type F3 (Fillion et al, 2007);
- relative motion between phases described by Ishii's correlation (Ishii, 1977);
- a generic Forster & Greif -like correlation (Forster and Greif, 1958), describing the wall-overheating with respect to saturation, in the regime of completely developed nucleate boiling;
- Groeneveld-95 correlation (Groeneveld, 1995) for computing the critical flux and the critical thermal flux ratio (CTFR).

A large variety of transient tests were performed in the BFBT facility in order to measure the transient void fraction distributions as functions of pressure, flow, and power changes representative of operational transients important for reactor safety, namely turbine trip without bypass, one pump trip, re-circulation pump tripped, and malfunction of pressure control system (pressure increase). The void fraction distributions were measured using the X-ray densitometer as indicated in Figure 1.2.
Three *system responses* have been considered for data assimilation and calibration, namely the time-dependent cross-sectional (transversal) averaged transient void distributions measured with the X-ray densitometers at three axial elevations: $h_1 = (3708 - 978)\ mm$, $h_2 = (h_1 - 1024)\ mm$, $h_3 = (h_2 - 1024)\ mm$ (see Figure 1.2). These responses are denoted by R_1, R_2, and R_3, respectively, and are depicted in

Figure 1.3. Although the BFBT experimental setup provides experimental data at every 0.02 seconds, such extremely narrow time intervals will not be necessary for the purpose of data assimilation and model calibration; a subset of these measurements will provide sufficiently accurate experimental information for this purpose.

The outlet pressure, flow rate, and power undergo large variations during the TTWB test. However, sensitivity analysis (Badea et al, 2012) has indicated that the influences of these parameters on the void fraction distribution can be neglected beyond 8 seconds after the occurrence of a perturbation in these parameters. These features permitted to limit the number of significant model parameters to 1260 for the experiments used for predictive modeling. The relative standard deviation for the flow rate was considered to be 4%, while the nominal relative standard deviation for the outlet pressure was taken as 2%. The standard deviations for the void fraction distributions were taken to be 3.5%, independent of time and space.

The BERRU-SMS predictive modeling procedure was applied individually to each of the void fraction responses R_1, R_2, and R_3. The following values of the corresponding consistency indicators were obtained: $\chi_1^2 = 473.39 \div 500 = 0.95$, $\chi_2^2 = 243.35 \div 500 = 0.49$ and $\chi_3^2 = 17.52 \div 500 = 3.5 \times 10^{-3}$, respectively. The white lines in Figure 1.3 depict the best estimate responses, which fall, as expected, between the originally measured and computed values. The accompanying predicted standard deviations for the three individually calibrated responses are depicted in Figure 1.4, which underscores the significant uncertainty reduction after data assimilation and model calibration. As predicted by the very small numerical value of χ_3^2, the largest inconsistency between the measured and the computational information occurred for the response R_3, which was only slightly improved by the predictive modeling procedure. This is because the computational and experimental results do not agree well in the regions of low void fractions. Additional sensitivity analysis and predictive modeling results are provided in the work of Badea et al (2012).

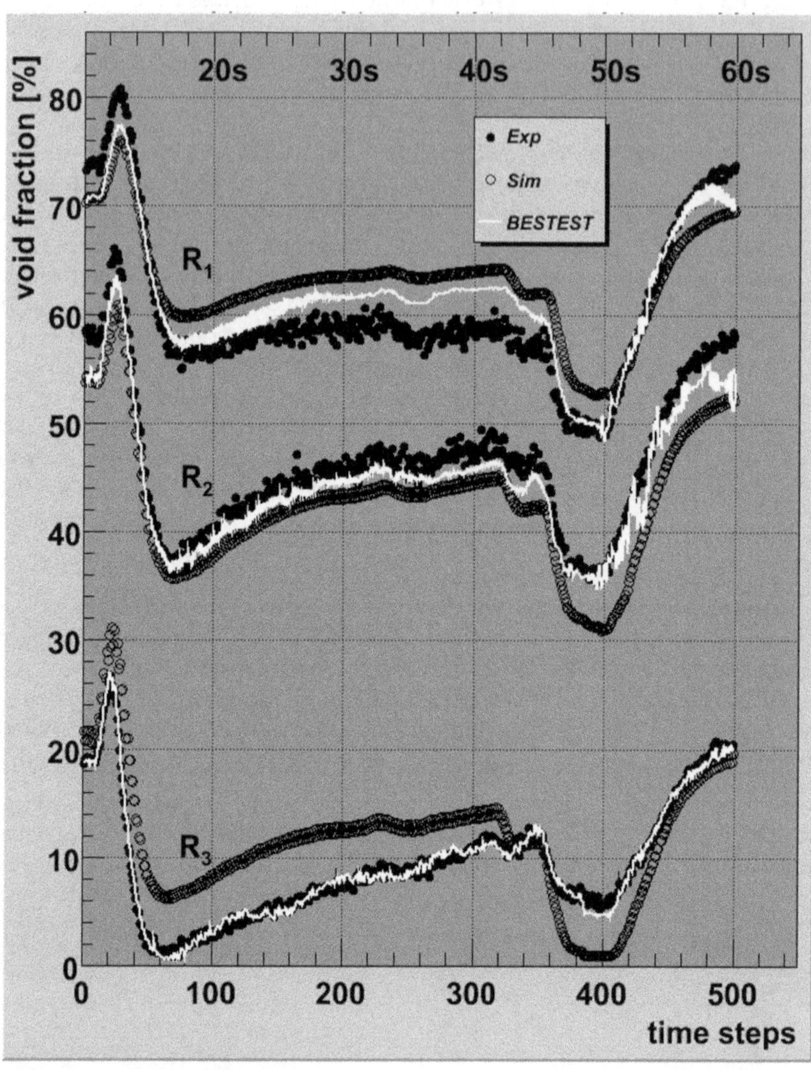

***Figure 1.3:** Experimental, simulated and best estimate void fraction distributions after data assimilation and individual calibration of responses.*

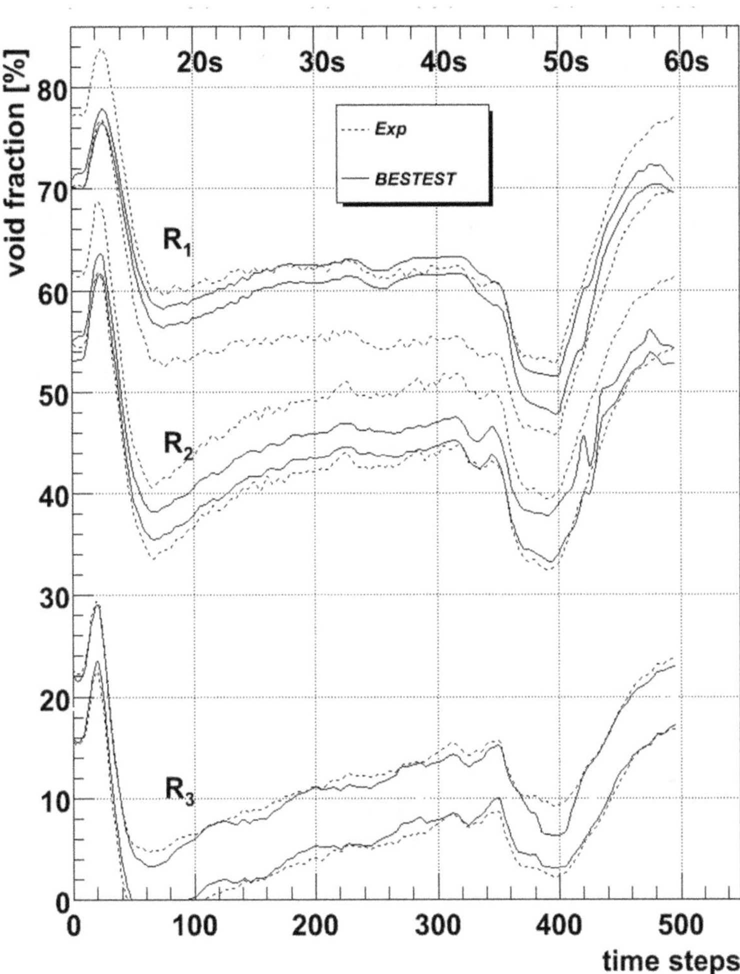

Figure 1.4: Error bands for experimental and best estimate void fraction distributions after data assimilation and individual calibration of responses.

1.4 BERRU-SMS APPLICATIONS TO LIQUID SODIUM FLOW EXPERIMENTS AND COMPUTATIONS

This Section presents two illustrative applications of the BERRU-SMS predictive modeling methodology to reduce predicted uncertainties in fluid dynamics and heat transfer processes involving liquid sodium. The results presented herein were originally obtained by Arslan and Cacuci (2014), who applied the BERRU-SMS methodology to the commercially available computational tool ANSYS CFX https://www.ansys.com/customer/center/documentation) in conjunction with experiments involving liquid sodium performed in the TEFLU (Knebel, 1993) and TEGENA (Moeller, 1989) experimental facilities.

1.4.1 BERRU-SMS Predictive Modeling of Turbulent Velocity Distributions in the TEFLU Liquid Sodium Experimental Facility

The TEFLU experimental facility was designed to investigate liquid-sodium flows with mixed convection. The TEFLU test section comprises an axially symmetric tube (for vertical flow) with an inner diameter of $D = 11.0\ cm$ and a total length of $L = 2.146 m$; a vertical section through this tube is schematically shown in Figure 1.5. The nozzle contains 158 holes, with a diameter $d = 7.2 mm$, configured in equilateral triangles with a total height of $l = 12.0 cm$ and a pitch of $p_{pitch} = 8.2 mm$. The nozzle's position is adjustable in the axial direction. The measurements of the mean velocity and temperature distributions are recorded simultaneously with MPPs (miniature permanent magnet potential probes). These probes can perform spatially-dependent measurements in the range $1.4 \le x/d \le 140$, where x/d is the distance of the corresponding axial measurement point to the nozzle diameter. Various flow characteristics are created by adjusting the velocities and temperatures at the inlet part of the test section. The jet-flow velocity $\overline{U}_i$ and temperature $\overline{T}_i$ can be changed independently from the co-flow velocity $\overline{U}_a$ and co-flow temperature $\overline{T}_a$. Heat losses in the inner pipe are reduced by evacuation and by using double-walled tubes.

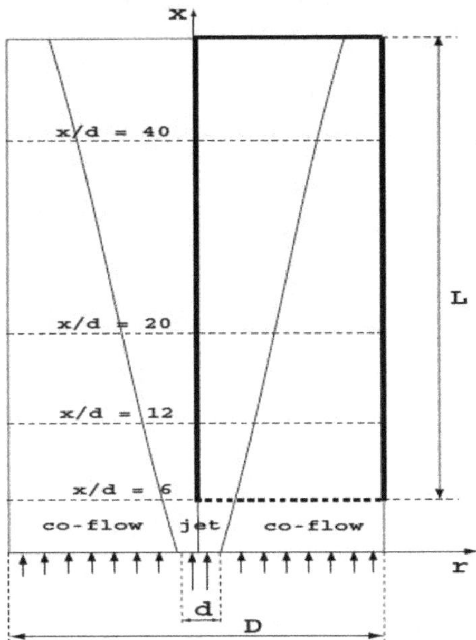

Figure 1.5: TEFLU test section

The TEFLU test section was modeled with CFX using a fine-scale mixed grid with predominantly tetragonal elements. The SAS SST (Scale Simulation Adaptive Shear Stress Transport) turbulence model provided in CFX was used for these numerical simulations. The responses (measured and computed) considered in this predictive modeling application were: (i) the r-dependent buoyant jet at $x/d = 6.0$ and (ii) r-dependent buoyant jet at $x/d = 20.0$. The relative standard deviations of the r-dependent measurements at $x/d = 6.0$ were also evaluated to be 5%, while the relative standard deviations of the r-dependent measurements at $x/d = 20.0$ were evaluated to be 7%. A total of 343 measurements were considered for the application of the predictive modeling methodology. All of these measurements were uncorrelated in time and space. The model parameters that displayed the largest relative sensitivities are listed in the first column of Table 1.4; their nominal values and corresponding standard deviations are listed in the second column of this table. The predicted best-estimate nominal parameter values, computed using Eq. (1.70), and their accompanying reduced predicted standard deviations, computed using Eq. (1.73), are presented in the 3rd and 4th columns of Table 1.4. The model parameters controlling the inlet velocities of the co-flow and jet-flow are the ones most affected by the BERRU-SMS predictive modeling procedure; the respective results are presented in the bottom two rows of Table 1.4. Due to their high sensitivities, the predicted best-estimate relative uncertainties (± 1 std. deviation) accompanying these parameters are reduced the most. On the other hand, the predicted best-estimate uncertainties for the other parameters are reduced only minimally from their original values; this is due to their very small sensitivities. The a posteriori predicted parameter-parameter correlations and response-parameter correlations, computed using Eqs. (1.73) and (1.80), respectively, are very small (under 1%), and are therefore not reported here. The results for the predicted best-estimate nominal values for the responses, along with their reduced predicted uncertainties (± 1 predicted standard deviations), are shown in Figures 1.6 and 1.7.

Table 1.4: CFX parameters (before and after calibration) for the buoyant-jet analysis

Parameter	Nominal Values ±1std. dev.	r-dependent jet at x/d=6; pred. values ±1std. dev.	r-dependent jet at x/d=20; pred. values ±1std. dev.
Co-flow inlet velocity [m/s]	0.10 ± 5%	0.110 ± 1.16%	0.110 ± 2.53%
Jet-flow Inlet velocity [m/s]	0.43 ± 5%	0.39 ± 1.70%	0.41 ± 2.74%
Eddy viscosity ν_t $[Pa \cdot s]$	0.010583 ± 5%	0.015473 ± 4.90%	0.016078 ± 4.78%
CFX coefficient β^*	0.09 ± 4%	0.09 ± 3.99%	0.09 ± 3.98%
Turbulent Schmidt number $(k-\omega)$ CFX model	0.90 ± 4%	0.89 ± 4.00%	0.90 ± 3.99%
Turbulent Schmidt number $(k-\varepsilon)$ CFX model	1.00 ± 4%	0.99 ± 4.00%	1.00 ± 3.99%
CFX SAS1 Coefficient	1.25 ± 4%	1.24 ± 3.99%	1.25 ± 3.99%
CFX SAS2 Coefficient	1.76 ± 4%	1.75 ± 4.00%	1.76 ± 3.99%
CFX SAS3 Coefficient	3.00 ± 4%	3.00 ± 3.99%	3.00 ± 3.99%
Turbulent Schmidt number $(k-\omega-\varepsilon)$ CFX model	1.17647 ± 4%	1.1775 ± 4.00%	1.1802 ± 3.99%
Dynamic viscosity η $[Pa \cdot s]$	344E-6 ± 5%	344E-6 ± 5.00%	346E-6 ± 4.97%

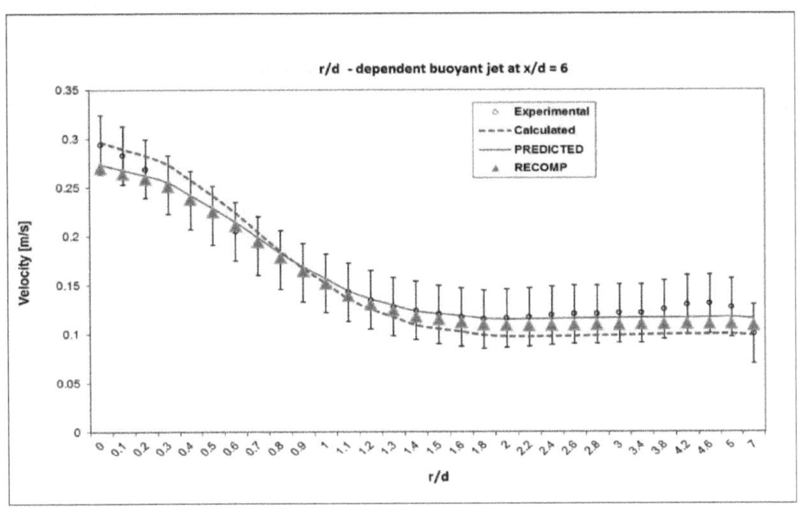

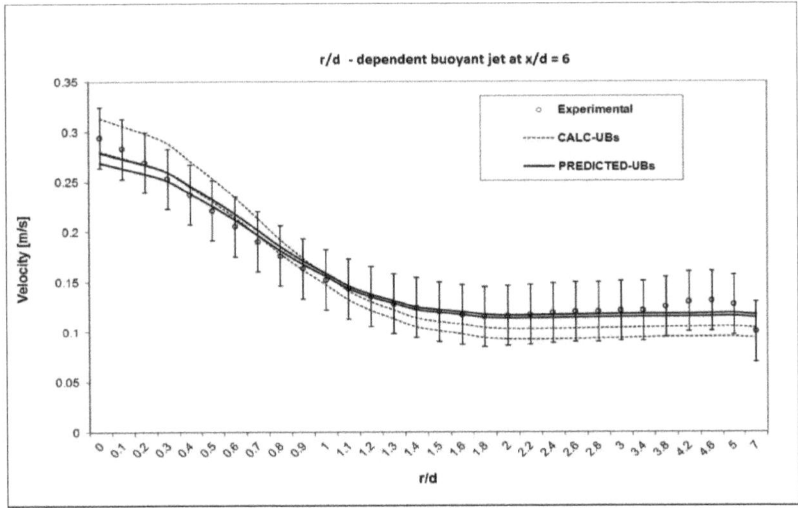

Figure 1.6. TOP: Experimental measurements with error bars, CFX-computation, predicted and recalculated best-estimate r ***-dependent velocity of the buoyant jet at*** $x/d = 6.0$

BOTTOM: Experimental measurements with (± 1) ***standard deviation bars, computed*** (± 1) ***standard deviations, and predicted best-estimate*** (± 1) ***standard deviations (UBs: upper and lower uncertainty bands)***

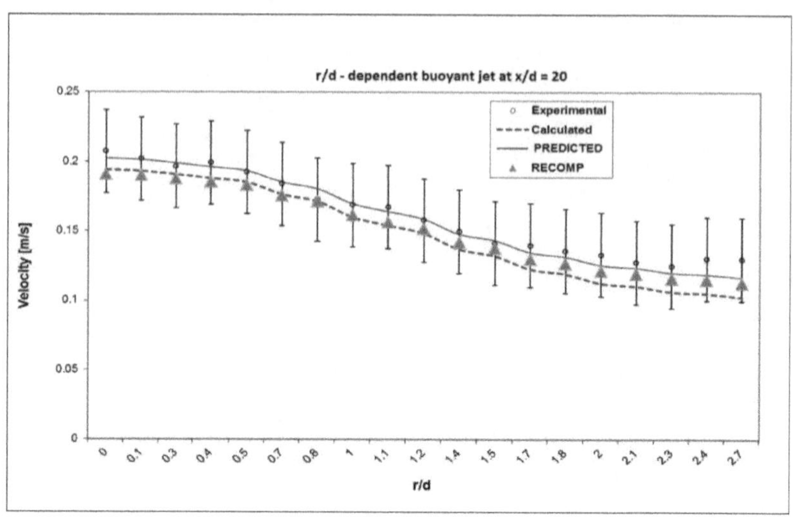

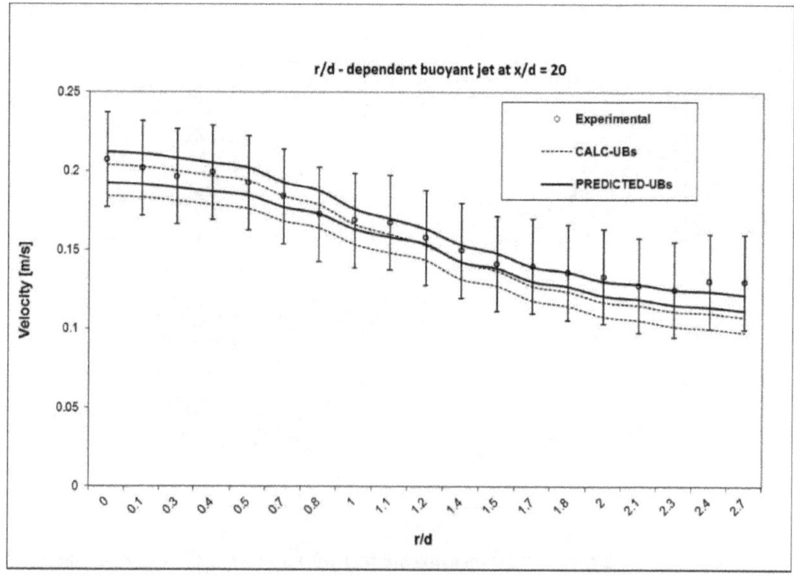

Figure 1.7. ***TOP: Experimental measurements with error bars, CFX-computation, predicted and recalculated best-estimate*** r ***-dependent velocity of the buoyant jet at*** $x/d = 24.0$ **.**

BOTTOM: Experimental measurements with (± 1) ***standard deviation bars, computed*** (± 1) ***standard deviations, and predicted best-estimate*** (± 1) ***standard deviations (UBs: upper and lower uncertainty bands)***

The top plot in Figure 1.6 depicts selected experimental measurements of the radially-dependent velocity of the buoyant jet at the axial position $x/d = 6.0$, together with their corresponding error bars. Also presented (with dashed lines) are the CFX-computations, using the nominal parameter values shown in the first column of Table 1.4. These CFX computations over-predict the experimental values for all radial positions $r/d < 0.75$, agree well with the experimental values in the relatively narrow range $0.75 < r/d < 0.95$, and under-predict the experimental values for the range of radial positions $r/d > 0.95$. The predicted best-estimate nominal values for the calibrated (adjusted) responses computed using Eq. (1.75) are also presented (depicted by a continuous line) in the top plot of Figure 1.6. In contradistinction to the CFX computations, these predicted response values agree very well with the experimental values, over the entire range of radial values $0.1 < r/d < 4.7$. Finally, the triangular-shaped points in the top plot of Figure 1.6 represent the recomputed responses using CFX, after having replaced the nominal parameter values in CFX with the predicted best-estimate parameter values shown in the fourth column of Table 1.4. The recomputed responses are very close to the predicted ones, indicating that the effects of the calibration of the of inlet velocities are significant.

The bottom plot in Figure 1.6 presents the results for the computational uncertainties (i.e., the "± 1 standard deviation"), obtained by using the "sandwich" formula shown in Eq. (1.61), along with the predicted best-estimate response uncertainties (± 1 standard deviation) computed using Eq. (1.78). The "± 1 standard deviations" are denoted as "UBs" (uncertainty bands). By comparison to both the original experimental standard deviation, and also the computed response standard deviation, the predicted standard deviation is strikingly smaller, and also considerably more consistent with the measurement uncertainty bands than the computed response uncertainty bands.

Although the full predicted response-response covariance matrix has been computed using Eq. (1.78), the off-diagonal elements are all less than 1%, and are therefore not presented here. The consistency indicator for the r-dependent velocity of the buoyant jet at $x/d = 6.0$ was computed using Eq. (1.101) and was found to have the value $\chi^2/N_r = 1.57$ (for $N_r = 30$ measurements). This result indicates that the CFX computations were a priori reasonably consistent with the experimental data, given the respective parameter and response uncertainties.

Figure 1.7 presents results for the predictive modeling of the radially-dependent velocity of the buoyant jet at the axial position $x/d = 24.0$. The top plot in Figure 1.7 depicts selected experimental measurements of this velocity, together with their corresponding "± 1 standard deviations". Also presented (with dashed lines) are the CFX-computations, using the nominal parameter values shown in the first column of Table 1.4. These CFX computations under-predict the experimental values for all radial positions. The predicted best-estimate nominal values for the calibrated (adjusted) responses, computed using Eq. (1.75), are also presented (depicted by a continuous line) in the top plot of Figure 1.7. In contradistinction to the CFX computations, these predicted response values agree very well the experimental values, practically over the entire range of radial values. Finally, the triangular-shaped points in the top plot of Figure 1.7 represent the recomputed responses using CFX with the predicted best-estimate parameter values shown in the fifth column of Table 1.4, rather than with the original nominal parameter values. Just as for the results presented in Figure 1.6, the recomputed responses agree very well with the predicted ones, indicating that the effects of the calibration of the of inlet velocities are significant.

The bottom plot in Figure 1.7 presents the results for the computational uncertainties (i.e., the "±1 standard deviations"), obtained by using the "sandwich" formula shown in Eq. (1.61), along with the predicted best-estimate response uncertainties (±1 standard deviations) computed using Eq. (1.78). By comparison to both the original experimental standard deviation, and also the computed response standard deviation, the predicted standard deviation is significantly smaller, and more consistent with the measurement uncertainty bands than the computed response uncertainty bands.

The "±1 standard deviations" are denoted as "UBs" (uncertainty bands). Although the full predicted response-response covariance matrix has been computed using Eq. (1.78), the off-diagonal elements are all less than 1%, and are therefore not presented here. The consistency indicator for the r-dependent velocity of the buoyant jet at $x/d = 24.0$ was computed using Eq. (1.101) and was found to have the value $\chi^2/N_r = 0.72$ (for $N_r = 19$ measurements). This result indicates that the CFX-computations were a priori reasonably consistent with the experimental data, given the respective parameter and response uncertainties.

The largest reduction of the predicted response uncertainties occurs for the buoyant-jet velocity distribution at $x/d = 6.0$ (cf., Figure 1.6). This is due to the fact that the original CFX-computations are partially over-predictive and partially under-predictive by comparison to the corresponding experimental results, which is in contrast to the situation presented in Figure 1.7, where the CFX-computations consistently under-predicted the experimental results (for $x/d = 24.0$).

1.4.2 BERRU-SMS Predictive Modeling of Coupled Fluid Dynamics and Heat Transfer Processes in the TEGENA Liquid Sodium Experimental Facility

The TEGENA experimental facility is depicted in Figure 1.8. The test section comprises a rectangular channel of total length $2.485\,m$, which is divided into three segments. Counting from the bottom to the top, the first segment comprises a pre-inlet (2.0 m) and an inlet (1.288 m) for the liquid sodium flow, both of which are not heated. The second segment comprises a heating zone, and the last section serves as the outlet for the flow. The heated rods are 2.5 cm in diameter, and are held in place by spacers in the form of pins ("ferrules"). These spacers are attached to the perimeters of the rods to ensure a radial-separation distance of 3.7 cm between all rods. The maximum heat flux on the rod-surface is 60 W/cm^2 at a maximum power of 475 KW. The liquid-sodium temperatures vary between 523 K and 673 K. The maximum temperature rise in the liquid sodium is limited to 100 K. The mean flow velocities range from 0.1 m/s to 2.4 m/s. These velocities give rise to Reynolds numbers (Re) in the range of Re = 3700 to 76000. The temperature profiles are measured by transverse and longitudinal displacements of probes. Depending on the specific experiments, the corresponding locations are at 29 mm (designated "ME6" in Figure1. 8) or at 31.5 mm (not shown in Figure 1.8) before the end of the test section.

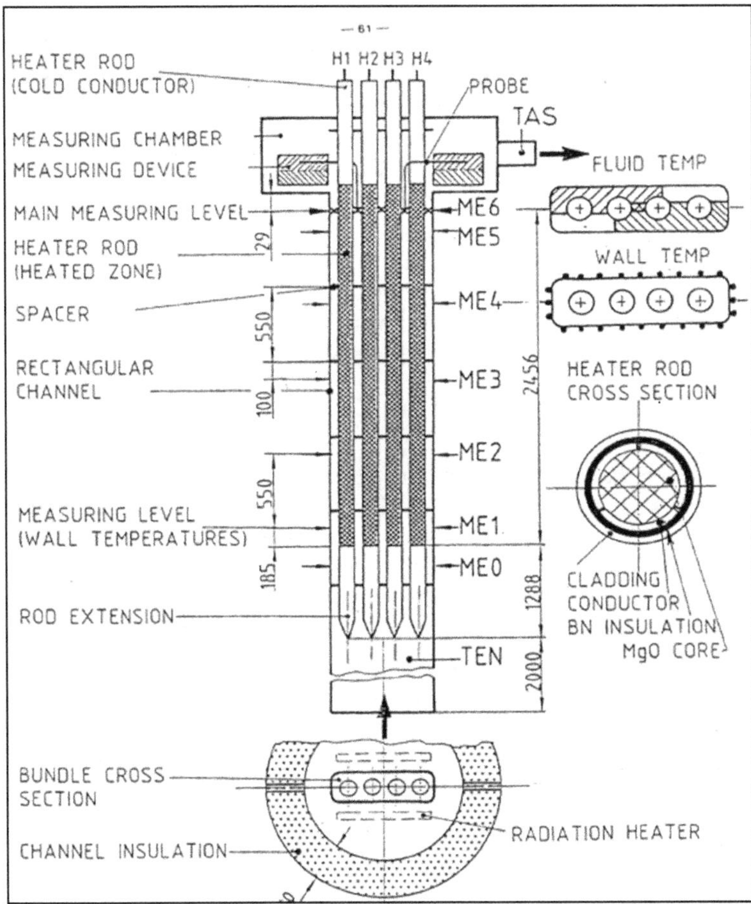

Figure 1.8: The TEGENA Experimental facility

The TEGENA experiment considered for the BERRU-SMS predictive modeling application presented in this Subsection was performed under the following conditions: average inlet velocity = 1.91 m/s; Reynolds number = 60100; mass flow rate

= 3.12 kg/s; average sodium temperature = 257.98 $°C$; heating power for rod H1 = 49.38 W/cm^2; heating power for rod H2 = 49.00 W/cm^2; heating power for rod H3 = 48.42 W/cm^2; heating power for rod H4 = 49.74 W/cm^2. The fluid temperature distribution was measured at 17 locations in the vertical (y-) direction, in the heated zone between heaters H3 and H4. The relative standard deviations for these measurements were quantified to be 1%, and the measurements were not correlated in space and time.

The TEGENA experiment was modeled with ANSYS CFX using a fine hexahedral mesh to enable the computation of the nominal values of axial temperature distribution within the 29 mm heated rod section, which was subsequently used to compute sensitivities (to CFX parameters, initial and boundary conditions) of the temperatures at the measurement locations. The relative response sensitivities were the largest for the eleven CFX parameters listed in the first column of Table 1.5. The second column of this table presents the nominal values together with the respective relative standard deviations. The third column in Table 1.5 presents the predicted best-estimate nominal parameter values, computed using Eq. (1.70), and their accompanying reduced predicted uncertainties, computed using Eq. (1.73). Notably, the BERRU-SMS modeling procedure has reduced considerably the predicted best-estimate uncertainties (± 1 std. deviation) for all of the CFX parameters presented in Table 1.5. In particular, the predicted best-estimate uncertainties for the heat fluxes are a factor or 3 smaller than the original uncertainties. The a posteriori predicted parameter-parameter correlations, and response-parameter correlations, computed using Eqs. (1.76) and (1.80), are very small (under 1%), and are therefore not reported here.

The results of applying the BERRU-SMS predictive modeling methodology to obtain the predicted responses and their reduced uncertainties using CFX-computations and the TEGENA experimental results are presented in Figure 1.9. The top plot in Figure 1.9 depicts selected experimentally measured values of the temperature distribution in TEGENA experiment, together with the corresponding error bars. Also presented (with dashed lines) are the CFX-calculations, using the nominal parameter values listed in the second column of Table 1.5. These spatially (y)-dependent CFX calculations over-predict systematically the experimental values for all values of the vertical spatial variable y. The predicted best-estimate nominal values for the calibrated (adjusted) responses, computed using Eq. (1.75), are also presented (depicted by a continuous line) in the top plot of Figure 1.9. These predicted response values are significantly closer to the experimental values than the CFX computations. Finally, the triangular-shaped points in the top plot of Figure 1.9 represent the recomputed responses using CFX using the predicted best-estimate parameter values shown in the third column of Table 1.5. The recomputed responses are closer to the nominal CFX-calculations rather than to the predicted ones, confirming the fact that, due to the small sensitivities or the responses to parameters, the original parameter values were adjusted very little by the predictive modeling formulas (already indicated in Table 1.5). The closer agreement between the measurements and the predicted (rather than the recalculated) response values stems from the fact that the predicted responses take the measurements directly into account in Eq. (1.75), whereas the recalculations using adjusted parameter values are influenced less strongly (comparatively) by the measurements.

The bottom plot in Figure 1.9 presents the results for the computational uncertainties (i.e., the "± 1 standard deviations"), obtained by using the "sandwich" formula shown in Eq. (1.61), along with the predicted best-estimate response uncertainties computed using Eq. (1.78). The "± 1 standard deviations" are denoted as "UBs" (uncertainty bands). Although the full predicted response-response covariance matrix has been computed using Eq. (1.71), the off-diagonal elements are all less than 1%, and are therefore not presented here. It is noteworthy that the predicted (± 1) standard deviations straddle the nominal measurement values, bringing a significant improvement (reduction) over both the computed and the measured response uncertainties.

Table 1.5: Nominal values and predicted best-estimate nominal values for CFX model parameters (with $\pm$ 1 rel. std. dev.) for TEGENA Experiment F04Q12B.DAT

Parameter	**Experiment F04Q12B.DAT**	
	Nominal Value $\pm$ 1 rel. std. dev.	**Predicted Value $\pm$ 1 rel. std. dev.**
Heat flux H1 [W/cm^2]	49.38 ± 7%	49.32 ± 1.97%
Heat flux H2 [W/cm^2]	49.00 ± 7%	48.94 ± 1.97%
Heat flux H3 [W/cm^2]	48.42 ± 7%	48.32 ± 1.89%
Heat flux H4 [W/cm^2]	49.74 ± 7%	48.64 ± 1.89%
$C_{\varepsilon,1}$ from the $k-\varepsilon$ Model	1.44 ± 7%	1.38 ± 2.88%
Inlet velocity [m/s]	1.91 ± 7%	1.91 ± 3.37%
$C_{\varepsilon,2}$ from the $k-\varepsilon$ Model	1.92 ± 7%	1.90 ± 3.92%
Turbulent Schmidt Number from ε correlation	1.00 ± 7%	0.99 ± 3.99%
Turbulent Schmidt Number from k correlation	1.30 ± 7%	1.30 ± 3.99%
C_μ from the $k-\varepsilon$ Model	0.09 ± 7%	0.089 ± 4.00%
Turbulent Prandtl Number	0.90 ± 7%	0.88 ± 5.53%

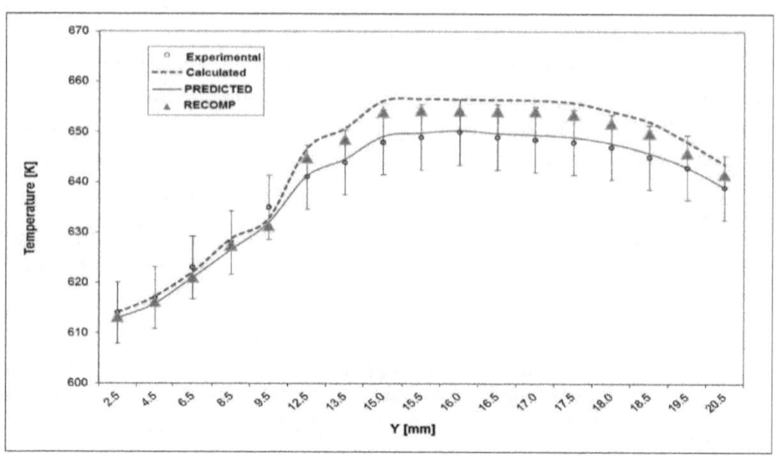

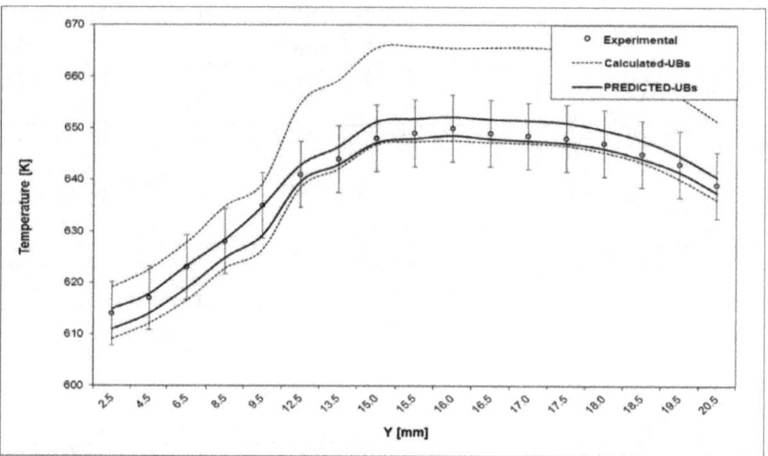

Figure 1.9: TOP: Experimental measurements with error bars, CFX-computation, predicted and recalculated best-estimate responses.

BOTTOM: Experimental measurements with (± 1) ***standard deviations bars, computed*** (± 1) ***standard deviations, and predicted best-estimate*** (± 1) ***standard deviations (UBs: upper/lower uncertainty bands).***

1.5 BERRU-SMS APPLICATION TO A COUPLED NEUTRON-KINETICS / THERMAL HYDRAULICS BENCHMARK

The predictive modeling methodology described in Section 1.2 is applied in this Section to achieve a substantial (up to 50%) reduction of uncertainties in the predicted total transient power in the boiling water reactor (BWR) Turbine Trip 2 (BWR-TT2) benchmark. The material presented in this Section is based on the work by Badea and Cacuci (2017).

A turbine trip (TT) in a BWR is characterized by a sudden closure of the turbine stop valve (TSV), and can be initiated by malfunctions affecting the turbine or the auxiliary nuclear system. A TT transient in a BWR is considered to be one of the most complex design basis accident scenarios because it involves the reactor core, the high pressure coolant boundary, associated valves and piping in highly complex interactions with variables changing very rapidly. The TT benchmark (Solis et al., 2001) defined by the OECD and the US Nuclear Regulatory Commission is based on an experiment with closure of the turbine stop valve which was carried out in 1977 in the nuclear power plant Peach Bottom 2. In the experiment, the closure of the valve caused a pressure wave which propagated with attenuation into the reactor core. The condensation of the steam in the reactor core caused by the increase of pressure led to a positive reactivity insertion. The subsequent rise of power was limited by the Doppler feedback and the insertion of the control rods, which started at 0.75 seconds and ended at 3.83 seconds into the transient, causing a decline of the total reactor power below the initial level. The safety, relief, and the bypass system valves helped releasing the steam production and limiting the pressure increase in the nuclear plant.

The BWR TT benchmark was published in four volumes as OECD/NEA reports (Solis et al., 2001, Akdeniz et al., 2005; 2006; 2010). The benchmark was divided into 3 phases. Phase I was used for validation of the thermo-hydraulic model of the system for a given power release in the core. In Phase II, three-dimensional core reactor physics (neutronics) computations were performed for given thermal-hydraulic boundary conditions. Coupled core neutronics/thermal-hydraulics computations were carried out for the selected experiment and four extreme scenarios in the Phase III. For the TT2 transient test, the dynamic measurements were registered with a high-speed digital acquisition system capable of sampling 150 signals every 6 milliseconds. The power distribution in the reactor core was measured using plant's local in-core flux detectors. Special fast response pressure and differential pressure transducers were installed in addition to the existing instruments in the plant's nuclear steam supply system.

The most important phenomena during the TT2-transient experiment occurred in the first 5 seconds after the initiating event. Therefore, the transient predictive modeling of TT2 presented in this Section is also limited to the first 5 seconds after the initiating event. The transient system response of interest is the time-dependent total reactor power, which is influenced mainly by the pressure in the reactor core and the insertion of control rods. Figure 1.10 depicts the transient pressure at the outlet of the reactor core (right) and the transient flow rate (left), as provided by Akdeniz et al. (2005). The evolution of these transient boundary conditions affects the reactor power most significantly during the first second after the initiation of the transient, after which the insertion of the control rods reduces the power to levels that are below those at the beginning of the transient. The transient TH boundary conditions, flow rate and outlet pressure will be also considered as model parameters within the BERRU-SMS predictive modeling application.

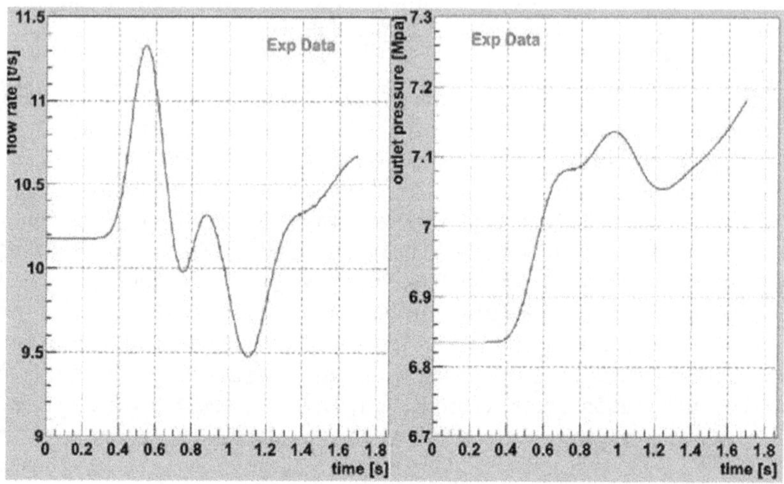

Figure 1.10: Transient boundary conditions: flow rate (left) and pressure (right) at the outlet of the reactor core.

The neutron conservation balance in the reactor core of the BWR-TT benchmark is modeled (Grundmann et al., 2004) using the DYN3D neutron kinetics code, which solves two-group (fast and thermal) three-dimensional neutron diffusion equations with up to six groups of precursors for delayed neutrons, using nodal expansion methods while assuming that the macroscopic cross sections are spatially constant in a node. The nodal expansion for Cartesian geometry uses transverse integration over all combinations of two directions of the rectangular node, yielding three one-dimensional diffusion equations, which are solved for the transverse integrated fluxes of the nodes in the three directions x, y, z. In each energy group, the one-dimensional equations are solved using flux expansions in polynomials up to 2nd-order and exponential solutions of the homogeneous equations. The fission source in the fast group, the scattering source in the thermal group, and the leakage terms are approximated by polynomials. The reactor core is divided into horizontal slices and the "nodes" (for the "nodal expansion") are spatially-homogenized assemblies in each slice. The neutron group constants are assumed to be spatially constant in each node. An implicit finite difference scheme together with an exponential transformation technique is used for the time integration of equations. Since the diffusion equation for the homogenized assembly is not equivalent to the original heterogeneous problem, the homogenized results are improved by introducing assembly discontinuity factors (ADF) defined (for each energy group) as the ratio of the average values of heterogeneous and homogeneous neutron fluxes at the radial fuel assembly boundaries. These ADFs are taken into account by modified interface conditions between the nodes. Homogenized two-group cross sections are used for the given fuel-element composition, nodal burnup, and the actual reactor operation conditions as characterized by the node-averaged values of fuel temperature, moderator temperature, moderator boron concentration and moderator density, which are calculated by DYN3D itself. Control rod presence or absence is also taken into account. The average value of the power densities at time t in a node n is computed using the expression $P^n(t)=\sum_{g=1}^{2}\varepsilon_g^n\Sigma_{f,g}^n(t)\bar{\Phi}_g^n(t)$, where ε_g^n denotes the energy release per fission. Gamma heating of the coolant is considered in the thermal-hydraulic model FLOCAL (discussed further below) which is coupled to DYN3D.

The radial geometry of the reactor core's computational model is shown in Figure 1.11; this geometry comprises 764 square fuel assemblies and 124 square reflector assemblies, each having a side of 15.24 cm. The reactor core is divided axially into 26 layers (24 core layers, with a reflector on the top and one at the bottom). The total active core height is 365.76 cm.

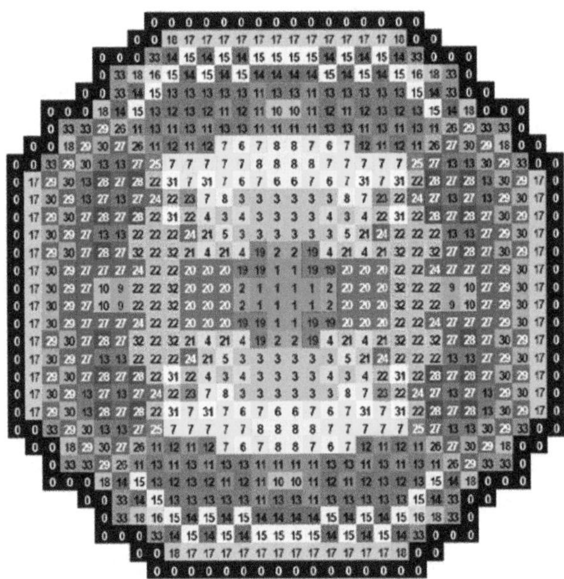

Figure 1.11: Transversal section of the TT benchmark reactor core computational model showing 888 homogenized assemblies.

Computation of the time-dependent variation of the total reactor power characterizing the BWR-TT2 benchmark is performed by tightly coupling DYN3D to the one-dimensional core thermal-hydraulics module FLOCAL using one channel per assembly in the axial direction (for a total of 888 thermal-hydraulic channels). The

FLOCAL module simulates the thermo-hydraulics phenomena within the reactor's core under steady-state and transient conditions based on following principles:

(A) The core is represented by one-dimensional parallel coolant channels, each channel comprising a fuel rod and the corresponding coolant flow. Each coolant channel is connected to one or more fuel assemblies describing the average conditions. The channels are isolated, i.e. no cross flow between the channels is taken into account. The channels are coupled only over common boundary conditions.

(B) One- or two-phase flow can be considered. The description of two-phase flow is based on four differential balance equations for mass, momentum and energy of the mixture and mass balance of the vapor phase. The two-phase model is closed by the following additional assumptions:
(i) One of the phases (vapor or liquid) is in saturation conditions. Its thermo-dynamic state is defined by the equation of state at saturation line.
(ii) The phase velocities are related to each other by a quasi-stationary phase slip model; an algebraic equation for the phase slip ratio is applied. In particular, the phase slip ratio, S, in the Molochnikov boiling model (1982) included in DYN3D has the expression $S = 1 + \left(P_2 + P_1\beta^2\right)/Fr^{1/4} \cdot \left(1 - p/p_{crit}\right) \cdot \sqrt{1 - \exp\left(-20\beta\right)}$, where β denotes the vapor volumetric flow ratio, Fr denotes the Froude number, p_{crit} denotes the critical pressure; the parameters $P_1 = 1.5$ and $P_2 = 0.6$ were considered to be imprecisely known and hence were explicitly considered in the application of the BERRU-SMS predictive modeling methodology.

(C) The thermo-hydraulic model is closed by constitutive relations for the slip ratio, one and two-phase frictional pressure losses, evaporation and condensation rates, heat transfer correlations and thermo-physical properties of the phases.

(D) A fuel rod model that simulates one-dimensional heat transfer in fuel, cladding, and within the gas gap between them is included. This model is coupled to the flow model via heat flux at the cladding surface facing the fluid.

(E) A model of boron acid tracking within the coolant channels is included by solving boron acid transport equation for each channel.

At steady-state, the solution of the eigenvalue neutron diffusion equation is used to compute power densities, which are in turn used to solve the fuel rod heat conduction and two-phase thermal-hydraulics equations. The node averaged values thus obtained for the thermal-hydraulics parameters, fuel temperature, coolant density, coolant temperature, and boron concentration are subsequently used to compute the macroscopic nodal cross sections, in conjunction with the DYN3D cross section library. The cross sections thus obtained are used in an iterative procedure for computing the flux and power density distribution in the core, and corresponding thermal-hydraulics temperatures, void fractions, etc. The iterations are terminated using convergence criteria for the reactor's fission source and effective multiplication factor. The iterative transient calculations are similar to the steady state iterations, except that the neutron kinetics iterations require a much smaller time step for convergence than the thermal-hydraulics ones. Therefore, the time steps for neutron kinetics (Δt_{NK}) and thermal-hydraulics (Δt_{TH}) are automatically controlled, requiring that $\Delta t_{NK} \leq \Delta t_{TH}$.

The coupled netronics/thermal-hydraulics equations were solved using the nominal values of the macroscopic cross sections, the nominal values of the transient boundary conditions plotted in Figure 1.10, and the nominal values of the Molochnikov parameters to obtain the nominal values of the computed responses, which are displayed in Figure 1.12 (in blue). The experimental measurements of the time-dependent power variation over the entire duration of the transient are also presented (in red) in Figure 1.12, along with the "close-up" of the power variation during the first 1.7 seconds into the transient, when the power displays the largest and fastest variations. This time-segment of the power variation is modeled by computing the power at 284 instances in the interval from the start of the transient (at 0 seconds) up to 1.7 seconds, taken at equally-spaced intervals of 6 milliseconds each. The power values computed --and also measured-- at these 284 equally spaced "time nodes" will subsequently provide 284 "model responses" for applying the BERRU-SMS predictive modeling methodology, including sensitivity and uncertainty analysis, data assimilation, model calibration, and reducing uncertainties in the best-estimate predictions.

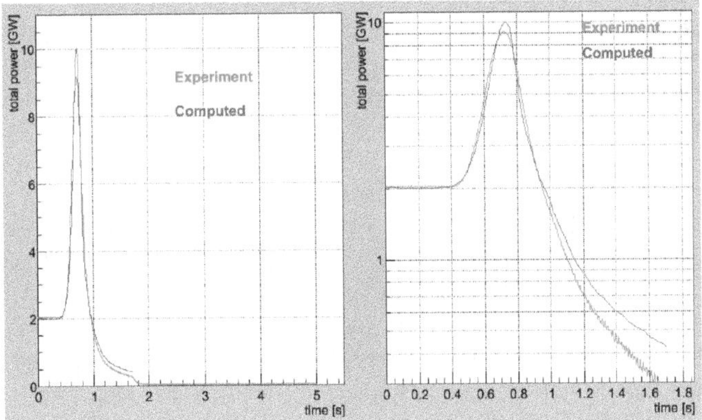

Figure 1.12. ***Left: experimental (red) and simulated (blue) power variation in the BWR-TT2 benchmark.***
Right: logarithmic close-up of the 284 computed powers, at equally-spaced time intervals of 6 milliseconds, during the transient's first 1.7 seconds, which defines the responses of interest for predictive modeling.

A total of 6660 imprecisely known model parameters have been calibrated by applying the BERRU-SMS predictive modeling methodology. These model parameters can be grouped in three broad categories, as follows:

(a) the flow rate and outlet pressure depicted in Figure 1.10, discretized at 284 instances in time, thus providing 568 imprecisely known model parameters which will generically illustrate the calibration of transient boundary conditions using the BWR-TT experimental data;

(b) parameters P_1 and P_2 in the Molochnikov (1982) phase-slip model in the FLOCAL module, which will be used to illustrate the use of the BWR-TT experimental data for calibrating parameters in thermal-hydraulics correlations.

(c) 6090 imprecisely known macroscopic cross section stemming from the two-group absorption, diffusion, and fission macroscopic cross sections of the 435 "materials" and 435 "rodded materials" (a total of 5 220 parameters stemming from: "2 groups" times "3 types of cross sections" times "2 x 435 materials") and scattering cross sections from the fast group #1 to the thermal group #2 (a total of 870 parameters stemming from "2 x 435 materials"). These model parameters will be used in the predictive modelling methodology to illustrate the calibration of macroscopic cross sections using the BWR-TT experimental data.

The exhaustive sensitivity analysis performed (Badea and Cacuci, 2017) on this (BWR-TT) benchmark revealed that the largest relative sensitivities of the power response are with respect to the time-dependent outlet pressure, indicating that the power is strongly influenced by perturbations in the void (steam) fraction/density/moderation efficiency in the moderator. Several typical sensitivity profiles, corresponding to perturbations in the outlet pressure at 0.42s, 0.84s, and 1.26s, are plotted on the left-side diagram in Figure 1.13. These relative sensitivities attain their largest values before the initiation of reactor scram by insertion of the control rods at 0.85 seconds into the transient. After the scram, the power level becomes somewhat less sensitive to perturbations in the outlet pressure, even though the respective sensitivities still remain large, as depicted by the sensitivity to the perturbation initiated at 1.26 seconds. The full matrix of sensitivities of the power response (at all 284 time instances spanning the time interval from 0 to 1.7 seconds) to perturbations in the outlet pressure (also considered at all 284 time instances spanning the time interval from 0 to 1.7 seconds) are displayed as "points" in Figure 1.13 (right). The logarithmic scale on the left-vertical axis presents values *sign(w)log10(abs(w))* for a relative sensitivity having the value *w%*. In order to preserve the one-to-one characteristic of the graphical representation for (significant) sensitivities with *abs(w)*>1 only, values of *abs(w)* below 1.0 (i.e., relative sensitivities in the interval -1% to +1%, including 0.0) are displayed as zeros, although they were exactly computed and used in subsequent computations. A relative sensitivity in the interval -1% to +1% thus corresponds to the value "zero" on the right-vertical colored axis in this diagram. Due to causality (i.e., $S^{\nu\mu}=0 \;\; when \;\; \nu<\mu$), the sensitivities at time instance on the left triangular part of the matrix of sensitivities are all zero, and are not displayed on the chosen logarithmic scale.

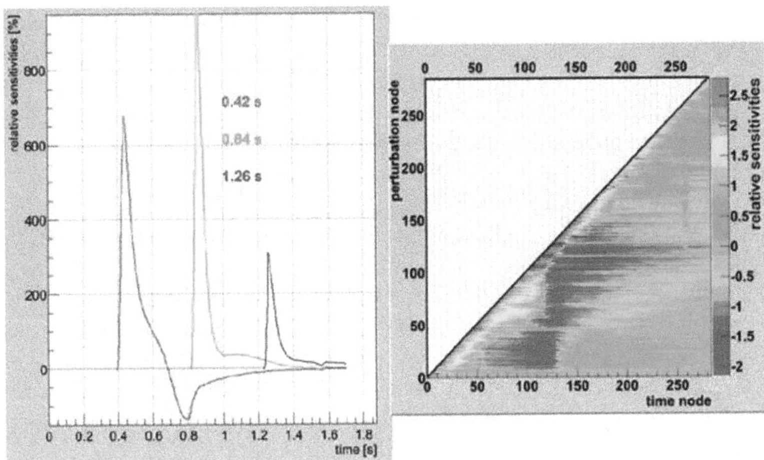

Figure 1.13: ***Relative sensitivities of the transient total power of the reactor to the outlet pressure (boundary condition).***
Left: corresponding to perturbations induced at the time nodes 0.42, 0.84, 1.26 s.
Right: corresponding to perturbations induced at each of the 284 time nodes.

The next largest relative sensitivities of the total reactor power are to the parameters P_1 and P_2 in the Molochnikov (1982) phase-slip model in the FLOCAL module. These sensitivities are presented in Figure 1.14, which indicates that they attain values of up to 120% at 0.7 seconds into the transient, and are also negative, which indicates that an increase in either or both of these parameters would cause the power response to decrease.

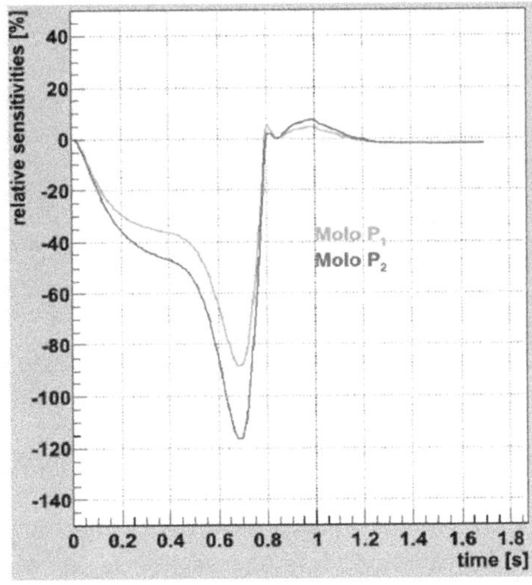

Figure 1.14: ***Relative sensitivities of the response (total power) to the 2 parameters of the Molochnikov (1982) phase-slip model (in green and blue).***

The relative sensitivities of the power response to perturbation in the time-dependent flow rate are presented in Figure 1.15, which shows that these relative sensitivities attain values up to 60%. Several typical sensitivity profiles, corresponding to perturbations in the outlet pressure at 0.42s, 0.84s, and 1.26s, are plotted on the left-side diagram in Figure 1.15. Just as for the sensitivities of the power to perturbation in the outlet pressure displayed in Figure 1.13, the relative sensitivities of the power to the flow rate attain larger values before the initiation of reactor scram by insertion of the control rods after the 0.85s. After the scram, the power level becomes increasingly less sensitive to perturbations in the flow rate. The full matrix of sensitivities of the power response (at all 284 time instances spanning the time interval from 0 to 1.7 seconds) to perturbations in the flow (also considered at all 284 time instances spanning the time interval from 0 to 1.7 seconds) are displayed as "points" in Figure 1.15 (right). The logarithmic scale on the right-vertical colored axis presents values *sign(w)log10(abs(w))* for a relative sensitivity having the value *w%*. The convention used for Figure 1.13 is also used for Figure 1.15 (right).

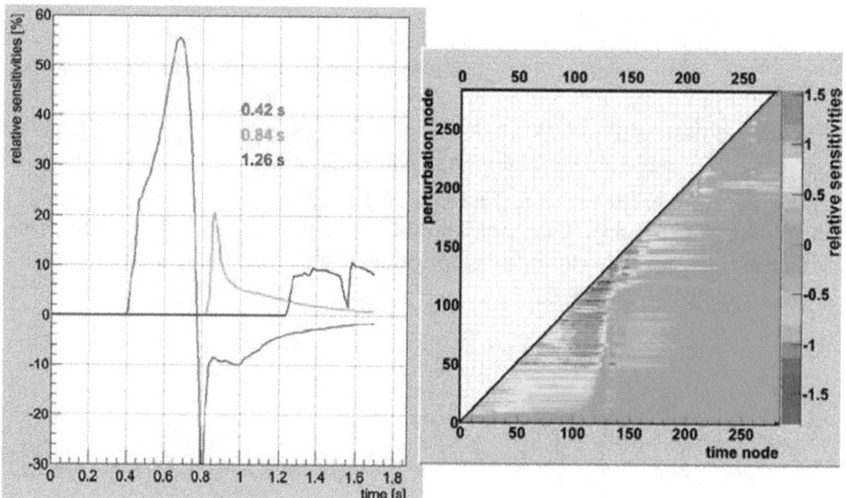

Figure 1.15: Relative sensitivities of the transient total power of the reactor to the flow rate (boundary condition).
Left: corresponding to perturbations induced at the time instances (nodes) 0.42, 0.84, 1.26 s.
Right: corresponding to perturbations induced at each of the 284 time instances (nodes).

The relative sensitivities (in %) of the transient total power of the reactor to the total number (6090) of macroscopic cross sections are displayed in Figures 1.16 through 1.19. The indexing of the 6090 macroscopic cross sections is provided in Table 1.6 and is employed in the plots of response sensitivities presented in Figures 1.16 through 1.19.

Table 1.6: Indexing of the 6090 macroscopic cross-sections considered as imprecisely known model (system) parameters, used in Figures 1.16 through 1.19.

MCS (macroscopic cross-sections)	un-rodded materials (no control rod)	rodded materials (including control rod)
absorption of fast neutrons (Fig. 1.16, left)	0001-0435 Top row	3046-3480 Bottom row
fission with fast neutrons (Fig. 1.17, left)	0436-0870 Top row	3481-3915 Bottom row
diffusion of fast neutrons (Fig. 1.18, left)	0871-1305 Top row	3916-4350 Bottom row
absorption of thermal neutrons (Fig. 1.16, right)	1306-1740 Top row	4351-4785 Bottom row
fission with thermal neutrons (Fig. 1.17, right)	1741-2175 Top row	4786-5220 Bottom row
diffusion of thermal neutrons (Fig. 1.18, right)	2176-2610 Top row	5221-5655 Bottom row
scattering from the fast into thermal group (Fig. 1.19)	2611-3045 Left row	5656-6090 Right row

Figure 1.16 displays the relative sensitivities (in %) of the transient total power of the reactor to the *absorption* cross sections of the 435 "materials" (top row) and 435 "rodded materials" (bottom row), corresponding to the fast energy group (left column) and thermal energy group (right column), respectively. The magnitude of these sensitivities varies according to the material type, amount of material, and also the material's location in the reactor. The relative sensitivities of the power response in the 435 "materials" (top row) are generally larger (up to 60%) than those for the 435 "rodded materials" (bottom row), where the absolute values of most sensitivities are below 20%. On the other hand, the magnitudes of the relative sensitivities in the fast energy group are comparable to those in the thermal group. In particular, the "stripes" at the levels 337-360 and 385-408 on the Oy axis in Figures 1.16 through 1.19 are caused by fuel assemblies of the type 15 and 17 (Solis, 2001), which have assembly design of type 3 initial fuel (comprising: 26 fuel rods with enrichment 2.94% 235U and no Gd2O3; 11 fuel rods with enrichment 1.94% 235U and no Gd2O3; 6 fuel rods with enrichment 1.69% 235U and no Gd2O3; 1 rod with enrichment 1.33% 235U and no Gd2O3; 3 rods with enrichment 2.93% 235U and 3% Gd2O3; 2 rods with enrichment 2.93% 235U and 4% Gd2O3; 1 rod with enrichment 1.94% 235U and 4% Gd2O3).

Figure 1.17 displays the relative sensitivities (in %) of the transient total power of the reactor to the *fission* cross sections of the 435 "materials" (top row) and 435 "rodded materials" (bottom row), corresponding to the fast energy group (left column) and thermal energy group (right column), respectively. These relative sensitivities are much smaller, by a factor of 10, than the sensitivities of the power response to the absorption cross sections. Figure 1.18 displays the relative sensitivities (in %) of the transient total power of the reactor to the *diffusion* cross sections of the 435 "materials" (top row) and 435 "rodded materials" (bottom row), corresponding to fast (left) and thermal (right) energy groups. These sensitivities are an even smaller, by yet another order of magnitude then the sensitivities to the fission cross sections.

Figure 1.19 presents the relative sensitivities (in %) of the transient total power of the reactor to the *scattering* cross sections from the fast into the thermal energy group for the 435 "materials" (left) and 435 "rodded materials" (right). These sensitivities are also relatively small, having absolute values of less than 15% for the "un-rodded materials" and less than 3% for the "rodded materials".

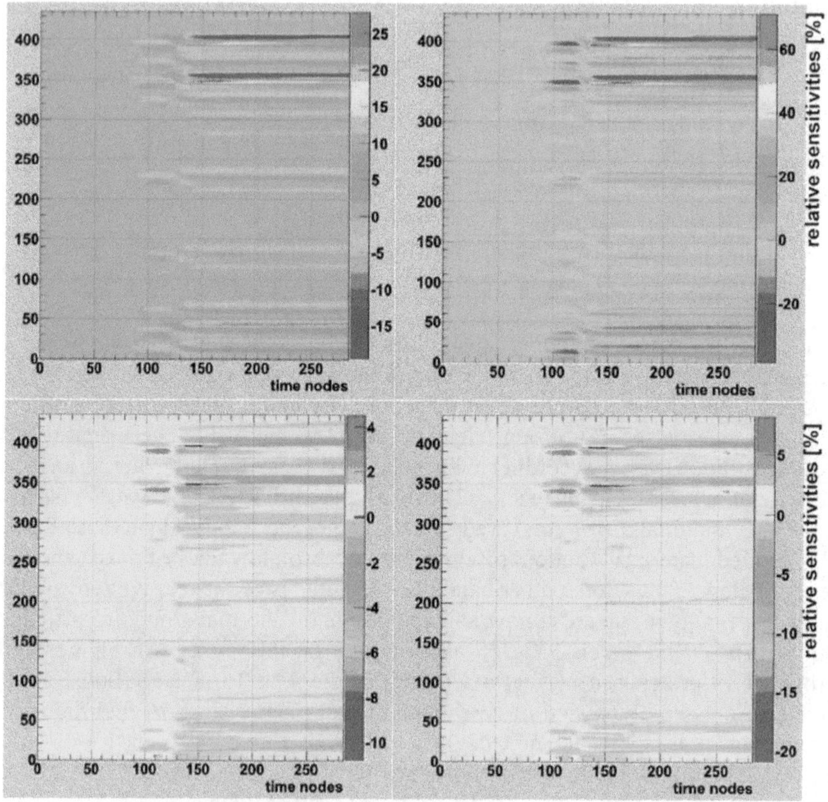

Figure 1.16. Relative sensitivities (in %) of the transient total power of the reactor to the absorption cross sections of the 435 "materials" (top row) and 435 "rodded materials" (bottom row), corresponding to fast (left) and thermal (right) energy groups.

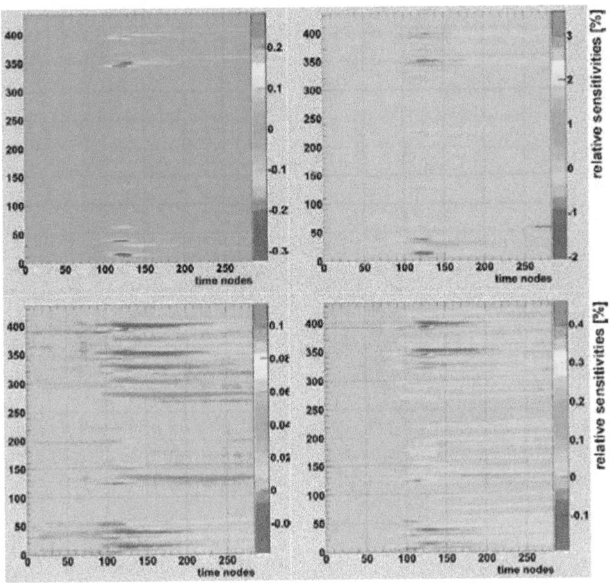

Figure 1.17: Relative sensitivities (in %) of the transient total power of the reactor to the fission cross sections of the 435 "materials" (top row) and 435 "rodded materials" (bottom row), corresponding to fast (left) and thermal (right) energy groups.

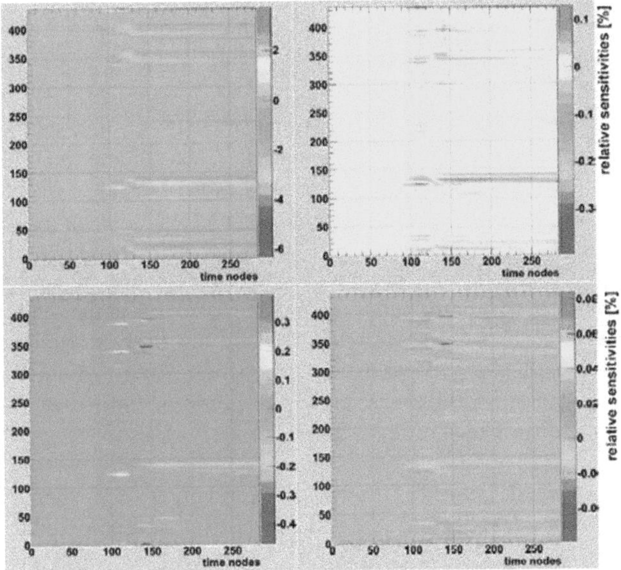

Figure 1.18: Relative sensitivities (in %) of the transient total power of the reactor to the diffusion cross sections of the 435 "materials" (top row) and 435 "rodded materials" (bottom row), corresponding to fast (left) and thermal (right) energy groups.

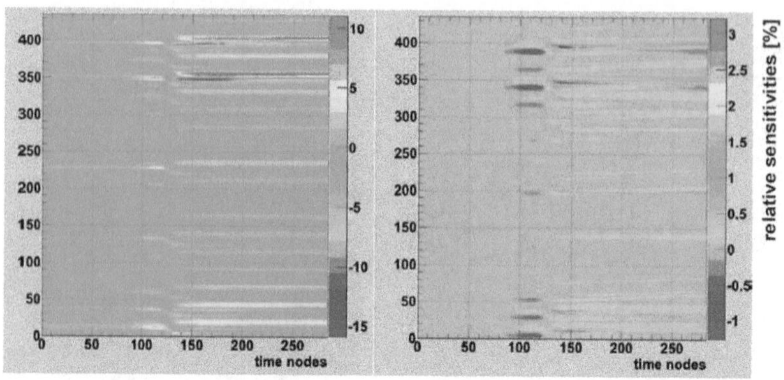

***Figure 1.19:** Relative sensitivities (in %) of the transient total power of the reactor to the scattering cross sections from the fast into the thermal energy group of the 435 "materials" (left) and 435 "rodded materials" (right).*

The sensitivities of the time-dependent power response to perturbations in the model parameters, which were computed in the previous sub-section, are used in Eq. (1.61) to compute the time-dependent correlation matrix of the response due to the uncertainties in the model parameters. Based on information available in the open literature (too extensive to be referenced here), the imprecisely known model parameter are considered to have the following relative standard deviations:

(i) The relative standard deviation of each of the parameters P_1 and P_2 in the phase-slip model of Molochnikov (1982) is considered to be 3%.

(ii) The relative standard deviation of each of the 3045 macroscopic cross sections for the un-rodded materials is considered to be 1.5%.

(iii) The relative standard deviation of each of the 3045 macroscopic cross sections for the rodded materials (i.e., with the control rods inserted) is considered to be 2.5%.

(iv) The time-dependent relative standard deviation of the time-dependent boundary conditions (pressure and flow rate at the output of the reactor core) is considered to be 0.75%, uniformly at each time instance.

(v) The time-dependent relative standard deviation of the time-dependent measured total reactor power response is considered to be 4%, uniformly at each time instance.

The respective nominal values of the computed and measured responses, the relative standard deviations described above, and the sensitivities computed in the previous sub-section are used in Eq. (1.101) to obtain the value $\chi^2 = 0.96$ for the resulting consistency estimator; this value indicates that the computational and experimental information considered is consistent.

The best-estimate predicted nominal value of the time-dependent power response is computed using Eq. (1.98), and is depicted in Figure 1.20 (left) along with the experimentally measured and computed nominal values of this response. The best estimate predicted nominal values are seen to fall in between the originally computed response and the experimental values (as would be expected by having applied the predictive modeling formulas to consistent information), being considerably closer to the more precisely known (i.e., with smaller standard deviations) experimental data.

The best estimate predicted covariance matrix of the time-dependent predicted power response is computed using Eq. (1.78). The square roots of the diagonal elements of the relative correlation matrix resulting from applying Eq. (1.78) are the predicted standard deviation of the predicted power response; these predicted standard deviations are plotted on the right-side of Figure 1.20, together with the standard deviations of the computed power response [i.e., the square roots of the diagonal elements of the relative correlation matrix shown in Eq. (1.61)] and the relative standard deviations of the measured responses. It is noted from these graphs that the predicted relative standard deviations are reduced by a factor of two by comparison to the experimental relative standard deviations and even more (by a factor of up to 8) by comparison to the computed relative standard deviation. The complete time-dependent relative correlation matrices of the computed and predicted best estimate power response, cf. Eqs. (1.61) and (1.78), are displayed in Figure 1.21, which highlights the significant overall reduction in the predicted versus computed errors.

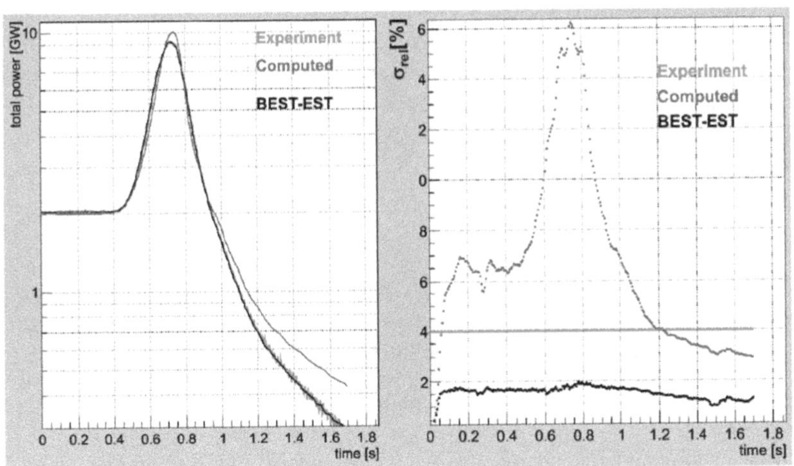

Figure 1.20: Left: experimental (red), simulated (blue) and best-estimate (black) transient power of the BWR for the TT2 benchmark.
Right: experimental (red), computed (blue) and best-estimate (black) relative standard deviations of the transient response.

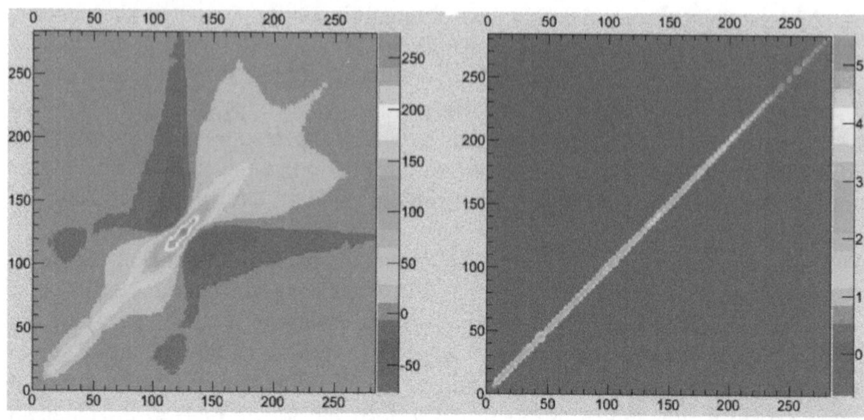

Figure 1.21: Nominal (left) and best-estimate predicted (right) relative correlations of the time dependent power response; the units on the colored scale are (%)^2.

As shown in the previous sub-section, the power response is most sensitive to perturbations in the time-dependent outlet pressure, outlet flow rate, and the parameters P_1 and P_2 in the phase-slip model of Molochnikov (1982). It is therefore expected that the predictive modeling methodology will reduce the predicted uncertainties in these model parameters by larger amounts than for parameters, such as the macroscopic cross sections, to which the power response is less sensitive. This expectation is confirmed by the actual results presented in Table 1.7, as well as in Figures 1.22 and 1.23. Thus, Table 1.7 lists the nominal and best-estimate values of the parameters P_1 and P_2 in the phase-slip model of Molochnikov (1982), along with the corresponding initial and best estimate predicted relative standard deviations. The BERRU-SMS predictive modeling methodology has calibrated the predicted optimal nominal values and has reduced the corresponding predicted relative standard deviations by up to 14%.

Table 1.7. Nominal and best-estimate predicted values of parameters P_1 and P_2 in the Molochnikov (1982) phase-slip model, with the corresponding initial and best estimated predicted relative standard deviations.

Parameter	Nominal Value	Best Estimate Predicted Value	Initial Rel. Std. Dev.	Best Estimate Predicted Rel. Std. Dev
P_1	1.5	1.465680	3.0%	2.77026%
P_2	0.6	0.583977	3.0%	2.57194%

The best-estimate predicted time-dependent flow rate and outlet pressure boundary conditions are compared with the corresponding initial nominal values for these quantities in the top row of Figure 1.22. Since the initial relative standard deviations were small (0.75%), the nominal values of the flow rate and outlet pressure are only slightly calibrated. Nevertheless, due to the influence of the high sensitivities of the power response to these time-dependent boundary conditions, the BERRU-SMS predictive modeling procedure reduces very significantly, by a factor of over 7, the predicted best-estimate relative standard deviations, as depicted in the bottom row in Figure 1.22.

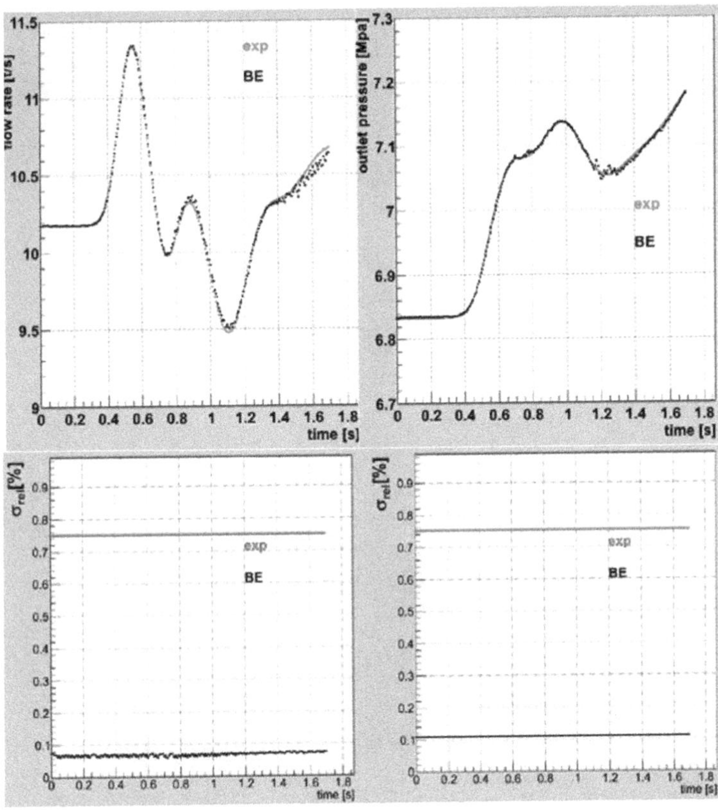

Figure 1.22: Experimental and best estimate predicted values for the time-dependent flow rate (top left) and outlet pressure (top right) and their corresponding experimental and best estimate predicted relative standard deviations (bottom row).

The application of the BERRU-SMS predictive modeling methodology also reduces the predicted uncertainties and provides best-estimate predicted values for the 6090 macroscopic cross-sections considered in the benchmark model. Figure 1.23 shows the relative adjustment (calibration) of the best-estimate macroscopic cross sections normalized to their initial nominal values. The indexing on the *Ox*-axis follows the convention presented in Table 1.6. The adjustments (calibrations) of the macroscopic cross sections are relatively small (up to 4%), which is expected in view of the relatively small corresponding sensitivities displayed in Figures 1.16 through 1.19. Nevertheless, the correspondences between the magnitude of the sensitivities displayed in Figures 1.16 through 1.19 and the magnitudes of the corresponding adjustments are evident: the larger the magnitudes of the sensitivities, the larger the adjustments. With the help of the indexing described in Table 1.6, it becomes clear that the following materials display "large sensitivities and corresponding large calibrations":

(i) the materials labeled 1306-1740 in Table 1.6, corresponding to absorption of the thermal neutrons in un-rodded materials;

(ii) the materials labeled 1-435 in Table 1.6, corresponding to absorption of fast neutrons in un-rodded materials;

(iii) the materials labeled in Table 1.6, corresponding to scattering from the fast into thermal group in un-rodded materials;

(iv) the materials labeled 4351-4785, 3046-3480 and 5656-6090 in Table 1.6, which corresponding to the materials described in items (i), (ii) and (iii) above but for "rodded materials."

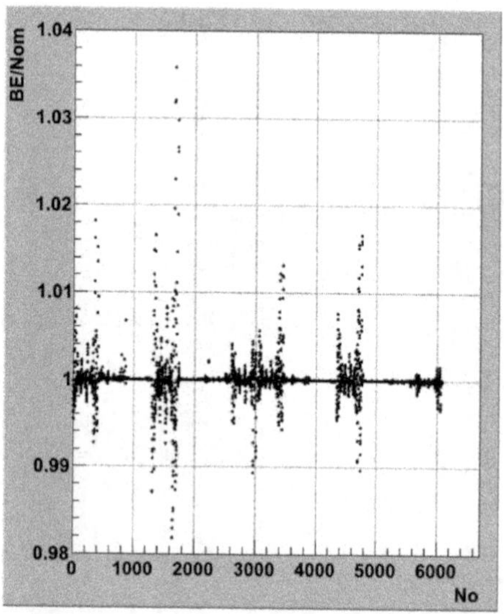

Figure 1.23: Best-estimate relative (to the initial values) adjustments for the various cross sections.

Figure 1.24 displays the reduction of the relative standard deviations of macroscopic cross sections (listed in Table 1.6) that characterize the un-rodded (left) and rodded materials (right), respectively. To facilitate the interpretation of the predicted uncertainties, the predicted standard deviations for the un-rodded material were normalized to the initial (nominal) values of the respective cross sections. Recall that the initial relative standard deviations were 1.5% for the un-rodded materials and 2.5% for rodded materials, respectively. Hence, the results displayed in Figure 1.23 reflect the (relatively small, up to 3%) reduction of the predicted standard deviation from the corresponding nominal (initial) values of 1.5% and 2.5%, respectively. Recalling the labelling of cross sections presented in Table 1.6, the results displayed in Figure 1.24 indicate that the largest reductions in the respective predicted standard deviations occur for the most adjusted cross sections in Figure 1.23. These comparatively larger reductions in the respective predicted uncertainties are due the correspondingly larger relative sensitivities of the power response to the aforementioned macroscopic cross sections.

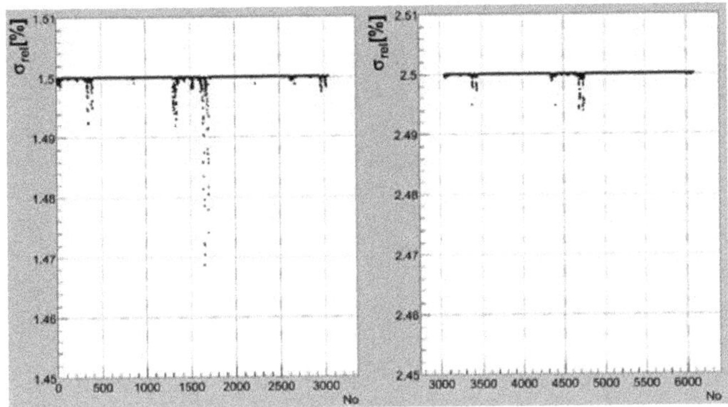

Figure 1.24: Best estimate predicted relative standard deviations for un-rodded (left) and rodded (right) materials.

In summary, the results presented in this Section have shown that the application of the BERRU-SMS methodology has simultaneously calibrated the time-dependent reactor power response for the coupled neutron kinetics/thermal-hydraulics numerical simulations of the BWR-TT2 benchmark together with 6660 imprecisely known model parameters (6090 macroscopic cross sections, and 570 thermal-hydraulics parameters involved in modeling the phase-slip correlation, transient outlet pressure, and total mass flow). The BERRU-SMS predictive modeling methodology has yielded best-estimate predicted values for all of the benchmark's parameters and power, achieving, in particular, substantial (up to 50%) reductions in the predicted standard deviation of the optimally predicted benchmark's power response.

2. BERRU-SMS FORWARD AND INVERSE PREDICTIVE MODELING APPLIED TO A SPENT FUEL DISSOLVER SYSTEM

Abstract.
This Chapter presents the application of the BERRU-SMS methodology to perform forward and inverse predictive modeling of a paradigm dissolver model that is representative for material separations and international safeguards activities. This dissolver model comprises eight active compartments in which the time-dependent nonlinear differential equations modeling the physical and chemical processes comprise sixteen time-dependent spatially dependent state functions and 635 model parameters related to the model's equation of state and inflow conditions. The most important response for the dissolver model is the computed nitric acid concentration in the compartment furthest away from the inlet, where measurements are available for comparisons. The sensitivities to all model parameters of the acid concentration at each of the instances in time when measurements were performed are computed exactly and efficiently using the adjoint sensitivity analysis method for nonlinear systems (ASAM). When using the traditional "ASAM for functional-type responses," 307 adjoint computations are needed for computing *exactly* the sensitivities of the time-dependent nitric concentration response in compartment #1, at the measured time instances, to all model parameters. Alternatively, the "ASAM for operator-type responses" developed by Cacuci (1981b) reduces the number of required adjoint computations from 307 to 17, with an overall loss of accuracy of less than 0.05% over the entire time-span of 10.5 hours (307 time steps). Similarly significant reductions in the number of computations when using the "adjoint sensitivity analysis methodology for operator-type responses" were also obtained when computing the sensitivities of the time-dependent responses (nitric acid concentrations) in the other dissolver compartments, e.g., compartments #4 and #7. These sensitivities are subsequently used for uncertainty analysis and predictive modeling, combining the computational results with experiments performed solely in the compartment furthest from the inlet. The application of the BERRU-SMS predictive modeling methodology yields optimally calibrated values for all 635 model parameters, with reduced predicted uncertainties, as well as optimal ("best-estimate") predicted values for the acid concentrations, also with reduced predicted uncertainties. Notably, even though the experimental data pertains solely to the compartment furthest from the inlet (where the data was measured), the predictive modeling methodology actually improves the predictions and reduces the predicted uncertainties not only in the compartment in which the data was actually measured, but throughout the entire dissolver, including the compartment furthest from the measurements (i.e., at the inlet). This is because the predictive modeling methodology combines and transmits information simultaneously over the entire phase-space, comprising all time steps and spatial locations.

D. G. Cacuci, *BERRU Predictive Modeling*,
https://doi.org/10.1007/978-3-662-58395-1_2

This Chapter also presents an application of the BERRU-SMS predictive modeling methodology in the *inverse* mode to determine, within a tight a priori specified convergence criterion and overall accuracy, an unknown time-dependent boundary condition (specifically: the time-dependent inlet acid concentration) for the dissolver model by using measurements of the state function (specifically: the time-dependent acid concentration) at a specified location (specifically: in the dissolver's compartment furthest from the inlet). The unknown time-dependent boundary condition is described by 635 unknown discrete scalar parameters. In general, the maximum entropy principle which underlies the BERRU-SMS predictive modeling methodology enables the construction of an intrinsic regularizing metric for solving any inverse problem. Specifically for the dissolver model, the unknown time-dependent boundary condition is predicted by the BERRU-SMS methodology within an a priori selected convergence criterion, without user intervention and/or introduction of arbitrary "regularization parameters."

2.1 INTRODUCTION

This Chapter illustrates the application of the BERRU-SMS methodology presented in Chapter 1 to perform predictive modeling of a paradigm dissolver model, which can serve as a case study due to its applicability to nuclear safeguard and nonproliferation activities. The material presented in this Chapter relies on the work reported by Peltz and Cacuci (2016a, 2016b), Peltz et al (2016), and Cacuci et al (2016). The dissolver is a mechanical and chemical module that produces feed stock for the chemical separation processes employed in the "head end" segment of an aqueous nuclear fuel reprocessing facility. The aim of research and development associated with nuclear nonproliferation is to understand and accurately exploit observables from signatures generated by processes and activities where a priori opportunities to measure relevant surrogate environments, collect statistically significant measurements, and calibrate instruments on real environments may be quite limited. This situation underscores the importance of understanding the limitations of computational tools and measured data when inferring or predicting characteristics of the underlying processes and activities, which, in turn, underscores the need for robust and scientifically well-founded predictive modeling methodologies.

The specific dissolver model analyzed in this work was originally developed by Lewis and Weber (1980). This dissolver model has been selected due to its applicability to material separations and for its potential role in diversion activities associated with proliferation and international safeguards. The dissolver resembles a rotating drum, comprising of eight active compartments in which the solids and liquids flow in opposite directions, and includes a ninth compartment used for rinsing. This material presented in this Chapter will concentrate on the flow of liquids, which are most relevant to material separation, and hence to non-proliferation objectives.

The original equations used by Lewis and Weber (1980) are presented in Section 2.2, but they will only serve as the starting point for developing a modified set of equations for describing the transient physical and chemical processes of interest to nonproliferation. The modified set of equations developed in Section 2.2 for the dissolver model is significantly more efficient computationally, and is also suitable for subsequent analyses (sensitivity and uncertainty quantification, predictive modeling, etc.). The modified dissolver model comprises sixteen spatially dependent state functions and 635 model parameters related to the model's equation of state and inflow conditions. In particular, the most important response for the dissolver model is the computed nitric acid in the compartment furthest away from the inlet, because this is the location where measurements (unique in the open literature) were performed by Lewis and Weber (1980), at 307 time instances, t_i, $i = 1,...,I = 307$, over a period of 10.5 hours.

The development of the "adjoint sensitivity dissolver model" using the adjoint sensitivity analysis method for nonlinear systems conceived by Cacuci (1981a, 1981b, 2003) is detailed in Section 2.3. The solution of this adjoint model enables the efficient computation of the first-order response sensitivities (i.e., functional derivatives) to all model parameters of the acid concentrations at each instance in time where response measurements are considered. The importance of the various sensitivities for the respective response at various time instances is also analyzed in Section 2.3 within the context of the underlying dissolver physics.

Section 2.4 uses the results described in Section 2.3 for performing comprehensive quantitative uncertainty analysis of the acid concentrations within the dissolver, and subsequently using the BERRU-SMS predictive modeling methodology developed in Chapter 1 for assimilating the available experimental information to calibrate the dissolver's model parameters and reduce the predicted uncertainties in the acid concentrations. The BERRU-SMS predictive modeling methodology yields optimally calibrated values for all model parameters, with reduced predicted uncertainties, as well as optimal ("best-estimate") predicted values for the model responses (in this "case study": the time-dependent acid concentrations in the dissolver's compartments), also with reduced predicted uncertainties. Notably, even though the experimental data pertains solely to the compartment furthest from the inlet, where the data was measured, the application of the BERRU-SMS methodology actually improves the predicted values and reduces their predicted uncertainties not only in the compartment in which the data was actually measured, but throughout the entire dissolver, including the compartment furthest from the measurements (i.e., at the inlet). This is because the forward and inverse BERRU-SMS predictive modeling methodology combines and transmits information simultaneously over the entire phase-space, comprising all time steps and spatial locations.

Section 2.5 demonstrates the application in the "inverse mode" of the BERRU-SMS predictive modeling to determine a time-dependent boundary condition within an a priori selected convergence criterion, without user intervention and/or introduction of arbitrary "regularization parameters," as would be required by all of the currently used procedures for analyzing "inverse problems."

2.2 MATHEMATICAL MODELING OF ROTARY DISSOLVER START-UP

The starting point for developing the dissolver model to be analyzed in this Chapter is provided by the rotary dissolver for used nuclear fuel developed originally by Lewis and Weber (1980), which is depicted in Figures 2.1 and 2.2. Only the liquid flow will be analyzed in this illustrative predictive modeling application. The liquid flows through the dissolver's eight compartments, labeled using the superscript $k = 1,...,8$, as depicted in Figure 2.2. Compartment 9 is used for rinsing, and is not relevant to the flow.

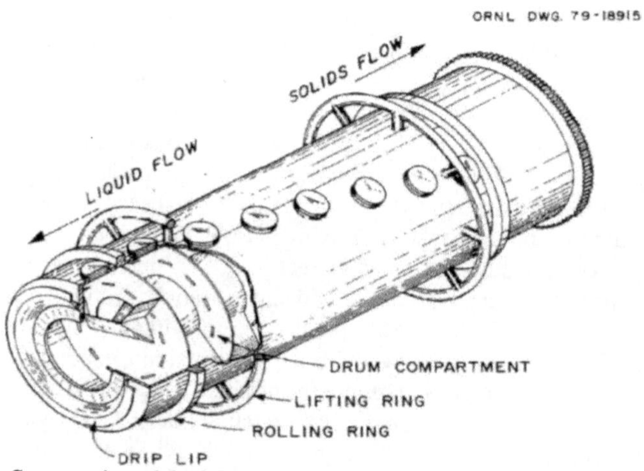

Figure 2.1: Cutaway view of the 0.5-t/d compartmented rotary dissolver drum (Lewis and Weber, 1980).

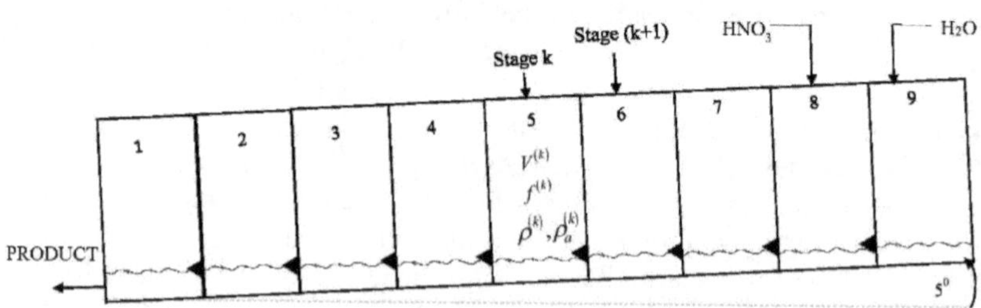

Figure 2.2: Liquid flow diagram for the compartmented rotary dissolver (Lewis and Weber, 1980)

The start-up conditions for the dissolver involve a non-ideal mixture of nitric acid and water at ambient conditions. The time and spatial variation of the physical and chemical processes occurring within the dissolver were originally modeled by Lewis and Weber (1980) by means of the following coupled nonlinear first-order time-dependent differential equations:

(A) The equations modeling conservation of mass:

Total mass:

$$\frac{d}{dt}\left[\rho^{(k)}V^{(k)}\right]=-\rho^{(k)}(t)f^{(k)}(t)+\rho^{(k+1)}(t)f^{(k+1)}(t),\ k=1,\dots,7,\ 0<t\le t_f,\quad(2.1)$$

Acid mass:

$$\frac{d}{dt}\left[\rho_a^{(k)}V^{(k)}\right]=-\rho_a^{(k)}(t)f^{(k)}(t)+\rho_a^{(k+1)}(t)f^{(k+1)}(t),\ k=1,\dots,7,\ 0<t\le t_f,\quad(2.2)$$

Total mass in compartment #8:

$$\frac{d}{dt}\left[\rho^{(8)}V^{(8)}\right]=-\rho^{(8)}(t)f^{(8)}(t)+\dot{m}^{(in)}(t),\ 0<t\le t_f,\quad(2.3)$$

Acid mass in compartment #8:

$$\frac{d}{dt}\left[\rho_a^{(8)}V^{(8)}\right]=-\rho_a^{(8)}(t)f^{(8)}(t)+\rho_a^{in}(t)f^{(in)}(t),\ 0<t\le t_f.\quad(2.4)$$

The quantities appearing in the above equations are defined as follows: (i) the index $k=1,\dots,8$ denotes the respective dissolver compartment; (ii) $V^{(k)}(t)$ denotes the volume of the liquid phase, in units of liters $[\ell]$; (iii) $\rho^{(k)}(t)$ denotes the volumetric mass density of the liquid phase, in units of gram/liter $[g/\ell]$; (iv) $\rho_a^{(k)}(t)$ denotes the volumetric mass concentration of nitric acid of the liquid phase, in units of $[g/\ell]$ in the solution; (v) $f^{(k)}(t)$ denotes the volumetric flow rate of the liquid mixture, in units of liter/hour $[\ell/h]$; and (vi) $\dot{m}^{(in)}(t)$ denotes the liquid solution mass rate in-flow in units of gram/hour $[g/h]$.

(B) Resistance to fluid flow through the compartments:

$$\frac{d}{dt}\left[V^{(k)}(t)\right]=-C\left(V^{(k)}\right)+f^{(k+1)}(t),\ k=1,\dots,7,\ 0<t\le t_f,\quad(2.5)$$

$$\frac{d}{dt}\left[V^{(8)}(t)\right] = -C\left(V^{(8)}\right) + f^{(in)}(t),\ 0 < t \le t_f, \tag{2.6}$$

where

$$C\left(V^{(k)}\right) \triangleq \begin{cases} \left(\dfrac{V^{(k)} - V_0}{G}\right)^p [\ell/h], & if\ V^{(k)}(t) - V_0 > 0,\ k = 1,...,8, \\ 0, & otherwise. \end{cases} \tag{2.7}$$

In the above relation, the scalar quantities G, V_0 and p are experimentally determined parameters, having nominal (mean) values and estimated relative standard deviations as presented in Table 2.1, below. Due to counter-flow conditions in the dissolver, the flow-inlet parameters $\dot{m}^{(in)}(t)$, $\rho_a^{(in)}(t)$, and $f^{(in)}(t)$ appear in compartment $k = 8$, as modeled by Eqs.(2.3), (2.4) and (2.6), rather than in compartment $k = 1$.

(C) Equation of state, which is needed to complete Eqs. (2.1) through (2.7):

$$\rho^{(k)}(t) = 63a\rho_a^{(k)}(t) + b,\ k = 1,...,8, \tag{2.8}$$

where a and b are experimentally determined scalar parameters with nominal (mean) values and estimated relative standard deviations presented in Table 2.1, below. The time-dependent variations of the inlet mass flow rate of solution, $\dot{m}^{(in)}(t)$, and inlet nitric mass concentration, $\rho_a^{(in)}(t)$, are presented in Figures 2.3 and 2.4, respectively. The estimated relative standard deviations of $\rho_a^{(in)}(t)$ and $\dot{m}^{(in)}(t)$ are also presented in Table 2.1.

Table 2.1: Nominal (mean) values and corresponding standard deviations for model parameters.

Parameter	$\rho_a^{(in)}(t)$	$\dot{m}^{(in)}(t)$	a	b	V_0	p	G
Nominal value	See Fig. 2.4	See Fig. 2.3	0.48916	1001.2 $[g/\ell]$	4.8 $[\ell]$	2.7	0.201941 $[\ell]$
Standard deviation	20%	10%	10%	10%	10%	10%	10%

The time-dependent nominal value of the inflow volumetric flow rate, $f^{(in)}(t)$, is obtained from the following expression:

$$f^{(in)}(t) = \dot{m}^{(in)}(t)/\rho^{(k)}(t) = \dot{m}^{(in)}(t)\left[63a\rho_a^{(in)}(t)+b\right]^{-1}, \tag{2.9}$$

which uses the equation of state, the inflow mass rate from Figure 2.3, and the time dependent nitric acid mass concentration data from Figure 2.4. In particular, the initial nominal value of $f^{(in)}(t)$ is $f^{(in)}(0) = 36.79\times 10^3/1001.2$ at $t = 0$.

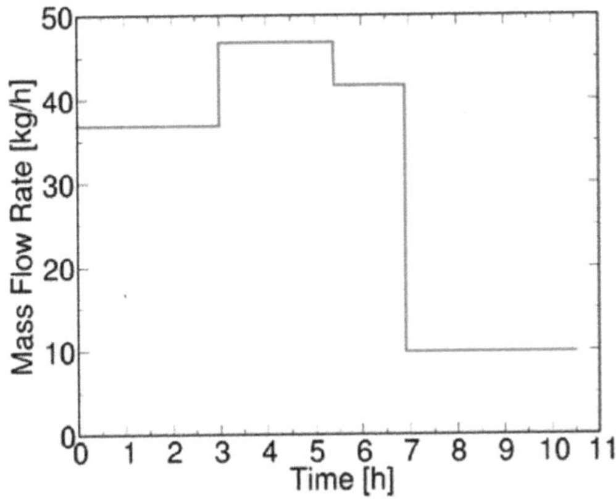

Figure 2.3. Time variation of the inlet mass flow rate of solution, $\dot{m}^{(in)}(t)/1000$ [kg/h].

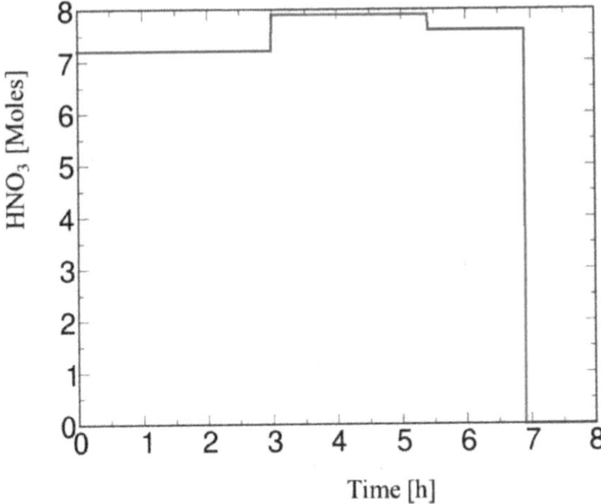

Figure 2.4: Time variation of the inlet nitric acid mass concentration, $\rho_a^{(in)}(t)/63$ [*moles*].

The original set of equations developed by Lewis and Weber (1980), namely Eqs. (2.1) through (2.6) are cumbersome for computations. Therefore, they will be transformed into a modified set of equations suitable for sensitivity analysis, uncertainty quantification and predictive modeling, by using the equation of state (2.8) to eliminate the volumetric mass density of the liquid phase, $\rho^{(k)}(t)$, from Eqs. (2.1) through (2.6). After some algebra, the modified set of equations describing the dissolver model is obtained in the following form:

$$V^{(k)}(t)\frac{d}{dt}\left[\rho_a^{(k)}\right]+\left[\rho_a^{(k)}(t)-\rho_a^{(k+1)}(t)\right]C\left[V^{(k+1)}(t)\right]=0,\quad k=1,\dots,7,\ 0<t\le t_f, \tag{2.10}$$

$$V^{(8)}(t)\frac{d}{dt}\left[\rho_a^{(8)}\right]+\rho_a^{(8)}(t)f^{(in)}(t)=\rho_a^{in}(t)f^{(in)}(t),\ 0<t\le t_f. \tag{2.11}$$

$$\frac{d}{dt}\left[V^{(k)}(t)\right]=-C\left[V^{(k)}(t)\right]+C\left[V^{(k+1)}(t)\right],\quad k=1,\dots,7,\ 0<t\le t_f, \tag{2.12}$$

$$\frac{d}{dt}\left[V^{(8)}(t)\right]=-C\left[V^{(8)}(t)\right]+f^{(in)}(t),\ 0<t\le t_f. \tag{2.13}$$

The initial conditions for Eqs. (2.10) through (2.13) are as follows:

$$\rho_a^{(k)}(0)=0.0,\ V^{(k)}(0)=V_0^{(k)},\quad k=1,\dots,8. \tag{2.14}$$

The compatibility condition for a fully developed initial flow implies that $\frac{d}{dt}V^{(k)}(0)=0,\ k=1,\dots,8$; in turn, this condition implies that

$$V_0^{(k)}\triangleq G\left[f^{(in)}(0)\right]^{1/p}+V_0,\quad k=1,\dots,8. \tag{2.15}$$

The nitric acid concentration in compartment #1, $\rho_a^{(1)}(t)$, was measured by Lewis and Weber (1980) at 307 time instances, $t_i,\ i=1,\dots,I=307$, over a period of 10.5 hours. The nominal values of these measurements are denoted as $\rho_{a,meas}^{(1)}(t_i)$, and are depicted using blue circles in Figure 2.5. Notably, these experimental results are unique in the open literature for a rotary dissolver. The relative standard deviation of each of these measurements has been estimated to be 5%. Figures 2.5 also depicts (red graph) the time-evolution of the normalized nominal value of the computed nitric acid concentration in compartment #1, $\rho_{a,nom}^{(1)}(t)$, which has been obtained by solving Eqs. (2.10) through (2.15) using a standard GMRES-solver (Saad and Schultz, 1986) and the nominal values for the model's parameters (as listed in Table 2.1). The agreement between the computed and experimentally measured values is remarkable.

Solving Eqs. (2.10) through (2.15) by using the nominal values for the model's parameters (as listed in Table 2.1) yields the time-dependent evolutions of the computed nominal values of the nitric acid concentrations in all of the compartments. In particular, the computed nominal values for $\rho_{a,nom}^{(1)}(t)$, $\rho_{a,nom}^{(4)}(t)$, and $\rho_{a,nom}^{(7)}(t)$, of the time-dependent acid concentrations in compartments #1 (furthest from the dissolver's inlet), #4 (in the dissolver's middle section), and #7 (closest to the dissolver's inlet), respectively, are depicted in Figure 2.6. The time evolutions of these concentrations are similar to each other, albeit time-delayed, as expected, and also resemble the time variation, depicted in Figure 2.4, of the inlet nitric acid mass concentration, $\rho_a^{(in)}(t)$.

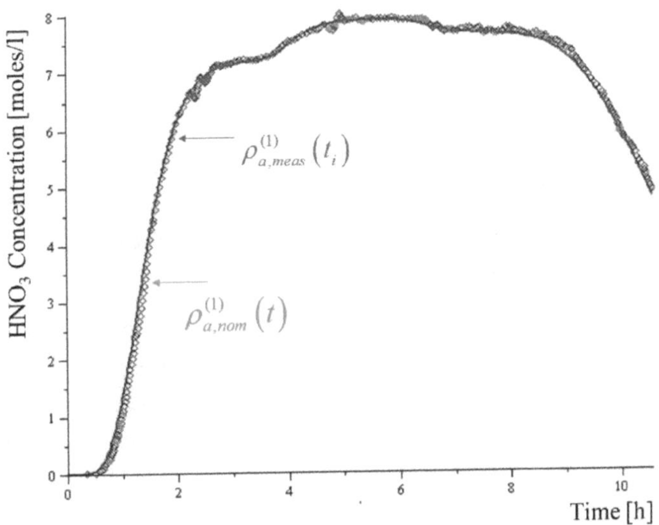

Figure 2.5: Red graph: time-evolution of the nominal values of the computed response of the nitric acid concentration, $\rho_{a,nom}^{(1)}(t)$.

Blue circles: experimentally measured nominal values, $\rho_{a,meas}^{(1)}(t_i)$, ***of the nitric acid concentration, at time instances*** t_i, $i = 1,\dots,I = 307$ ***[Lewis and Weber, 1980].***

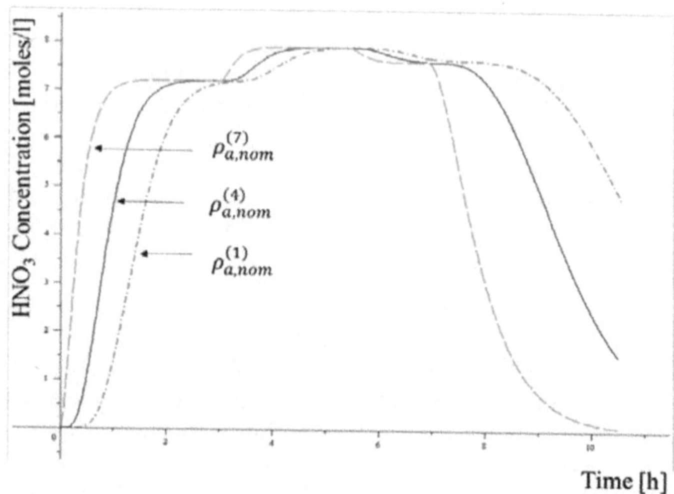

Figure 2.6: Time-evolution of the nominal values of the computed nitric acid concentrations $\rho_{a,nom}^{(1)}(t)$, $\rho_{a,nom}^{(4)}(t)$, $\rho_{a,nom}^{(7)}(t)$, in compartments #1, #4, and #7, respectively.

2.3 APPLICATION OF CACUCI'S ASAM FOR COMPUTING EFFICIENTLY AND EXACTLY THE FIRST-ORDER SENSITIVITIES OF TIME-DEPENDENT NITRIC ACID CONCENTRATIONS

The general adjoint sensitivity analysis methodology (ASAM) for nonlinear systems developed by Cacuci (1981a, b) considers a physical system represented mathematically by means of N_u coupled equations that are written in the following operator notation

$$\mathbf{N}\left[\mathbf{u}(\mathbf{x}),\boldsymbol{\alpha}(\mathbf{x})\right]=\mathbf{Q}\left[\boldsymbol{\alpha}(\mathbf{x})\right], \quad \mathbf{x}\in\Omega_x, \tag{2.16}$$

where:

1. $\mathbf{x}$ denotes the J_x-dimensional phase-space position vector for the primary system; $\mathbf{x}\in\Omega_x\subset\mathbb{R}^{J_x}$, where Ω_x is a subset of the J_x-dimensional real vector space $\mathbb{R}^{J_x}$;

2. $\mathbf{u}(\mathbf{x})\triangleq\left[u_1(\mathbf{x}),\ldots,u_{N_u}(\mathbf{x})\right]$ denotes a N_u-dimensional column vector whose components are the primary system's dependent (i.e., state) variables; $\mathbf{u}(\mathbf{x})\in\mathscr{E}_u$, where $\mathscr{E}_u$ is a normed linear space over the scalar field $\mathscr{F}$ of real numbers;

3. $\boldsymbol{\alpha}(\mathbf{x}) \triangleq \left[\alpha_1(\mathbf{x}),\ldots,\alpha_{N_\alpha}(\mathbf{x})\right]$ denotes a N_α-dimensional column vector whose components are the primary system's parameters; $\boldsymbol{\alpha} \in \mathscr{E}_\alpha$, where $\mathscr{E}_\alpha$ is also a normed linear space;
4. $\mathbf{Q}\left[\boldsymbol{\alpha}(\mathbf{x})\right] \triangleq \left[Q_1(\boldsymbol{\alpha}),\ldots,Q_{N_u}(\boldsymbol{\alpha})\right]$ denotes a N_u-dimensional column vector whose elements represent inhomogeneous source terms that depend either linearly or nonlinearly on $\boldsymbol{\alpha}$; $\mathbf{Q} \in \mathscr{E}_Q$, where $\mathscr{E}_Q$ is also a normed linear space; the components of $\mathbf{Q}$ may be operators, rather than just functions, acting on $\boldsymbol{\alpha}(\mathbf{x})$ and $\mathbf{x}$;
5. $\mathbf{N} \equiv \left[N_1(\mathbf{u},\boldsymbol{\alpha}),\ldots,N_{N_u}(\mathbf{u},\boldsymbol{\alpha})\right]$ denotes a N_u-component column vector whose components are operators (including differential, difference, integral, distributions, and/or infinite matrices) acting nonlinearly on $\mathbf{u}$ and $\boldsymbol{\alpha}$. In view of the definitions given above, $\mathbf{N}$ represents the mapping $\mathbf{N}:\mathscr{D} \subset \mathscr{E} \to \mathscr{E}_Q$, where $\mathscr{D} = \mathscr{D}_u \times \mathscr{D}_\alpha$, $\mathscr{D}_u \subset \mathscr{E}_u$, $\mathscr{D}_\alpha \subset \mathscr{E}_\alpha$, and $\mathscr{E} = \mathscr{E}_u \times \mathscr{E}_\alpha$. Note that an arbitrary element $\mathbf{e} \in \mathscr{E}$ is of the form $\mathbf{e} = (\mathbf{u},\boldsymbol{\alpha})$. Note also that even though the vector-valued quantities defined above were written out "horizontally" (for notational convenience), all vectors in this work are considered to be column vectors; transposition will be indicated by a dagger $(\dagger)$.

If differential operators appear in Eq. (2.16), then a corresponding set of boundary and/or initial conditions (which are essential to define $\mathscr{D}$) must also be given; these can be represented in operator form as

$$\left[\mathbf{B}(\mathbf{u},\boldsymbol{\alpha}) - \mathbf{A}(\boldsymbol{\alpha})\right]_{\partial\Omega_x} = \mathbf{0}, \quad \mathbf{x} \in \partial\Omega_x, \tag{2.17}$$

where $\partial\Omega_x$ denotes the boundary of Ω_x while $\mathbf{A}$ and $\mathbf{B}$ denote operators that act nonlinearly on $\mathbf{u}$ and $\boldsymbol{\alpha}$.

For the dissolver model analyzed in this work, Eq. (2.16) symbolically represents Eqs. (2.10) through (2.13), while Eq. (2.17) represents the initial conditions given in Eq. (2.14). The specific correspondences between the equations describing the dissolver model presented in Section 2.2 and the general notation introduced in Eqs. (2.16) and (2.17) are as follows: $\mathbf{x} \in \Omega_x \to t \in (0,t_f)$, $\partial\Omega_x \to \{t=0, t=t_f\}$,

$$\mathbf{u}(\mathbf{x}) \to \mathbf{u}(t) \triangleq \left[\rho_a^{(1)}(t),\ldots,\rho_a^{(8)}(t), V^{(1)}(t),\ldots,V^{(8)}(t)\right] \tag{2.18}$$

$$\alpha(\mathbf{x}) \to \alpha(t) \triangleq \begin{bmatrix} \rho_a^{(in)}(t_1),\ldots,\rho_a^{(in)}(t_{307}), \dot{m}^{(in)}(t_1),\ldots,\dot{m}^{(in)}(t_{307}), \\ \rho_a^{(1)}(0),\ldots,\rho_a^{(8)}(0), V^{(1)}(0),\ldots,V^{(8)}(0), a, b, V_0, p, G \end{bmatrix} \tag{2.19}$$

Thus, the dissolver model comprises 16 dependent variables (i.e., state functions) and 635 model parameters. The system response (i.e., result of interest) associated with the problem modeled by Eqs. (2.16) and (2.17) will be denoted as $R(\mathbf{u},\boldsymbol{\alpha})$, and will, in general, be a phase-space dependent operator that acts (linearly or non-linearly) on the system's state (vector-valued) function $\mathbf{u}$ and parameters $\boldsymbol{\alpha}$. Most generally, such a response can be represented in operator form as

$$R(\mathbf{u},\boldsymbol{\alpha}): \mathcal{D}_R \subset \mathcal{E} \to \mathcal{E}_R, \tag{2.20}$$

where $\mathcal{E}_R$ denotes another normed vector space. In particular, the measured and/or computed nitric acid concentration in a compartment k at a time-instance $t_i, i=1,...,I=307$ can be represented in the form

$$\rho_a^{(k)}(t_i) = \int_0^{t_f} \rho_a^{(k)}(t)\,\delta(t-t_i)\,dt; \quad i=1,...,I, \tag{2.21}$$

where $\delta(t-t_i)$ represents the customary Dirac-delta functional.

In general, the N_α model parameters α_i are usually derived from experiments and are therefore subject to uncertainties. The mean or nominal values of the model parameters are denoted as α_i^0; the vector of nominal vales will be denoted as $\boldsymbol{\alpha}^0 \triangleq (\alpha_1^0,...,\alpha_{N_\alpha}^0)^\dagger$. The nominal parameter values $\boldsymbol{\alpha}^0(\mathbf{x})$ are used in Eqs. (2.16) and (2.17) to obtain the nominal solution $\mathbf{u}^0(\mathbf{x})$ by solving

$$\mathbf{N}(\mathbf{u}^0,\boldsymbol{\alpha}^0) = \mathbf{Q}(\boldsymbol{\alpha}^0), \quad \mathbf{x} \in \Omega_x, \tag{2.22}$$

$$\mathbf{B}(\mathbf{u}^0,\boldsymbol{\alpha}^0) = \mathbf{A}(\boldsymbol{\alpha}^0), \quad \mathbf{x} \in \partial\Omega_x. \tag{2.23}$$

Equations (2.22) and (2.23) represent the "base-case" or nominal state of the physical system. The nominal solution, $\mathbf{u}^0(\mathbf{x})$, obtained by solving Eqs. (2.22) and (2.23) is used to compute the nominal value $R(\mathbf{e}^0)$ of the response $R(\mathbf{e})$, using the nominal values $\mathbf{e}^0 = (\mathbf{u}^0,\boldsymbol{\alpha}^0) \in \mathcal{E}$ of the model's state function and parameters. The response considered in Eq. (2.20) generally depends on the model parameters and the state variables. For this reason, arbitrary variations denoted as $\mathbf{h}_\alpha \triangleq (\delta\alpha_1,...,\delta\alpha_{N_\alpha}) \in \mathcal{E}_\alpha$, in the model parameters, and arbitrary variations, denoted as $\mathbf{h}_u \triangleq (\delta u_1,...,\delta u_{K_u}) \in \mathcal{E}_u$, in the state variables, will induce a variation δR in the response. As shown by Cacuci (1981a), the most general definition of the (first-order) sensitivity of the response $R(\mathbf{e})$ to variations $\mathbf{h} \triangleq (\mathbf{h}_u,\mathbf{h}_\alpha) \in \mathcal{E} = \mathcal{E}_u \times \mathcal{E}_\alpha$, in the system parameters is given by the Gâteaux- (G)-differential $\delta R(\mathbf{e}^0;\mathbf{h})$ of the response $R(\mathbf{e})$ at $\mathbf{e}^0 \triangleq (\mathbf{u}^0,\boldsymbol{\alpha}^0)$ with increment $\mathbf{h}$, which is defined as

$$\delta R(\mathbf{e}^0;\mathbf{h}) \triangleq \left\{\frac{d}{d\varepsilon}\left[R(\mathbf{e}^0+\varepsilon\mathbf{h})\right]\right\}_{\varepsilon=0} = \lim_{\varepsilon\to 0}\frac{R(\mathbf{e}^0+\varepsilon\mathbf{h})-R(\mathbf{e}^0)}{\varepsilon}, \tag{2.24}$$

for $\varepsilon \in \mathcal{F}$, and all (i.e., arbitrary) vectors $\mathbf{h}\in\mathcal{E}$. The symbol $\mathcal{F}$ denotes the underlying field of scalars. The sensitivity $\delta R(\mathbf{e}^0;\mathbf{h})$ is also an operator, defined on the same domain and with the same range as $R(\mathbf{e})$.

Since the system's state vector $\mathbf{u}$ and parameters $\boldsymbol{\alpha}$ are related to each other through Eqs. (2.16) and (2.17), it follows that $\mathbf{h}_u$ and $\mathbf{h}_\alpha$ are also related to each other. Therefore, the sensitivity $\delta R(\mathbf{e}^0;\mathbf{h})$ of $R(\mathbf{e})$ at $\mathbf{e}^0$ can be evaluated only after determining the vector of variations $\mathbf{h}_u$ in terms of the vector of parameter variations $\mathbf{h}_\alpha$. The first-order relationship between $\mathbf{h}_u$ and $\mathbf{h}_\alpha$ is determined by taking the G-differentials of Eqs. (2.16) and (2.17). Thus, taking the G-differential at $\mathbf{e}^0$ of Eq. (2.16) yields

$$\frac{d}{d\varepsilon}\left[\mathbf{N}(\mathbf{u}^0+\varepsilon\mathbf{h}_u,\boldsymbol{\alpha}^0+\varepsilon\mathbf{h}_\alpha)-\mathbf{Q}(\boldsymbol{\alpha}^0+\varepsilon\mathbf{h}_\alpha)\right]_{\varepsilon=0}=0, \tag{2.25}$$

or

$$\mathbf{N}'_u(\mathbf{e}^0)\mathbf{h}_u = \mathbf{Q}'_\alpha(\boldsymbol{\alpha}^0)\mathbf{h}_\alpha - \mathbf{N}'_\alpha(\mathbf{e}^0)\mathbf{h}_\alpha, \quad \mathbf{x}\in\boldsymbol{\Omega}_x, \tag{2.26}$$

where $\mathbf{N}'_u(\mathbf{e}^0)$ represents the partial G-derivative of $\mathbf{N}(\mathbf{e})$ with respect to $\mathbf{u}$, while $\mathbf{Q}'_\alpha(\boldsymbol{\alpha}^0)$ and $\mathbf{N}'_\alpha(\mathbf{e}^0)$ represent the partial G-derivatives with respect to $\boldsymbol{\alpha}$. Similarly, taking the G-differential at $\mathbf{e}^0$ of the boundary and initial conditions represented by Eq. (2.17) yields

$$\mathbf{B}'_u(\mathbf{e}^0)\mathbf{h}_u = \mathbf{A}'_\alpha(\boldsymbol{\alpha}^0)\mathbf{h}_\alpha - \mathbf{B}'_\alpha(\mathbf{e}^0)\mathbf{h}_\alpha, \quad \mathbf{x}\in\partial\Omega_x. \tag{2.27}$$

Equations (2.26) and (2.27) represent the "forward sensitivity equations". For a given vector of parameter variations $\mathbf{h}_\alpha$ around $\boldsymbol{\alpha}^0$, the forward sensitivity system represented by Eqs. (2.26) and (2.27) can be solved to obtain $\mathbf{h}_u$. Once $\mathbf{h}_u$ is available, it can be used in Eq. (2.24) to compute the sensitivity $\delta R(\mathbf{e}^0;\mathbf{h})$, for a given vector of parameter variations $\mathbf{h}_\alpha$. The direct computation of the response sensitivity

$\delta R\left(\mathbf{e}^0;\mathbf{h}\right)$ by using the ($\mathbf{h}_\alpha$-dependent) solution $\mathbf{h}_u$ of Eqs. (2.26) and (2.27) is called (Cacuci, 1980a) the Forward Sensitivity Analysis Procedure (FSAP).
For the dissolver model, the vector of parameter variations comprises the following 635 components:

$$\mathbf{h}_\alpha(t) \triangleq \begin{bmatrix} \Delta\rho_a^{(in)}(t_1),\ldots,\Delta\rho_a^{(in)}(t_{307}),\Delta\dot{m}^{(in)}(t_1),\ldots,\Delta\dot{m}^{(in)}(t_{307}), \\ \Delta\rho_{a0}^{(1)},\ldots,\Delta\rho_{a0}^{(8)},\Delta V_0^{(1)},\ldots,\Delta V_0^{(8)},\Delta a,\Delta b,\Delta V_0,\Delta G,\Delta p \end{bmatrix}. \tag{2.28}$$

The variations in the state functions are represented by the vector $\mathbf{h}_u(t) \triangleq \left[\mathbf{h}_\rho(t),\mathbf{h}_V(t)\right]$, with components defined as follows:

$$\mathbf{h}_\rho(t) \triangleq \left[\Delta\rho_a^{(1)}(t),\ldots,\Delta\rho_a^{(8)}(t)\right],\ \mathbf{h}_V(t) \triangleq \left[\Delta V^{(1)}(t),\ldots,\Delta V^{(8)}(t)\right]. \tag{2.29}$$

For example, if the model's response of interest is the nitric acid concentration in the compartment k, then the sensitivity of this response is

$$\delta R\left(\mathbf{e}^0;\mathbf{h}\right) \triangleq \frac{d}{d\varepsilon}\left[\rho_{a,nom}^{(k)} + \varepsilon\Delta\rho_a^{(k)}(t)\right]_{\varepsilon=0} = \Delta\rho_a^{(k)}(t). \tag{2.30}$$

The relations between $\mathbf{h}_\alpha(t)$ and $\mathbf{h}_u(t)$ are provided by the forward sensitivity system, cf. Eq. (2.26) and (2.27), which is obtained by computing the first-order Gateaux-differential of Eqs. (2.10) through (2.14) at the nominal parameter values $\boldsymbol{\alpha}^0(t)$ and nominal solution $\mathbf{u}^0(t)$. Thus, applying the definition of the G-differential to Eqs. (2.10) through (2.14) yields the following forward sensitivity system:

$$\begin{aligned} & V_{nom}^{(k)}(t)\frac{d\left[\Delta\rho_a^{(k)}(t)\right]}{dt} + C\left[V_{nom}^{(k+1)}(t)\right]\left[\Delta\rho_a^{(k)}(t)\right] - C\left[V_{nom}^{(k+1)}(t)\right]\left[\Delta\rho_a^{(k+1)}(t)\right] \\ & + \left[\frac{d\rho_a^{(k)}(t)}{dt}\right]_{nom}\left[\Delta V^{(k)}(t)\right] + \left[\rho_{a,nom}^{(k)}(t) - \rho_{a,nom}^{(k+1)}(t)\right]\left[\frac{dC\left[V^{(k+1)}(t)\right]}{dV^{(k+1)}(t)}\right]_{nom}\left[\Delta V^{(k+1)}(t)\right] \\ & = q_\rho^{(k)}(t), \quad k = 1,\ldots,7,\ 0 < t \le t_f; \end{aligned} \tag{2.31}$$

$$V_{nom}^{(8)}(t)\frac{d\left[\Delta\rho_a^{(8)}(t)\right]}{dt}+f_{nom}^{(in)}(t)\left[\Delta\rho_a^{(8)}(t)\right]
+\left[\frac{d\rho_a^{(8)}(t)}{dt}\right]_{nom}\left[\Delta V^{(8)}(t)\right]=q_\rho^{(8)}(t),\; 0<t\le t_f; \tag{2.32}$$

$$\frac{d\left[\Delta V^{(k)}(t)\right]}{dt}+\left[\frac{dC\left[V^{(k)}(t)\right]}{dV^{(k)}(t)}\right]_{nom}\left[\Delta V^{(k)}(t)\right]
-\left[\frac{dC\left[V^{(k+1)}(t)\right]}{dV^{(k+1)}(t)}\right]_{nom}\left[\Delta V^{(k+1)}(t)\right]=q_V^{(k)}(t),\quad k=1,\dots,7,\; 0<t\le t_f; \tag{2.33}$$

$$\frac{d\left[\Delta V^{(8)}(t)\right]}{dt}+\left[\frac{dC\left[V^{(8)}(t)\right]}{dV^{(8)}(t)}\right]_{nom}\left[\Delta V^{(8)}(t)\right]=q_V^{(8)}(t),\; 0<t\le t_f, \tag{2.34}$$

where the source terms on the right-sides of the above equations are defined as follows:

$$q_\rho^{(k)}(t)\triangleq\left[\rho_{a,nom}^{(k+1)}(t)-\rho_{a,nom}^{(k)}(t)\right]\left\{\left[\partial C\left(V^{(k+1)}\right)/\partial V_0\right]_{nom}\Delta V_0
+\left[\partial C\left(V^{(k+1)}\right)/\partial G\right]_{nom}\Delta G+\left[\partial C\left(V^{(k+1)}\right)/\partial p\right]_{nom}\Delta p\right\},\; k=1,\dots,7, \tag{2.35}$$

$$q_\rho^{(8)}(t)\triangleq\left[\rho_{a,nom}^{(in)}(t)-\rho_{a,nom}^{(8)}(t)\right]\Delta f^{(in)}(t)+f_{nom}^{(in)}(t)\left[\Delta\rho_a^{(in)}(t)\right], \tag{2.36}$$

$$q_V^{(k)}(t)\triangleq-\left[\partial C\left(V^{(k)}\right)/\partial V_0\right]_{nom}\Delta V_0-\left[\partial C\left(V^{(k)}\right)/\partial G\right]_{nom}\Delta G
-\left[\partial C\left(V^{(k)}\right)/\partial p\right]_{nom}\Delta p+\left[\partial C\left(V^{(k+1)}\right)/\partial V_0\right]_{nom}\Delta V_0
+\left[\partial C\left(V^{(k+1)}\right)/\partial G\right]_{nom}\Delta G+\left[\partial C\left(V^{(k+1)}\right)/\partial p\right]_{nom}\Delta p,\; k=1,\dots,7, \tag{2.37}$$

$$q_V^{(8)}(t)\triangleq-\left[\partial C\left(V^{(8)}\right)/\partial V_0\right]_{nom}\Delta V_0-\left[\partial C\left(V^{(8)}\right)/\partial G\right]_{nom}\Delta G
-\left[\partial C\left(V^{(8)}\right)/\partial p\right]_{nom}\Delta p+\Delta f^{(in)}(t), \tag{2.38}$$

with

$$\Delta f^{(in)}(t) \triangleq \frac{\Delta \dot{m}^{(in)}(t)}{63 a_{nom} \rho_{a,nom}^{(in)}(t) + b_{nom}} - \frac{\dot{m}_{nom}^{(in)}(t)\left\{(\Delta a) 63 \rho_{a,nom}^{(in)}(t) + 63 a_{nom}\left[\Delta \rho_{a}^{(in)}(t)\right] + \Delta b_{nom}\right\}}{\left[63 a_{nom} \rho_{a,nom}^{(in)}(t) + b_{nom}\right]^{2}}. \tag{2.39}$$

As generally represented by Eq. (2.27), the initial conditions for the differential equations (2.31) through (2.34) are obtained by taking the Gateaux-differential of the initial conditions given in Eq. (2.14) to obtain:

$$\begin{aligned}
&\Delta \rho_{a}^{(k)}(0) = \Delta \rho_{a0}^{(k)}, \quad k = 1, \ldots, 8, \\
&\Delta V^{(k)}(0) \equiv \Delta V_{0}^{(k)} = (\Delta G)\left[f_{nom}^{(in)}(0)\right]^{1/p} \\
&+\Delta\left(f^{(in)}(0)\right) \frac{G_{nom}}{p_{nom}}\left[f_{nom}^{(in)}(0)\right]^{(1-p_{nom})/p_{nom}} \\
&+(\Delta p) G_{nom}\left[f_{nom}^{(in)}(0)\right]^{1/p_{nom}} \ln\left[f_{nom}^{(in)}(0)\right] + \Delta V_{0}, \quad k = 1, \ldots, 8.
\end{aligned} \tag{2.40}$$

Note also the following relations:

$$\frac{\partial C\left[V^{(k)}(t)\right]}{\partial V_{0}} = \begin{cases} -\dfrac{p}{G}\left[\dfrac{V^{(k)}(t) - V_{0}}{G}\right]^{p-1}, & if\ V^{(k)}(t) - V_{0} > 0,\ k = 1, \ldots, 8, \\ 0, & otherwise. \end{cases} \tag{2.41}$$

$$\frac{\partial C\left[V^{(k)}(t)\right]}{\partial G} = \begin{cases} -\dfrac{p}{G}\left[\dfrac{V^{(k)}(t) - V_{0}}{G}\right]^{p}, & if\ V^{(k)}(t) - V_{0} > 0,\ k = 1, \ldots, 8, \\ 0, & otherwise. \end{cases} \tag{2.42}$$

$$\frac{\partial C\left[V^{(k)}(t)\right]}{\partial p} = \begin{cases} \left[\dfrac{V^{(k)}(t) - V_{0}}{G}\right]^{p} \ln\left[\dfrac{V^{(k)}(t) - V_{0}}{G}\right], & if\ V^{(k)}(t) - V_{0} > 0,\ k = 1, \ldots, 8, \\ 0, & otherwise. \end{cases} \tag{2.43}$$

The response sensitivity $\Delta \rho_{a}^{(k)}(t)$, cf. Eq. (2.30), can be computed after solving the forward sensitivity system given by Eqs. (2.31) through (2.34) repeatedly, for every possible parameter variation contained in the vector $\mathbf{h}_{\alpha}(t)$ defined in Eq. (2.28).

Furthermore, the variations $\Delta\rho^{(k)}(t)$ can be computed after obtaining the values of $\Delta\rho_a^{(k)}(t)$, by taking the first Gateaux-variation of the equation of state, cf. Eq. (2.8) to obtain the relation

$$\Delta\rho^{(k)}(t) = a_{nom}\left[\Delta\rho_a^{(k)}(t)\right] + \rho_{a,nom}^{(k)}(t)(\Delta a) + (\Delta b), \; k = 1,\ldots,8. \qquad (2.44)$$

From the standpoint of computational costs and effort, the FSAP requires, in general, $O(N_\alpha)$ large-scale computations. Specifically for the dissolver model, Eqs. (2.31) through (2.34) would need to be solved 635 times. It is therefore inefficient to employ the FSAP, because it is prohibitively expensive computationally to solve repeatedly the $\mathbf{h}_\alpha$-dependent FSE in order to determine $\mathbf{h}_u$ for all possible vectors $\mathbf{h}_\alpha$.

2.3.1 ASAM Computation of Nitric Acid Concentration Sensitivities at Specified Time Instances

In most practical situations, the number of model parameters exceeds significantly the number of functional responses of interest, i.e., $N_r \ll N_\alpha$. This is also the case for the dissolver model, which comprises 635 model parameters but only one primary response of interest, $\rho_a^{(1)}(t)$. In such cases, Cacuci (1981a, b) has shown that the (1st-Order) Adjoint Sensitivity Analysis Methodology (ASAM) is the most efficient method for computing exactly the first-order sensitivities, since it requires only $O(N_r)$ large-scale computations. The framework of the ASAM for computing the first-order G-differential $\delta R(\mathbf{e}^0;\mathbf{h})$ requires the spaces $\mathcal{E}_u$, $\mathcal{E}_Q$ and $\mathcal{E}_R$ to be Hilbert spaces, to enable the introduction of appropriately defined adjoint operators. Consequently these vector spaces will henceforth be considered to be Hilbert spaces, and will be denoted as $\mathcal{H}_u(\Omega_x)$, $\mathcal{H}_Q(\Omega_x)$ and $\mathcal{H}_R(\Omega_R)$, respectively. The elements of $\mathcal{H}_u(\Omega_x)$ and $\mathcal{H}_Q(\Omega_x)$ are vector-valued functions defined on the open set $\Omega_x \subset \mathbb{R}^{J_x}$, with smooth boundary $\partial\Omega_x$. The inner product of two vectors $\mathbf{u}^{(1)} \in \mathcal{H}_u(\Omega_x)$ and $\mathbf{u}^{(2)} \in \mathcal{H}_u(\Omega_x)$ will be denoted as $\left\langle \mathbf{u}^{(1)}, \mathbf{u}^{(2)} \right\rangle_u$, while the inner product on $\mathcal{H}_Q(\Omega_x)$ of two vectors $\mathbf{Q}^{(1)} \in \mathcal{H}_Q$ and $\mathbf{Q}^{(2)} \in \mathcal{H}_Q$ will be denoted as $\left\langle \mathbf{Q}^{(1)}, \mathbf{Q}^{(2)} \right\rangle_Q$. The elements of $\mathcal{H}_R(\Omega_R)$ are vector or scalar functions defined on the open set $\Omega_R \subset \mathbb{R}^{J_m}$, $1 \le m \le J_x$. The inner product on $\mathcal{H}_R(\Omega_R)$ of two vectors $\mathbf{R}^{(1)} \in \mathcal{H}_R$ and $\mathbf{R}^{(2)} \in \mathcal{H}_R$ will be denoted as $\left\langle \mathbf{R}^{(1)}, \mathbf{R}^{(2)} \right\rangle_R$.

The ASAM requires that $\delta R(\mathbf{e}^0;\mathbf{h})$ be linear in $\mathbf{h}$, which implies that $R(\mathbf{e})$ must satisfy a weak Lipschitz condition at $\mathbf{e}^0$, and also satisfy the following condition

$$R(\mathbf{e}^0+\varepsilon\mathbf{h}_1+\varepsilon\mathbf{h}_2)-R(\mathbf{e}^0+\varepsilon\mathbf{h}_1)-R(\mathbf{e}^0+\varepsilon\mathbf{h}_2)+R(\mathbf{e}^0)=o(\varepsilon);$$
$$\mathbf{h}_1,\mathbf{h}_2\in\mathcal{H}_u\times\mathcal{H}_\alpha;\quad \varepsilon\in\mathcal{F}. \tag{2.45}$$

If $R(\mathbf{e})$ satisfies the two conditions above, then the total response variation $\delta R(\mathbf{e}^0;\mathbf{h})$ is indeed linear in $\mathbf{h}$, a fact that will be highlighted by denoting it as $DR(\mathbf{e}^0;\mathbf{h})$. Consequently, $R(\mathbf{e})$ admits a total G-derivative at $\mathbf{e}^0=(\mathbf{u}^0,\boldsymbol{\alpha}^0)$, such that the relationship

$$DR(\mathbf{e}^0;\mathbf{h})=R'_u(\mathbf{e}^0)\mathbf{h}_u+R'_\alpha(\mathbf{e}^0)\mathbf{h}_\alpha \tag{2.46}$$

holds, where $R'_u(\mathbf{e}^0)$ and $R'_\alpha(\mathbf{e}^0)$ denote, respectively, the partial G-derivatives at $\mathbf{e}^0$ of $R(\mathbf{e})$ with respect to $\mathbf{u}$ and $\boldsymbol{\alpha}$. It is convenient to refer to the quantities $R'_u(\mathbf{e}^0)\mathbf{h}_u$ and $R'_\alpha(\mathbf{e}^0)\mathbf{h}_\alpha$ appearing in Eq. (2.46) as the "indirect effect term" and the "direct effect term," respectively. The operator $R'_u(\mathbf{e}^0)$ acts linearly on the vector of (arbitrary) variations $\mathbf{h}_u$, from $\mathcal{H}_u$ into $\mathcal{H}_R(\Omega_R)$, while the operator $R'_\alpha(\mathbf{e}^0)$ acts linearly on the vector of (arbitrary) variations $\mathbf{h}_\alpha$, from $\mathcal{H}_\alpha$ into $\mathcal{H}_R(\Omega_R)$.

When the response of interest is the nitric acid concentration, in any dissolver compartment k, $k=1,\ldots,8$, at some specified time instance, the operator $R(\mathbf{e})$ is a scalar-valued functional, which implies that the operator $R'_u(\mathbf{e}^0)\mathbf{h}_u$ is also a scalar-valued functional. In such cases, the operator $R'_u(\mathbf{e}^0)\mathbf{h}_u$ can be represented as an inner-product, as follows:

$$R'_u(\mathbf{e}^0)\mathbf{h}_u=\left\langle\nabla_u R(\mathbf{e}^0),\mathbf{h}_u\right\rangle_R, \tag{2.47}$$

where $\nabla_u R(\mathbf{e}^0)$ denotes the partial gradient of $R(\mathbf{e})$ with respect to $\mathbf{u}$ at $\mathbf{e}^0$. In particular, the total sensitivity of the nitric acid concentration in a compartment k at a time instance t_i can be represented in the form shown in Eq. (2.47) as follows:

$$\delta R\left(e^0;h\right) = D\rho_a^{(k)}\left(t_i\right) = \int_0^{t_f} \Delta\rho_a^{(k)}(t)\delta\left(t-t_i\right)dt = \left\langle \delta\left(t-t_i\right), \Delta\rho_a^{(k)}(t)\right\rangle_R; \quad (2.48)$$

$i = 1,\ldots,I.$

The general 1st-order Adjoint Sensitivity Analysis Methodology (ASAM) proceeds by defining the formal adjoint operator $\mathbf{N}^+\left(\mathbf{e}^0\right)$ of $\mathbf{N}'_u\left(\mathbf{e}^0\right)$, which is defined through the following relationship (in the respective Hilbert spaces) which is required to hold for an arbitrary vector $\boldsymbol{\psi}(\mathbf{x}) \in \mathcal{H}_Q$:

$$\left\langle \boldsymbol{\psi}, \mathbf{N}'_u\left(\mathbf{e}^0\right)\mathbf{h}_u \right\rangle_Q = \left\langle \mathbf{N}^+\left(\mathbf{e}^0\right)\boldsymbol{\psi}, \mathbf{h}_u \right\rangle_u + \left\{P\left(\mathbf{h}_u;\boldsymbol{\psi}\right)\right\}_{\partial\Omega_x} \quad (2.49)$$

The formal adjoint operator $\mathbf{N}^+\left(\mathbf{e}^0\right)$ is the $N_u \times N_u$ matrix $\mathbf{N}^+\left(\mathbf{e}^0\right) \triangleq \left[N_{ji}^+\left(\mathbf{e}^0\right)\right]$, $(i,j=1,\ldots,N_u)$, comprising elements $N_{ji}^+\left(\mathbf{e}^0\right)$ obtained by transposing the formal adjoints of the components of $\mathbf{N}'_u\left(\mathbf{e}^0\right)$, while $\left\{P\left(\mathbf{h}_u;\boldsymbol{\psi}\right)\right\}_{\partial\Omega_x}$ is the associated bilinear form evaluated on $\partial\Omega_x$. The domain of $\mathbf{N}^+\left(\mathbf{e}^0\right)$ is determined by selecting appropriate adjoint boundary conditions, represented here in operator form as

$$\mathbf{B}^+\left(\mathbf{e}^0\right)\boldsymbol{\psi}(\mathbf{x}) - \mathbf{A}^+\left(\boldsymbol{\alpha}^0\right) = \mathbf{0}, \ \mathbf{x} \in \partial\Omega_x. \quad (2.50)$$

The above boundary conditions for $\mathbf{N}^+\left(\mathbf{e}^0\right)$ are obtained by requiring that:

(a) They must be independent of $\mathbf{h}_u$ and $\mathbf{h}_\alpha$;

(b) The substitution of Eqs. (2.27) and (2.50) into the expression of $\left\{P\left(\mathbf{h}_u;\boldsymbol{\psi}\right)\right\}_{\partial\Omega_x}$ must cause all terms containing unknown values of $\mathbf{h}_u$ to vanish.

This selection of the boundary conditions for $\mathbf{N}^+\left(\mathbf{e}^0\right)$ reduces $\left\{P\left(\mathbf{h}_u;\boldsymbol{\psi}\right)\right\}_{\partial\Omega_x}$ to a quantity that contains boundary terms involving only known values of $\mathbf{h}_\alpha$, $\boldsymbol{\psi}$, and, possibly, $\boldsymbol{\alpha}^0$; this quantity will be denoted by $\hat{P}\left(\mathbf{h}_\alpha, \boldsymbol{\psi}; \boldsymbol{\alpha}^0\right)$. In general, $\hat{P}$ does not automatically vanish as a result of these manipulations, although it may do so in particular instances. In principle, $\hat{P}$ can be forced to vanish by considering extensions of $\mathbf{N}'_u\left(\mathbf{e}^0\right)$, in the operator sense, but this is seldom needed in practice.

Introducing now Eqs. (2.26), (2.27) and (2.50) into Eq. (2.49) and re-arranging the various terms yields

$$\left\langle \mathbf{N}^{+}\left(\mathbf{e}^{0}\right)\boldsymbol{\psi}, \mathbf{h}_{u}\right\rangle_{u} = \left\langle \boldsymbol{\psi}, \mathbf{Q}'_{\alpha}\left(\mathbf{e}^{0}\right)\mathbf{h}_{\alpha} - \mathbf{N}'_{\alpha}\left(\mathbf{e}^{0}\right)\mathbf{h}_{\alpha}\right\rangle_{Q} - \hat{P}\left(\mathbf{h}_{\alpha}, \boldsymbol{\psi}; \boldsymbol{\alpha}^{0}\right). \tag{2.51}$$

Note that the function $\boldsymbol{\psi}$ is still incompletely defined at this stage, since only the Hilbert space to which it belongs was specified thus far. Furthermore, note that the left-side of Eq. (2.51) is actually a linear functional of $\mathbf{h}_u$, and so is the right side of Eq. (2.47). Therefore, we can require that these two quantities be identical, i.e.,

$$\left\langle \mathbf{N}^{+}\left(\mathbf{e}^{0}\right)\boldsymbol{\psi}, \mathbf{h}_{u}\right\rangle_{u} = \left\langle \nabla_{u} R\left(\mathbf{e}^{0}\right), \mathbf{h}_{u}\right\rangle_{R}. \tag{2.52}$$

The Riesz Representation Theorem applied to Eq. (2.52) implies that $\boldsymbol{\psi}(\mathbf{x})$ is the unique weak solution of the adjoint system

$$\mathbf{N}^{+}\left(\mathbf{e}^{0}\right)\boldsymbol{\psi}(\mathbf{x}) = \nabla_{u} R\left(\mathbf{e}^{0}\right), \tag{2.53}$$

subject to the adjoint boundary conditions determined in Eq. (2.50). Note that the adjoint function $\boldsymbol{\psi}(\mathbf{x})$ is called a weak solution of Eq. (2.53) because $\nabla_u R\left(\mathbf{e}^0\right)$ may contain distributions (e.g., Heaviside and/or Dirac-delta functionals, as well as derivatives thereof), so the various Hilbert spaces introduced in the foregoing would become appropriately defined Sobolev spaces.

Equations (2.46), (2.47), (2.51) and (2.53) can now be used to obtain the following expression for the total sensitivity $D\mathrm{R}\left(\mathbf{e}^0;\mathbf{h}\right)$ of $\mathrm{R}(\mathbf{e})$ at $\mathbf{e}^0$:

$$\begin{aligned} D\mathrm{R}\left(\mathbf{e}^{0};\mathbf{h}\right) &= \mathrm{R}'_{\alpha}\left(\mathbf{e}^{0}\right)\mathbf{h}_{\alpha} + \left\langle \boldsymbol{\psi}, \mathbf{Q}'_{\alpha}\left(\mathbf{e}^{0}\right)\mathbf{h}_{\alpha} - \mathbf{N}'_{\alpha}\left(\mathbf{e}^{0}\right)\mathbf{h}_{\alpha}\right\rangle_{Q} - \hat{P}\left(\mathbf{h}_{\alpha}, \boldsymbol{\psi}; \boldsymbol{\alpha}^{0}\right) \\ &\triangleq \sum_{i=1}^{N_{\alpha}} S_{i}\left(\mathbf{e}^{0}\right)\left(\Delta\alpha_{i}\right), \end{aligned} \tag{2.54}$$

where the quantities $S_i\left(\mathbf{e}^0\right)$ denote the absolute sensitivities of the response $\mathrm{R}(\mathbf{e})$ at $\mathbf{e} = \mathbf{e}^0$. These absolute sensitivities coincide with the ordinary partial derivatives or the system response with respect to the model parameters, when all quantities are continuous. The relative sensitivities, denoted as $S_i^{rel}\left(\mathbf{e}^0\right)$, are obtained, as usual, by means of their non-dimensional definition

$$S_{i}^{rel}\left(\mathbf{e}^{0}\right) \triangleq S_{i}\left(\mathbf{e}^{0}\right)\alpha_{i}^{0} / R\left(\mathbf{e}^{0}\right). \tag{2.55}$$

As Eq. (2.54) indicates, the desired elimination of all unknown values of $\mathbf{h}_u$ from the expressions of the sensitivities $S_i\left(\mathbf{e}^0\right)$, $i=1,\ldots,N_\alpha$, of $\mathrm{R}\left(\mathbf{e}^0\right)$ at $\mathbf{e}^0$ has been accomplished. The sensitivities $S_i\left(\mathbf{e}^0\right)$ can therefore be computed by means of Eq. (2.55), after solving the adjoint sensitivity system consisting of Eqs. (2.53) and (2.50), once for each response, to obtain the corresponding adjoint function $\boldsymbol{\psi}(\mathbf{x})$. Recall that the FSAP requires solving the forward sensitivity system defined by Eqs. (2.26) and (2.27) for each parameter variation $\Delta\alpha_i$, $i=1,\ldots,N_\alpha$. Hence, using the ASAM instead of the FSAP is advantageous if the number of responses is less than the number of model parameters, which is definitely the case for the dissolver model analyzed in this work.

To apply the ASAM to the dissolver model, the forward sensitivity equations, cf., Eqs. (2.31) through (2.34), are written in matrix form as

$$\mathbf{N}'_u\left(\mathbf{e}^0\right)\mathbf{h}_u \triangleq \begin{bmatrix} \mathbf{N}_{11} & \mathbf{N}_{12} \\ \mathbf{N}_{21} & \mathbf{N}_{22} \end{bmatrix} \begin{bmatrix} \mathbf{h}_\rho(t) \\ \mathbf{h}_V(t) \end{bmatrix} = \begin{bmatrix} \mathbf{q}_\rho(t) \\ \mathbf{q}_V(t) \end{bmatrix}, \tag{2.56}$$

where

$$\mathbf{q}_\rho(t) \triangleq \left[q_\rho^{(1)}(t),\ldots,q_\rho^{(8)}(t)\right],\ \mathbf{q}_V(t) \triangleq \left[q_V^{(1)}(t),\ldots,q_V^{(8)}(t)\right], \tag{2.57}$$

$$\begin{aligned}
\mathbf{N}_{11}(t) &\triangleq \begin{bmatrix} a_{11}^{11} & a_{12}^{11} & 0 & . & 0 & 0 \\ 0 & a_{22}^{11} & a_{23}^{11} & . & 0 & 0 \\ . & . & . & . & . & . \\ 0 & 0 & 0 & . & a_{77}^{11} & a_{78}^{11} \\ 0 & 0 & 0 & . & 0 & a_{88}^{11} \end{bmatrix}, \\
\mathbf{N}_{12}(t) &\triangleq \begin{bmatrix} a_{11}^{12} & a_{12}^{12} & 0 & . & 0 & 0 \\ 0 & a_{22}^{12} & a_{23}^{12} & . & 0 & 0 \\ . & . & . & . & . & . \\ 0 & 0 & 0 & . & a_{77}^{12} & a_{78}^{12} \\ 0 & 0 & 0 & . & 0 & a_{88}^{12} \end{bmatrix},
\end{aligned} \tag{2.58}$$

$$\mathbf{N}_{22}(t) \triangleq \begin{bmatrix} a_{11}^{22} & a_{12}^{22} & 0 & . & 0 & 0 \\ 0 & a_{22}^{22} & a_{23}^{22} & . & 0 & 0 \\ . & . & . & . & . & . \\ 0 & 0 & 0 & . & a_{77}^{22} & a_{78}^{22} \\ 0 & 0 & 0 & . & 0 & a_{88}^{22} \end{bmatrix}, \ \mathbf{N}_{21} \triangleq [0]_{8\times 8}, \tag{2.59}$$

$$a_{ii}^{11}(t) \triangleq V_{nom}^{(i)}(t)\frac{d[*]}{dt} + C\left[V_{nom}^{(i+1)}(t)\right], \ i = 1,\ldots,7; \tag{2.60}$$

$$a_{88}^{11}(t) \triangleq V_{nom}^{(8)}(t)\frac{d[*]}{dt} + f_{nom}^{(in)}(t); \tag{2.61}$$

$$a_{i,i+1}^{11}(t) \triangleq -C\left[V_{nom}^{(i+1)}(t)\right], \ i = 1,\ldots,7; \tag{2.62}$$

$$a_{ii}^{12}(t) \triangleq \frac{d\rho_{a,nom}^{(i)}(t)}{dt}, \ i = 1,\ldots,8; \tag{2.63}$$

$$a_{i,i+1}^{12}(t) \triangleq \left[\rho_{a,nom}^{(i)}(t) - \rho_{a,nom}^{(i+1)}(t)\right]\left[\frac{dC\left(V^{(i+1)}\right)}{dV^{(i+1)}}\right]_{nom}, \ i = 1,\ldots,7; \tag{2.64}$$

$$a_{ii}^{22}(t) \triangleq \frac{d[*]}{dt} + \left[\frac{dC\left(V^{(i)}\right)}{dV^{(i)}}\right]_{nom}, \ i = 1,\ldots,8; \tag{2.65}$$

$$a_{i,i+1}^{22}(t) \triangleq -\left[\frac{dC\left(V^{(i+1)}\right)}{dV^{(i+1)}}\right]_{nom}, \ i = 1,\ldots,7;. \tag{2.66}$$

In view of Eq. (2.56), the Hilbert spaces $\mathcal{H}_u(\Omega_x)$ and $\mathcal{H}_Q(\Omega_x)$ are identical, comprising all square-integrable vector-valued functions of the form $\mathbf{f}(t) \triangleq \left[\mathbf{f}_\rho(t), \mathbf{f}_V(t)\right]$, $\mathbf{g}(t) \triangleq \left[\mathbf{g}_\rho(t), \mathbf{g}_V(t)\right]$, and endowed with the inner product defined as

$$\begin{aligned}\langle \mathbf{f}(t),\mathbf{g}(t)\rangle_u &\triangleq \int_0^{t_f}\left[\mathbf{f}_\rho(t)\mathbf{g}_\rho(t)+\mathbf{f}_V(t)\mathbf{g}_V(t)\right]dt \\ &= \sum_{i=1}^{8}\int_0^{t_f}\left[f_\rho^{(i)}(t)g_\rho^{(i)}(t)+f_V^{(i)}(t)g_V^{(i)}(t)\right]dt.\end{aligned} \tag{2.67}$$

The formal adjoint operator $\mathbf{N}^+\left(\mathbf{e}^0\right)\triangleq\left[N_{ji}^+\left(\mathbf{e}^0\right)\right]$, $(i,j=1,2)$, of the operator $\mathbf{N}'_u\left(\mathbf{e}^0\right)\mathbf{h}_u$ is derived by using Eq. (2.49) to obtain:

$$\mathbf{N}_{11}^+(t)\triangleq\begin{bmatrix} b_{11}^{11} & 0 & 0 & . & 0 & 0 \\ b_{21}^{11} & b_{22}^{11} & 0 & . & 0 & 0 \\ . & . & . & . & . & . \\ 0 & 0 & 0 & . & b_{77}^{11} & 0 \\ 0 & 0 & 0 & . & b_{87}^{11} & b_{88}^{11}\end{bmatrix},\ \mathbf{N}_{21}^+(t)\triangleq\begin{bmatrix} b_{11}^{21} & 0 & 0 & . & 0 & 0 \\ b_{21}^{21} & b_{22}^{21} & 0 & . & 0 & 0 \\ . & . & . & . & . & . \\ 0 & 0 & 0 & . & b_{77}^{21} & 0 \\ 0 & 0 & 0 & . & b_{87}^{21} & b_{88}^{21}\end{bmatrix}, \tag{2.68}$$

$$\mathbf{N}_{22}^+(t)\triangleq\begin{bmatrix} b_{11}^{22} & 0 & 0 & . & 0 & 0 \\ b_{21}^{22} & b_{22}^{22} & 0 & . & 0 & 0 \\ . & . & . & . & . & . \\ 0 & 0 & 0 & . & b_{77}^{22} & 0 \\ 0 & 0 & 0 & . & b_{87}^{22} & b_{88}^{22}\end{bmatrix},\ \mathbf{N}_{12}^+\triangleq[0]_{8\times 8}, \tag{2.69}$$

$$b_{ii}^{11}(t)\triangleq-\frac{d\left[V_{nom}^{(i)}(t)^*\right]}{dt}+C\left[V_{nom}^{(i+1)}(t)\right],\ i=1,\dots,7; \tag{2.70}$$

$$b_{88}^{11}(t)\triangleq-\frac{d\left[V_{nom}^{(8)}(t)^*\right]}{dt}+f_{nom}^{(in)}(t); \tag{2.71}$$

$$b_{i+1,i}^{11}(t)\triangleq-C\left[V_{nom}^{(i+1)}(t)\right],\ i=1,\dots,7; \tag{2.72}$$

$$b_{ii}^{21}(t)\triangleq\frac{d\rho_{a,nom}^{(i)}(t)}{dt},\ i=1,\dots,8; \tag{2.73}$$

$$b_{i+1,i}^{21}(t)\triangleq\left[\rho_{a,nom}^{(i)}(t)-\rho_{a,nom}^{(i+1)}(t)\right]\left[\frac{dC\left(V^{(i+1)}\right)}{dV^{(i+1)}}\right]_{nom},\ i=1,\dots,7; \tag{2.74}$$

$$b_{ii}^{22}(t) \triangleq -\frac{d[*]}{dt} + \left[\frac{dC\left(V^{(i)}\right)}{dV^{(i)}}\right]_{nom}, \; i = 1,\ldots,8; \tag{2.75}$$

$$b_{i+1,i}^{22}(t) \triangleq -\left[\frac{dC\left(V^{(i+1)}\right)}{dV^{(i+1)}}\right]_{nom}, \; i = 1,\ldots,7;. \tag{2.76}$$

The adjoint sensitivity system corresponding to Eq. (2.56) is derived by particularizing the general form represented by Eq. (2.53), to obtain:

$$-\frac{d\left[V_{nom}^{(1)}(t)\psi_{\rho}^{(1)}(t)\right]}{dt} + C\left[V_{nom}^{(2)}(t)\right]\psi_{\rho}^{(1)}(t) = w_{\rho}^{(1)}(t), \tag{2.77}$$

$$\begin{aligned} &-\frac{d\left[V_{nom}^{(k)}(t)\psi_{\rho}^{(k)}(t)\right]}{dt} + C\left[V_{nom}^{(k+1)}(t)\right]\psi_{\rho}^{(k)}(t) \\ &-C\left[V_{nom}^{(k)}(t)\right]\psi_{\rho}^{(k-1)}(t) = w_{\rho}^{(k)}(t), \; k = 2,\ldots,7; \end{aligned} \tag{2.78}$$

$$-\frac{d\left[V_{nom}^{(8)}(t)\psi_{\rho}^{(8)}(t)\right]}{dt} + f_{nom}^{(in)}(t)\psi_{\rho}^{(8)}(t) - C\left[V_{nom}^{(8)}(t)\right]\psi_{\rho}^{(7)}(t) = w_{\rho}^{(8)}(t), \tag{2.79}$$

$$-\frac{d\left[\psi_{V}^{(1)}(t)\right]}{dt} + \left[\frac{dC\left[V^{(1)}(t)\right]}{dV^{(1)}(t)}\right]_{nom}\psi_{V}^{(1)}(t) + \left[\frac{d\rho_{a,nom}^{(1)}(t)}{dt}\right]\psi_{\rho}^{(1)}(t) = w_{V}^{(1)}(t), \tag{2.80}$$

$$\begin{aligned} &-\frac{d\left[\psi_{V}^{(k)}(t)\right]}{dt} + \left[\frac{dC\left[V^{(k)}(t)\right]}{dV^{(k)}(t)}\right]_{nom}\psi_{V}^{(k)}(t) - \left[\frac{dC\left[V^{(k)}(t)\right]}{dV^{(k)}(t)}\right]_{nom}\psi_{V}^{(k-1)}(t) \\ &+\left[\frac{d\rho_{a,nom}^{(k)}(t)}{dt}\right]\psi_{\rho}^{(k)}(t) + \left[\rho_{a,nom}^{(k-1)}(t) - \rho_{a,nom}^{(k)}(t)\right]\left[\frac{dC\left[V^{(k)}(t)\right]}{dV^{(k)}(t)}\right]_{nom}\psi_{\rho}^{(k-1)}(t) \\ &= w_{V}^{(k)}(t), \; k = 2,\ldots,8;. \end{aligned} \tag{2.81}$$

Comparing the source terms $\mathbf{w}_{\rho} \triangleq \left[w_{\rho}^{(1)}(t),\ldots,w_{\rho}^{(8)}(t)\right]$ and $\mathbf{w}_{V} \triangleq \left[w_{V}^{(1)}(t),\ldots,w_{V}^{(8)}(t)\right]$ in Eqs. (2.77) through (2.81) with the right-side of the general expression of the adjoint sensitivity system defined in Eq. (2.53), it follows that

$$\nabla_u R(\mathbf{e}^0) \triangleq [\mathbf{w}_\rho, \mathbf{w}_V] = [w_\rho^{(1)}(t), \ldots, w_\rho^{(8)}, w_V^{(1)}(t), \ldots, w_V^{(8)}(t)]. \tag{2.82}$$

The actual expressions of the above source terms will be specified in the Sub-sections to follow, according to the particular response under investigation.
Equations (2.77) through (2.81) can be further simplified by using the original forward model, namely Eqs. (2.10) through (2.13), along with the following definitions,

$$D_\rho^{(k)}(t) \triangleq \frac{d\rho_{a,nom}^{(k)}(t)}{dt} = \frac{\rho_{a,nom}^{(k+1)}(t) - \rho_{a,nom}^{(k)}(t)}{V_{nom}^{(k)}(t)} C\left[V_{nom}^{(k+1)}(t)\right], \; k = 1, \ldots, 7; \tag{2.83}$$

$$D_\rho^{(8)}(t) \triangleq \frac{d\rho_{a,nom}^{(8)}(t)}{dt} = \frac{\rho_{a,nom}^{(in)}(t) - \rho_{a,nom}^{(8)}(t)}{V_{nom}^{(8)}(t)} f_{nom}^{(in)}(t), \tag{2.84}$$

$$D_V^{(k)}(t) \triangleq \left[\frac{dC\left[V^{(k)}(t)\right]}{dV^{(k)}(t)}\right]_{nom} = \begin{cases} \dfrac{p}{G}\left[\dfrac{V^{(k)}(t) - V_0}{G}\right]^{p-1}, & if \; V^{(k)}(t) - V_0 > 0, \; k = 1, \ldots, 8, \\ 0, & otherwise. \end{cases} \tag{2.85}$$

Following the introduction of the above relations, we obtain the following final form of the adjoint sensitivity system:

$$-V_{nom}^{(1)}(t)\frac{d\psi_\rho^{(1)}(t)}{dt} + C\left[V_{nom}^{(1)}(t)\right]\psi_\rho^{(1)}(t) = w_\rho^{(1)}(t), \; 0 < t < t_f, \tag{2.86}$$

$$-V_{nom}^{(k)}(t)\frac{d\psi_\rho^{(k)}(t)}{dt} + C\left[V_{nom}^{(k)}(t)\right]\left[\psi_\rho^{(k)}(t) - \psi_\rho^{(k-1)}(t)\right] = w_\rho^{(k)}(t), \tag{2.87}$$
$$0 < t < t_f, \; k = 2, \ldots, 8;$$

$$-\frac{d\psi_V^{(1)}(t)}{dt} + D_V^{(1)}(t)\psi_V^{(1)}(t) + D_\rho^{(1)}(t)\psi_\rho^{(1)}(t) = w_V^{(1)}(t), \; 0 < t < t_f, \tag{2.88}$$

$$-\frac{d\psi_V^{(k)}(t)}{dt}+D_V^{(k)}(t)\left[\psi_V^{(k)}(t)-\psi_V^{(k-1)}(t)\right]+D_\rho^{(k)}(t)\psi_\rho^{(k)}(t)$$
$$+\left[\rho_{a,nom}^{(k-1)}(t)-\rho_{a,nom}^{(k)}(t)\right]D_V^{(k)}(t)\psi_\rho^{(k-1)}(t)=w_V^{(k)}(t),\ k=2,\ldots,8,\ 0<t<t_f. \tag{2.89}$$

The boundary conditions will be determined next by inserting Eqs. (2.56), and (2.86) through (2.89) into Eq. (2.49) and performing the integration over time to obtain

$$\begin{aligned}
\{P(\mathbf{h}_u;\boldsymbol{\psi})\}_{\partial\Omega_x} &= \left\langle \mathbf{N}^+(\mathbf{e}^0)\boldsymbol{\psi},\mathbf{h}_u\right\rangle_u-\left\langle \boldsymbol{\psi},\mathbf{N}'_u(\mathbf{e}^0)\mathbf{h}_u\right\rangle_Q \\
&=\int_0^{t_f}\left[\mathbf{h}_\rho,\mathbf{h}_V\right]^\dagger\begin{bmatrix}\mathbf{N}_{11}^+ & \mathbf{N}_{12}^+\\ \mathbf{N}_{21}^+ & \mathbf{N}_{22}^+\end{bmatrix}\begin{bmatrix}\boldsymbol{\psi}_\rho(t)\\ \boldsymbol{\psi}_V(t)\end{bmatrix}dt \\
&-\int_0^{t_f}\left[\boldsymbol{\psi}_\rho,\boldsymbol{\psi}_V\right]^\dagger\begin{bmatrix}\mathbf{N}_{11} & \mathbf{N}_{12}\\ \mathbf{N}_{21} & \mathbf{N}_{22}\end{bmatrix}\begin{bmatrix}\mathbf{h}_\rho(t)\\ \mathbf{h}_V(t)\end{bmatrix}dt \\
&=\sum_{k=1}^{8}\left\{\psi_\rho^{(k)}(t)V_{nom}^{(k)}(t)\left[\Delta\rho_a^{(k)}(t)\right]+\psi_V^{(k)}(t)\left[\Delta V^{(k)}(t)\right]\right\}_{t=t_f} \\
&-\sum_{k=1}^{8}\left\{\psi_\rho^{(k)}(t)V_{nom}^{(k)}(t)\left[\Delta\rho_a^{(k)}(t)\right]+\psi_V^{(k)}(t)\left[\Delta V^{(k)}(t)\right]\right\}_{t=0}.
\end{aligned} \tag{2.90}$$

Since the values of $\Delta\rho_a^{(k)}(t)$, $\Delta V^{(k)}(t)$, $k=1,\ldots,8$, are known at $t=0$ but are unknown at the final time, $t=t_f$, these unknown values are eliminated by setting to zero the "final-time" values for the adjoint functions $\boldsymbol{\psi}\triangleq\left[\boldsymbol{\psi}_\rho,\boldsymbol{\psi}_V\right]^\dagger$, namely

$$\psi_\rho^{(k)}(t_f)=0,\ \psi_V^{(k)}(t_f)=0,\ k=1,\ldots,8. \tag{2.91}$$

Inserting the above "final-time" conditions into Eq. (2.90) reduces it to the following expression:

$$\hat{P}\left(\mathbf{h}_\alpha,\boldsymbol{\psi};\boldsymbol{\alpha}^0\right)=-\sum_{k=1}^{8}\left\{\psi_\rho^{(k)}(0)V_{nom}^{(k)}(0)\left[\Delta\rho_{a0}^{(k)}\right]+\psi_V^{(k)}(0)\left[\Delta V_0^{(k)}\right]\right\}. \tag{2.92}$$

It is important to note that the above expression contains the sensitivities of the response to the initial conditions, which will play a crucial role in verifying the accuracy of the computation of the adjoint functions discussed in subsequent sections of this work.

The "final-time" conditions in Eq. (2.91) clearly indicate that the adjoint sensitivity system is a "final-time problem" rather than an initial-value problem. It is useful to convert the adjoint sensitivity system from a "final-time" problem to an "initial-condition problem" as customarily required by solvers of ordinary differential equations. This can be accomplished by changing the independent variable t to another independent variable, τ , defined as follows:

$$\tau \triangleq t_f - t. \tag{2.93}$$

Introducing the above change of independent variable into Eqs. (2.86) through (2.89) transforms them into the following "computationally-suitable" form of the adjoint sensitivity system:

$$V_{nom}^{(1)}\left(t_f - \tau\right)\frac{d\psi_\rho^{(1)}(\tau)}{d\tau} + C\left[V_{nom}^{(1)}\left(t_f - \tau\right)\right]\psi_\rho^{(1)}(\tau) = w_\rho^{(1)}\left(t_f - \tau\right),\ 0 < \tau < t_f, \tag{2.94}$$

$$\begin{aligned} &V_{nom}^{(k)}\left(t_f - \tau\right)\frac{d\psi_\rho^{(k)}(\tau)}{d\tau} + C\left[V_{nom}^{(k)}\left(t_f - \tau\right)\right]\left[\psi_\rho^{(k)}(\tau) - \psi_\rho^{(k-1)}(\tau)\right] \\ &= w_\rho^{(k)}\left(t_f - \tau\right),\ 0 < \tau < t_f,\ k = 2,\ldots,8; \end{aligned} \tag{2.95}$$

$$\begin{aligned} &\frac{d\psi_V^{(1)}(\tau)}{d\tau} + D_V^{(1)}\left(t_f - \tau\right)\psi_V^{(1)}(\tau) + D_\rho^{(1)}\left(t_f - \tau\right)\psi_\rho^{(1)}(\tau) = w_V^{(1)}\left(t_f - \tau\right), \\ &0 < \tau < t_f, \end{aligned} \tag{2.96}$$

$$\begin{aligned} &\frac{d\psi_V^{(k)}(\tau)}{d\tau} + D_V^{(k)}\left(t_f - \tau\right)\left[\psi_V^{(k)}(\tau) - \psi_V^{(k-1)}(\tau)\right] + D_\rho^{(k)}\left(t_f - \tau\right)\psi_\rho^{(k)}(\tau) \\ &+\left[\rho_{a,nom}^{(k-1)}\left(t_f - \tau\right) - \rho_{a,nom}^{(k)}\left(t_f - \tau\right)\right]D_V^{(k)}\left(t_f - \tau\right)\psi_\rho^{(k-1)}(\tau) = w_V^{(k)}\left(t_f - \tau\right), \\ &k = 2,\ldots,8,\ 0 < \tau < t_f. \end{aligned} \tag{2.97}$$

The initial conditions for Eqs. (2.94) through (2.97) are

$$\psi_\rho^{(k)}(\tau = 0) = 0,\ \psi_V^{(k)}(\tau = 0) = 0,\ k = 1,\ldots,8. \tag{2.98}$$

The measured or computed nitric acid concentration in a compartment k at a time-instance $t_i, i = 1,\ldots, I = 307,$ can be represented in the form shown in Eq. (2.21). Applying the definition of the (first-order) Gateaux-differential shown in Eq. (2.30) to Eq. (2.21) yields the following expression for the total sensitivity of $\rho_a^{(k)}(t_i)$ to model parameter variations:

$$D\rho_a^{(k)}(t_i) \triangleq \frac{d}{d\varepsilon}\left\{\int_0^{t_f}\left[\rho_{a,nom}^{(k)}(t)+\varepsilon\Delta\rho_a^{(k)}(t)\right]\delta(t-t_i)\,dt\right\}_{\varepsilon=0}$$
$$=\int_0^{t_f}\left[\Delta\rho_a^{(k)}(t)\right]\delta(t-t_i)\,dt,\quad i=1,\ldots,\mathrm{I}. \tag{2.99}$$

As indicated by Eq. (2.99), the "direct effect term" $\mathrm{R}'_\alpha(\mathbf{e}^0)\mathbf{h}_\alpha$ is identically zero for this response. Comparing the above expression with the general expression of the adjoint sensitivity system given in Eq. (2.82), it follows that

$$w_\rho^{(k)}(t)=\delta(t-t_i),\; w_\rho^{(j)}(t)=0,\quad 1\le j\ne k\le 8;\; w_V^{(1)}(t)=\ldots=w_V^{(8)}(t)=0. \tag{2.100}$$

It further follows from Eq. (2.56) that

$$\left\langle \boldsymbol{\psi},\mathbf{Q}'_\alpha(\mathbf{e}^0)\mathbf{h}_\alpha-\mathbf{N}'_\alpha(\mathbf{e}^0)\mathbf{h}_\alpha\right\rangle_Q=\sum_{k=1}^{8}\int_0^{t_f}\psi_\rho^{(k)}(t)\,q_\rho^{(k)}(t)\,dt$$
$$+\sum_{k=1}^{8}\int_0^{t_f}\psi_V^{(k)}(t)\,q_V^{(k)}(t)\,dt, \tag{2.101}$$

where the quantities $q_\rho^{(k)}(t)$ and $q_V^{(k)}(t)$ were defined in Eqs. (2.35) through (2.38) . Replacing now Eqs. (2.92) and (2.101) into Eq. (2.99) transforms the latter into the following expression:

$$D\rho_a^{(k)}(t_i)=\sum_{k=1}^{8}\int_0^{t_f}\psi_\rho^{(k)}(t)\,q_\rho^{(k)}(t)\,dt+\sum_{k=1}^{8}\int_0^{t_f}\psi_V^{(k)}(t)\,q_V^{(k)}(t)\,dt$$
$$+\sum_{k=1}^{8}\left\{\psi_\rho^{(k)}(0)V_{nom}^{(k)}(0)\left[\Delta\rho_{a0}^{(k)}\right]+\psi_V^{(k)}(0)\left[\Delta V_0^{(k)}\right]\right\}. \tag{2.102}$$

As the above equation indicates, *the sensitivities* $D\rho_a^{(k)}(t_i)$ for all 635 system parameters *can be evaluated exactly and efficiently* as soon as the components of the adjoint function $\boldsymbol{\psi}\triangleq\left[\boldsymbol{\psi}_\rho,\boldsymbol{\psi}_V\right]^\dagger$ have been computed by solving *once* the adjoint sensitivity system defined in Eqs. (2.94) through (2.98). This is because the adjoint sensitivity system, namely Eqs. (2.94) through (2.98), are independent of parameter variations, in contradistinction to the forward sensitivity system, cf., Eqs. (2.31) through (2.40). The expression in Eq. (2.102) indicates that the following relations hold:

$$
\begin{aligned}
\frac{\partial R(t_i)}{\partial \rho_{a0}^{(k)}} &\triangleq \left[\frac{\partial \rho_a^{(1)}(t_i)}{\partial \rho_{a0}^{(k)}}\right] = \psi_\rho^{(k)}(0) V_{nom}^{(k)}(0), \\
\frac{\partial R(t_i)}{\partial V_0^{(k)}} &\triangleq \left[\frac{\partial \rho_a^{(1)}(t_i)}{\partial V_0^{(k)}}\right] = \psi_V^{(k)}(0),\ k = 1, \dots, 8.
\end{aligned}
\tag{2.103}
$$

Therefore, the values of $\psi_V^{(k)}(0)$ are actually the response sensitivities to the initial conditions $V_0^{(k)}$, while the values of $\psi_\rho^{(k)}(0)$ are obtained from Eq. (2.102) by dividing the respective sensitivities through $V_{nom}^{(k)}(0)$. The above sensitivities can be independently computed using recalculations with perturbed initial conditions, i.e.,

$$
\begin{aligned}
&\frac{1}{V_{nom}^{(k)}(0)} \frac{\partial R(t_i)}{\partial \rho_{a0}^{(k)}} \triangleq \frac{1}{V_{nom}^{(k)}(0)} \left[\frac{\partial \rho_a^{(1)}(t_i)}{\partial \rho_{a0}^{(k)}}\right] = \psi_\rho^{(k)}(t=0) \\
&\cong \frac{\rho_a^{(1)}\left(t_i, \rho_{a0}^{(k)} + \Delta \rho_{a0}^{(k)}\right) - \rho_a^{(1)}\left(t_i, \Delta \rho_{a0}^{(k)}\right)}{\Delta \rho_{a0}^{(k)}},
\end{aligned}
\tag{2.104}
$$

and

$$
\frac{\partial R(t_i)}{\partial V_0^{(k)}} \triangleq \frac{\partial \rho_a^{(1)}(t_i)}{\partial V_0^{(k)}} = \psi_V^{(k)}(t=0) \cong \frac{\rho_a^{(1)}\left(t_i, V_0^{(k)} + \Delta V_0^{(k)}\right) - \rho_a^{(1)}\left(t_i, V_0^{(k)}\right)}{\Delta V_0^{(k)}}. \tag{2.105}
$$

Note that the values of the adjoint functions $\psi_\rho^{(k)}(t=0)$ and $\psi_V^{(k)}(t=0)$ are obtained only after computing the solution of the complete adjoint sensitivity system, i.e., after "running" the adjoint sensitivity system through the entire time-period under consideration. Hence, the relations given in Eqs. (2.104) and (2.105) provide a stringent numerical "verification" of the solution accuracy of the adjoint sensitivity system. Such verifications have been performed using Eqs. (2.104) and (2.105), obtaining very close agreement (within less than 0.1%) between the respective sensitivities, computed by the ASAM (on the one hand) and direct re-computations (on the other hand).

2.3.2 ASAM Computation of Time-Dependent Sensitivities of Nitric Acid Concentration to Model Parameters

A response that is a function-valued operator, rather than a scalar-valued functional, does not naturally provide an inner product. Cacuci (1981b, 2003) has provided a definitive adjoint sensitivity analysis methodology for computing the sensitivities of operator-valued responses to model parameters. The application of this methodology to the dissolver model will be illustrated in this Sub-section, showing that the number of required adjoint computations for computing all of the sensitivities of the time-dependent responses over the entire time-span is reduced drastically --from 307 to 17-- with an overall loss of accuracy of less than 0.01% over the entire time-span.

Applying the methodology of Cacuci (1981b) to compute the sensitivities of an operator-valued response $\mathbf{R}(\mathbf{e})$ requires the introduction of an orthonormal basis for the Hilbert space $\mathcal{H}_R(\Omega_R)$ that enables the representation of the operator-valued $\mathbf{R}(\mathbf{e})$ by a spectral (generalized Fourier) expansion. Thus, consider that the set $\{\boldsymbol{\varphi}_n(\mathbf{x})\}$, $n \in N$ (where the index set N may be finite or infinite) is an orthonormal basis for the Hilbert space $\mathcal{H}_R(\Omega_R)$, so that $\mathbf{R}(\mathbf{e})$ can be represented by the spectral (generalized Fourier) expansion

$$\mathbf{R}(\mathbf{e}^0) = \sum_{n \in N} \left\langle \mathbf{R}(\mathbf{e}^0), \boldsymbol{\varphi}_n(\mathbf{x}) \right\rangle_R \boldsymbol{\varphi}_n(\mathbf{x}). \tag{2.106}$$

Using the above spectral expansion in Eq. (2.24) yields

$$\begin{aligned} D\mathbf{R}(\mathbf{e}^0;\mathbf{h}) &\triangleq \left\{ \frac{d}{d\varepsilon}\left[\mathbf{R}(\mathbf{e}^0+\varepsilon\mathbf{h})\right]\right\}_{\varepsilon=0} = \left\{ \frac{d}{d\varepsilon}\left[\sum_{n\in N}\left\langle \mathbf{R}(\mathbf{e}^0+\varepsilon\mathbf{h}),\boldsymbol{\varphi}_n\right\rangle_R \boldsymbol{\varphi}_n\right]\right\}_{\varepsilon=0} \\ &= \sum_{n\in N}\left\langle \mathbf{R}'_u(\mathbf{e}^0)\mathbf{h}_u,\boldsymbol{\varphi}_n\right\rangle_R \boldsymbol{\varphi}_n + \sum_{n\in N}\left\langle \mathbf{R}'_\alpha(\mathbf{e}^0)\mathbf{h}_\alpha,\boldsymbol{\varphi}_n\right\rangle_R \boldsymbol{\varphi}_n \\ &= \mathbf{R}'_\alpha(\mathbf{e}^0)\mathbf{h}_\alpha + \sum_{n\in N}\left\langle \mathbf{R}'_u(\mathbf{e}^0)\mathbf{h}_u,\boldsymbol{\varphi}_n\right\rangle_R \boldsymbol{\varphi}_n. \end{aligned} \tag{2.107}$$

Actually, the "direct effect term,"

$$\mathbf{R}'_\alpha(\mathbf{e}^0)\mathbf{h}_\alpha \triangleq \sum_{n\in N}\left\langle \mathbf{R}'_\alpha(\mathbf{e}^0)\mathbf{h}_\alpha,\boldsymbol{\varphi}_n\right\rangle_R \boldsymbol{\varphi}_n \tag{2.108}$$

can be computed directly from Eq. (2.24) without needing to evaluate its spectral expansion. On the other hand, the generalized Fourier coefficients $\left\langle \mathbf{R}'_u(\mathbf{e}^0)\mathbf{h}_u,\boldsymbol{\varphi}_n\right\rangle_R$ in the spectral expansion of the indirect-effect term, namely

$$\mathbf{R}'_u(\mathbf{e}^0)\mathbf{h}_u \triangleq \sum_{n\in N}\left\langle \mathbf{R}'_u(\mathbf{e}^0)\mathbf{h}_u,\boldsymbol{\varphi}_n\right\rangle_R \boldsymbol{\varphi}_n \tag{2.109}$$

are linear functionals of $\mathbf{h}_u$ and therefore provide the appropriate inner product for introducing the corresponding adjoint sensitivity system. For this purpose, we introduce the formal adjoint operator, denoted as $\mathbf{N}^+(\mathbf{e}^0)$, of the operator $\mathbf{N}'_u(\mathbf{e}^0)$ through the following relationship (in the respective Hilbert spaces) which holds for an arbitrary vector $\boldsymbol{\psi}_n \in \mathcal{H}_Q(\Omega_x)$:

$$\left\langle \boldsymbol{\psi}_n, \mathbf{N}'_u(\mathbf{e}^0)\mathbf{h}_u \right\rangle_Q = \left\langle \mathbf{N}^+(\mathbf{e}^0)\boldsymbol{\psi}_n, \mathbf{h}_u \right\rangle_u + \left\{ P(\mathbf{h}_u;\boldsymbol{\psi}_n) \right\}_{\partial\Omega_x}. \tag{2.110}$$

As before, the formal adjoint operator $\mathbf{N}^+(\mathbf{e}^0)$ is the $N_u \times N_u$ matrix $\mathbf{N}^+(\mathbf{e}^0) \triangleq \left[N_{ji}^+(\mathbf{e}^0) \right]$, $(i,j = 1,\ldots,N_u)$, comprising elements $N_{ji}^+(\mathbf{e}^0)$ obtained by transposing the formal adjoints of the components of $\mathbf{N}'_u(\mathbf{e}^0)$, while $\left\{ P(\mathbf{h}_u;\boldsymbol{\psi}_n) \right\}_{\partial\Omega_x}$ denotes the associated bilinear form evaluated on $\partial\Omega_x$. The domain of $\mathbf{N}^+(\mathbf{e}^0)$ is determined by selecting appropriate adjoint boundary conditions, represented here in operator form as

$$\mathbf{B}^+(\mathbf{e}^0)\boldsymbol{\psi}_n(\mathbf{x}) - \mathbf{A}^+(\boldsymbol{\alpha}^0) = \mathbf{0}, \quad \mathbf{x} \in \partial\Omega_x. \tag{2.111}$$

The above boundary conditions for $\mathbf{N}^+(\mathbf{e}^0)$ are obtained by requiring that:

(a) They must be independent of $\mathbf{h}_u$ and $\mathbf{h}_\alpha$;
(b) The substitution of Eqs. (2.111) and (2.27) into the expression of $\left\{ P(\mathbf{h}_u;\boldsymbol{\psi}_n) \right\}_{\partial\Omega_x}$ in Eq. (2.110) must cause all terms containing unknown values of $\mathbf{h}_u$ to vanish.

The above selection of the adjoint boundary conditions for $\mathbf{N}^+(\mathbf{e}^0)$ reduces $\left\{ P(\mathbf{h}_u;\boldsymbol{\psi}_n) \right\}_{\partial\Omega_x}$ to a quantity that contains boundary terms involving only known values of $\mathbf{h}_\alpha$, $\boldsymbol{\psi}_n$, and, possibly, $\boldsymbol{\alpha}^0$; this quantity will be denoted by $\hat{P}(\mathbf{h}_\alpha, \boldsymbol{\psi}_n; \boldsymbol{\alpha}^0)$. In general, $\hat{P}$ does not automatically vanish as a result of these manipulations, although it may do so in particular instances. In principle, $\hat{P}$ can be forced to vanish by considering extensions of $\mathbf{N}'_u(\mathbf{e}^0)$, in the operator sense, but such an action is seldom needed in practice.

Using Eq. (2.26) to replace the quantity $\mathbf{N}'_u(\mathbf{e}^0)\mathbf{h}_u$ on the left-side of Eq. (2.110), and re-arranging terms transforms Eq. (2.110) into the form

$$\left\langle \mathbf{N}^{+}\left(\mathbf{e}^{0}\right)\boldsymbol{\psi}_{n}, \mathbf{h}_{u}\right\rangle_{u} = \left\langle \boldsymbol{\psi}_{n}, \mathbf{Q}'_{\alpha}\left(\mathbf{e}^{0}\right)\mathbf{h}_{\alpha} - \mathbf{N}'_{\alpha}\left(\mathbf{e}^{0}\right)\mathbf{h}_{\alpha}\right\rangle_{Q} - \hat{P}\left(\mathbf{h}_{\alpha}, \boldsymbol{\psi}_{n}; \boldsymbol{\alpha}^{0}\right). \quad (2.112)$$

The function $\boldsymbol{\psi}_n$ is still incompletely defined, in that only the Hilbert space to which it belongs has been specified thus far. Since the left-side of Eq. (2.112) is actually a linear functional of $\mathbf{h}_u$, as is each of the coefficients $\left\langle \mathbf{R}'_u\left(\mathbf{e}^{0}\right)\mathbf{h}_u, \boldsymbol{\varphi}_n\right\rangle_R$ in the spectral expansion on the right side of Eq. (2.109), it is possible to require that these two quantities be identical, i.e.,

$$\left\langle \mathbf{N}^{+}\left(\mathbf{e}^{0}\right)\boldsymbol{\psi}_{n}, \mathbf{h}_{u}\right\rangle_{u} = \left\langle \mathbf{R}'_{u}\left(\mathbf{e}^{0}\right)\mathbf{h}_{u}, \boldsymbol{\varphi}_{n}\right\rangle_{R} = \left\langle \mathbf{M}\left(\mathbf{e}^{0}\right)\boldsymbol{\varphi}_{n}, \mathbf{h}_{u}\right\rangle_{u}. \quad (2.113)$$

In the rightmost equality in Eq. (2.113), the operator $\mathbf{M}\left(\mathbf{e}^{0}\right)$ denotes the adjoint of the partial Gateaux-derivative $\mathbf{R}'_u\left(\mathbf{e}^{0}\right)$, including any boundary terms that might arise when defining it. The Riesz representation theorem assures that the three inner products shown in Eq. (2.113) represent uniquely the same linear functional of $\mathbf{h}_u(\mathbf{x})$. The first and third inner products in Eq. (2.113) also complete the definition of the adjoint function $\boldsymbol{\psi}_n(\mathbf{x})$ since they imply that $\boldsymbol{\psi}_n(\mathbf{x})$ is the unique solution of the following adjoint sensitivity system

$$\mathbf{N}^{+}\left(\mathbf{e}^{0}\right)\boldsymbol{\psi}_{n}(\mathbf{x}) = \mathbf{M}\left(\mathbf{e}^{0}\right)\boldsymbol{\varphi}_{n}(\mathbf{x}), \; n \in N, \quad (2.114)$$

subject to the adjoint boundary conditions determined in Eq. (2.111). Equations (2.112) and (2.113) can now be used in Eq. (2.107) to obtain the following expression for the total sensitivity $D\mathbf{R}\left(\mathbf{e}^{0};\mathbf{h}\right)$ of $\mathbf{R}(\mathbf{e})$ at $\mathbf{e}^{0}$:

$$\begin{aligned} D\mathbf{R}\left(\mathbf{e}^{0};\mathbf{h}\right) &= \mathbf{R}'_{\alpha}\left(\mathbf{e}^{0}\right)\mathbf{h}_{\alpha} + \sum_{n\in N}\left\{\left\langle \boldsymbol{\psi}_{n}, \mathbf{Q}'_{\alpha}\left(\mathbf{e}^{0}\right)\mathbf{h}_{\alpha} - \mathbf{N}'_{\alpha}\left(\mathbf{e}^{0}\right)\mathbf{h}_{\alpha}\right\rangle_{Q} - \hat{P}\left(\mathbf{h}_{\alpha}, \boldsymbol{\psi}_{n}; \boldsymbol{\alpha}^{0}\right)\right\}\boldsymbol{\varphi}_{n} \\ &\triangleq \sum_{i=1}^{N_{\alpha}} S_{i}\left(\mathbf{e}^{0}\right)\left(\Delta\alpha_{i}\right), \end{aligned} \quad (2.115)$$

where the quantities $S_i\left(\mathbf{e}^{0}\right)$ denote the absolute sensitivities of the response $\mathbf{R}(\mathbf{e})$ at $\mathbf{e} = \mathbf{e}^{0}$; these quantities coincide with the corresponding ordinary partial derivatives when all quantities involved in their computation are continuous. As Eq. (2.115) indicates, the desired elimination of all unknown values of $\mathbf{h}_u$ from the expressions of the sensitivities $S_i(\mathbf{e}^0)$, $i=1,\ldots,N_\alpha$, of $\mathbf{R}(\mathbf{e}^0)$ at $\mathbf{e}^0$ has been accomplished. The sensitivities $S_i(\mathbf{e}^0)$ can therefore be computed by means of Eq. (2.115), after solving the adjoint sensitivity system, consisting of Eqs. (2.111) and (2.114), to obtain the adjoint function $\boldsymbol{\psi}_n$ for each $n\in N$. Thus, the adjoint system must be solved anew, with a different source on the right-side of Eq. (2.114), for each $n\in N$. From the foregoing considerations, it is evident that the orthonormal basis $\{\varphi_n(\mathbf{x})\}$, $n\in N$, must be chosen such as to minimize an a priori user-selected error criterion, to ensure that the spectral expansion in Eq. (2.106) represents the known nominal value of $\mathbf{R}(\mathbf{e}^0)$ within the selected error criterion with a minimal number, N, of terms in the expansion. The selection of this error criterion and of the basis $\{\varphi_n\}$, $n\in N$, are problem-dependent issues but the procedures and considerations for performing this selection (e.g., using classical Fourier expansion, orthogonal and/or chaos polynomials, wavelets, collocation, pseudo-spectral methods, etc.) are well-known.

The methodology leading to Eq. (2.115) will be illustrated next by applying it to compute the sensitivities of the time-dependent nitric acid concentration, $\rho_a^{(1)}(t)$, in compartment #1. For this purpose, the Legendre polynomials will be chosen to serve as the orthonormal basis for the spectral representation of $\rho_a^{(1)}(t)$. Recall that the Nth-order spectral expansion, $f_N(x)$, of a function $f(x)$, $x\in[-1,1]$, using Legendre polynomials, $P_n(x)$, is defined as

$$f_N(x)=\sum_{n=0}^{N}a_nP_n(x), \tag{2.116}$$

where $P_n(x)$ denotes the Legendre polynomial of order n, and where the coefficients a_n are defined as

$$a_n\triangleq\frac{2n+1}{2}\int_{-1}^{1}f(x)P_n(x)dx,\quad n=0,1,\ldots,N. \tag{2.117}$$

The Legendre polynomials satisfy several well-known recursion relationships; the relationship below can be conveniently used for the numerical computations, to avoid undue accumulation and magnification of round-off errors:

$$P_{n+1}(x)=2xP_n(x)-P_{n-1}(x)-\frac{xP_n(x)-P_{n-1}(x)}{n+1}. \tag{2.118}$$

Recall also that the Legendre polynomials satisfy the orthogonality relation

$$\int_{-1}^{1} P_n(x)P_n(x)\,dx = \frac{2}{2n+1}\delta_{mn} \tag{2.119}$$

where δ_{mn} represents the Kronecker delta functional, defined as $\delta_{mn}=1,\ m=n$ and $\delta_{mn}=0,\ m\neq n$.

Since the time-dependent response of interest, namely the nitric acid concentration in the first compartment, $\rho_a^{(1)}(t)$, is defined over the time interval $t\in[0,t_f]$, it follows that the interval $x\in[-1,1]$ must be shifted to the interval $t\in[0,t_f]$ in order to obtain the corresponding spectral expansion for $\rho_a^{(1)}(t)$. The correspondence between the independent variables $t\in[0,t_f]$ and $x\in[-1,1]$ is provided by the relationships

$$t=(x+1)t_f/2, \quad x=(2t/t_f-1). \tag{2.120}$$

Denoting the Nth-order spectral expansion $\rho_a^{(1)}(t)$ by $\rho_{a,S}^{(1)}(t)$, where the subscript "S" indicates "spectral," it follows from Eqs. (2.116), (2.117) and (2.120) that

$$\rho_{a,S}^{(1)}(t)=\sum_{n=0}^{N} a_n P_n(2t/t_f-1), \quad 0<t\leq t_f, \tag{2.121}$$

with

$$a_n \triangleq \frac{2n+1}{t_f}\int_0^{t_f} \rho_a^{(1)}(t)\,P_n(2t/t_f-1)\,dt, \quad n=0,1,\ldots,N. \tag{2.122}$$

For the shifted Legendre polynomials $P_n(2t/t_f-1)$, the "orthogonality relation" expressed by Eq. (2.119) takes on the following form:

$$\int_0^{t_f} P_n(2t/t_f-1)P_n(2t/t_f-1)\,dt=\frac{\delta_{mn}}{2n+1}. \tag{2.123}$$

Using the shifted polynomials $P_n(2t/t_f-1)$, the nitric acid concentration $\rho_a^{(1)}(t)$ is approximated by $\rho_{a,S}^{(1)}(t)$ within a maximum global relative error of less than 0.01%, by using $N=17$ in the spectral expansion in Eq. (2.123). This excellent approximation is depicted in Figure 2.7, and it ensures similarly accurate computations of the response sensitivities, as will be shown in the sequel, after establishing the corresponding adjoint sensitivity system. Similarly, the time-dependent acid concentration in compartment #4 can be represented, within a maximum global relative error of less than 0.01%, using 21 Legendre polynomials, while the time-dependent acid concentration in compartment #7 can be represented, within a maximum global relative error of less than 0.01%, using 29 Legendre polynomials. These representations are shown in Figures 2.8 and 2.9, respectively.

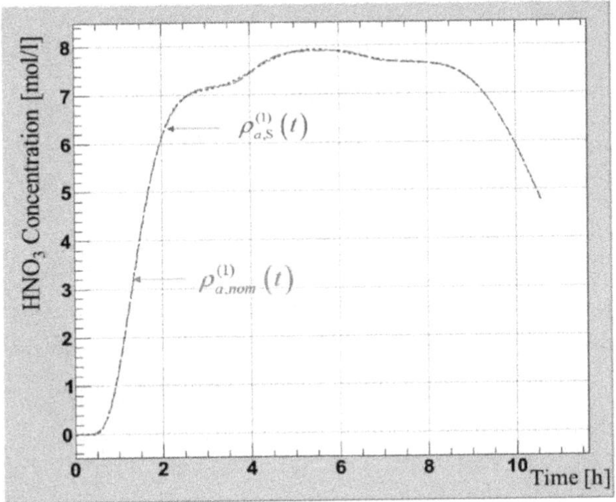

Figure 2.7: ***Time-dependent behavior of the exact nominal value of the nitric acid concentration in compartment #1,*** $\rho_{a,nom}^{(1)}(t)$***, and its spectral representation,*** $\rho_{a,S}^{(1)}(t)$***, using the first 17 Legendre polynomials*** $(N=16)$.

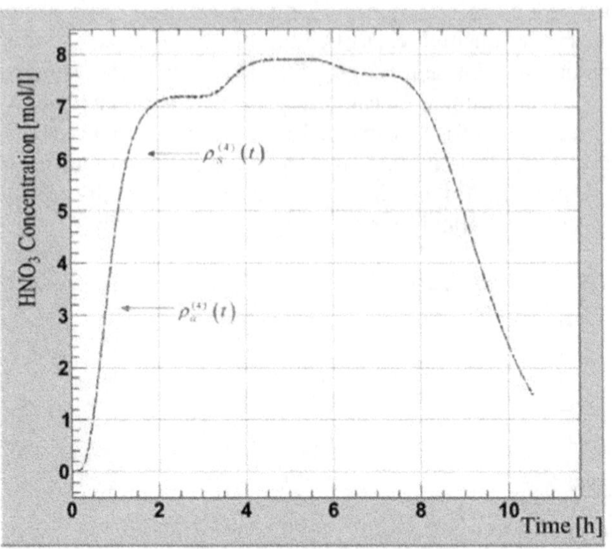

***Figure 2.8:** Time-dependent behavior of the exact nominal value of the nitric acid concentration in compartment #4, $\rho_{a,nom}^{(4)}(t)$, and its spectral representation, $\rho_{a,S}^{(4)}(t)$, using the first 21 Legendre polynomials $(N=20)$.*

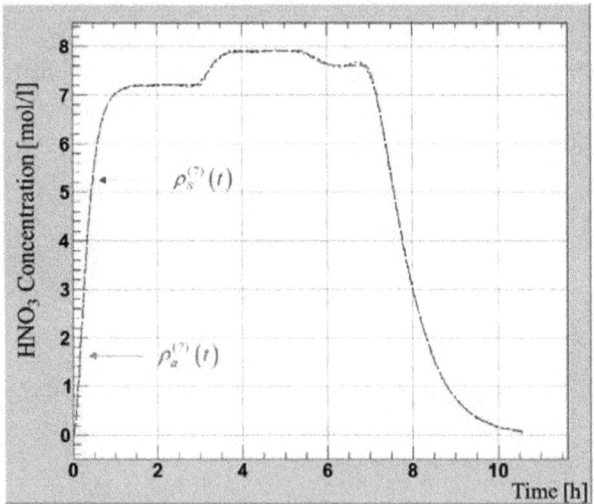

***Figure 2.9:** Time-dependent behavior of the exact nominal value of the nitric acid concentration in compartment #7, $\rho_{a,nom}^{(7)}(t)$, and its spectral representation, $\rho_{a,S}^{(7)}(t)$, using the first 29 Legendre polynomials $(N=28)$.*

For the particular case of the response $\rho_{a,S}^{(1)}(t)$ defined in Eq. (2.121), Eq. (2.115) takes on the following particular form:

$$D\rho_{a,S}^{(1)}(t)=\sum_{n=0}^{N}\left[\frac{2n+1}{t_f}\int_0^{t_f}\Delta\rho_{a,S}^{(1)}(t)\,P_n\left(2t/t_f-1\right)dt\right]P_n\left(2t/t_f-1\right)$$
$$=\sum_{n=0}^{N}\left\langle \Delta\rho_{a,S}^{(1)}(t),\frac{2n+1}{t_f}P_n\left(2t/t_f-1\right)\right\rangle_R P_n\left(2t/t_f-1\right). \quad (2.124)$$

Comparing the right-side of Eq. (2.124) with the right-most side of Eq. (2.113) and keeping in mind Eq. (2.114), it follows that the sources for the adjoint system are $\left[(2n+1)/t_f\right]P_n\left(2t/t_f-1\right)$ for the equation involving the adjoint function $\psi_{\rho,n}^{(1)}(t)$, $n=0,1,\ldots,N=17$; and zero for the other equations. Consequently, the corresponding adjoint sensitivity system becomes

- For each $n=0,1,\ldots,N=16$, solve:

$$V_{nom}^{(1)}\left(t_f-\tau\right)\frac{d\psi_{\rho,n}^{(1)}(\tau)}{d\tau}+C\left[V_{nom}^{(1)}\left(t_f-\tau\right)\right]\psi_{\rho,n}^{(1)}(\tau)$$
$$=\frac{2n+1}{t_f}P_n\left(1-2\tau/t_f\right),\quad 0<\tau<t_f, \quad (2.125)$$

$$V_{nom}^{(k)}\left(t_f-\tau\right)\frac{d\psi_{\rho,n}^{(k)}(\tau)}{d\tau}+C\left[V_{nom}^{(k)}\left(t_f-\tau\right)\right]\left[\psi_{\rho,n}^{(k)}(\tau)-\psi_{\rho,n}^{(k-1)}(\tau)\right]=0,$$
$$k=2,\ldots,7;\ 0<\tau<t_f, \quad (2.126)$$

$$\frac{d\psi_{V,n}^{(1)}(\tau)}{d\tau}+D_V^{(1)}\left(t_f-\tau\right)\psi_{V,n}^{(1)}(\tau)+D_\rho^{(1)}\left(t_f-\tau\right)\psi_{\rho,n}^{(1)}(\tau)=0,\ 0<\tau<t_f, \quad (2.127)$$

$$\frac{d\psi_{V,n}^{(k)}(\tau)}{d\tau}+D_V^{(k)}\left(t_f-\tau\right)\left[\psi_{V,n}^{(k)}(\tau)-\psi_{V,n}^{(k-1)}(\tau)\right]+D_\rho^{(k)}\left(t_f-\tau\right)\psi_{\rho,n}^{(k)}(\tau)$$
$$+\left[\rho_{a,nom}^{(k-1)}\left(t_f-\tau\right)-\rho_{a,nom}^{(k)}\left(t_f-\tau\right)\right]D_V^{(k)}\left(t_f-\tau\right)\psi_{\rho,n}^{(k-1)}(\tau)=0,$$
$$k=2,\ldots,8,\ 0<\tau<t_f, \quad (2.128)$$

subject to the initial conditions

$$\psi_{\rho,n}^{(k)}(\tau=0)=0,\ \ \psi_{V,n}^{(k)}(\tau=0)=0,\ \ k=1,\ldots,8. \quad (2.129)$$

Furthermore, it follows that the general expression of the response sensitivities, cf. Eq. (2.115), takes on the following particular form for computing the sensitivities $D\rho_{a,S}^{(1)}(t)$:

$$D\rho_{a,S}^{(1)}(t) = \sum_{n=0}^{N=16} Da_n(\mathbf{h}_\alpha, \boldsymbol{\psi}; \boldsymbol{\alpha}^0) P_n(2t/t_f - 1), \tag{2.130}$$

where

$$\begin{aligned} Da_n(\mathbf{h}_\alpha, \boldsymbol{\psi}; \boldsymbol{\alpha}^0) &= \sum_{k=1}^{8} \int_0^{t_f} \psi_{\rho,n}^{(k)}(t) q_\rho^{(k)}(t)\, dt + \sum_{k=1}^{8} \int_0^{t_f} \psi_{V,n}^{(k)}(t) q_V^{(k)}(t)\, dt \\ &+ \sum_{k=1}^{8} \left\{ \psi_{\rho,n}^{(k)}(0) V_{nom}^{(k)}(0) \left[\Delta\rho_{a0}^{(k)}\right] + \psi_{V,n}^{(k)}(0) \left[\Delta V_0^{(k)}\right] \right\}, \quad n = 0,1,\ldots,N = 16. \end{aligned} \tag{2.131}$$

The sensitivities $D\rho_{a,S}^{(4)}(t)$ and $D\rho_{a,S}^{(7)}(t)$ are computed by using a similar procedure as outlined above for computing the sensitivities $D\rho_{a,S}^{(1)}(t)$. Thus, the adjoint sensitivity system for computing the sensitivities of $\rho_{a,S}^{(4)}(t)$ is obtained as follows:

- For each $n = 0,1,\ldots,N = 20$, solve:

$$V_{nom}^{(1)}(t_f - \tau) \frac{d\psi_{\rho,n}^{(1)}(\tau)}{d\tau} + C\left[V_{nom}^{(1)}(t_f - \tau)\right] \psi_{\rho,n}^{(1)}(\tau) = 0, \; 0 < \tau < t_f, \tag{2.132}$$

$$\begin{aligned} &V_{nom}^{(k)}(t_f - \tau) \frac{d\psi_{\rho,n}^{(k)}(\tau)}{d\tau} + C\left[V_{nom}^{(k)}(t_f - \tau)\right]\left[\psi_{\rho,n}^{(k)}(\tau) - \psi_{\rho,n}^{(k-1)}(\tau)\right] = 0, \\ &\quad k = 2,3,5,6,7; \; 0 < \tau < t_f, \end{aligned} \tag{2.133}$$

$$\begin{aligned} &V_{nom}^{(k)}(t_f - \tau) \frac{d\psi_{\rho,n}^{(k)}(\tau)}{d\tau} + C\left[V_{nom}^{(k)}(t_f - \tau)\right]\left[\psi_{\rho,n}^{(k)}(\tau) - \psi_{\rho,n}^{(k-1)}(\tau)\right] \\ &\quad = \frac{2n+1}{t_f} P_n(1 - 2\tau/t_f), \; k = 4; \; 0 < \tau < t_f, \end{aligned} \tag{2.134}$$

$$\frac{d\psi_{V,n}^{(1)}(\tau)}{d\tau} + D_V^{(1)}(t_f - \tau)\psi_{V,n}^{(1)}(\tau) + D_\rho^{(1)}(t_f - \tau)\psi_{\rho,n}^{(1)}(\tau) = 0, \; 0 < \tau < t_f, \tag{2.135}$$

$$\frac{d\psi_{V,n}^{(k)}(\tau)}{d\tau} + D_V^{(k)}(t_f - \tau)\left[\psi_{V,n}^{(k)}(\tau) - \psi_{V,n}^{(k-1)}(\tau)\right]$$
$$+ D_\rho^{(k)}(t_f - \tau)\psi_{\rho,n}^{(k)}(\tau) \tag{2.136}$$
$$+\left[\rho_{a,nom}^{(k-1)}(t_f - \tau) - \rho_{a,nom}^{(k)}(t_f - \tau)\right]D_V^{(k)}(t_f - \tau)\psi_{\rho,n}^{(k-1)}(\tau) = 0,$$
$$k = 2,\ldots,8,\ 0 < \tau < t_f,$$

subject to the initial conditions

$$\psi_{\rho,n}^{(k)}(\tau = 0) = 0,\ \ \psi_{V,n}^{(k)}(\tau = 0) = 0,\ \ k = 1,\ldots,8. \tag{2.137}$$

The expression obtained for the total time-dependent sensitivity of $\rho_{a,S}^{(4)}(t)$ to model parameters is as follows:

$$D\rho_{a,S}^{(4)}(t) = \sum_{n=0}^{N=20} Da_n\left(\mathbf{h}_\alpha, \boldsymbol{\psi}; \boldsymbol{\alpha}^0\right)P_n\left(2t/t_f - 1\right), \tag{2.138}$$

where the expression of $Da_n\left(\mathbf{h}_\alpha, \boldsymbol{\psi}; \boldsymbol{\alpha}^0\right)$ remains formally the same as shown in Eq. (2.131), except that the adjoint functions $\psi_{\rho,n}^{(k)}(t)$ and $\psi_{V,n}^{(k)}(t)$ are now the solutions of Eqs. (2.132) through (2.137).
The expression of the total time-dependent sensitivity of $\rho_{a,S}^{(7)}(t)$ is computed similarly, except that the corresponding summation formula becomes

$$D\rho_{a,S}^{(7)}(t) = \sum_{n=0}^{N=28} Da_n\left(\mathbf{h}_\alpha, \boldsymbol{\psi}; \boldsymbol{\alpha}^0\right)P_n\left(2t/t_f - 1\right), \tag{2.139}$$

where the expression of $Da_n\left(\mathbf{h}_\alpha, \boldsymbol{\psi}; \boldsymbol{\alpha}^0\right)$ remains formally the same as shown in Eq. (2.131), except that the adjoint functions $\psi_{\rho,n}^{(k)}(t)$ and $\psi_{V,n}^{(k)}(t)$ are now the solutions of the adjoint sensitivity system given below:

➢ For each $n = 0, 1, \ldots, N = 28$, solve:

$$V_{nom}^{(1)}(t_f - \tau)\frac{d\psi_{\rho,n}^{(1)}(\tau)}{d\tau} + C\left[V_{nom}^{(1)}(t_f - \tau)\right]\psi_{\rho,n}^{(1)}(\tau) = 0,\ 0 < \tau < t_f, \tag{2.140}$$

$$V_{nom}^{(k)}\left(t_f-\tau\right)\frac{d\psi_{\rho,n}^{(k)}(\tau)}{d\tau}+C\left[V_{nom}^{(k)}\left(t_f-\tau\right)\right]\left[\psi_{\rho,n}^{(k)}(\tau)-\psi_{\rho,n}^{(k-1)}(\tau)\right]=0,\quad k=2,3,4,5,6;\ 0<\tau<t_f, \tag{2.141}$$

$$V_{nom}^{(k)}\left(t_f-\tau\right)\frac{d\psi_{\rho,n}^{(k)}(\tau)}{d\tau}+C\left[V_{nom}^{(k)}\left(t_f-\tau\right)\right]\left[\psi_{\rho,n}^{(k)}(\tau)-\psi_{\rho,n}^{(k-1)}(\tau)\right]=\frac{2n+1}{t_f}P_n\left(1-2\tau/t_f\right),\ k=7;\ 0<\tau<t_f, \tag{2.142}$$

$$\frac{d\psi_{V,n}^{(1)}(\tau)}{d\tau}+D_V^{(1)}\left(t_f-\tau\right)\psi_{V,n}^{(1)}(\tau)+D_\rho^{(1)}\left(t_f-\tau\right)\psi_{\rho,n}^{(1)}(\tau)=0,\ 0<\tau<t_f, \tag{2.143}$$

$$\begin{aligned}&\frac{d\psi_{V,n}^{(k)}(\tau)}{d\tau}+D_V^{(k)}\left(t_f-\tau\right)\left[\psi_{V,n}^{(k)}(\tau)-\psi_{V,n}^{(k-1)}(\tau)\right]\\&+D_\rho^{(k)}\left(t_f-\tau\right)\psi_{\rho,n}^{(k)}(\tau)\\&+\left[\rho_{a,nom}^{(k-1)}\left(t_f-\tau\right)-\rho_{a,nom}^{(k)}\left(t_f-\tau\right)\right]D_V^{(k)}\left(t_f-\tau\right)\psi_{\rho,n}^{(k-1)}(\tau)=0,\\&k=2,\ldots,8,\ 0<\tau<t_f,\end{aligned} \tag{2.144}$$

subject to the initial conditions

$$\psi_{\rho,n}^{(k)}(\tau=0)=0,\ \psi_{V,n}^{(k)}(\tau=0)=0,\ k=1,\ldots,8. \tag{2.145}$$

The absolute and relative sensitivities of $\rho_{a,nom}^{(1)}(t_i)$, $\rho_{a,nom}^{(4)}(t_i)$, and $\rho_{a,nom}^{(7)}(t_i)$ to the scalar parameter a are presented in Figures 2.10 and 2.11, respectively. To enable the comparison of the respective relative sensitivities, the absolute sensitivities were normalized, arbitrarily but consistently, to be the acid concentration after 60 minutes into the transient. The main features of these sensitivities are as follows:

- The sensitivities of the responses to each of these parameters is localized in time.
- All of these sensitivities are negative, meaning that an increase in the magnitude of the parameter a will induce a decrease in the magnitude of the respective response.
- After reaching a minimum (or maximum in absolute value), all of these sensitivities decay quickly to zero, for the remaining duration of the transient.
- The earliest (in time) impact of the respective sensitivity is on compartment #7, closest to the inlet; the impact of the sensitivity/disturbance propagates, in time, towards the last compartment.
- The largest impact of the each of these sensitivities is on the compartment #1, which is located furthest from the inlet.

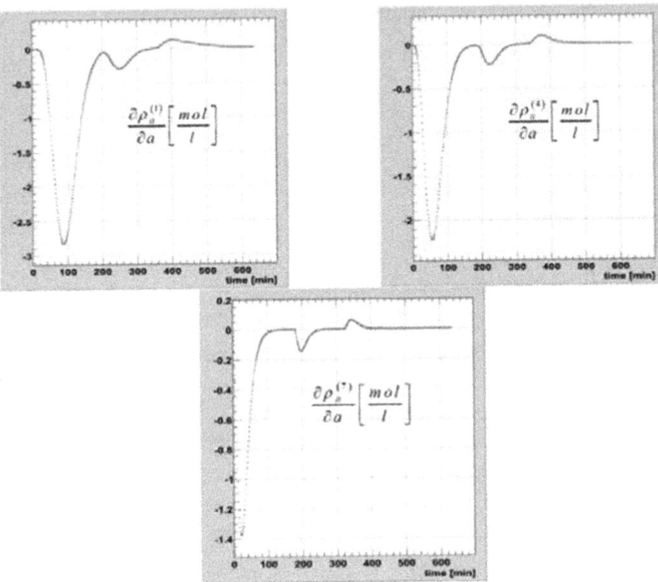

Figure 2.10: Absolute sensitivities of $\rho^{(1)}_{a,nom}(t_i)$, $\rho^{(4)}_{a,nom}(t_i)$, *and* $\rho^{(7)}_{a,nom}(t_i)$ *to the scalar parameter* a.

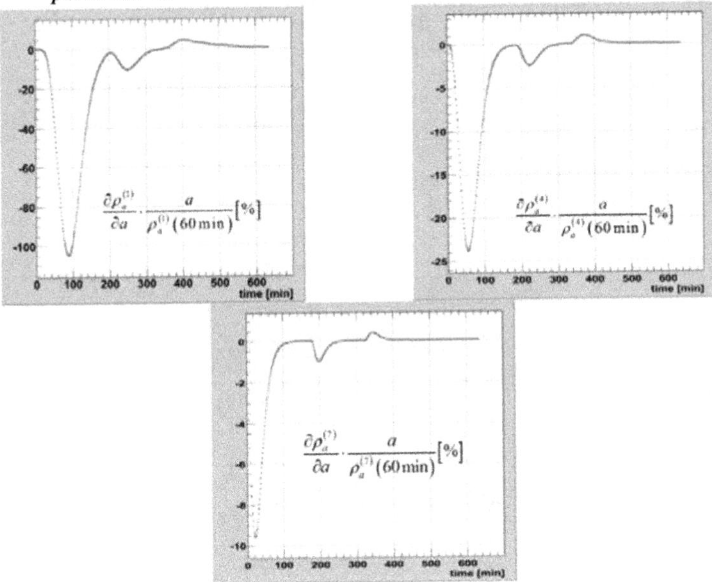

Figure 2.11: Relative sensitivities of $\rho^{(1)}_{a,nom}(t_i)$, $\rho^{(4)}_{a,nom}(t_i)$, *and* $\rho^{(7)}_{a,nom}(t_i)$ *to the scalar parameter* a.

The absolute and relative sensitivities of $\rho_{a,nom}^{(1)}(t_i)$, $\rho_{a,nom}^{(4)}(t_i)$, and $\rho_{a,nom}^{(7)}(t_i)$ to the scalar parameters V_0, b, and G are presented in Figures 2.12 through 2.17. The response sensitivities to these parameters show similar features, as follows:

- In time, the sensitivities of the responses to each of these parameters undergo first a minimum having large negative values, then display a plateau around zero, followed by a rise to a maximum; this maximum decays towards zero in the first compartment, but does not "have sufficient time" to do the same in the last compartment, where the respective sensitivity attains a relatively high positive value at the end of the transient.
- The earliest (in time) impact of these sensitivities is on compartment #7, which is located closest to the inlet; the impact of the sensitivity/disturbance propagates, in time, towards the last compartment.
- The largest impact, in absolute value, of the each of these sensitivities is on compartment #1, which is located furthest from the inlet.

To facilitate the comparison of these relative sensitivities, the respective normalizations were chosen, arbitrarily but consistently, to be the acid concentration after 60 minutes into the transient. Comparing the various relative sensitivities reveals that their magnitudes are largest for the parameters V_0 and b, and smaller for G, indicating that the impact of comparable uncertainties in V_0 and b will have a higher impact on the response uncertainties than those for G.

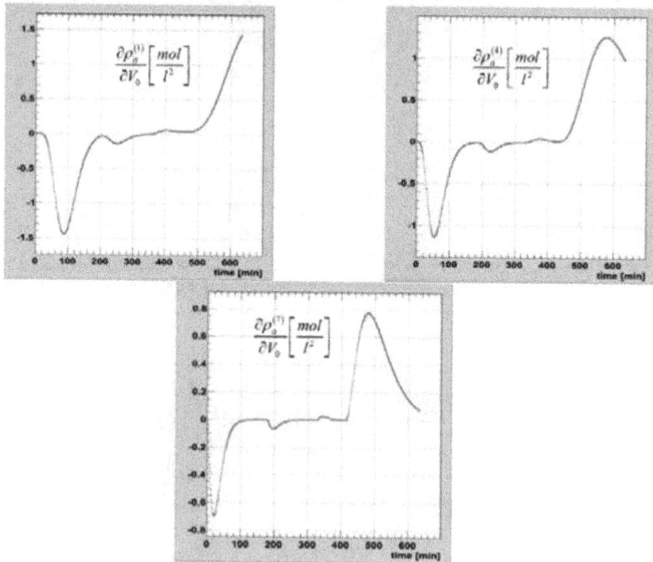

Figure 2.12: Absolute sensitivities of $\rho_{a,nom}^{(1)}(t_i)$, $\rho_{a,nom}^{(4)}(t_i)$, and $\rho_{a,nom}^{(7)}(t_i)$ to the scalar parameter V_0.

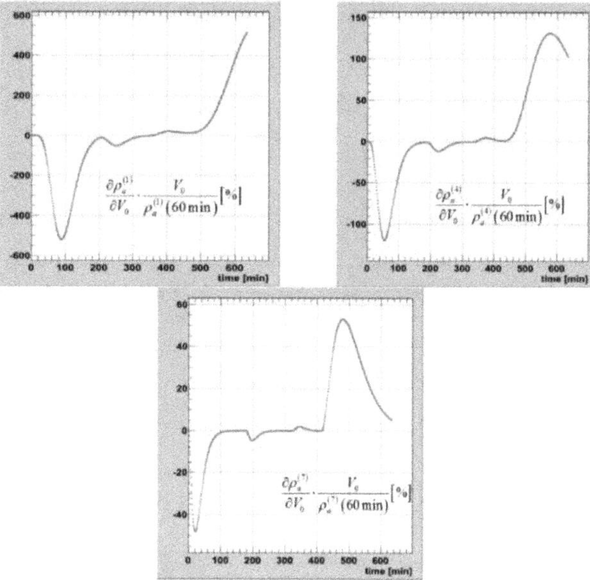

Figure 2.13: Relative sensitivities of $\rho_{a,nom}^{(1)}(t_i)$, $\rho_{a,nom}^{(4)}(t_i)$, ***and*** $\rho_{a,nom}^{(7)}(t_i)$ ***to the scalar parameter*** V_0.

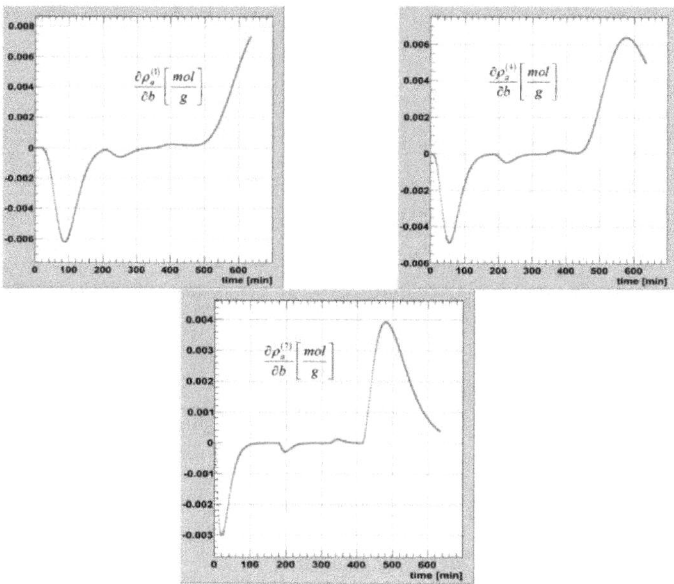

Figure 2.14: Absolute sensitivities of $\rho_{a,nom}^{(1)}(t_i)$, $\rho_{a,nom}^{(4)}(t_i)$, ***and*** $\rho_{a,nom}^{(7)}(t_i)$ ***to the scalar parameter*** b.

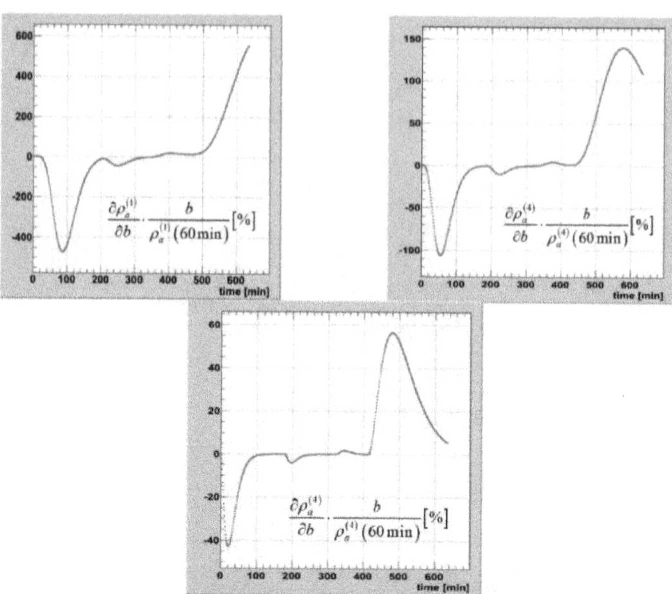

Figure 2.15: Relative sensitivities of $\rho^{(1)}_{a,nom}(t_i)$, $\rho^{(4)}_{a,nom}(t_i)$, ***and*** $\rho^{(7)}_{a,nom}(t_i)$ ***to the scalar parameter*** b.

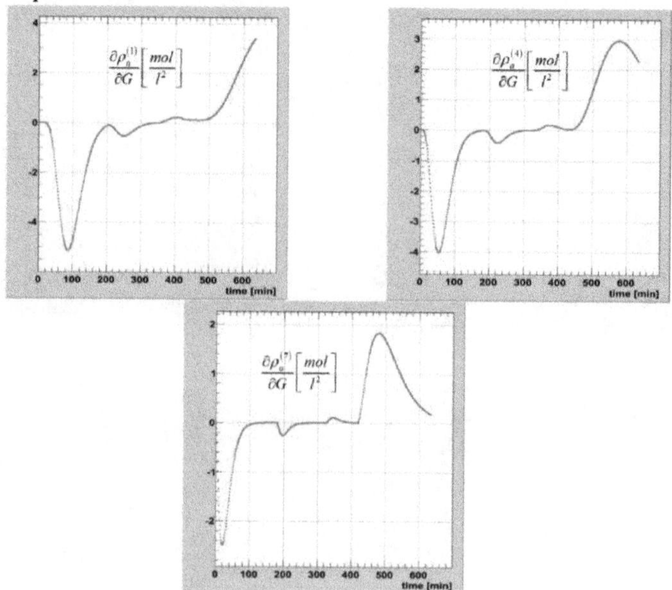

Figure 2.16: Absolute sensitivities of $\rho^{(1)}_{a,nom}(t_i)$, $\rho^{(4)}_{a,nom}(t_i)$, ***and*** $\rho^{(7)}_{a,nom}(t_i)$ ***to the scalar parameter*** G.

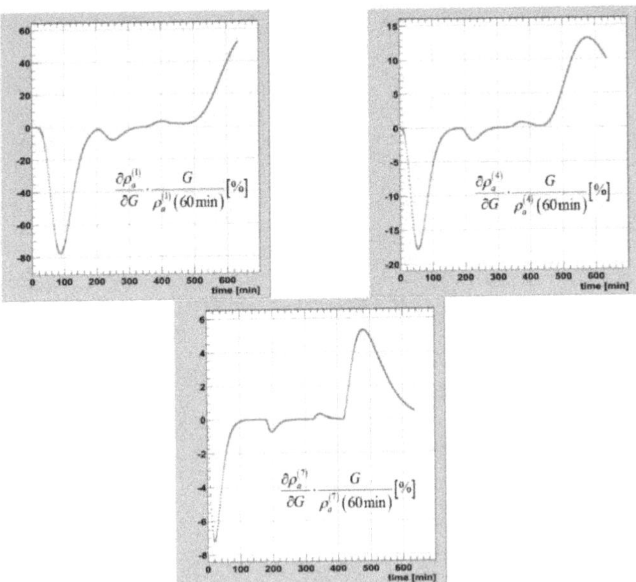

Figure 2.17: Relative sensitivities of $\rho_{a,nom}^{(1)}(t_i)$, $\rho_{a,nom}^{(4)}(t_i)$, ***and*** $\rho_{a,nom}^{(7)}(t_i)$ ***to the scalar parameter*** G.

The sensitivities of $\rho_{a,nom}^{(1)}(t_i)$, $\rho_{a,nom}^{(4)}(t_i)$, and $\rho_{a,nom}^{(7)}(t_i)$ to the scalar parameter p are depicted in Figures 2.18 and 2.19. Comparing Figures 2.18 and 2.19 to Figures 2.12 through 2.17 indicates that the sensitivities of $\rho_{a,nom}^{(1)}(t_i)$, $\rho_{a,nom}^{(4)}(t_i)$, and $\rho_{a,nom}^{(7)}(t_i)$ to the scalar parameter p behave in the opposite manner to the behavior of the respective response sensitivities to the parameters V_0, b and G, namely:

- In time, the sensitivities of the respective responses to parameter p first rise to a maximum having positive values, then display a plateau around zero, followed by a decay to a minimum having negative values. This minimum subsequently rises in time towards zero in the first compartment, but does not "have sufficient time" to do the same in the last compartment, where the respective sensitivity remains negative at the end of the transient.
- The earliest (in time) impact of the respective responses sensitivity to parameter p is on compartment #7, which is located closest to the inlet; the impact of the sensitivity/disturbance propagates, in time, towards the last compartment.
- The largest impact, in absolute value, of each of these response sensitivities to the parameter p occurs in compartment #1, which is located furthest from the inlet.

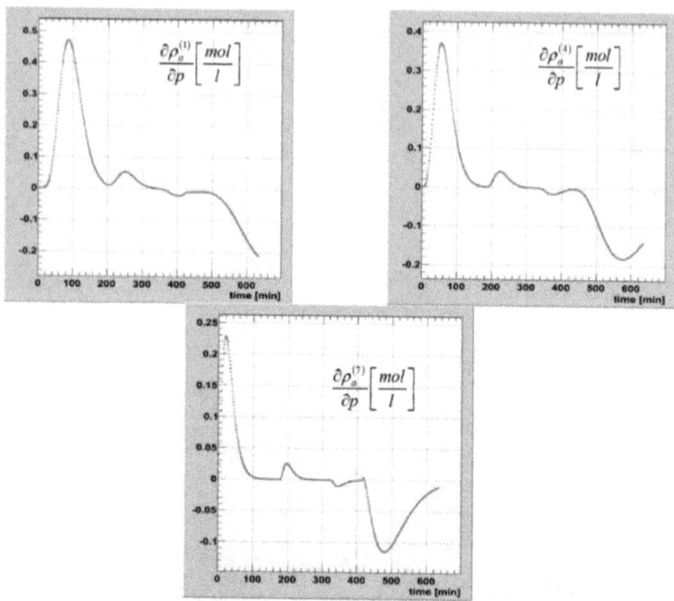

Figure 2.18: Absolute sensitivities of $\rho_{a,nom}^{(1)}(t_i)$, $\rho_{a,nom}^{(4)}(t_i)$, ***and*** $\rho_{a,nom}^{(7)}(t_i)$ ***to the scalar parameter*** p.

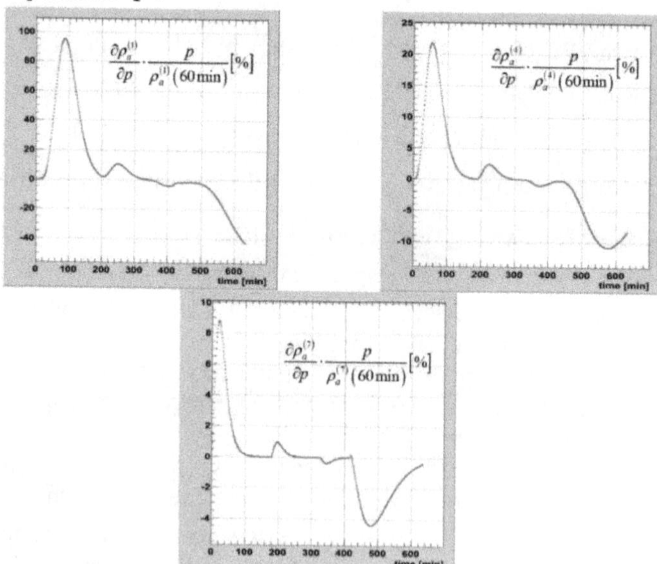

Figure 2.19: Relative sensitivities of $\rho_{a,nom}^{(1)}(t_i)$, $\rho_{a,nom}^{(4)}(t_i)$, ***and*** $\rho_{a,nom}^{(7)}(t_i)$ ***to the scalar parameter*** p.

Although *all* of the sensitivities of $\rho_{a,nom}^{(1)}(t_i)$, $\rho_{a,nom}^{(4)}(t_i)$, and $\rho_{a,nom}^{(7)}(t_i)$ to the scalar parameters $\dot{m}^{(in)}(t_i)$, $i = 1,...,I = 307$, have been computed, they are too numerous to be all displayed. The major trends of these sensitivities are displayed in Figures 2.20 through 2.25, which display the relative sensitivities of $\rho_{a,nom}^{(1)}(t_i)$, $\rho_{a,nom}^{(4)}(t_i)$, and $\rho_{a,nom}^{(7)}(t_i)$ to the scalar parameters $\dot{m}^{(in)}(t_i)$, at representative time instances.

As Figure 2.20 indicates, the relative sensitivities of all of the considered responses to a perturbation in $\dot{m}^{(in)}(t_i)$ that occurs in the early stages of the transient (e.g., at $t_i = 31$ minutes) have values greater than 1, thereby indicating that such a parameter change would induce significant changes in the respective responses. Recall that a relative sensitivity of 1 implies that a 1% change in the respective parameter would induce a 1% change in the respective response. The response furthest from the inlet, namely $\rho_{a,nom}^{(1)}(t_i)$, displays the largest relative sensitivities to the parameter $\dot{m}^{(in)}(t_i)$. The closer the compartments to the inlet, the increasingly smaller the corresponding sensitivities, with the response in the compartment closest to the inlet, namely $\rho_{a,nom}^{(7)}(t_i)$, displaying the smallest sensitivities, and hence being the least affected by uncertainties in this parameter. The compartment (#7) closest to the inlet responds first, while the compartment (#1) furthest from the inlet responds last. All of these sensitivities are positive.

As Figure 2.21 indicates, the relative sensitivities to the parameter $\dot{m}^{(in)}(t_i)$ at $t_i = 240$ minutes continue to display the same trend as above, remaining all positive while decreasing proportionally from the inlet to the outlet. All of them have become comparatively smaller than those shown in Figure 2.20 (i.e., with respect to $\dot{m}^{(in)}(t_i)$ at $t_i = 31$ minutes).

As Figure 2.3 shows, the inlet mass rate flow, $\dot{m}^{(in)}(t)/1000$, of nitric acid decreases at 325 minutes (5.42 hours) from the value of 46.83 to 41.67 $[\text{kg/h}]$. This (negative) change is reflected in Figure 2.22, which shows that the response closest to the change, namely the acid concentration $\rho_{a,nom}^{(7)}(t_i)$ in compartment #7, has already undergone a sign change in its derivative, $\partial\rho_{a,nom}^{(7)}(t_i)/\partial\dot{m}^{(in)}(t_i)$, from positive to negative. This discontinuous change causes some very minor (of the order of 10^{-5}) oscillations around zero in $\partial\rho_{a,nom}^{(7)}(t_i)/\partial\dot{m}^{(in)}(t_i)$, which propagate at later times. Since the sensitivities will be "integrated over time" when ultimately used, these small oscillations are inconsequential. The negative change in $\dot{m}^{(in)}(t)$ is seen to affect all compartments "downstream" from the inlet, up to and including compartment #4, which is just in the process of changing signs from positive (for times earlier than 325 minutes) to negative (for times later than 325 minutes), displaying the sign discontinuity precisely at 325 minutes, as depicted by the graph involving $\partial \rho_{a,nom}^{(4)}(325\,\text{min}) / \partial \dot{m}^{(in)}(325\,\text{min})$. The sensitivities of the acid concentrations in compartments #3, #2, and #1 have remained positive, thus indicating that the disturbance in the inlet mass flow rate has not yet affected the acid concentrations in these compartments (which are located further away from the disturbance). One time-step (1 minute) later, at 326 minutes, the graphs in Figure 2.23 show that the disturbance in the inlet mass flow rate, $\dot{m}^{(in)}(t)$, has just reached the furthest compartment from the inlet, namely compartment #1, where the sensitivity $\partial \rho_{a,nom}^{(1)}(t_i) / \partial \dot{m}^{(in)}(t_i)$ is just in the process of changing signs from positive (for times prior to the disturbance) to negative (for times after the occurrence of the disturbance), with the discontinuity in the derivative $\partial \rho_{a,nom}^{(1)}(t_i) / \partial \dot{m}^{(in)}(t_i)$ occurring at the time of the disturbance (325 minutes). The corresponding sensitivities in all of the other compartments are negative, reflecting the effect of the disturbance in $\dot{m}^{(in)}(t)$. Figure 2.24 shows that the relative sensitivities to the parameter $\dot{m}^{(in)}(t_i)$ at $t_i = 360$ minutes remain negative and small in absolute value. The largest (albeit small) impact of a change in $\dot{m}^{(in)}(t_i)$ is on the response furthest from the outlet, while the smallest impact is on the response in the compartment closest to the inlet. Figure 2.25 shows that at $t_i = 540$ minutes, the relative sensitivities to the parameter $\dot{m}^{(in)}(t_i)$ continue to remain negative, but their absolute values have increased in time, becoming significantly larger than those with respect to the parameter $\dot{m}^{(in)}(t_i)$ at $t_i = 360$ minutes. At all times, all of the sensitivities depicted in Figures 2.20 through 2.25 are sharply localized around the respective instance in time, and are zero otherwise.

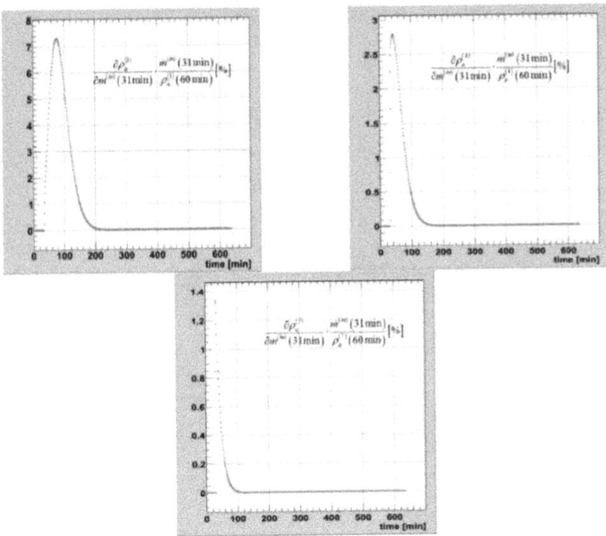

Figure 2.20: Relative sensitivities of $\rho_{a,nom}^{(1)}(t_i)$, $\rho_{a,nom}^{(4)}(t_i)$, and $\rho_{a,nom}^{(7)}(t_i)$ to the scalar parameter $\dot{m}^{(in)}(t_i)$ at $t_i = 31$ minutes.

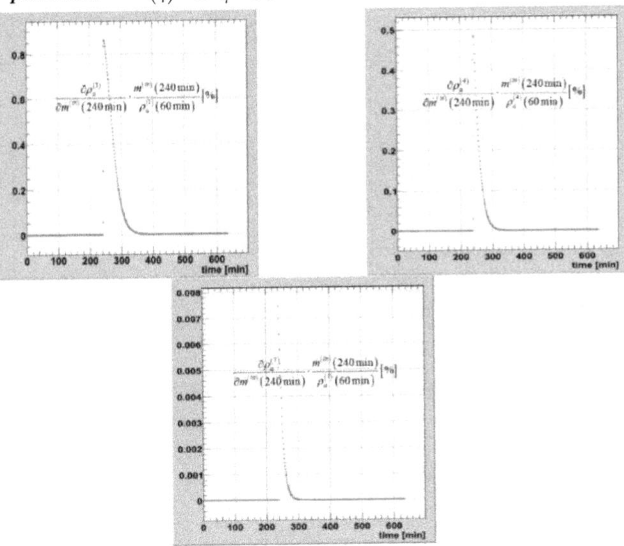

Figure 2.21: Relative sensitivities of $\rho_{a,nom}^{(1)}(t_i)$, $\rho_{a,nom}^{(4)}(t_i)$, and $\rho_{a,nom}^{(7)}(t_i)$ to the scalar parameter $\dot{m}^{(in)}(t_i)$ at $t_i = 240$ minutes.

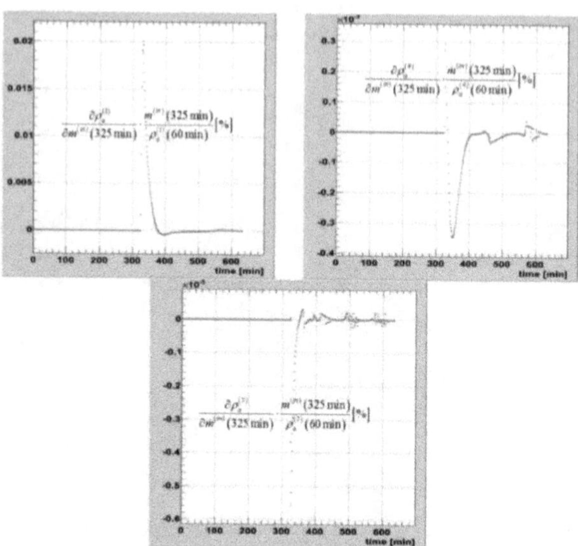

Figure 2.22: Relative sensitivities of $\rho^{(1)}_{a,nom}(t_i)$, $\rho^{(4)}_{a,nom}(t_i)$, ***and*** $\rho^{(7)}_{a,nom}(t_i)$ ***to the scalar parameter*** $\dot{m}^{(in)}(t_i)$ ***at*** $t_i = 325$ ***minutes.***

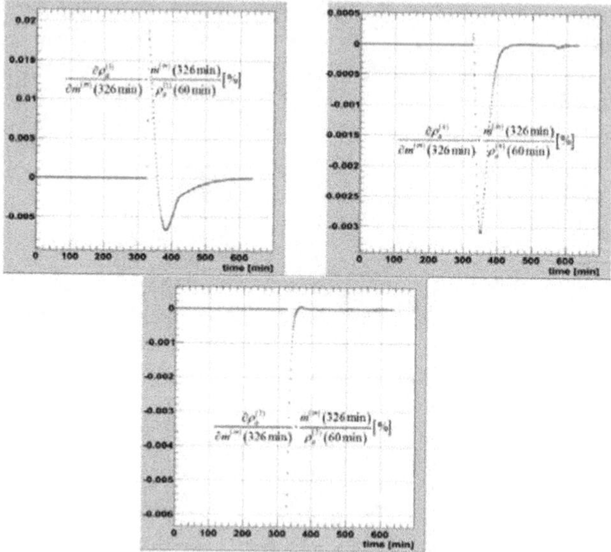

Figure 2.23: Relative sensitivities of $\rho^{(1)}_{a,nom}(t_i)$, $\rho^{(4)}_{a,nom}(t_i)$, ***and*** $\rho^{(7)}_{a,nom}(t_i)$ ***to the scalar parameter*** $\dot{m}^{(in)}(t_i)$ ***at*** $t_i = 326$ ***minutes.***

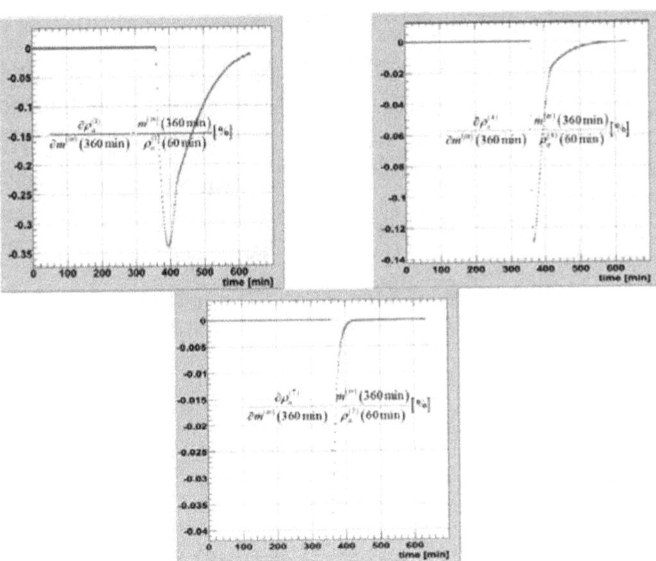

Figure 2.24: Relative sensitivities of $\rho^{(1)}_{a,nom}(t_i)$, $\rho^{(4)}_{a,nom}(t_i)$, and $\rho^{(7)}_{a,nom}(t_i)$ to the scalar parameter $\dot{m}^{(in)}(t_i)$ at $t_i = 360$ minutes.

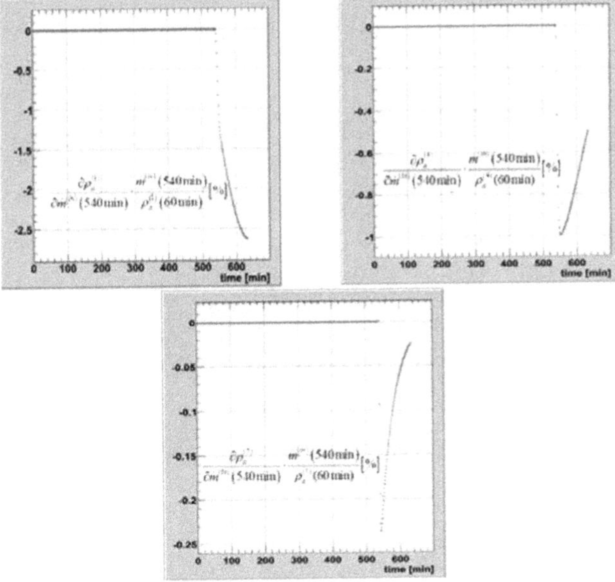

Figure 2.25: Relative sensitivities of $\rho^{(1)}_{a,nom}(t_i)$, $\rho^{(4)}_{a,nom}(t_i)$, and $\rho^{(7)}_{a,nom}(t_i)$ to the scalar parameter $\dot{m}^{(in)}(t_i)$ at $t_i = 540$ minutes.

Although all of the sensitivities of $\rho_{a,nom}^{(1)}(t_i)$, $\rho_{a,nom}^{(4)}(t_i)$, and $\rho_{a,nom}^{(7)}(t_i)$ to the scalar parameters $\rho_a^{(in)}(t_i)$, $i = 1,...,I = 307$, have also been computed for subsequent use, they are too numerous to be all displayed. The major trends are typified by the time-variation of the relative sensitivities of $\rho_{a,nom}^{(1)}(t_i)$, $\rho_{a,nom}^{(4)}(t_i)$, and $\rho_{a,nom}^{(7)}(t_i)$ at the representatively selected time instances, which are displayed in Figures 2.26 through 2.29. These figures indicate that all of the relative sensitivities to the parameter $\rho_a^{(in)}(t_i)$ are significant (i.e., larger than 1) and positive. Recall from Figure 2.4 that $\rho_a^{(in)}(t_i) = 0$ beyond 6.91 hours (415 minutes).

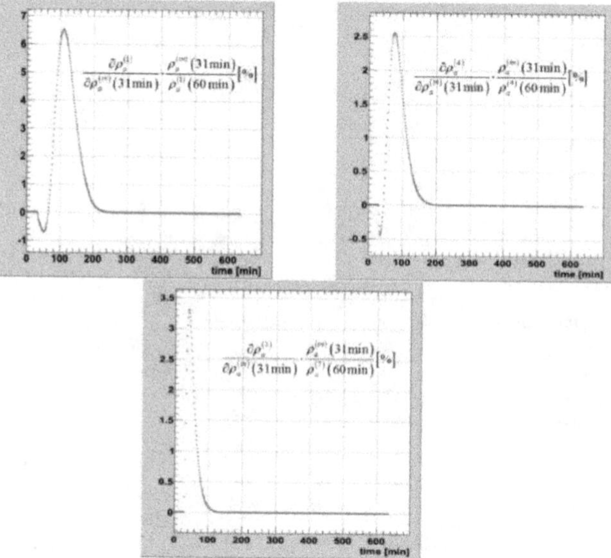

Figure 2.26: Relative sensitivities of $\rho_{a,nom}^{(1)}(t_i)$, $\rho_{a,nom}^{(4)}(t_i)$, and $\rho_{a,nom}^{(7)}(t_i)$ to the scalar parameter $\rho_a^{(in)}(t_i)$ at $t_i = 31$ minutes.

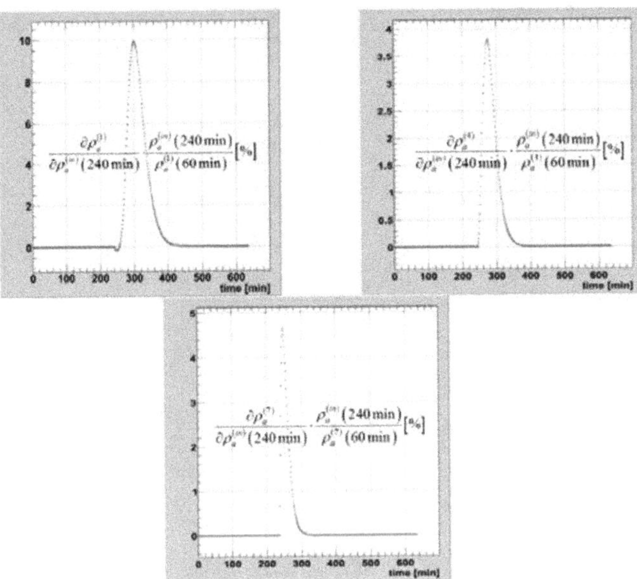

Figure 2.27: Relative sensitivities of $\rho_{a,nom}^{(1)}(t_i)$, $\rho_{a,nom}^{(4)}(t_i)$, ***and*** $\rho_{a,nom}^{(7)}(t_i)$ ***to the scalar parameter*** $\rho_{a}^{(in)}(t_i)$ ***at*** $t_i = 240$ ***minutes.***

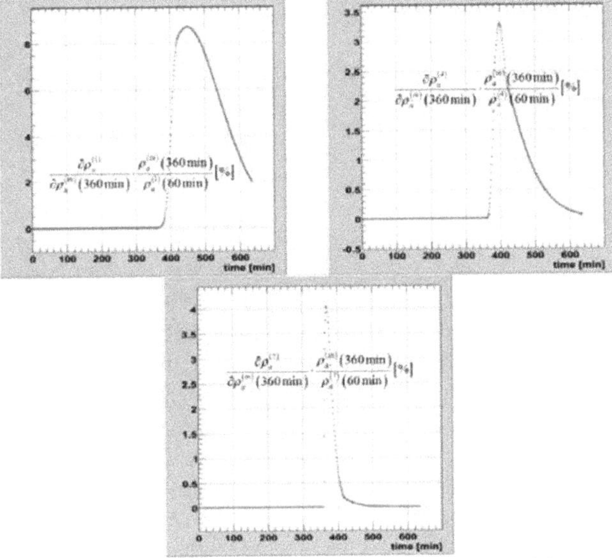

Figure 2.28: Relative sensitivities of $\rho_{a,nom}^{(1)}(t_i)$, $\rho_{a,nom}^{(4)}(t_i)$, ***and*** $\rho_{a,nom}^{(7)}(t_i)$ ***to the scalar parameter*** $\rho_{a}^{(in)}(t_i)$ ***at*** $t_i = 360$ ***minutes.***

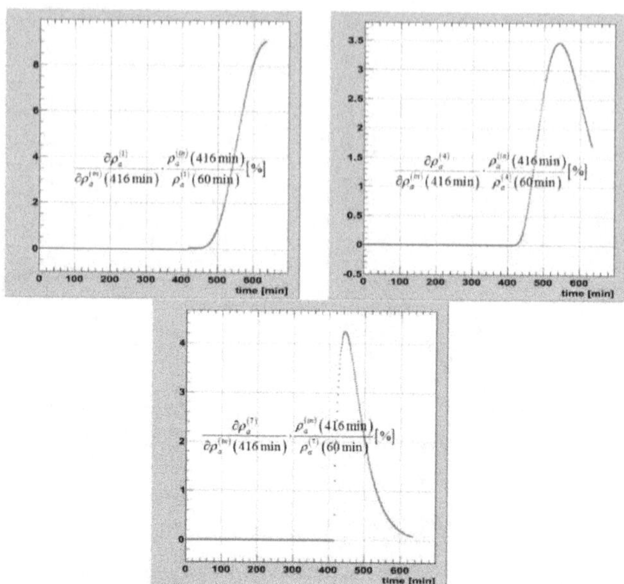

Figure 2.29: Relative sensitivities of** $\rho^{(1)}_{a,nom}(t_i)$, $\rho^{(4)}_{a,nom}(t_i)$, **and** $\rho^{(7)}_{a,nom}(t_i)$ **to the scalar parameter** $\rho^{(in)}_{a}(t_i)$ **at** $t_i = 416$ **minutes.

2.4 BERRU-SMS PREDICTIVE MODELING OF DISSOLVER START-UP

Using the adjoint sensitivity analysis methodology developed by Cacuci (1981a, 1981b, 2003), the first-order sensitivities of the nitric acid distribution within the dissolver with respect to the 635 model parameters related to the dissolver's equation of state and inflow conditions were computed in the previous Section. They will be used in this Section, in conjunction with the BERRU-SMS predictive modeling methodology presented in Chapter 1, to perform predictive modeling methodology of the dissolver model, commencing by quantifying the uncertainties in the acid concentration ("system responses") in various dissolver compartments, arising from uncertainties in the model parameters. Subsequently, the acid concentration's sensitivities will be used in this Section within the predictive modeling methodology to combine the computational results with the experimental data of Lewis and Weber (1980), which were performed solely in the compartment furthest from the inlet. The results of applying the BERRU-SMS methodology yields optimally calibrated values for all 635 model parameters, with reduced predicted uncertainties, as well as optimal ("best-estimate") predicted values for the acid concentrations, also with reduced predicted uncertainties. Notably, even though the experimental data pertains solely to the compartment furthest from the inlet (where the data was measured), the predictive modeling methodology actually improves the predictions and reduces the predicted uncertainties not only in the compartment in which the data was actually measured, but throughout the entire dissolver, including the compartment furthest from the measurements (i.e., at the inlet). This is because the BERRU-SMS methodology combines and transmits information simultaneously over the entire phase-space, comprising all time steps and spatial locations, thereby calibrating simultaneously all model parameters and responses, over all spatial locations and over the entire time interval under consideration.

Table 2.2 and Figures 2.30 through 2.33 present the results of applying Eqs. (1.70) and (1.73) of the BERRU-SMS predictive modeling methodology to obtain optimally predicted "best-estimate" nominal values along with reduced predicted variances ("uncertainties") for the scalar model parameters involved in the equation of state. For convenience, the equations describing the final BERRU-SMS results are reproduced below:

$$\left(\boldsymbol{\alpha}^{be}\right)^{\nu}=\left(\boldsymbol{\alpha}^{0}\right)^{\nu}+\sum_{\mu=1}^{N_t}\left\{\left[\mathbf{C}_{\alpha r}^{\nu\mu}-\sum_{\rho=1}^{\mu}\mathbf{C}_{\alpha}^{\nu\rho}\left(\mathbf{S}^{\dagger}\right)^{\mu\rho}\right]\left[\sum_{\eta=1}^{N_t}\mathbf{K}_{d}^{\mu\eta}\mathbf{d}^{\eta}\right]\right\},\quad \nu\in J_t \tag{2.146}$$

$$\left(\mathbf{C}_{\alpha}^{be}\right)^{\nu\mu}=\mathbf{C}_{\alpha}^{\nu\mu}-\sum_{\eta=1}^{N_t}\sum_{\rho=1}^{N_t}\left[\mathbf{C}_{\alpha r}^{\nu\rho}-\sum_{\pi=1}^{\rho}\mathbf{C}_{\alpha}^{\nu\pi}\left(\mathbf{S}^{\dagger}\right)^{\rho\pi}\right]\mathbf{K}_{d}^{\rho\eta}\left[\mathbf{C}_{r\alpha}^{\eta\mu}-\sum_{\pi=1}^{\eta}\mathbf{S}^{\eta\pi}\mathbf{C}_{\alpha}^{\pi\mu}\right]. \tag{2.147}$$

As Table 2.2 indicates, the uncertainties for these parameters are reduced from initially 10% to values as low as 4.5%. The uncertainty reduction is proportional to the sensitivity of the responses (i.e., acid concentrations) to the respective parameters. The predicted optimal values were also calibrated accordingly, as shown in Table 2.2, differing from their original nominal values.

Table 2.2: Initial and Predicted Nominal Values and Standard Deviations for the Scalar Model Parameters

Scalar Parameters	Nom. Values	Predicted Values	Nom. Rel. Std. Dev.	Pred. Rel. Std. Dev.
#1: a	0.48916	0.50621	10%	7.67834 %
#2: b	$1001.2[g/\ell]$	$948.7\ [g/\ell]$	10%	4.54535%
#3: V_0	$4.8[\ell]$	$5.123\,[\ell]$	10%	4.97098%
#4: G	$0.20194\,[\ell]$	$0.20591\,[\ell]$	10%	9.82085%
#5: p	2.7	2.61256	10%	9.44417%

Figure 2.30 displays the initial correlation matrix for the initially uncorrelated scalar parameters listed in Table 2.2, all having relative standard deviations of 10%. The numbers on the vertical axis are in units of $(\%)^2$, so the numbers shown are to be multiplied by 10^{-4}, while the numbers on the two horizontal axes correspond to the numbering of the parameters indicated in Table 2.2 The results after having applied the BERRU-SMS methodology are displayed in Figure 2.31, which depicts the predicted correlation matrix for the scalar parameters listed in Table 2.2. Figure 2.31 indicates that the application of Eq. (2.147) induces non-zero correlations among several of the parameters, notably between parameters #4 and #5 (G and p) and, to a lesser extent, between parameters #2 and #3 (b and V_0). The diagonal values in Figure 2.31 are the predicted variances, i.e., the squares of the values shown in the last column of Table 2.2.

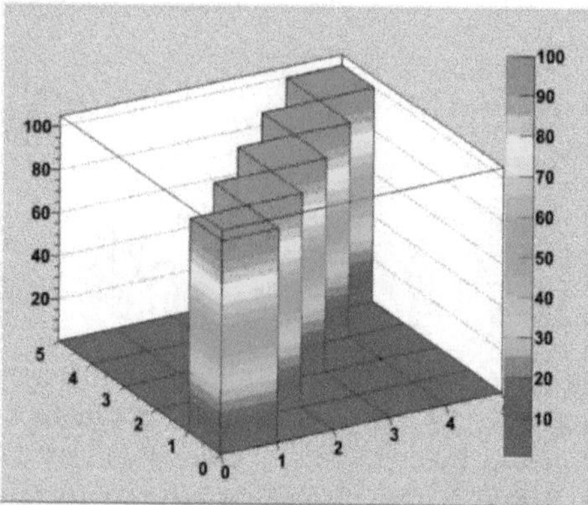

Figure 2.30: Initial correlation matrix for the scalar parameters listed in Table 2.2

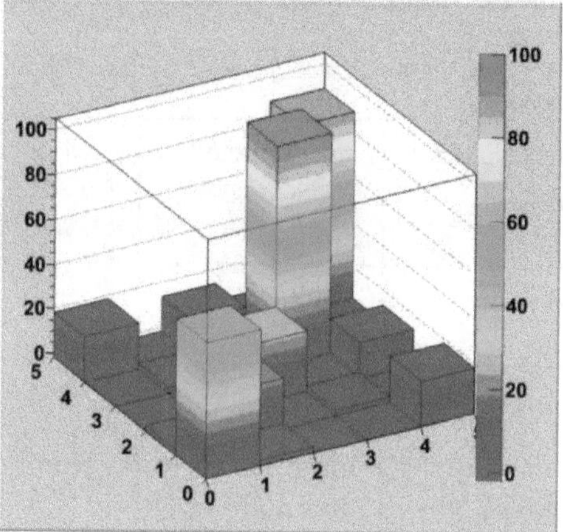

Figure 2.31: Predicted correlation matrix, $\left(C_{\alpha}^{be}\right)^{\nu\mu}$, *for the scalar parameters listed in Table 2.2*

The results of applying Eqs. (2.146) and (2.147) to the time dependent inlet acid concentration, $\rho_a^{(in)}(t)$ are depicted in Figures 2.32 and 2.33, respectively. The time-dependent calibration of the nominal value $\rho_a^{(in)}(t)$ is relatively small, and so is the reduction in the corresponding time-dependent standard deviation, from the initial value of $\sigma\left[\rho_a^{(in)}(t)\right]=20\%$.

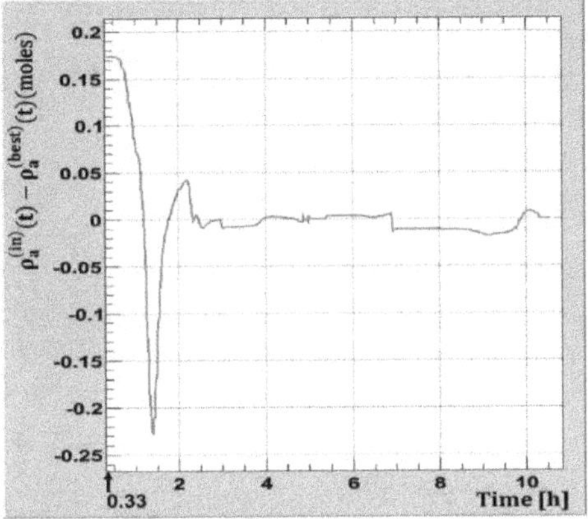

Figure 2.32: Time-dependent behavior of the difference between the nominal value, $\rho_a^{(in)}(t)$, and the optimally predicted "best estimate" value, $\rho_a^{(best)}(t)$, for the inlet acid concentration.

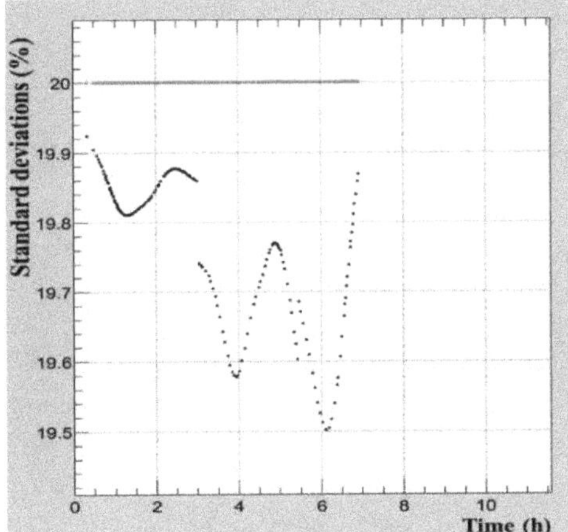

Figure 2.33: Time-dependent behavior of the original relative standard deviation $\sigma\left[\rho_a^{(in)}(t)\right]=20\%$ (in red) and the optimally predicted "best estimate" relative standard deviation $\sigma\left[\rho_a^{(best)}(t)\right]$ (in black), for the inlet acid concentration.

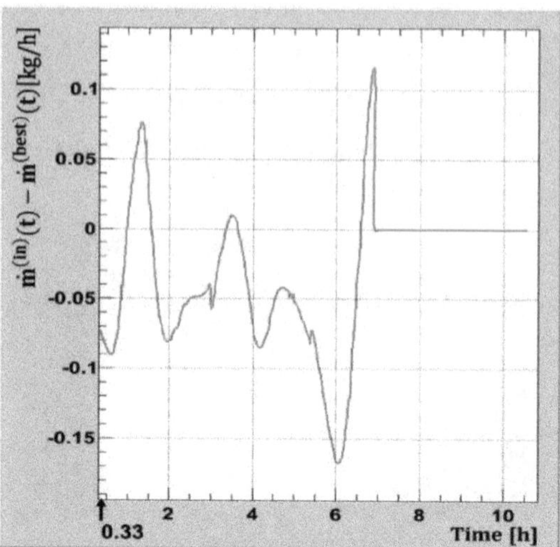

Figure 2.34: Time-dependent behavior of the difference between the nominal value, $\dot{m}^{(in)}(t)$, and the optimally predicted "best estimate" value, $\dot{m}^{(best)}(t)$, for the inlet mass flow rate.

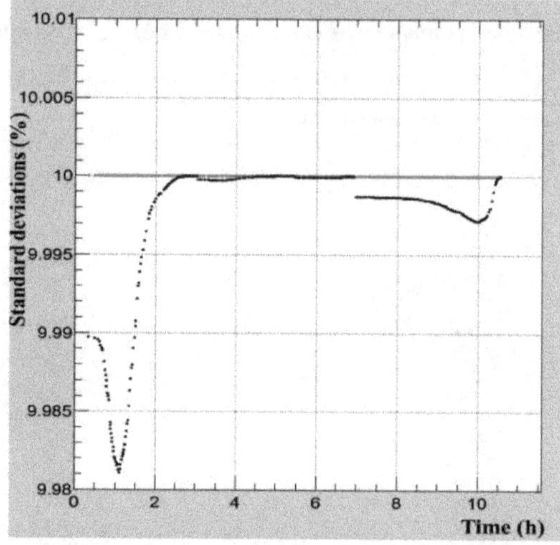

Figure 2.35. Time-dependent behavior of original relative standard deviation, $\sigma\left[\dot{m}^{(in)}(t)\right] = 10\%$, and the optimally predicted "best estimate" $\sigma\left[\dot{m}^{(best)}(t)\right]$, for the inlet mass flow rate.

The results of applying Eqs. (2.146) and (2.147) to the time-dependent mass flow rate, $\dot{m}^{(in)}(t)$ are depicted in Figure 2.34 and 2.35, respectively. The time-dependent calibration of the nominal value $\dot{m}^{(in)}(t)$ is also relatively small, and so is the reduction in the corresponding time-dependent standard deviation, from the initial value of $\sigma\left[\dot{m}^{(in)}(t)\right]=10\%$.

The best-estimate predicted nominal values for the nitric acid concentration are obtained by using Eq. (1.75) of Chapter 1, which is reproduced below, for convenience:

$$\left(\mathbf{r}^{be}\right)^{\nu}=\left(\mathbf{r}_{m}\right)^{\nu}+\sum_{\mu=1}^{N_t}\left\{\left[\mathbf{C}_{x}^{\nu\mu}-\sum_{\rho=1}^{\mu}\mathbf{C}_{r\alpha}^{\nu\rho}\left(\mathbf{S}^{\dagger}\right)^{\mu\rho}\right]\left[\sum_{\eta=1}^{N_t}\mathbf{K}_{d}^{\mu\eta}\mathbf{d}^{\eta}\right]\right\},\quad \nu\in J_t\,. \tag{2.148}$$

Figure 2.36 presents the computed, experimental, and best-estimate predicted nominal values for the nitric acid concentration in compartment #1. All of these values are in close agreement with one another.

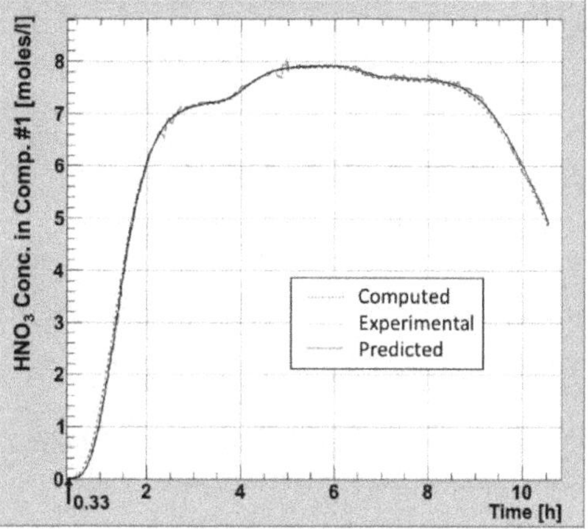

Figure 2.36: Computed, experimental, and best estimate predicted nominal values for the nitric acid concentration in compartment #1.

The full covariance matrix, which arises due to uncertainties in the model parameters, of the *computed* acid concentration in any compartment are obtained by using Eq. (1.61) of Chapter 1, which is reproduced, for convenience, below:

$$\mathbf{C}_{rc}^{\nu\mu}=\sum_{\eta=1}^{\nu}\sum_{\rho=1}^{\mu}\mathbf{S}^{\nu\eta}\mathbf{C}_{\alpha}^{\eta\rho}\left(\mathbf{S}^{\mu\rho}\right)^{\dagger}=\left(\mathbf{C}_{rc}^{\mu\nu}\right)^{\dagger};\quad \nu,\mu\in J_t\,. \tag{2.149}$$

The result of using Eq. (2.149) for the computed acid concentration in compartment #1 is depicted in Figure 2.37. As can be noted from this figure, the computed responses in the early stages of the transient, between time instances 5 through 75, are strongly (up to -0.86 [moles/l]^2) anti-correlated in time with the responses computed towards the end of the transient, between time instances 266-307. At other time instances, the responses are weakly correlated, except for the responses between time instances 5 through 60, which are strongly (up to 0.86 [moles/l]^2) correlated to each other, and again at the end of the transient, between time instances 260 through 307, when they again become strongly correlated. Variances of 0.86 [moles/l]^2, as noticed at the end of the transient, correspond to relative standard deviations of about 20%.

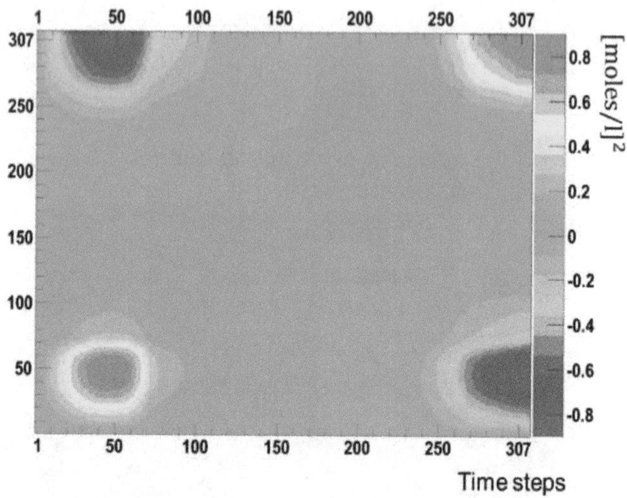

Figure 2.37: Time-dependent computed correlation matrix (arising from parameter uncertainties), $C_{rc}^{\nu\mu}$, for the nitric acid concentration in compartment #1.

The predicted best estimate response correlations are obtained by using Eq. (1.78) of Chapter 1, which is reproduced, for convenience, below:

$$\left(\mathbf{C}_r^{be}\right)^{\nu\mu} = \mathbf{C}_m^{\nu\mu} - \sum_{\eta=1}^{N_t}\sum_{\rho=1}^{N_t}\left[\mathbf{C}_m^{\nu\rho} - \sum_{\pi=1}^{\rho}\mathbf{C}_{r\alpha}^{\nu\pi}\left(\mathbf{S}^{\dagger}\right)^{\rho\pi}\right]\mathbf{K}_d^{\rho\eta}\left[\mathbf{C}_m^{\eta\mu} - \sum_{\pi=1}^{\eta}\mathbf{S}^{\eta\pi}\mathbf{C}_{\alpha r}^{\pi\mu}\right]. \quad (2.150)$$

The predicted best estimate response correlations for the nitric acid concentration in compartment #1, obtained by using Eq. (2.150), are depicted in Figure 2.38. As indicated in this figure, all of the best-estimate correlations, including the predicted standard deviations, are significantly reduced and rendered uniform. The corresponding (+/-) one-standard deviations are plotted in Figure 2.39, which depicts the behavior in time of the measured response standard deviation (5%), the computed response standard deviation (i.e., the diagonal elements of Eq. (2.149) stemming from uncertainties in the model parameters), and the best-estimate predicted response standard deviation obtained from the diagonal elements of Eq. (2.150). It is evident from Figure 2.39 that the "predicted best-estimate" response standard deviation is smaller than either the "measured" standard deviation or the "computed" one, for the entire time-interval under consideration.

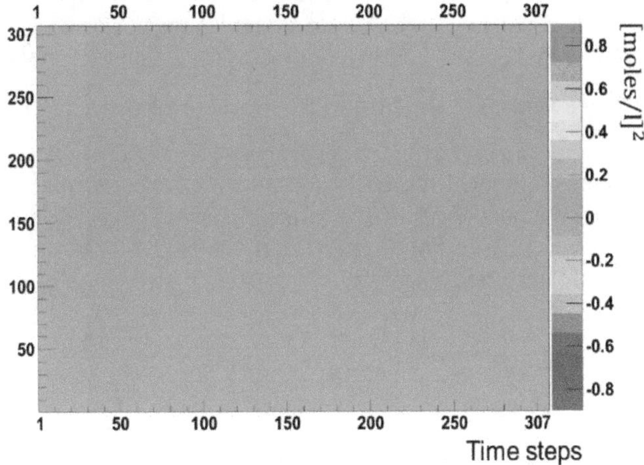

Figure 2.38: *Time-dependent best-estimate predicted correlation matrix,* $\left(C_r^{be}\right)^{\nu\mu}$ *, for the nitric acid concentration in compartment #1.*

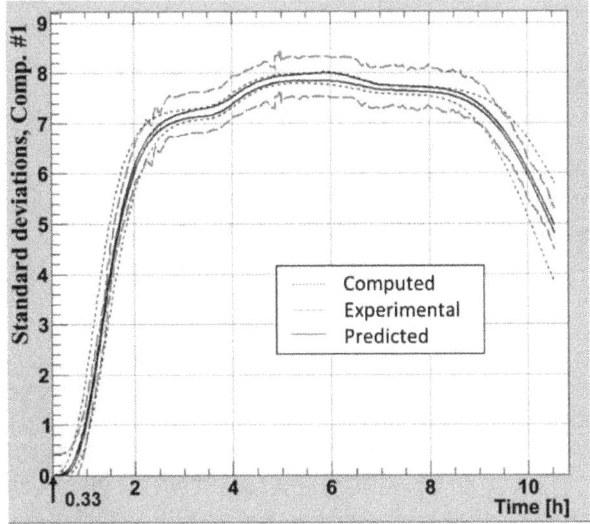

Figure 2.39: *Computed, experimental, and best estimate predicted (+/-) absolute standard deviations for the nitric acid concentration in compartment #1.*

Even though no measurements were performed in the dissolver compartments #2 through #8, the nominal values of the "best-estimate" responses, $\left(\boldsymbol{r}^{be}\right)^{\nu}$, in these compartments can be computed by using the calibrated best estimate parameter values $\left(\boldsymbol{\alpha}^{be}\right)^{\nu}$. In this vein, the best-estimate predicted parameter values for all 635 model parameters (as presented in Table 2.2 and depicted in Figures 2.32 and 2.34) together with their reduced predicted uncertainties (as presented in Table 2.2 and depicted Figures in 2.33 and 2.35) were used to re-compute the nominal values of the best-estimate responses, $\left(\boldsymbol{r}^{be}\right)^{\nu}$, which turned to be very close to the originally computed nominal values. Furthermore, the best-estimate predicted uncertainties in the best-estimate computed responses for which no experimental data is a priori available can be obtained by using the "propagation of errors" formula given in Eq. (2.149) but using the best estimated parameter values and their corresponding best-estimate standard deviations, namely:

$$\left(\mathbf{C}_{r}^{be}\right)^{\nu\mu}=\sum_{\eta=1}^{\nu}\sum_{\rho=1}^{\mu}\left[\mathbf{S}^{\nu\eta}\right]^{be}\left[\mathbf{C}_{\alpha}^{\eta\rho}\right]^{be}\left[\left(\mathbf{S}^{\mu\rho}\right)^{\dagger}\right]^{be};\quad \nu,\mu=1,\dots,N_{t}\,. \tag{2.151}$$

As will be shown in the following, the computation of the best-estimate uncertainties using Eq. (2.151) for the compartments in which no measurements were performed indeed underwent reductions, in all compartments, by comparison to the originally computed uncertainties. Typical results will be presented in Figures 2.40 through 2.45, for compartment #4 (in the middle of the dissolver) and for compartment #7. The uncertainty reductions in the other compartments are not presented here because they can be obtained by interpolating linearly between the results presented for compartments #1, #4, and #7.

The covariance matrix of the *computed* acid concentration in compartment #4, obtained using Eq. (2.149), is depicted in Figure 2.40. As can be noted from this figure, the computed responses in the early stages of the transient, between time instances 5 through 30, are anti-correlated in time with the responses computed towards the end of the transient, between time instances 266 through 307. The anti-correlations for the acid concentration in compartment #4 are similar to the time-dependent response anti-correlations in compartment #1. The acid concentration responses in compartment #4 are less strongly correlated at other time instances, except for the responses between the initial stages of the transient (i.e., time instances 1 through 50) and again at the end of the transient (i.e., time instances 260 through 307), when they are positively correlated, with variances reaching as high as 0.6 [moles/l]^2. This value corresponds to an absolute standard deviation of 0.77 [moles/l], which in turn corresponds to a relative standard deviation of over 50%, which is quite large. Overall, the time-correlations for the acid concentration in compartment #4 are similar to, but stronger than, the time-dependent response correlations in compartment #1.

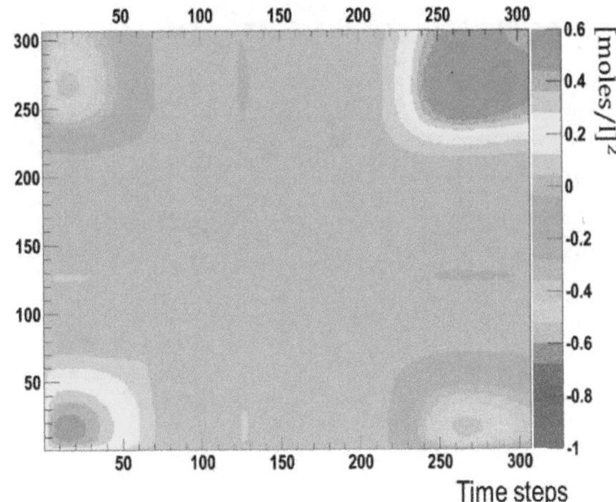

***Figure 2.40:** Time-dependent computed correlation matrix (arising from parameter uncertainties), $C_{rc}^{\nu\mu}$, for the nitric acid concentration in compartment #4.*

The predicted best estimate response correlations for the acid concentration in compartment #4 are obtained by using Eq. (2.151), and are depicted in Figure 2.41. As indicated in this figure, all of the best-estimate correlations, including the predicted standard deviations, are drastically reduced and rendered much more uniform. The corresponding (+/-) one-standard deviations are plotted in Figure 2.42, which depicts the behavior in time of the computed response standard deviation (i.e., the diagonal elements of Eq. (2.149), which stem from uncertainties in the model parameters) and the best-estimate predicted response standard deviation obtained from the diagonal elements of the matrix obtained using Eq. (2.151). It is evident from Figures 2.41 and 2.42 that the "predicted best-estimate" response standard deviation is considerably smaller than the "computed" one, for the entire time-interval under consideration.

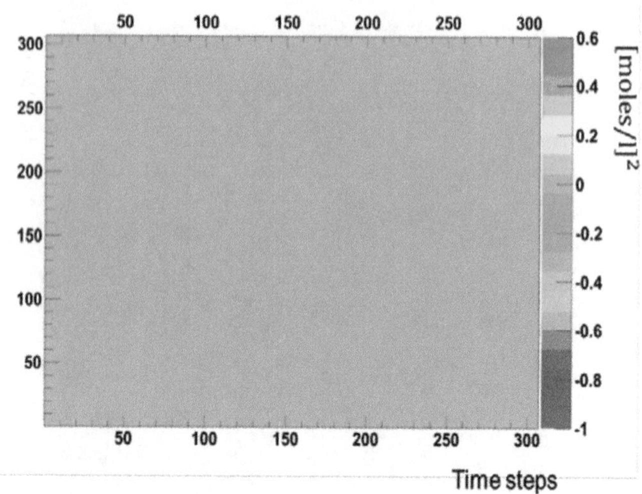

Figure 2.41: Time-dependent best-estimate predicted correlation matrix, $\left(C_r^{be}\right)^{\nu\mu}$, for the nitric acid concentration in compartment #4.

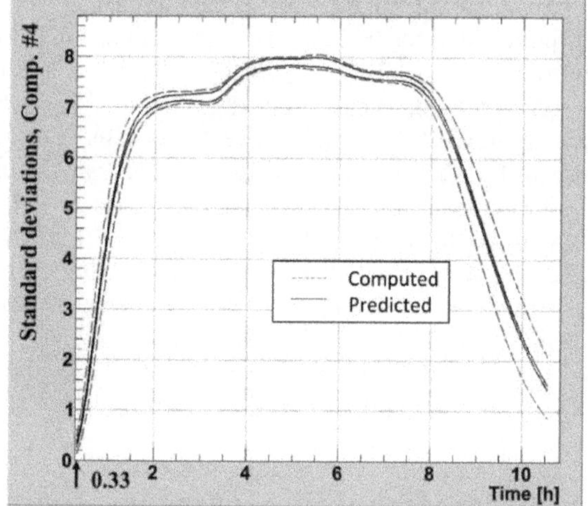

Figure 2.42: Computed and best estimate predicted absolute standard deviations (+/-) for the nitric acid concentration in compartment #4.

The covariance matrix of the *computed* acid concentration in compartment #7, obtained using Eq. (2.149), is depicted in Figure 2.43, which displays an "island" of anti-correlated responses between time instances 1 through 10 and responses at time instances 220 through 260, as well as an "island" of positively correlated acid concentrations among the time instances 220 through 260. Although the absolute values of the overall uncertainties are smaller in this compartment, by comparison to the other compartments, their relative values are actually larger than in the other compartments. For example, the largest variance of the acid concentration in compartment being 0.2 $[\text{moles/l}]^2$, which occurs during the time instances 220 through 240; this variance corresponds to a relative standard deviation of 90%, as can be deduced from Figure 2.45. The predicted best estimate correlations for the acid concentration in compartment #7 are obtained by using Eq. (2.151) and are depicted in Figure 2.44. As indicated in this figure, all of the best-estimate correlations for the acid concentration in compartment #7, including the predicted standard deviations, are drastically reduced and rendered much more uniform. The corresponding (+/-) one-standard deviations are plotted in Figure 2.45, which depicts the behavior in time of the computed response standard deviation (i.e., the diagonal elements of Eq. (2.149) stemming from uncertainties in the model parameters) and the best-estimate predicted response standard deviation obtained using Eq. (2.151). It is evident from Figure 2.45 that the "predicted best-estimate" response standard deviation for the acid concentration in compartment #7 is considerably smaller than the "computed" one, over the entire time-interval under consideration.

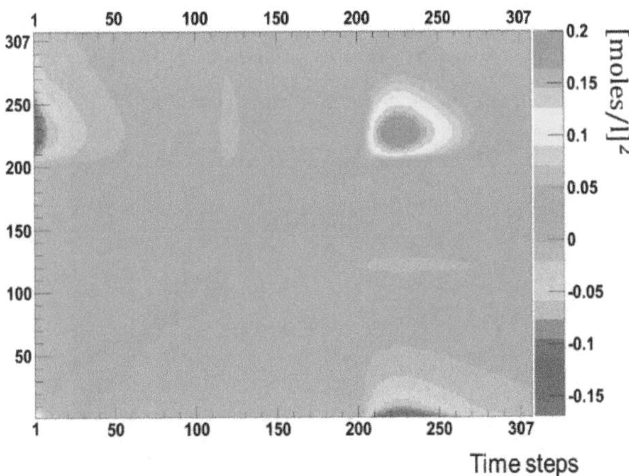

Figure 2:43: Time-dependent computed correlation matrix (arising from parameter uncertainties), $C_{rc}^{\nu\mu}$, for the nitric acid concentration in compartment #7.

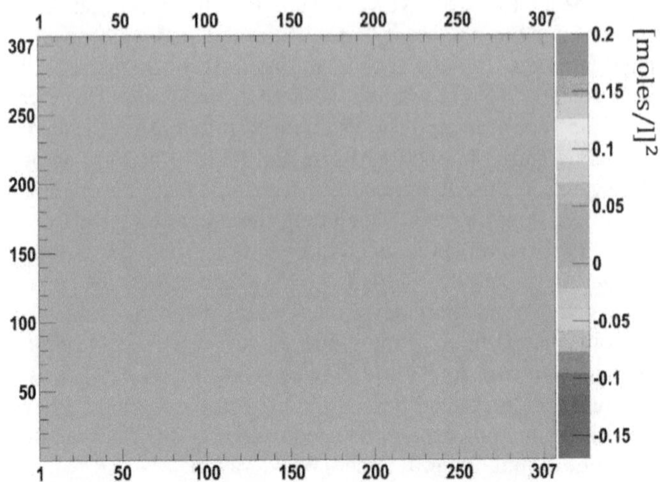

Figure 2.44: Time-dependent best-estimate predicted correlation matrix, $\left(C_r^{be}\right)^{\nu\mu}$, *for the nitric acid concentration in compartment #7.*

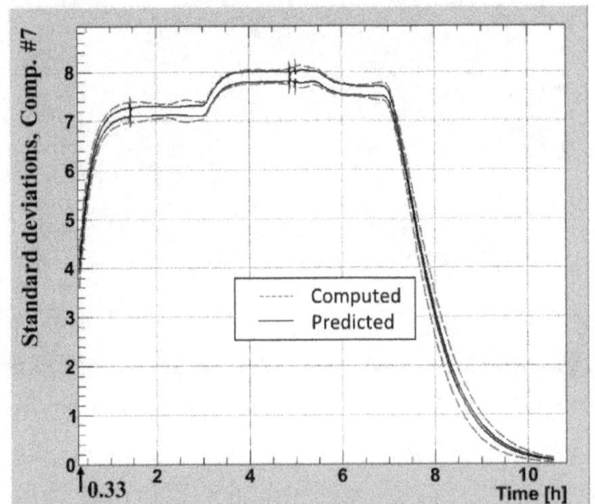

Figure 2.45: Computed (blue graph) and best estimate predicted (black graph) absolute standard deviations (+/-) for the nitric acid concentration in compartment #7.

The results presented in the forgoing highlight the very beneficial effects of the comprehensive framework of the BERRU-SMS predictive modeling methodology presented in Chapter 1, which considers the entire phase-space of parameters and responses simultaneously over the entire time interval of interest. In particular, this unique feature makes it possible to "spread out" the positive effects of having performed measurements in one region of the dissolver (in this case, in compartment

#1) to reduce significantly the predicted uncertainties in the acid concentration not only in the compartment where measurements were performed but also in all of the other compartments, where measurements were lacking.

2.5 BERRU-SMS INVERSE PREDICTIVE MODELING OF DISSOLVER START-UP

Many measurement problems are "inverse" to the "forward" problem in that they seek to determine the properties of a source term in a medium and/or the size of the medium or properties of the medium from measurements of quantities that depend on the unknown state-variables in the medium. Inverse problems are ill-posed (admitting non-unique solutions) and ill-conditioned, unstable to small errors or perturbations that are inherently affecting both the model parameters and the experimental measurements. Cacuci (2014a) has highlighted the ill-posed nature of inverse problems by presenting a paradigm inverse neutron diffusion problem that admits an exact solution on paper, yet the amplification of "noise" makes the exact solution non-computable. The methods for solving such inverse problems can be categorized as "explicit" or "implicit". The historically older "explicit" methods explicitly attempt to manipulate the forward model in conjunction with measurements in order to estimate explicitly the unknown source and/or other unknown characteristics of the medium. The current state-of-the-art practice relies on implicit methods, which combine measurements with repeated solutions of the direct problem obtained with different values of the unknowns, iterating until an a priori selected functional, usually representing a user-defined "goodness of fit" between measurements and direct computations, is reduced to a value deemed to be "acceptable" by the user. The introduction of the user defined "goodness of fit" functional comprises ad-hoc parameters which "regularize" the inverse problem at hand, as typified by the methods described by Tichonov (1963), Levenberg-Marquardt (1944/1963), and Tarantola (2005). The fundamental difficulties associated with inverse problems affect profoundly the numerical methods for solving them, particularly in the presence of errors, including numerical ones. Consequently, using different methods for solving an inverse problem produce different results, depending on the user-defined assumptions employed to "regularize" and solve the inverse respective problem.

The BERRU-SMS methodology can be used not only in the "forward predictive mode," as has been used in the applications presented thus far, but can also be used in the "inverse predictive mode" to determine from measurements the value of imprecisely known model parameters, within an a priori specified accuracy or convergence criterion, by employing Eqs. (2.146) through (2.151). It is important to note that these equations do not contain any arbitrary, user-defined, parameters for "regularizing" the inverse problem. This Section illustrates the "inverse predictive modeling" application of the BERRU-SMS methodology to determine, within an a priori specified convergence criterion and overall accuracy, an imprecisely known time-dependent boundary condition (specifically: the time-dependent inlet acid concentration) for the dissolver model "case study" by using measurements of the state function (specifically: the time-dependent acid concentration) at a specified location (specifically: in the dissolver's compartment furthest from the inlet). The imprecisely known time-dependent boundary condition is described by 635 imprecisely known discrete scalar parameters.

2.5.1 *Preliminary Computation of Acid Concentration Distribution in the Dissolver Using an Initial Guess for the Inlet Acid Concentration*

In this illustrative application, the time-dependent inlet acid concentration, which is a "time-dependent inlet boundary condition." will be considered to be unknown. The BERRU-SMS methodology presented in Section 2.1 will be applied in the inverse predictive mode to predict the time-dependent inlet acid concentration within an a priori specified convergence error, by using the measurements of the acid concentration in compartment #1, which are depicted in Figure 2.5. Recall that compartment #1 is the furthest from the dissolver's inlet. For computational purposes, 1291 model parameters will be considered to be the components of the vector $\boldsymbol{\alpha}$ defined below:

$$\boldsymbol{\alpha} \triangleq \left[\rho_a^{(in)}(t_1),...,\rho_a^{(in)}(t_{635}),\dot{m}^{(in)}(t_1),...,\dot{m}^{(in)}(t_{635}),\rho_{a0}^{(1)},...,\rho_{a0}^{(8)},V_0^{(1)},...,V_0^{(8)},a,b,V_0,G,p\right] \tag{2.152}$$

For the inverse modeling prediction of the "unknown" inlet acid concentrations $\rho_a^{(in)}(t_1),...,\rho_a^{(in)}(t_{635})$, the other model parameters are considered to be known within negligible errors: the model parameters a,b,V_0,G,p have nominal values as provided in Table 2.1 (within negligible errors); the time-dependent inlet mass flow rate $\dot{m}^{(in)}(t)$ behaves (within negligible errors) as depicted in Figure 2.3. The initial conditions $\rho_{a0}^{(1)},...,\rho_{a0}^{(8)}$, $V_0^{(1)},...,V_0^{(8)}$ are also considered to be known within negligible errors. The experimentally measured values shown in Figure 2.5 are considered to be known within a relative standard deviation of 1%. Thus, using the nominal values and standard deviations of all of these "known parameters," the predicative modeling methodology will be employed (in the "inverse predictive mode") in the remainder of this Section into predict (within 0.1%, as will be required in the sequel) the time-dependent behavior of the inlet acid concentration, $\rho_a^{(in)}(t)$, which actually evolves as depicted in Figure 2.4. Since the inlet acid concentration is categorized as a (time-dependent) model *parameter*, as opposed to a model *response*, it is to be expected that model parameters will ultimately be predicted by Eq. (2.146), in conjunction with the iterative application of not only this equation but also of Eqs. (2.147) through (2.151). These computations will be described in the remainder of this Section.

To initiate the predictive modeling computations indicated in Eqs. (2.146) through (2.151), it is necessary to perform a preliminary "base-case" computation to obtain preliminary "base-case" values for the acid concentrations, $\rho_a^{(k)}(t)$, and the volumes of the liquid phase, $V^{(k)}(t)$, in all of the dissolver compartments $k = 1,...,8$. Such a preliminary "base-case" computation can be performed by using "nominal values" estimated by "expert opinion" for the unknown time-dependent inlet acid concentration, $\rho_a^{(in)}(t)$. The only a priori information related to the behavior of $\rho_a^{(in)}(t)$ is provided by the measurements $\rho_{a,meas}^{(1)}(t_i)$ depicted in Figure 2.5 Therefore, these measurements will be used in Eqs. (2.146) through (2.151) as "expert opinion values" for the unknown "starting-values" of the inlet acid concentration $\rho_a^{(in)}(t)$, to carry out a preliminary "base-case" computation of . $\rho_a^{(k)}(t)$, $k = 1,...,8$. The results provided by this "expert opinion base-case" computation are depicted in Figure 2.46; for brevity, only the acid concentrations in compartments #1, 4, and 7 are shown.

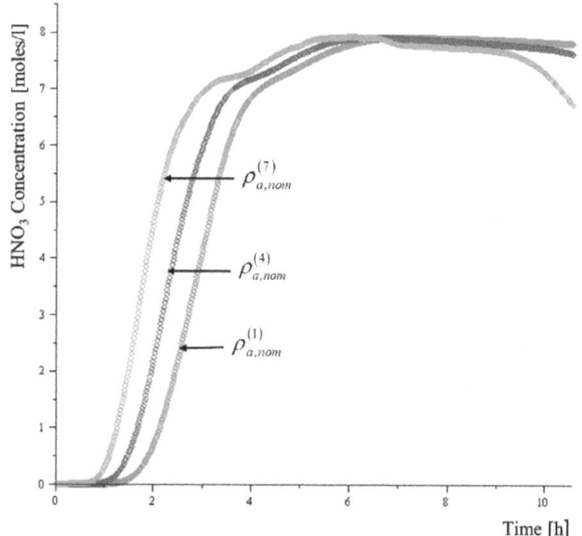

Figure 2.46: Preliminary "expert opinion base-case" computations of the time-dependent nitric acid concentrations $[\text{mol/L}]$ ***in compartments #1, #4, and #7.***

Comparing now the experimental measurements, $\rho_{a,meas}^{(1)}(t_i)$, given in Figure 2.5, with the preliminary "base-case" computed results (depicted in Figure 2.46) for $\rho_a^{(1)}(t)$ obtained using the "expert opinion" values for the inlet acid concentration indicates that there is a time-lag of about 140 minutes between the phenomena occurring at the dissolver's inlet and those occurring at the dissolver's other end, namely in compartment #1. In other words, the effects of the time-dependent inlet acid concentration take about 140 minutes to "propagate" to compartment #1. This information will be used in the inverse modeling of the inlet acid concentration, which will be performed following the iterative procedure described in the next sub-section.

2.5.2 *Iterative Inverse Modeling of the Time-Dependent Evolution of the Inlet Acid Concentration*

Using the results and discussion presented in the previous sub-section, the preliminary "base-case" values obtained for $\rho_a^{(7)}(t)$, as depicted in Figure 2.46, are shifted by 140 minutes, and are subsequently used as the values of the "time-dependent inlet acid concentration" to be replaced into Eqs. (2.10) through (2.14) for the first iterative computation of the forward functions $\left\{\rho_a^{(k)}(t)\right\}^{(1)}$ and $\left\{V^{(k)}(t)\right\}^{(1)}$, $k = 1,\ldots,8$; the superscript "(1)" indicates "iteration #1" . These forward functions are then used in the adjoint dissolver model to compute the "1st-iteration values" of the adjoint functions $\left\{\psi_\rho^{(k)}(t)\right\}^{(1)}$ and $\left\{\psi_V^{(k)}(t)\right\}^{(1)}$. Subsequently, these adjoint functions are to compute the "1st-iteration values" of the sensitivities $\left\{D\rho_a^{(k)}(t_i)\right\}^{(1)}$. All of these "1st-iteration values" are finally used together with the measured response values depicted in Figure 2.5, assuming a 1% standard deviation for these measurements, in the predictive modeling Eqs. (2.146) through (2.151) to obtain the "1st-iteration predicted best-estimate values" for the predicted responses and the predicted model parameters. The results for the "predicted best-estimate parameter values after the 1st-iteration," $\left\{\left(\boldsymbol{\alpha}^{be}\right)^{\nu}\right\}^{(1)}$, include the predicted nominal values of all 1291 model parameters, but only the values for the time-dependent inlet acid concentration are affected. The "inverse predictive modeling" methodology will not calibrate the nominal values for the other parameters because these nominal values were considered to be very well known (i.e., within negligible uncertainties). The results included in $\left\{\left(\boldsymbol{\alpha}^{be}\right)^{\nu}\right\}^{(1)}$ for the "first-iteration predicted values for the time-dependent inlet acid concentration $\left\{\rho_a^{(in)}(t)\right\}^{(1)}$", which we are seeking to determine in the "inverse predictive mode", are depicted in Figure 2.47 using the label "iteration 1". The above procedure is repeated ("iterated") until the maximum difference between the time-dependent predicted inlet acid concentration converges within 0.1%, for the entire time interval, i.e., $\max_t \left|1-\left\{\rho_a^{(in)}(t)\right\}^{(J-1)} / \left\{\rho_a^{(in)}(t)\right\}^{(J)}\right| < 0.1\%$; the convergence criterion of "1%" was selected based on "expert opinion" regarding the accuracy that can be expected of measurements of the time-dependent inlet acid concentration over the duration of 10.5 hours. This inequality is reached after $J=5$ iterations, at which stage the inverse predictive modeling iterations are considered to have converged. The results obtained after the 3rd- and 5th-iteration, respectively, are also depicted in Figure 2.47. To facilitate the comparison between the various iterations and the desired result, Figure 2.47 also presents the actual time-dependency of the inlet acid concentration (which was considered to be unknown for the purposes of this "inverse predictive modeling" illustrative example). The results depicted in Figure 2.47 indicate that the various iterations converge reasonably well towards the exact time-dependent inlet acid concentration $\rho_a^{(in)}(t)$ discontinuous at three time instances, taking on constant values between the respective discontinuities, and becoming discontinuously zero after 7 hours, in a manner very similar to a Heaviside step-function. It is known that such step-functions are notoriously difficult to approximate by continuous functions, and every such approximation will display oscillations around discontinuities. In particular, the largest point-wise discrepancies between the "inverse predicted values" and the exact values are clustered at the beginning and the end of the transient period, which is not surprising in view of the step-like time-distribution of the exact inlet acid concentration. These discrepancies are irrelevant, however, for the time-integrated inlet acid concentration $\left\{\int_0^{t_f} \rho_a^{(in)}(t)\,dt\right\}^{(5)}$, predicted after the 5th-iteration, which differs by less than 1% of the time-integral of the exact inlet acid concentration.

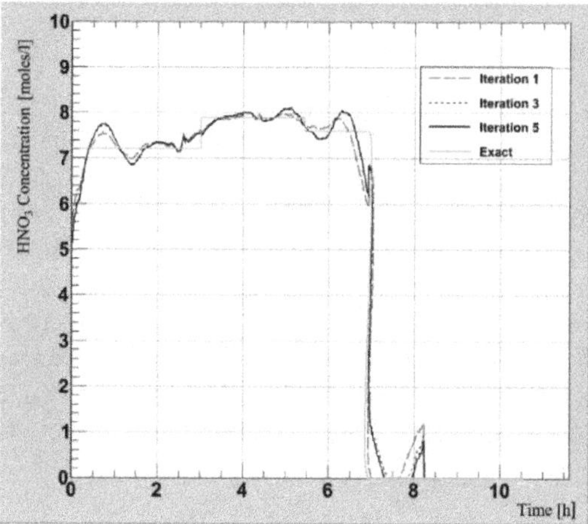

Figure 2.47: Values predicted for the inlet acid concentration, $\left\{\rho_a^{(in)}(t)\right\}^{(1)}$, $\left\{\rho_a^{(in)}(t)\right\}^{(3)}$, $\left\{\rho_a^{(in)}(t)\right\}^{(5)}$ ***in*** $[\mathrm{mol/L}]$ ***after the inverse predictive modeling iterations #1, #3, and #5, respectively, and their comparison to the actual time-dependence of the inlet nitric acid concentration.***

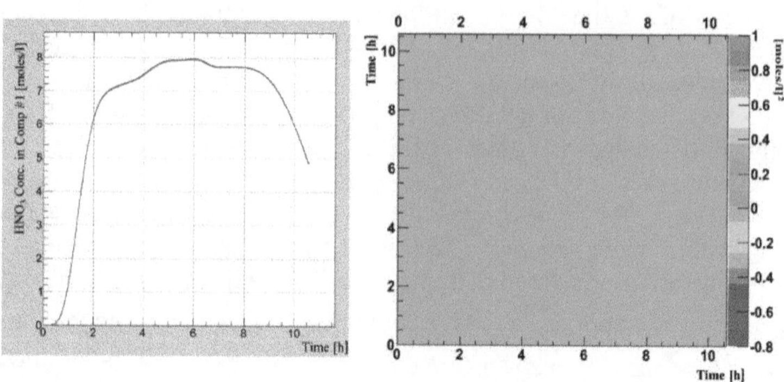

Figure 2.48: Left: predicted nominal time-dependent acid concentration $[\mathrm{mol/L}]$ ***in compartment #1 after the 5th iteration,*** $\left\{\rho_a^{(1)}(t)\right\}^{(5)}$***, with one standard deviation error bands;***

Right: the accompanying predicted covariance matrix for $\left\{\rho_a^{(1)}(t)\right\}^{(5)}$.

The results from the last (5th) iteration for the predicted nominal value of the acid concentration, $\left\{\rho_a^{(1)}(t)\right\}^{(5)}$, in compartment #1, and the accompanying predicted covariance matrix for this concentration are depicted in Figure 2.48. The result for $\left\{\rho_a^{(1)}(t)\right\}^{(5)}$ is within 1% of the experimentally measured results depicted in Figure 2.5, over the entire duration (10.5 hours) of the transient, with a standard deviation of less than 2%.

The results produced by the 5th-iteration of the inverse BERRU-SMS predictive modeling methodology also yield predicted values for the nominal values of the acid concentrations in all of the other dissolver compartments. Furthermore, using the results from this 5th-iteration in Eq. (2.151) provides the covariance matrix ("uncertainties") for the predicted acid concentration responses in the dissolver's compartments where measurements are not available. In particular, the predicted results for compartment #4 (in the middle of the dissolver) and compartment #7 (closest to the inlet), along with the accompanying covariance matrices, are depicted in Figures 2.49 and 2.50.

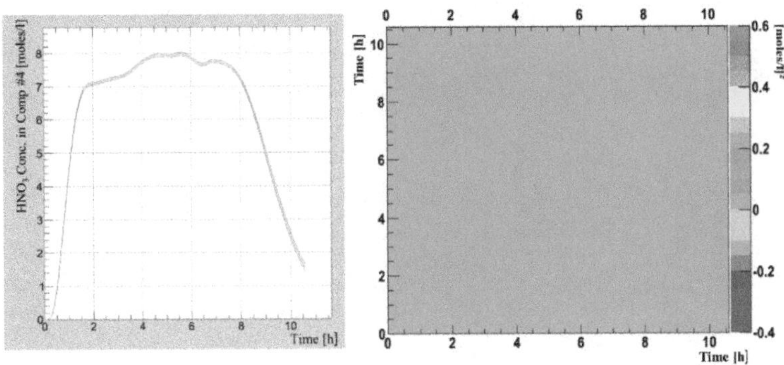

Figure 2.49: Left: predicted nominal time-dependent acid concentration $[\text{mol/L}]$ ***in compartment #4 after the 5th iteration,*** $\left\{\rho_a^{(4)}(t)\right\}^{(5)}$***, with one standard deviation error bands;***

Right: the accompanying predicted covariance matrix for $\left\{\rho_a^{(4)}(t)\right\}^{(5)}$.

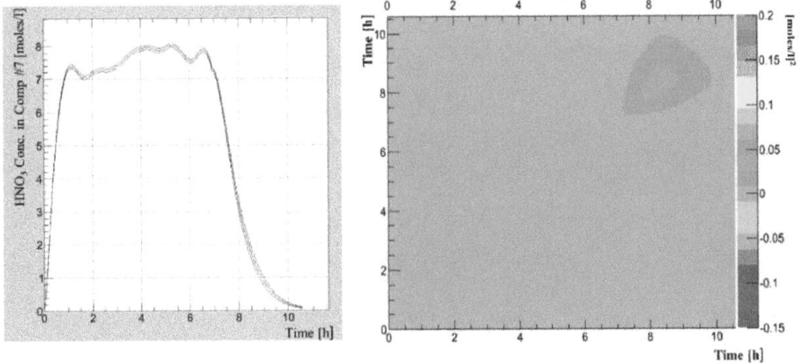

Figure 2.50: Left: predicted nominal time-dependent acid concentration $[\text{mol/L}]$ ***in compartment #7 after the 5th iteration,*** $\left\{\rho_a^{(7)}(t)\right\}^{(5)}$***, with one standard deviation error bands;***

Right: the accompanying predicted covariance matrix for $\left\{\rho_a^{(7)}(t)\right\}^{(5)}$.

On the other hand, the exact results for the actual time-dependent distributions of acid concentrations within the dissolver and, in particular, for compartments #1, #4 and #7, produced by the actual inlet acid concentration distributions are actually available from the results presented in Section 2.2. In particular, the results obtained in Figures 2.48 through 2.50 for the time-dependent acid concentrations in the respective compartments predicted by the "inverse predictive modeling" of the inlet acid concentration can be compared with the exact results displayed in Figure 2.6. These comparisons are presented in Figures 2.51 through 2.53.

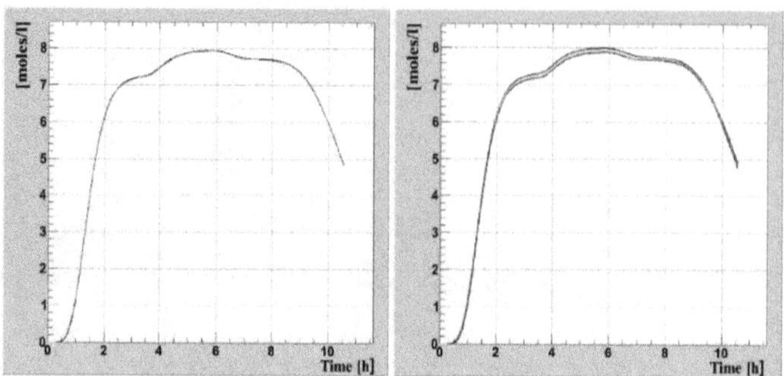

Figure 2.51: (Left) Black graph: forward predicted nominal value of the acid concentration $[\mathrm{mol/L}]$ in compartment #1, from Figure 2.6. Green graph: inverse predicted nominal value of the acid concentration in compartment #1, from Figure 2.48. (Right) Black graphs: forward predicted one standard deviation error bands of the acid concentration in compartment #1, from Figure 2.6. Green graphs: inverse predicted one standard deviation error bands of the acid concentration in compartment #1, from Figure 2.48.

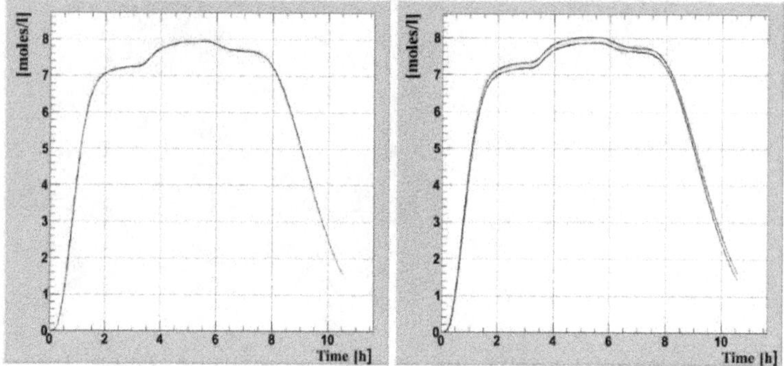

Figure 2.52: (Left) Black graph: forward predicted nominal value of the acid concentration $[\mathrm{mol/L}]$ in compartment #4, from Figure 2.6. Green graph: inverse predicted nominal value of the acid concentration in compartment #4, from Figure 2.49. (Right) Black graphs: forward predicted one standard deviation error bands of the acid concentration in compartment #4, from Figure 2.6. Green graphs: inverse predicted one standard deviation error bands of the acid concentration in compartment #4, from Figure 2.49.

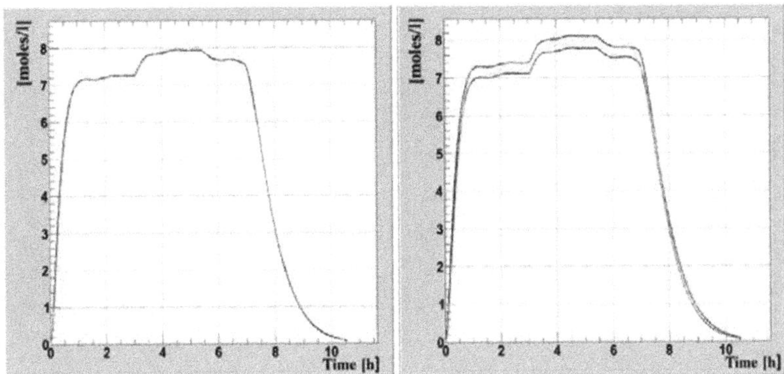

***Figure 2.53:** (Left) Black graph: forward predicted nominal value of the acid concentration* $[\mathrm{mol/L}]$ *in compartment #7, from Figure 2.6. Green graph: inverse predicted nominal value of the acid concentration in compartment #7, from Figure 2.50. (Right) Black graphs: forward predicted one standard deviation error bands of the acid concentration in compartment #7, from Figure 2.6. Green graphs: inverse predicted one standard deviation error bands of the acid concentration in compartment #7, from Figure 2.50.*

Figures 2.51 through 2.53 indicate a very good agreement, within the respective standard deviations, between the exact forward predictions (i.e., those obtained using the known inlet acid concentration) produced by the BERRU-SMS predictive modeling methodology presented in Chapter 1 and the corresponding values for the acid concentrations obtained using the BERRU-SMS methodology in the inverse predictive mode. This close agreement indicates that the effects of the less-than-perfect inverse prediction of the time-dependent inlet acid concentration (time-dependent boundary condition) have very little effect on predicting the responses of interest, namely the time-dependent acid concentrations in the various compartments, and, in particular, in compartment #1, where the experimental measurements are available. Notably, the use of the maximum entropy principle enables the forward and inverse BERRU-SMS predictive modeling methodology presented in Chapter 1 to construct an intrinsic regularizing metric for solving any inverse problem without user intervention and/or introduction of arbitrary "regularization parameters," as other inverse-problem procedures need to do (Tichonov, 1963; Levenberg-Marquardt, 1944/1063; Tarantola, 2005).

3. BERRU-CMS PREDICTIVE MODELING OF COUPLED MULTIPHYSICS SYSTEMS

Abstract

This Chapter presents the "BERRU-CMS" methodology, which extends the BERRU-SMS predictive modeling methodology presented in Chapter 1 to predictive modeling of *coupled* (as opposed to "single") multi-physics systems. The BERRU-CMS methodology is build on the same principles as the BERRU-SMS predictive modeling methodology that was presented in Chapter 1. Thus, the maximum entropy principle is used to construct an optimal approximation of the unknown a priori distribution based on a priori known mean values and uncertainties characterizing the parameters and responses for both multi-physics models. This "maximum entropy"-approximate a priori distribution is combined, using Bayes' theorem, with the "likelihood" provided by the multi-physics simulation models. Subsequently, the posterior distribution thus obtained is evaluated using the saddle-point method to obtain analytical expressions for the optimally predicted values for the multi-physics models parameters and responses along with corresponding reduced uncertainties. Noteworthy, the predictive modeling methodology for the coupled systems is constructed such that the systems can be considered sequentially rather than simultaneously, while preserving exactly the same results as if the systems were treated simultaneously. Consequently, very large coupled systems, which could perhaps exceed available computational resources if treated simultaneously, can be treated with the BERRU-CMS methodology presented in this work sequentially, without any loss of generality or information while requiring just the resources that would be needed if the systems were treated separately.

3.1 MATHEMATICAL FRAMEWORK

3.1.1 A Priori Information for Two Multi-Physics Models

Consider a multi-physics model, henceforth called "Model A" comprising N_α system (model) parameters α_n. Model A is used to compute results, henceforth called responses, which can also be measured experimentally. Consider now a second physical system, henceforth called "Model B," comprising N_β system (model) parameters β_m , and which is also used to compute responses that can be measured experimentally. Model A and Model B are considered to be coupled to each other. In the analysis and design of nuclear reactors, for example, Model A may comprise the neutron transport and nuclide depletion equations which are coupled to Model B, which would computing the thermal-hydraulics conservation (mass, momentum,

D. G. Cacuci, *BERRU Predictive Modeling*,
https://doi.org/10.1007/978-3-662-58395-1_3

energy) equations describing the flow of coolant and heat transfer through the reactor's materials.

Consider next that there are N_r experimentally measured responses r_i associated mostly, but not necessarily exclusively, with Model A. Furthermore, consider also that there are N_q experimentally measured responses q_j associated mostly, but not necessarily exclusively, with Model B. For example, measurement of reaction rates and power (or flux) distributions could be considered to be responses of type r_i, while measurements of flow rates and temperature distributions could be considered responses of type q_j. Furthermore-, cross sections can be considered to be model parameters of type α_n, while heat transfer correlations can be considered model parameters of type β_m . Parameters modeling the geometry of the system (e.g., rod and assembly dimensions, core dimensions), for example, could be considered to belong to either type of model parameters (i.e., either α_n or β_m), since they affect both the neutron transport equation and the thermal-hydraulics conservation equations.

In practice, the values of the parameters α_n and β_m are determined experimentally. Therefore, these parameters cannot be known exactly, but can be considered to behave stochastically, obeying some probability distribution function which (for large-scale systems, as customarily encountered in practice) is unknown. Such stochastic quantities are usually called variates; thus, the parameters α_n and β_m, as well as the measured responses r_i and q_j, are variates. To simplify the mathematical derivations to follow in this Section, the model parameters α_n will be considered to constitute the components of the N_α-dimensional column vector $\boldsymbol{\alpha}$ of, defined as follows:

$$\boldsymbol{\alpha} \triangleq \left(\alpha_1,\ldots,\alpha_{N_\alpha}\right)^\dagger, \tag{3.1}$$

while the model parameters β_m will be considered to constitute the components of the N_β-dimensional column vector $\boldsymbol{\beta}$ defined as

$$\boldsymbol{\beta} \triangleq \left(\beta_1,\ldots,\beta_{N_\beta}\right)^\dagger. \tag{3.2}$$

By convention, all of the vectors considered in this work (e.g., $\boldsymbol{\alpha}$ and $\boldsymbol{\beta}$) are column vectors; a dagger ($\dagger$) will be used to denote "transposition." Similarly, the N_r experimentally measured responses r_i will be considered to be components of the N_r -dimensional column vector

$$\mathbf{r} \triangleq \left(r_1,\ldots,r_{N_r}\right)^{\dagger}, \tag{3.3}$$

while the N_q experimentally measured responses q_j will be considered to be components of the N_q-dimensional column vector

$$\mathbf{q} \triangleq \left(q_1,\ldots,q_{N_q}\right)^{\dagger}. \tag{3.4}$$

Most generally, the parameters α_n and β_m, as well as the responses r_i and q_j can be considered to obey some a priori probability distribution function $P(\boldsymbol{\alpha},\boldsymbol{\beta},\mathbf{r},\mathbf{q})$. For large-scale systems, as customarily encountered in practice, the probability distribution $P(\boldsymbol{\alpha},\boldsymbol{\beta},\mathbf{r},\mathbf{q})$ cannot possibly be known. The information usually available in practice comprises the mean values of the model parameters and responses together with the corresponding uncertainties (standard deviations and, occasionally, correlations) computed about the respective mean values. For notational simplicity, angular brackets will be usedto denote the integral, $\langle f \rangle$, of the quantity $f(\boldsymbol{\alpha},\boldsymbol{\beta},\mathbf{r},\mathbf{q})$ over the joint probability distribution $P(\boldsymbol{\alpha},\boldsymbol{\beta},\mathbf{r},\mathbf{q})$; $\langle f \rangle$ is defined as

$$\langle f \rangle \triangleq \int f(\boldsymbol{\alpha},\boldsymbol{\beta},\mathbf{r},\mathbf{q}) P(\boldsymbol{\alpha},\boldsymbol{\beta},\mathbf{r},\mathbf{q}) d\boldsymbol{\alpha}\, d\boldsymbol{\beta}\, d\mathbf{r}\, d\mathbf{q}. \tag{3.5}$$

Using the above convention, the mean values of the model parameters α_n will be denote as $\alpha_n^0 \triangleq \langle \alpha_n \rangle$, and will be considered to constitute the components of the vector $\boldsymbol{\alpha}^0$ defined as

$$\boldsymbol{\alpha}^0 \triangleq \left(\alpha_1^0,\ldots,\alpha_{N_q}^0\right)^{\dagger}. \tag{3.6}$$

Similarly, the mean values of the parameters β_n are considered to be known, and will be denoted as $\beta_n^0 \triangleq \langle \beta_n \rangle$. These mean values are considered to be the components of the vector $\boldsymbol{\beta}^0$ defined as

$$\boldsymbol{\beta}^0 \triangleq \left(\beta_1^0,\ldots,\beta_{N_\beta}^0\right)^{\dagger}. \tag{3.7}$$

The parameters' second-order central moments, namely the standard deviations and correlations, will also be considered to be known. For the parameters α_n, the second-order central moments are considered to be the components of covariance matrices $\mathbf{C}_{\alpha\alpha}^{(N_\alpha \times N_\alpha)}$ defined as

$$\mathbf{C}_{\alpha\alpha}^{(N_\alpha \times N_\alpha)} \triangleq \left[\operatorname{cov}\left(\alpha_i,\alpha_j\right)\right]_{N_\alpha \times N_\alpha} \triangleq \left\langle \left(\alpha_i - \alpha_i^0\right)\left(\alpha_j - \alpha_j^0\right)\right\rangle_{N_\alpha \times N_\alpha}; \; i,j = 1,\ldots,N_\alpha, \tag{3.8}$$

while the second-order central moments (i.e., the standard deviations and correlations) for the parameters β_m form covariance matrices $\mathbf{C}_{\beta\beta}^{(N_\beta \times N_\beta)}$ defined as

$$\mathbf{C}_{\beta\beta}^{(N_\beta \times N_\beta)} \triangleq \left[\operatorname{cov}\left(\beta_i,\beta_j\right)\right]_{N_\beta \times N_\beta} \triangleq \left\langle \left(\beta_i - \beta_i^0\right)\left(\beta_j - \beta_j^0\right)\right\rangle_{N_\beta \times N_\beta}; \; i,j = 1,\ldots,N_\beta. \tag{3.9}$$

In general, the components of the vectors $\boldsymbol{\alpha}$ and $\boldsymbol{\beta}$ may be correlated. The correlations among the parameters $\boldsymbol{\alpha}$ and $\boldsymbol{\beta}$ are quantified by correlation matrices $\mathbf{C}_{\alpha\beta}^{(N_\alpha \times N_\beta)}$ defined as

$$\mathbf{C}_{\alpha\beta}^{(N_\alpha \times N_\beta)} \triangleq \left\langle \left(\boldsymbol{\alpha} - \boldsymbol{\alpha}^0\right)\left(\boldsymbol{\beta} - \boldsymbol{\beta}^0\right)^\dagger \right\rangle = \left[\mathbf{C}_{\beta\alpha}^{(N_\beta \times N_\alpha)}\right]^\dagger. \tag{3.10}$$

These experimentally measured responses are also considered to be characterized by known mean measured values and measured variances and covariances. Thus, for the N_r experimentally measured responses r_i, the mean measured values will be denoted as r_i^m, and will be considered to constitute the components of the vector $\mathbf{r}^m$ defined as

$$\mathbf{r}^m \triangleq \left(r_1^m,\ldots,r_{N_r}^m\right)^\dagger, \;\; r_i^m \triangleq \left\langle r_i \right\rangle, \; i = 1,\ldots,N_r, \tag{3.11}$$

while the corresponding measured covariance matrix, denoted as $\mathbf{C}_{rr}^{(N_r \times N_r)}$, is defined as

$$\mathbf{C}_{rr}^{(N_r \times N_r)} \triangleq \left\langle \left(r_i - r_i^m\right)\left(r_j - r_j^m\right)\right\rangle_{N_r \times N_r}, \; i,j = 1,\ldots,N_r. \tag{3.12}$$

Similarly, the N_q experimentally measured responses q_j are characterized by mean measured values, denoted as q_j^m, which constitute the components of the vector $\mathbf{q}^m$ defined as

$$\mathbf{q}^m \triangleq \left(q_1^m,\ldots,q_{N_q}^m\right)^\dagger, \;\; q_j^m \triangleq \left\langle q_j \right\rangle, \; j = 1,\ldots,N_q, \tag{3.13}$$

and by the measured covariance matrix $\mathbf{C}_{qq}^{(N_q \times N_q)}$ defined as

$$\mathbf{C}_{qq}^{(N_q \times N_q)} \triangleq \left\langle \left(q_i - q_i^m\right)\left(q_j - q_j^m\right) \right\rangle_{N_q \times N_q}, \quad i,j = 1,\ldots,N_q. \tag{3.14}$$

Furthermore, the responses $\mathbf{r}$ and $\mathbf{q}$ may also be correlated; such correlations would be quantified by correlation matrices defined as

$$\mathbf{C}_{rq}^{(N_r \times N_q)} \triangleq \left\langle \left(\mathbf{r} - \mathbf{r}^m\right)\left(\mathbf{q} - \mathbf{q}^m\right)^\dagger \right\rangle = \left[\mathbf{C}_{qr}^{(N_q \times N_r)}\right]^\dagger . \tag{3.15}$$

In the most general case, correlations my also exist among all parameters and responses. Such correlations would be quantified through matrices defined as follows:

$$\mathbf{C}_{\alpha r}^{(N_\alpha \times N_r)} \triangleq \left\langle \left(\boldsymbol{\alpha} - \boldsymbol{\alpha}^0\right)\left(\mathbf{r} - \mathbf{r}^m\right)^\dagger \right\rangle = \left[\mathbf{C}_{r\alpha}^{(N_r \times N_\alpha)}\right]^\dagger , \tag{3.16}$$

$$\mathbf{C}_{\alpha q}^{(N_\alpha \times N_q)} \triangleq \left\langle \left(\boldsymbol{\alpha} - \boldsymbol{\alpha}^0\right)\left(\mathbf{q} - \mathbf{q}^m\right)^\dagger \right\rangle = \left[\mathbf{C}_{q\alpha}^{(N_q \times N_\alpha)}\right]^\dagger , \tag{3.17}$$

$$\mathbf{C}_{\beta r}^{(N_\beta \times N_r)} \triangleq \left\langle \left(\boldsymbol{\beta} - \boldsymbol{\beta}^0\right)\left(\mathbf{r} - \mathbf{r}^m\right)^\dagger \right\rangle = \left[\mathbf{C}_{r\beta}^{(N_r \times N_\beta)}\right]^\dagger , \tag{3.18}$$

$$\mathbf{C}_{\beta q}^{(N_\beta \times N_q)} \triangleq \left\langle \left(\boldsymbol{\beta} - \boldsymbol{\beta}^0\right)\left(\mathbf{q} - \mathbf{q}^m\right)^\dagger \right\rangle = \left[\mathbf{C}_{q\beta}^{(N_q \times N_\beta)}\right]^\dagger . \tag{3.19}$$

3.1.2 Construction of the A Priori Distribution Function $p(\boldsymbol{\alpha},\boldsymbol{\beta},\mathbf{r},\mathbf{q})$ as the Maximum Entropy Principle Approximation of the True but Unknown A Priori Distribution Function $P(\boldsymbol{\alpha},\boldsymbol{\beta},\mathbf{r},\mathbf{q})$

The quantities defined in Eqs (3.1) through (3.19) constitute the prior information regarding the uncertain parameters and measured responses in the two-model multi-physics system considered in the previous Section. This prior information prescribes the means (i.e., the first-order moments) and covariances (i.e., the second-order moments) of an otherwise unknown distribution function $p(\boldsymbol{\alpha},\boldsymbol{\beta},\mathbf{r},\mathbf{q})$. Mathematically, these means and covariances are functionals of $p(\boldsymbol{\alpha},\boldsymbol{\beta},\mathbf{r},\mathbf{q})$, having the generic form

$$\langle F_k \rangle \triangleq \int p(\mathbf{x}) F_k(\mathbf{x}) d\mathbf{x}, \quad \mathbf{x} \triangleq (\boldsymbol{\alpha},\boldsymbol{\beta},\mathbf{r},\mathbf{q}), \; d\mathbf{x} \triangleq d\boldsymbol{\alpha}\, d\boldsymbol{\beta}\, d\mathbf{r}\, d\mathbf{q}, \; k = 1,2,\ldots,K, \tag{3.20}$$

with $F_k(\mathbf{x})$ representing, in turn, the quantities: $(\alpha_n - \alpha_n^0)$, $(\beta_n - \beta_n^0)$, $(r_n - r_n^m)$, $(q_n - q_n^m)$, $(\alpha_i - \alpha_i^0)(\alpha_j - \alpha_j^0)$, $(\beta_i - \beta_i^0)(\beta_j - \beta_j^0)$, $(r_i - r_i^m)(r_j - r_j^m)$, $(q_i - q_i^m)(q_j - q_j^m)$, $(\alpha_i - \alpha_i^0)(\beta_j - \beta_j^0)$, $(\alpha_i - \alpha_i^0)(r_j - r_j^m)$, $(\alpha_i - \alpha_i^0)(q_j - q_j^m)$, $(\beta_i - \beta_i^0)(r_j - r_j^m)$, $(\beta_i - \beta_i^0)(q_j - q_j^m)$, and $(r_i - r_i^m)(q_j - q_j^m)$.

The total number of first- and second-order moments is

$$\begin{aligned} K \triangleq\, & N_\alpha + N_\beta + N_r + N_q + N_\alpha^2 + N_\beta^2 + N_r^2 + N_q^2 \\ & + (N_\alpha \times N_\beta) + (N_\alpha \times N_r) + (N_\alpha \times N_q) \\ & + (N_\beta \times N_r) + (N_\beta \times N_q) + (N_r \times N_q). \end{aligned} \tag{3.21}$$

The maximum entropy formalism enables the determination of an approximate probability distribution function, denoted here as $p(\mathbf{x})$, which approximates the true but unknown distribution $P(\mathbf{x})$ by maximizing over $p(\mathbf{x})$ the Shannon information entropy (Shannon, 1948), defined as

$$S \triangleq -\int d\mathbf{x}\, p(\mathbf{x}) \ln \frac{p(\mathbf{x})}{m(\mathbf{x})}, \tag{3.22}$$

where $m(\mathbf{x})$ is a prior density that ensures form invariance under change of variable, while satisfying the constraints given in Eq. (3.20). This maximum entropy principle insures that the approximate distribution function $p(\mathbf{x})$ maximizes the optimal compatibility with the available information, namely the constraints given in Eq. (3.20), while simultaneously ensuring minimal spurious information content. Maximizing the information entropy S over $p(\mathbf{x})$ subject to the constraints expressed by Eq. (3.20) constitutes a variational problem that can be solved by using the method of Lagrange multipliers to obtain a member of the exponential family, namely

$$p(\mathbf{x}) = \frac{1}{Z} m(\mathbf{x}) \exp\left[-\sum_k \lambda_k F_k(\mathbf{x}) \right], \tag{3.23}$$

where the quantities λ_k are the Lagrange multipliers. The normalization constant Z in Eq.(3.23). is defined as

$$Z \triangleq \int d\mathbf{x}\, m(\mathbf{x}) \exp\left[-\sum_k \lambda_k F_k(\mathbf{x}) \right]. \tag{3.24}$$

The Lagrange multipliers λ_k can be determined directly from the constraints [i.e., using Eqs. (3.20) and (3.23) or from the equivalent equations

$$\langle F_k \rangle = -\frac{\partial}{\partial \lambda_k} \ln Z \ , \qquad k = 1, 2, \ldots, K \ , \tag{3.25}$$

which are more convenient to compute if Z can be expressed as an analytic function of the Lagrange parameters.

In the case of discrete distributions, if only the alternatives can be enumerated but the macroscopic data $\langle F_k \rangle$ are not known, then $m(\mathbf{x}) = 1$, and the maximum entropy algorithm described in the foregoing yields the uniform distribution, as would be required by the principle of insufficient reason. Therefore, the maximum entropy principle can be considered to be a far-reaching generalization of the principle of insufficient reason, ranging from discrete alternatives with no other information given, to cases with given global or macroscopic information, and also encompassing continuous distributions. Physicists will recognize the maximum entropy algorithm described above as the essence of the Gibbs-formalism for statistical mechanics, where Z is the partition function (or sum over states), carrying all information about the possible states of the system, from which the expected macroscopic parameters can be obtained by differentiation with respect to the Lagrange multipliers. If only the possible energies of a system and the average energy (i.e., the temperature) are given, one finds Gibbs' canonical ensemble, with probabilities proportional to the Boltzmann factors $\exp(-\lambda E_j)$, the Lagrange multiplier λ being essentially the inverse temperature. If, in addition, the average particle number is given, one finds the grand-canonical ensemble, with a second Lagrange multiplier equal to the chemical potential, an so on.

Performing the respective (lengthy but straightforward) computations indicated in Eq (3.25), solving the resulting system of equation for the Lagrange multipliers λ_k, and replacing the resulting expressions in Eq. (3.23) leads to the following expression for $p(\mathbf{x})$

$$p(\mathbf{x} \mid \langle \mathbf{x} \rangle, \mathbf{C})\, d\mathbf{x} = \frac{\exp\left[-\frac{1}{2} (\mathbf{x} - \langle \mathbf{x} \rangle)^{\dagger} \mathbf{C}^{-1} (\mathbf{x} - \langle \mathbf{x} \rangle) \right] d\mathbf{x},}{\sqrt{\det(2\pi \mathbf{C})}}, \quad -\infty < x_j < \infty \ , \tag{3.26}$$

where the dagger ($\dagger$) denotes transposition (Hermitian conjugation of real vectors and matrices), and the matrix $\mathbf{C}$ is defined as

$$\mathbf{C} \simeq \begin{pmatrix} \mathbf{C}_{\alpha\alpha} & \mathbf{C}_{\alpha\beta} & \mathbf{C}_{\alpha r} & \mathbf{C}_{\alpha q} \\ \mathbf{C}_{\beta\alpha} & \mathbf{C}_{\beta\beta} & \mathbf{C}_{\beta r} & \mathbf{C}_{\beta q} \\ \mathbf{C}_{r\alpha} & \mathbf{C}_{r\beta} & \mathbf{C}_{rr} & \mathbf{C}_{rq} \\ \mathbf{C}_{q\alpha} & \mathbf{C}_{q\beta} & \mathbf{C}_{qr} & \mathbf{C}_{qq} \end{pmatrix}, \; with \; \mathbf{x} \triangleq \begin{pmatrix} \boldsymbol{\alpha} \\ \boldsymbol{\beta} \\ \mathbf{r} \\ \mathbf{q} \end{pmatrix}, \; \langle \mathbf{x} \rangle \triangleq \begin{pmatrix} \boldsymbol{\alpha}^0 \\ \boldsymbol{\beta}^0 \\ \mathbf{r}^m \\ \mathbf{q}^m \end{pmatrix}. \tag{3.27}$$

Thus, the foregoing considerations show that, when only mean values and covariances are known, the maximum entropy algorithm yields the Gaussian probability distribution shown in Eq.(3.27) as the most objective probability distribution consistent with the available information. Although all of the above results are valid for $-\infty < x_j < \infty$, these results can also be used for $0 < x_j < \infty$ after introduction of a logarithmic scale (which leads to lognormal distributions on the original scale).
Gaussian distributions are often considered appropriate only if many independent random deviations act together so that the central limit theorem is applicable. At other times, Gaussian distributions are invoked for more convenience, with accompanying warnings about consequences if the true distribution is not Gaussian. The maximum entropy principle cannot eliminate these consequences, but it reassures the data user who is given only mean values and their (co)variances that the corresponding Gaussian is the best choice for all further inferences, whatever the unknown true distribution may happen to be. In contrast to the central limit theorem, the maximum entropy principle is also valid for correlated data.

3.1.3 Construction of the A Posteriori Predicted Mean Values and Covariances for the Given Models (Likelihood Function) and Maximum Entropy Prior Distribution

Consider next that the coupled Models A and B are used to compute the $(N_r + N_q)$ experimentally measured responses. These computed responses will be denoted as $\mathbf{r}^c(\boldsymbol{\alpha},\boldsymbol{\beta}) \triangleq (r_1^c,...,r_{N_r}^c)^\dagger$, and $\mathbf{q}^c(\boldsymbol{\alpha},\boldsymbol{\beta}) \triangleq (q_1^c,...,q_{N_q}^c)^\dagger$, respectively, where the superscript "c" indicates "computed." In principle, the computed responses may depend on some or all of the components of $\boldsymbol{\alpha}$ and $\boldsymbol{\beta}$. Consequently, $\mathbf{r}^c(\boldsymbol{\alpha},\boldsymbol{\beta})$ and $\mathbf{q}^c(\boldsymbol{\alpha},\boldsymbol{\beta})$ are also variates, characterized by probability distribution functions, which cannot, in general, be obtained in explicitly closed forms.
The next step is to combine the experimental and computational information in order to obtain the posterior distribution of $\mathbf{x} \triangleq (\boldsymbol{\alpha},\boldsymbol{\beta},\mathbf{r},\mathbf{q})$. This combination is rigorously performed by using Bayes' theorem, in which the (maximum entropy) prior is the Gaussian distribution computed in Eq. (3.26), while the likelihood is provided by the computational models $\mathbf{r}^c(\boldsymbol{\alpha},\boldsymbol{\beta})$ and $\mathbf{q}^c(\boldsymbol{\alpha},\boldsymbol{\beta})$. When the numerical and/or

modeling errors are not explicitly taken into account, but are considered to be amenable to treatment via uncertain model parameters that are included among the components of $\boldsymbol{\alpha}$, the computational models are considered to be "hard constraints" of the form

$$\mathbf{r} = \mathbf{r}^{c}(\boldsymbol{\alpha},\boldsymbol{\beta}), \;\; \mathbf{q} = \mathbf{q}^{c}(\boldsymbol{\alpha},\boldsymbol{\beta}) \, . \tag{3.28}$$

It is clear the posterior distribution, which consists of the prior given in Eq (3.26) together with the likelihood expressed by Eq. (3.28), cannot be computed exactly. Nevertheless, the main contribution to the posterior distribution, and, in particular, the main contributions to the posterior distribution's means and covariances, can be obtained by applying the saddle-point method to evaluate the Gaussian prior in Eq. (3.26) subject to the constraints expressed by Eq. (3.28). As is well known, the saddle-point is the point where the gradient of exponent of the Gaussian prior in Eq.(3.26) vanishes subject to the constraints in Eq. (3.28). The method of Lagrange multipliers can be used to determine this saddle-point, by setting to zero the (partial) gradients with respect to $\boldsymbol{\alpha},\boldsymbol{\beta},\mathbf{r},\mathbf{q}$ of the following functional:

$$\begin{aligned} P(\boldsymbol{\alpha},\boldsymbol{\beta},\mathbf{r},\mathbf{q}) &\triangleq -\frac{1}{2}(\mathbf{x}-\langle\mathbf{x}\rangle)^{\dagger}\mathbf{C}^{-1}(\mathbf{x}-\langle\mathbf{x}\rangle) \\ &+\boldsymbol{\lambda}_r^{\dagger}\left[\mathbf{r}-\mathbf{r}^{c}(\boldsymbol{\alpha},\boldsymbol{\beta})\right]+\boldsymbol{\lambda}_q^{\dagger}\left[\mathbf{q}-\mathbf{q}^{c}(\boldsymbol{\alpha},\boldsymbol{\beta})\right], \end{aligned} \tag{3.29}$$

where $\boldsymbol{\lambda}_r$ and $\boldsymbol{\lambda}_q$ are vectors of (yet undetermined) Lagrange multipliers of sizes N_r and N_q, respectively. Thus, the saddle point of $P(\boldsymbol{\alpha},\boldsymbol{\beta},\mathbf{r},\mathbf{q})$ is attained at $\mathbf{x}^{pred} \triangleq \left(\boldsymbol{\alpha}^{pred},\boldsymbol{\beta}^{pred},\mathbf{r}^{pred},\mathbf{q}^{pred}\right)$ where the following conditions are simultaneously fulfilled:

$$\nabla_{\lambda_r} P = \mathbf{r}-\mathbf{r}^{c}(\boldsymbol{\alpha},\boldsymbol{\beta}) = \mathbf{0}; \quad \nabla_{\lambda_q} P = \mathbf{q}-\mathbf{q}^{c}(\boldsymbol{\alpha},\boldsymbol{\beta}) = \mathbf{0}; \tag{3.30}$$

$$\nabla_{\alpha} P = \mathbf{0}; \; \nabla_{\beta} P = \mathbf{0}; \; \nabla_{r} P = \mathbf{0}; \; \nabla_{q} P = \mathbf{0}. \tag{3.31}$$

The conditions expressed in Eq. (3.30) simply ensure that the saddle-point will satisfy the constraints imposed by the numerical simulation Models A and B. On the other hand, the conditions imposed in Eq. (3.31) can be written in block-matrix form as

$$\begin{pmatrix} \boldsymbol{\alpha}^{pred}-\boldsymbol{\alpha}^{0} \\ \boldsymbol{\beta}^{pred}-\boldsymbol{\beta}^{0} \\ \mathbf{r}^{pred}-\mathbf{r}^{m} \\ \mathbf{q}^{pred}-\mathbf{q}^{m} \end{pmatrix} = \begin{pmatrix} \mathbf{C}_{\alpha\alpha} & \mathbf{C}_{\alpha\beta} & \mathbf{C}_{\alpha r} & \mathbf{C}_{\alpha q} \\ \mathbf{C}_{\alpha\beta}^{\dagger} & \mathbf{C}_{\beta\beta} & \mathbf{C}_{\beta r} & \mathbf{C}_{\beta q} \\ \mathbf{C}_{\alpha r}^{\dagger} & \mathbf{C}_{\beta r}^{\dagger} & \mathbf{C}_{rr} & \mathbf{C}_{rq} \\ \mathbf{C}_{\alpha q}^{\dagger} & \mathbf{C}_{\beta q}^{\dagger} & \mathbf{C}_{rq}^{\dagger} & \mathbf{C}_{qq} \end{pmatrix} \begin{pmatrix} -\mathbf{S}_{r\alpha}^{\dagger}\boldsymbol{\lambda}_r-\mathbf{S}_{q\alpha}^{\dagger}\boldsymbol{\lambda}_q \\ -\mathbf{S}_{r\beta}^{\dagger}\boldsymbol{\lambda}_r-\mathbf{S}_{q\beta}^{\dagger}\boldsymbol{\lambda}_q \\ \boldsymbol{\lambda}_r \\ \boldsymbol{\lambda}_q \end{pmatrix} \tag{3.32}$$

where the matrices $\mathbf{S}_{r\alpha}\left(\boldsymbol{\alpha}^0,\boldsymbol{\beta}^0\right)$, $\mathbf{S}_{r\beta}\left(\boldsymbol{\alpha}^0,\boldsymbol{\beta}^0\right)$, $\mathbf{S}_{q\alpha}\left(\boldsymbol{\alpha}^0,\boldsymbol{\beta}^0\right)$, and $\mathbf{S}_{q\beta}\left(\boldsymbol{\alpha}^0,\boldsymbol{\beta}^0\right)$ comprise first-order response-derivatives with respect to the model parameters, computed at the nominal parameter values $\left(\boldsymbol{\alpha}^0,\boldsymbol{\beta}^0\right)$, and are defined as follows:

$$\mathbf{S}_{r\alpha}^{N_r\times N_\alpha} \triangleq \begin{bmatrix} \frac{\partial r_1}{\partial \alpha_1} & \cdots & \frac{\partial r_1}{\partial \alpha_{N_\alpha}} \\ \vdots & \ddots & \vdots \\ \frac{\partial r_{N_r}}{\partial \alpha_1} & \cdots & \frac{\partial r_{N_r}}{\partial \alpha_{N_\alpha}} \end{bmatrix}, \quad \mathbf{S}_{r\beta}^{N_r\times N_\beta} \triangleq \begin{bmatrix} \frac{\partial r_1}{\partial \beta_1} & \cdots & \frac{\partial r_1}{\partial \beta_{N_\beta}} \\ \vdots & \ddots & \vdots \\ \frac{\partial r_{N_r}}{\partial \beta_1} & \cdots & \frac{\partial r_{N_r}}{\partial \beta_{N_\beta}} \end{bmatrix}, \tag{3.33}$$

$$\mathbf{S}_{q\alpha}^{N_q\times N_\alpha} \triangleq \begin{bmatrix} \frac{\partial q_1}{\partial \alpha_1} & \cdots & \frac{\partial q_1}{\partial \alpha_{N_\alpha}} \\ \vdots & \ddots & \vdots \\ \frac{\partial q_{N_q}}{\partial \alpha_1} & \cdots & \frac{\partial q_{N_q}}{\partial \alpha_{N_\alpha}} \end{bmatrix}, \quad \mathbf{S}_{q\beta}^{N_q\times N_\beta} \triangleq \begin{bmatrix} \frac{\partial q_1}{\partial \beta_1} & \cdots & \frac{\partial q_1}{\partial \beta_{N_\beta}} \\ \vdots & \ddots & \vdots \\ \frac{\partial q_{N_q}}{\partial \beta_1} & \cdots & \frac{\partial q_{N_q}}{\partial \beta_{N_\beta}} \end{bmatrix}. \tag{3.34}$$

In component form, Eq. (3.32) is equivalent to the following relations:

$$\left(\boldsymbol{\alpha}^{pred}-\boldsymbol{\alpha}^0\right)=-\mathbf{C}_{\alpha\alpha}\left(\mathbf{S}_{r\alpha}^\dagger\boldsymbol{\lambda}_r+\mathbf{S}_{q\alpha}^\dagger\boldsymbol{\lambda}_q\right)-\mathbf{C}_{\alpha\beta}\left(\mathbf{S}_{r\beta}^\dagger\boldsymbol{\lambda}_r+\mathbf{S}_{q\beta}^\dagger\boldsymbol{\lambda}_q\right)+\mathbf{C}_{\alpha r}\boldsymbol{\lambda}_r+\mathbf{C}_{\alpha q}\boldsymbol{\lambda}_q \tag{3.35}$$

$$\left(\boldsymbol{\beta}^{pred}-\boldsymbol{\beta}^0\right)=-\mathbf{C}_{\alpha\beta}^\dagger\left(\mathbf{S}_{r\alpha}^\dagger\boldsymbol{\lambda}_r+\mathbf{S}_{q\alpha}^\dagger\boldsymbol{\lambda}_q\right)-\mathbf{C}_{\beta\beta}\left(\mathbf{S}_{r\beta}^\dagger\boldsymbol{\lambda}_r+\mathbf{S}_{q\beta}^\dagger\boldsymbol{\lambda}_q\right)+\mathbf{C}_{\beta r}\boldsymbol{\lambda}_r+\mathbf{C}_{\beta q}\boldsymbol{\lambda}_q \tag{3.36}$$

$$\left(\mathbf{r}^{pred}-\mathbf{r}^0\right)=-\mathbf{C}_{\alpha r}^\dagger\left(\mathbf{S}_{r\alpha}^\dagger\boldsymbol{\lambda}_r+\mathbf{S}_{q\alpha}^\dagger\boldsymbol{\lambda}_q\right)-\mathbf{C}_{\beta r}^\dagger\left(\mathbf{S}_{r\beta}^\dagger\boldsymbol{\lambda}_r+\mathbf{S}_{q\beta}^\dagger\boldsymbol{\lambda}_q\right)+\mathbf{C}_{rr}\boldsymbol{\lambda}_r+\mathbf{C}_{rq}\boldsymbol{\lambda}_q \tag{3.37}$$

$$\left(\mathbf{q}^{pred}-\mathbf{q}^0\right)=-\mathbf{C}_{\alpha q}^\dagger\left(\mathbf{S}_{r\alpha}^\dagger\boldsymbol{\lambda}_r+\mathbf{S}_{q\alpha}^\dagger\boldsymbol{\lambda}_q\right)-\mathbf{C}_{\beta q}^\dagger\left(\mathbf{S}_{r\beta}^\dagger\boldsymbol{\lambda}_r+\mathbf{S}_{q\beta}^\dagger\boldsymbol{\lambda}_q\right)+\mathbf{C}_{rq}^\dagger\boldsymbol{\lambda}_r+\mathbf{C}_{qq}\boldsymbol{\lambda}_q \tag{3.38}$$

Note that no approximations have been introduced thus far, so that Eqs. (3.35) through (3.38) are exact for the a priori information considered to be known (i.e., known means and covariance matrices for the parameters and measured responses). On the other hand, these equations cannot be used to compute the optimally predicted mean values for the parameters and responses, since the Lagrange multipliers $\boldsymbol{\lambda}_r$ and $\boldsymbol{\lambda}_q$ are still undetermined. Two additional relations are needed to determine these Lagrange multipliers. These relations are obtained by considering the model responses as explicit functions of the model parameters.

To first-order in the parameter variations the model responses $\mathbf{r}$ (for Model A) and $\mathbf{q}$ (for Model B) would be linear functions of the parameter variations of the form

$$\mathbf{r} = \mathbf{r}^c\left(\boldsymbol{\alpha}^0, \boldsymbol{\beta}^0\right) + \mathbf{S}_{r\alpha}\left(\boldsymbol{\alpha} - \boldsymbol{\alpha}^0\right) + \mathbf{S}_{r\beta}\left(\boldsymbol{\beta} - \boldsymbol{\beta}^0\right) + \textit{higher order terms}, \tag{3.39}$$

$$\mathbf{q} = \mathbf{q}^c\left(\boldsymbol{\alpha}^0, \boldsymbol{\beta}^0\right) + \mathbf{S}_{q\alpha}\left(\boldsymbol{\alpha} - \boldsymbol{\alpha}^0\right) + \mathbf{S}_{q\beta}\left(\boldsymbol{\beta} - \boldsymbol{\beta}^0\right) + \textit{higher order terms}. \tag{3.40}$$

In particular, for the predicted parameter values $\boldsymbol{\alpha}^{pred}$ and $\boldsymbol{\beta}^{pred}$, the responses predicted by the linearized models would be given the following expressions:

$$\mathbf{r}^{pred} = \mathbf{r}^c\left(\boldsymbol{\alpha}^0, \boldsymbol{\beta}^0\right) + \mathbf{S}_{r\alpha}\left(\boldsymbol{\alpha}^{pred} - \boldsymbol{\alpha}^0\right) + \mathbf{S}_{r\beta}\left(\boldsymbol{\beta}^{pred} - \boldsymbol{\beta}^0\right) + \textit{higher order terms}, \tag{3.41}$$

$$\mathbf{q}^{pred} = \mathbf{q}^c\left(\boldsymbol{\alpha}^0, \boldsymbol{\beta}^0\right) + \mathbf{S}_{q\alpha}\left(\boldsymbol{\alpha}^{pred} - \boldsymbol{\alpha}^0\right) + \mathbf{S}_{q\beta}\left(\boldsymbol{\beta}^{pred} - \boldsymbol{\beta}^0\right) + \textit{higher order terms}. \tag{3.42}$$

The following intermediate steps are now performed in order to eliminate the Lagrange multipliers: (i) replace $\mathbf{r}^{pred}$ and $\mathbf{q}^{pred}$ from Eqs.(3.41) and (3.42) into Eqs.(3.35) through (3.38) to obtain a system of four equations for the four unknowns $\left(\boldsymbol{\alpha}^{pred}, \boldsymbol{\beta}^{pred}, \boldsymbol{\lambda}_r, \boldsymbol{\lambda}_q\right)$; (ii) from this system, eliminate the quantities $\left(\boldsymbol{\alpha}^{pred} - \boldsymbol{\alpha}^0\right)$ and $\left(\boldsymbol{\beta}^{pred} - \boldsymbol{\beta}^0\right)$; and (iii) re-arrange the resulting equations to obtain the following coupled equations for the Lagrange multipliers:

$$\begin{bmatrix} \mathbf{D}_{rr} & \mathbf{D}_{rq} \\ \mathbf{D}_{qr} & \mathbf{D}_{qq} \end{bmatrix} \begin{bmatrix} \boldsymbol{\lambda}_r \\ \boldsymbol{\lambda}_q \end{bmatrix} = \begin{bmatrix} \mathbf{r}^d\left(\boldsymbol{\alpha}^0, \boldsymbol{\beta}^0\right) \\ \mathbf{q}^d\left(\boldsymbol{\alpha}^0, \boldsymbol{\beta}^0\right) \end{bmatrix}, \tag{3.43}$$

where the block-matrix of known quantities on the left-side, and the block-vector of known quantities on the right-side of the above equations are defined as follows:

$$\begin{aligned} \mathbf{D}_{rr} \triangleq\; & \mathbf{S}_{r\alpha}\left(\mathbf{C}_{\alpha\alpha}\mathbf{S}_{r\alpha}^{\dagger} + \mathbf{C}_{\alpha\beta}\mathbf{S}_{r\beta}^{\dagger} - \mathbf{C}_{\alpha r}\right) + \mathbf{S}_{r\beta}\left(\mathbf{C}_{\alpha\beta}^{\dagger}\mathbf{S}_{r\alpha}^{\dagger} + \mathbf{C}_{\beta\beta}\mathbf{S}_{r\beta}^{\dagger} - \mathbf{C}_{\beta r}\right) \\ & -\mathbf{C}_{\alpha r}^{\dagger}\mathbf{S}_{r\alpha}^{\dagger} - \mathbf{C}_{\beta r}^{\dagger}\mathbf{S}_{r\beta}^{\dagger} + \mathbf{C}_{rr}, \end{aligned} \tag{3.44}$$

$$\begin{aligned} \mathbf{D}_{rq} \triangleq\; & \mathbf{S}_{r\alpha}\left(\mathbf{C}_{\alpha\alpha}\mathbf{S}_{q\alpha}^{\dagger} + \mathbf{C}_{\alpha\beta}\mathbf{S}_{q\beta}^{\dagger} - \mathbf{C}_{\alpha q}\right) + \mathbf{S}_{r\beta}\left(\mathbf{C}_{\alpha\beta}^{\dagger}\mathbf{S}_{q\alpha}^{\dagger} + \mathbf{C}_{\beta\beta}\mathbf{S}_{q\beta}^{\dagger} - \mathbf{C}_{\beta q}\right) \\ & -\mathbf{C}_{\alpha r}^{\dagger}\mathbf{S}_{q\alpha}^{\dagger} - \mathbf{C}_{\beta r}^{\dagger}\mathbf{S}_{q\beta}^{\dagger} + \mathbf{C}_{rq}, \end{aligned} \tag{3.45}$$

$$\begin{aligned} \mathbf{D}_{qr} \triangleq\; & \mathbf{S}_{q\alpha}\left(\mathbf{C}_{\alpha\alpha}\mathbf{S}_{r\alpha}^{\dagger} + \mathbf{C}_{\alpha\beta}\mathbf{S}_{r\beta}^{\dagger} - \mathbf{C}_{\alpha r}\right) + \mathbf{S}_{q\beta}\left(\mathbf{C}_{\alpha\beta}^{\dagger}\mathbf{S}_{r\alpha}^{\dagger} + \mathbf{C}_{\beta\beta}\mathbf{S}_{r\beta}^{\dagger} - \mathbf{C}_{\beta r}\right) \\ & -\mathbf{C}_{\alpha q}^{\dagger}\mathbf{S}_{r\alpha}^{\dagger} - \mathbf{C}_{\beta q}^{\dagger}\mathbf{S}_{r\beta}^{\dagger} + \mathbf{C}_{rq}^{\dagger} = \mathbf{D}_{rq}^{\dagger}, \end{aligned} \tag{3.46}$$

$$\begin{aligned} \mathbf{D}_{qq} \triangleq\; & \mathbf{S}_{q\alpha}\left(\mathbf{C}_{\alpha\alpha}\mathbf{S}_{q\alpha}^{\dagger} + \mathbf{C}_{\alpha\beta}\mathbf{S}_{q\beta}^{\dagger} - \mathbf{C}_{\alpha q}\right) + \mathbf{S}_{q\beta}\left(\mathbf{C}_{\alpha\beta}^{\dagger}\mathbf{S}_{q\alpha}^{\dagger} + \mathbf{C}_{\beta\beta}\mathbf{S}_{q\beta}^{\dagger} - \mathbf{C}_{\beta q}\right) \\ & -\mathbf{C}_{\alpha q}^{\dagger}\mathbf{S}_{q\alpha}^{\dagger} - \mathbf{C}_{\beta q}^{\dagger}\mathbf{S}_{\beta q}^{\dagger} + \mathbf{C}_{qq}, \end{aligned} \tag{3.47}$$

$$\mathbf{r}^d\left(\boldsymbol{\alpha}^0,\boldsymbol{\beta}^0\right) \triangleq \mathbf{r}^c\left(\boldsymbol{\alpha}^0,\boldsymbol{\beta}^0\right) - \mathbf{r}^m; \quad \mathbf{q}^d\left(\boldsymbol{\alpha}^0,\boldsymbol{\beta}^0\right) \triangleq \mathbf{q}^c\left(\boldsymbol{\alpha}^0,\boldsymbol{\beta}^0\right) - \mathbf{q}^m. \tag{3.48}$$

Note that the vectors $\mathbf{r}^d\left(\boldsymbol{\alpha}^0,\boldsymbol{\beta}^0\right)$ and $\mathbf{q}^d\left(\boldsymbol{\alpha}^0,\boldsymbol{\beta}^0\right)$ measure the differences ("deviations") between the computed and measured responses. Note also that the matrices defined in Eqs. (3.44) through (3.47) have the following dimensions: $\dim \mathbf{D}_{rr} = \left(N_r \times N_r\right)$; $\dim \mathbf{D}_{rq} = \left(N_r \times N_q\right)$; $\dim \mathbf{D}_{qr} = \mathbf{D}_{rq}^{\dagger} = \left(N_q \times N_r\right)$; and $\dim \mathbf{D}_{qq} = \left(N_q \times N_q\right)$, and are shown in Appendix 3.A to have the following physical meanings:

(a) The matrix $\mathbf{D}_{rr}$ is actually the covariance matrix of the vector of response "deviations" for Model A, i.e.,

$$\mathbf{D}_{rr} = \left\langle \mathbf{r}^d\left(\boldsymbol{\alpha}^0,\boldsymbol{\beta}^0\right)\left[\mathbf{r}^d\left(\boldsymbol{\alpha}^0,\boldsymbol{\beta}^0\right)\right]^{\dagger} \right\rangle; \tag{3.49}$$

(b) The matrix $\mathbf{D}_{qq}$ is actually the covariance matrix of the vector of response "deviations" for Model B, i.e.,

$$\mathbf{D}_{qq} = \left\langle \mathbf{q}^d\left(\boldsymbol{\alpha}^0,\boldsymbol{\beta}^0\right)\left[\mathbf{q}^d\left(\boldsymbol{\alpha}^0,\boldsymbol{\beta}^0\right)\right]^{\dagger} \right\rangle; \tag{3.50}$$

(c) The matrix $\mathbf{D}_{rq} = \mathbf{D}_{rq}^{\dagger}$ is actually the correlation matrix between the vector of response "deviations" for Model A and Model B, i.e.,

$$\mathbf{D}_{rq} = \left\langle \mathbf{q}^d\left(\boldsymbol{\alpha}^0,\boldsymbol{\beta}^0\right)\left[\mathbf{r}^d\left(\boldsymbol{\alpha}^0,\boldsymbol{\beta}^0\right)\right]^{\dagger} \right\rangle; \; \mathbf{D}_{qr} = \left\langle \mathbf{r}^d\left(\boldsymbol{\alpha}^0,\boldsymbol{\beta}^0\right)\left[\mathbf{q}^d\left(\boldsymbol{\alpha}^0,\boldsymbol{\beta}^0\right)\right]^{\dagger} \right\rangle. \tag{3.51}$$

The Lagrange multipliers $\boldsymbol{\lambda}_r$ and $\boldsymbol{\lambda}_q$ are obtained by solving Eq. (3.41), which requires the inverse of the matrix

$$\mathbf{D} \triangleq \begin{bmatrix} \mathbf{D}_{rr} & \mathbf{D}_{rq} \\ \mathbf{D}_{rq}^{\dagger} & \mathbf{D}_{qq} \end{bmatrix}. \tag{3.52}$$

The above matrix can be inverted by partitioning it to obtain

$$\mathbf{D}^{-1} \triangleq \begin{bmatrix} \mathbf{D}_{11} & \mathbf{D}_{12} \\ \mathbf{D}_{12}^{\dagger} & \mathbf{D}_{22} \end{bmatrix}, \tag{3.53}$$

where

$$\mathbf{D}_{11} \triangleq \mathbf{D}_{rr}^{-1} + \mathbf{D}_{rr}^{-1}\mathbf{D}_{rq}\mathbf{D}_{22}\mathbf{D}_{rq}^{\dagger}\mathbf{D}_{rr}^{-1}, \tag{3.54}$$

$$\mathbf{D}_{12} \triangleq -\mathbf{D}_{rr}^{-1}\mathbf{D}_{rq}\mathbf{D}_{22}, \tag{3.55}$$

$$\mathbf{D}_{12}^{\dagger} \triangleq -\mathbf{D}_{22}\mathbf{D}_{rq}^{\dagger}\mathbf{D}_{rr}^{-1}, \tag{3.56}$$

$$\mathbf{D}_{22} \triangleq \left(\mathbf{D}_{qq} - \mathbf{D}_{rq}^{\dagger}\mathbf{D}_{rr}^{-1}\mathbf{D}_{rq}\right)^{-1}. \tag{3.57}$$

The expressions of $\boldsymbol{\lambda}_r$ and $\boldsymbol{\lambda}_q$ obtained by solving Eq.(3.43) are subsequently replaced in Eqs.(3.35) through (3.38) to obtain the following expressions for the optimally predicted values of model parameters and responses:

1. The best-estimate (optimal) values, $\boldsymbol{\alpha}^{pred}$, for the predicted nominal (mean) values for the parameters of Model A:

$$\boldsymbol{\alpha}^{pred} = \boldsymbol{\alpha}^{0} - \left[\mathbf{X}_{\alpha}\mathbf{D}_{11} + \mathbf{Y}_{\alpha}\mathbf{D}_{12}^{\dagger}\right]\mathbf{r}^{d}\left(\boldsymbol{\alpha}^{0},\boldsymbol{\beta}^{0}\right) - \left[\mathbf{X}_{\alpha}\mathbf{D}_{12} + \mathbf{Y}_{\alpha}\mathbf{D}_{22}\right]\mathbf{q}^{d}\left(\boldsymbol{\alpha}^{0},\boldsymbol{\beta}^{0}\right), \tag{3.58}$$

2. The best-estimate (optimal) values, $\boldsymbol{\beta}^{pred}$, for the predicted nominal (mean) values for the parameters of Model B:

$$\boldsymbol{\beta}^{pred} = \boldsymbol{\beta}^{0} - \left[\mathbf{X}_{\beta}\mathbf{D}_{11} + \mathbf{Y}_{\beta}\mathbf{D}_{12}^{\dagger}\right]\mathbf{r}^{d}\left(\boldsymbol{\alpha}^{0},\boldsymbol{\beta}^{0}\right) - \left[\mathbf{X}_{\beta}\mathbf{D}_{12} + \mathbf{Y}_{\beta}\mathbf{D}_{22}\right]\mathbf{q}^{d}\left(\boldsymbol{\alpha}^{0},\boldsymbol{\beta}^{0}\right), \tag{3.59}$$

3. The best-estimate (optimal) values, $\mathbf{r}^{pred}$, for the predicted nominal (mean) values for the responses of Model A:

$$\mathbf{r}^{pred} = \mathbf{r}^{m} - \left[\mathbf{X}_{r}\mathbf{D}_{11} + \mathbf{Y}_{r}\mathbf{D}_{12}^{\dagger}\right]\mathbf{r}^{d}\left(\boldsymbol{\alpha}^{0},\boldsymbol{\beta}^{0}\right) - \left[\mathbf{X}_{r}\mathbf{D}_{12} + \mathbf{Y}_{r}\mathbf{D}_{22}\right]\mathbf{q}^{d}\left(\boldsymbol{\alpha}^{0},\boldsymbol{\beta}^{0}\right), \tag{3.60}$$

4. The best-estimate (optimal) values, $\mathbf{q}^{pred}$, for the predicted nominal (mean) values for the responses of Model B:

$$\mathbf{q}^{pred} = \mathbf{q}^{m} - \left[\mathbf{X}_{q}\mathbf{D}_{11} + \mathbf{Y}_{q}\mathbf{D}_{12}^{\dagger}\right]\mathbf{r}^{d}\left(\boldsymbol{\alpha}^{0},\boldsymbol{\beta}^{0}\right) - \left[\mathbf{X}_{q}\mathbf{D}_{12} + \mathbf{Y}_{q}\mathbf{D}_{22}\right]\mathbf{q}^{d}\left(\boldsymbol{\alpha}^{0},\boldsymbol{\beta}^{0}\right), \tag{3.61}$$

where

$$\mathbf{X}_{\alpha} \triangleq \mathbf{C}_{\alpha\alpha}\mathbf{S}_{r\alpha}^{\dagger} + \mathbf{C}_{\alpha\beta}\mathbf{S}_{r\beta}^{\dagger} - \mathbf{C}_{\alpha r}, \tag{3.62}$$

$$\mathbf{Y}_{\alpha} \triangleq \mathbf{C}_{\alpha\alpha}\mathbf{S}_{q\alpha}^{\dagger} + \mathbf{C}_{\alpha\beta}\mathbf{S}_{q\beta}^{\dagger} - \mathbf{C}_{\alpha q}, \tag{3.63}$$

$$\mathbf{X}_{\beta} \triangleq \mathbf{C}_{\alpha\beta}^{\dagger}\mathbf{S}_{r\alpha}^{\dagger} + \mathbf{C}_{\beta\beta}\mathbf{S}_{r\beta}^{\dagger} - \mathbf{C}_{\beta r}, \tag{3.64}$$

$$\mathbf{Y}_{\beta} \triangleq \mathbf{C}_{\beta\alpha}\mathbf{S}_{q\alpha}^{\dagger} + \mathbf{C}_{\beta\beta}\mathbf{S}_{q\beta}^{\dagger} - \mathbf{C}_{\beta q}, \tag{3.65}$$

$$\mathbf{X}_{r} \triangleq \mathbf{C}_{\alpha r}^{\dagger}\mathbf{S}_{r\alpha}^{\dagger} + \mathbf{C}_{\beta r}^{\dagger}\mathbf{S}_{r\beta}^{\dagger} - \mathbf{C}_{rr}, \tag{3.66}$$

$$\mathbf{Y}_{r} \triangleq \mathbf{C}_{\alpha r}^{\dagger}\mathbf{S}_{q\alpha}^{\dagger} + \mathbf{C}_{\beta r}^{\dagger}\mathbf{S}_{q\beta}^{\dagger} - \mathbf{C}_{rq}^{\dagger}, \tag{3.67}$$

$$\mathbf{X}_{q} \triangleq \mathbf{C}_{\alpha q}^{\dagger}\mathbf{S}_{r\alpha}^{\dagger} + \mathbf{C}_{\beta q}^{\dagger}\mathbf{S}_{r\beta}^{\dagger} - \mathbf{C}_{rq}^{\dagger}, \tag{3.68}$$

$$\mathbf{Y}_{q} \triangleq \mathbf{C}_{\alpha q}^{\dagger}\mathbf{S}_{q\alpha}^{\dagger} + \mathbf{C}_{\beta q}^{\dagger}\mathbf{S}_{q\beta}^{\dagger} - \mathbf{C}_{qq}. \tag{3.69}$$

The computations of the optimal predicted covariance matrices for the responses and parameters involve tedious algebra which is presented in Appendix 3.B. The final results of these algebraic computations are presented below:

5. The predicted optimal covariance matrix $\mathbf{C}_{\alpha\alpha}^{pred}$ for the parameters $\boldsymbol{\alpha}$ of Model A:

$$\begin{aligned}\mathbf{C}_{\alpha\alpha}^{pred} &\triangleq \left\langle \left(\boldsymbol{\alpha} - \boldsymbol{\alpha}^{pred}\right)\left(\boldsymbol{\alpha} - \boldsymbol{\alpha}^{pred}\right)^{\dagger} \right\rangle \\ &= \mathbf{C}_{\alpha\alpha} - \left[\mathbf{X}_{\alpha}\left(\mathbf{D}_{11}\mathbf{X}_{\alpha}^{\dagger} + \mathbf{D}_{12}\mathbf{Y}_{\alpha}^{\dagger}\right) + \mathbf{Y}_{\alpha}\left(\mathbf{D}_{21}\mathbf{X}_{\alpha}^{\dagger} + \mathbf{D}_{22}\mathbf{Y}_{\alpha}^{\dagger}\right)\right];\end{aligned} \tag{3.70}$$

6. The predicted covariance matrix $\mathbf{C}_{rr}^{pred}$ for the responses $\mathbf{r}$ of Model A:

$$\begin{aligned}\mathbf{C}_{rr}^{pred} &\triangleq \left\langle \left(\mathbf{r} - \mathbf{r}^{pred}\right)\left(\mathbf{r} - \mathbf{r}^{pred}\right)^{\dagger} \right\rangle \\ &= \mathbf{C}_{rr} - \left[\mathbf{X}_{r}\left(\mathbf{D}_{11}\mathbf{X}_{r}^{\dagger} + \mathbf{D}_{12}\mathbf{Y}_{r}^{\dagger}\right) + \mathbf{Y}_{r}\left(\mathbf{D}_{21}\mathbf{X}_{r}^{\dagger} + \mathbf{D}_{22}\mathbf{Y}_{r}^{\dagger}\right)\right];\end{aligned} \tag{3.71}$$

7. The predicted correlation matrix $\mathbf{C}_{\alpha r}^{pred}$ for the parameters $\boldsymbol{\alpha}$ and $\mathbf{r}$ responses of Model A:

$$\begin{aligned}\mathbf{C}_{\alpha r}^{pred} &\triangleq \left\langle \left(\boldsymbol{\alpha} - \boldsymbol{\alpha}^{pred}\right)\left(\mathbf{r} - \mathbf{r}^{pred}\right)^{\dagger} \right\rangle \\ &= \mathbf{C}_{\alpha r} - \left[\mathbf{X}_{\alpha}\left(\mathbf{D}_{11}\mathbf{X}_{r}^{\dagger} + \mathbf{D}_{12}\mathbf{Y}_{r}^{\dagger}\right) + \mathbf{Y}_{\alpha}\left(\mathbf{D}_{21}\mathbf{X}_{r}^{\dagger} + \mathbf{D}_{22}\mathbf{Y}_{r}^{\dagger}\right)\right];\end{aligned} \tag{3.72}$$

8. The predicted covariance matrix $\mathbf{C}_{\beta\beta}^{pred}$ for the parameters $\boldsymbol{\beta}$ of Model B:

$$\mathbf{C}_{\beta\beta}^{pred} \triangleq \left\langle \left(\boldsymbol{\beta}-\boldsymbol{\beta}^{pred}\right)\left(\boldsymbol{\beta}-\boldsymbol{\beta}^{pred}\right)^{\dagger} \right\rangle$$
$$= \mathbf{C}_{\beta\beta} - \left[\mathbf{X}_{\beta}\left(\mathbf{D}_{11}\mathbf{X}_{\beta}^{\dagger}+\mathbf{D}_{12}\mathbf{Y}_{\beta}^{\dagger}\right)+\mathbf{Y}_{\beta}\left(\mathbf{D}_{21}\mathbf{X}_{\beta}^{\dagger}+\mathbf{D}_{22}\mathbf{Y}_{\beta}^{\dagger}\right)\right]; \tag{3.73}$$

9. The predicted covariance matrix $\mathbf{C}_{qq}^{pred}$ for the responses $\mathbf{q}$ of Model B:

$$\mathbf{C}_{qq}^{pred} \triangleq \left\langle \left(\mathbf{q}-\mathbf{q}^{pred}\right)\left(\mathbf{q}-\mathbf{q}^{pred}\right)^{\dagger} \right\rangle$$
$$= \mathbf{C}_{qq} - \left[\mathbf{X}_{q}\left(\mathbf{D}_{11}\mathbf{X}_{q}^{\dagger}+\mathbf{D}_{12}\mathbf{Y}_{q}^{\dagger}\right)+\mathbf{Y}_{q}\left(\mathbf{D}_{21}\mathbf{X}_{q}^{\dagger}+\mathbf{D}_{22}\mathbf{Y}_{q}^{\dagger}\right)\right]; \tag{3.74}$$

10. The predicted correlation matrix $\mathbf{C}_{\beta q}^{pred}$ for the parameters $\boldsymbol{\beta}$ and the responses $\mathbf{q}$ of Model B:

$$\mathbf{C}_{\beta q}^{opt} \triangleq \left\langle \left(\boldsymbol{\beta}-\boldsymbol{\beta}^{pred}\right)\left(\mathbf{q}-\mathbf{q}^{pred}\right)^{\dagger} \right\rangle$$
$$= \mathbf{C}_{\beta q} - \left[\mathbf{X}_{\beta}\left(\mathbf{D}_{11}\mathbf{X}_{q}^{\dagger}+\mathbf{D}_{12}\mathbf{Y}_{q}^{\dagger}\right)+\mathbf{Y}_{\beta}\left(\mathbf{D}_{21}\mathbf{X}_{q}^{\dagger}+\mathbf{D}_{22}\mathbf{Y}_{q}^{\dagger}\right)\right]; \tag{3.75}$$

11. The predicted correlation matrix $\mathbf{C}_{\alpha\beta}^{pred}$ for the parameters $\boldsymbol{\alpha}$ of Model A and the parameters $\boldsymbol{\beta}$ of Model B:

$$\mathbf{C}_{\alpha\beta}^{pred} \triangleq \left\langle \left(\boldsymbol{\alpha}-\boldsymbol{\alpha}^{pred}\right)\left(\boldsymbol{\beta}-\boldsymbol{\beta}^{pred}\right)^{\dagger} \right\rangle$$
$$= \mathbf{C}_{\alpha\beta} - \left[\mathbf{X}_{\alpha}\left(\mathbf{D}_{11}\mathbf{X}_{\beta}^{\dagger}+\mathbf{D}_{12}\mathbf{Y}_{\beta}^{\dagger}\right)+\mathbf{Y}_{\alpha}\left(\mathbf{D}_{21}\mathbf{X}_{\beta}^{\dagger}+\mathbf{D}_{22}\mathbf{Y}_{\beta}^{\dagger}\right)\right]; \tag{3.76}$$

12. The predicted correlation matrix $\mathbf{C}_{\alpha q}^{pred}$ for the parameters $\boldsymbol{\alpha}$ of Model A and the responses $\mathbf{q}$ of Model B:

$$\mathbf{C}_{\alpha q}^{pred} \triangleq \left\langle \left(\boldsymbol{\alpha}-\boldsymbol{\alpha}^{pred}\right)\left(\mathbf{q}-\mathbf{q}^{pred}\right)^{\dagger} \right\rangle$$
$$= \mathbf{C}_{\alpha q} - \left[\mathbf{X}_{\alpha}\left(\mathbf{D}_{11}\mathbf{X}_{q}^{\dagger}+\mathbf{D}_{12}\mathbf{Y}_{q}^{\dagger}\right)+\mathbf{Y}_{\alpha}\left(\mathbf{D}_{21}\mathbf{X}_{q}^{\dagger}+\mathbf{D}_{22}\mathbf{Y}_{q}^{\dagger}\right)\right]; \tag{3.77}$$

13. The predicted correlation matrix $\mathbf{C}_{\beta r}^{pred}$ for the parameters $\boldsymbol{\beta}$ of Model B and the responses $\mathbf{r}$ of Model A:

$$\mathbf{C}_{\beta r}^{pred} \triangleq \left\langle \left(\boldsymbol{\beta}-\boldsymbol{\beta}^{pred}\right)\left(\mathbf{r}-\mathbf{r}^{pred}\right)^{\dagger} \right\rangle$$
$$= \mathbf{C}_{\beta r} - \left[\mathbf{X}_{\beta}\left(\mathbf{D}_{11}\mathbf{X}_{r}^{\dagger}+\mathbf{D}_{12}\mathbf{Y}_{r}^{\dagger}\right)+\mathbf{Y}_{\beta}\left(\mathbf{D}_{21}\mathbf{X}_{r}^{\dagger}+\mathbf{D}_{22}\mathbf{Y}_{r}^{\dagger}\right)\right]; \tag{3.78}$$

14. The predicted correlation matrix $\mathbf{C}_{rq}^{pred}$ for the responses $\mathbf{r}$ of Model A and the responses $\mathbf{q}$ of Model B:

$$\begin{aligned}\mathbf{C}_{rq}^{pred} &\triangleq \left\langle \left(\mathbf{r}-\mathbf{r}^{pred}\right)\left(\mathbf{q}-\mathbf{q}^{pred}\right)^{\dagger}\right\rangle \\ &= \mathbf{C}_{rq} - \left[\mathbf{X}_{r}\left(\mathbf{D}_{11}\mathbf{X}_{q}^{\dagger}+\mathbf{D}_{12}\mathbf{Y}_{q}^{\dagger}\right)+\mathbf{Y}_{r}\left(\mathbf{D}_{21}\mathbf{X}_{q}^{\dagger}+\mathbf{D}_{22}\mathbf{Y}_{q}^{\dagger}\right)\right].\end{aligned} \quad (3.79)$$

Note also that, to first-order in response sensitivities, the covariance matrices of the computed responses arising from the uncertainties in the model parameters can be computed from Eqs.(3.39) and (3.40), respectively, to obtain:

$$\begin{aligned}\mathbf{C}_{rr}^{comp} &\triangleq \left\langle \left[\mathbf{r}-\mathbf{r}^{c}\left(\boldsymbol{\alpha}^{0},\boldsymbol{\beta}^{0}\right)\right]\left[\mathbf{r}-\mathbf{r}^{c}\left(\boldsymbol{\alpha}^{0},\boldsymbol{\beta}^{0}\right)\right]^{\dagger}\right\rangle \\ &= \mathbf{S}_{r\alpha}\mathbf{C}_{\alpha\alpha}\mathbf{S}_{r\alpha}^{\dagger} + 2\mathbf{S}_{r\alpha}\mathbf{C}_{\alpha\beta}\mathbf{S}_{r\beta}^{\dagger} + \mathbf{S}_{r\beta}\mathbf{C}_{\beta\beta}\mathbf{S}_{r\beta}^{\dagger},\end{aligned} \quad (3.80)$$

$$\begin{aligned}\mathbf{C}_{qq}^{comp} &\triangleq \left\langle \left[\mathbf{q}-\mathbf{q}^{c}\left(\boldsymbol{\alpha}^{0},\boldsymbol{\beta}^{0}\right)\right]\left[\mathbf{q}-\mathbf{q}^{c}\left(\boldsymbol{\alpha}^{0},\boldsymbol{\beta}^{0}\right)\right]^{\dagger}\right\rangle \\ &= \mathbf{S}_{q\alpha}\mathbf{C}_{\alpha\alpha}\mathbf{S}_{q\alpha}^{\dagger} + 2\mathbf{S}_{q\alpha}\mathbf{C}_{\alpha\beta}\mathbf{S}_{q\beta}^{\dagger} + \mathbf{S}_{q\beta}\mathbf{C}_{\beta\beta}\mathbf{S}_{q\beta}^{\dagger},\end{aligned} \quad (3.81)$$

$$\begin{aligned}\mathbf{C}_{rq}^{comp} &\triangleq \left\langle \left[\mathbf{r}-\mathbf{r}^{c}\left(\boldsymbol{\alpha}^{0},\boldsymbol{\beta}^{0}\right)\right]\left[\mathbf{q}-\mathbf{q}^{c}\left(\boldsymbol{\alpha}^{0},\boldsymbol{\beta}^{0}\right)\right]^{\dagger}\right\rangle \\ &= \mathbf{S}_{r\alpha}\mathbf{C}_{\alpha\alpha}\mathbf{S}_{q\alpha}^{\dagger} + \mathbf{S}_{r\alpha}\mathbf{C}_{\alpha\beta}\mathbf{S}_{q\alpha}^{\dagger} + \mathbf{S}_{r\beta}\mathbf{C}_{\alpha\beta}^{\dagger}\mathbf{S}_{q\alpha}^{\dagger} + \mathbf{S}_{r\beta}\mathbf{C}_{\beta\beta}\mathbf{S}_{q\beta}^{\dagger}.\end{aligned} \quad (3.82)$$

3.1.4 Construction of the A Posteriori Predicted Consistency Metrics for Model Validation

At the saddle-point $\left(\boldsymbol{\alpha}^{pred},\boldsymbol{\beta}^{pred},\mathbf{r}^{pred},\mathbf{q}^{pred}\right)$, the functional $P\left(\boldsymbol{\alpha},\boldsymbol{\beta},\mathbf{r},\mathbf{q}\right)$ defined in Eq. (3.29), and the first-order computational model equations become

$$P^{\min} = \begin{pmatrix} \boldsymbol{\alpha}^{pred}-\boldsymbol{\alpha}^{0} \\ \boldsymbol{\beta}^{pred}-\boldsymbol{\beta}^{0} \\ \mathbf{r}^{pred}-\mathbf{r}^{m} \\ \mathbf{q}^{pred}-\mathbf{q}^{m} \end{pmatrix}^{\dagger} \mathbf{C}^{-1} \begin{pmatrix} \boldsymbol{\alpha}^{pred}-\boldsymbol{\alpha}^{0} \\ \boldsymbol{\beta}^{pred}-\boldsymbol{\beta}^{0} \\ \mathbf{r}^{pred}-\mathbf{r}^{m} \\ \mathbf{q}^{pred}-\mathbf{q}^{m} \end{pmatrix}, \quad (3.83)$$

$$\mathbf{r}^{pred} = \mathbf{r}^{c}\left(\boldsymbol{\alpha}^{0},\boldsymbol{\beta}^{0}\right) + \mathbf{S}_{r\alpha}\left(\boldsymbol{\alpha}^{pred}-\boldsymbol{\alpha}^{0}\right) + \mathbf{S}_{r\beta}\left(\boldsymbol{\beta}^{pred}-\boldsymbol{\beta}^{0}\right) = \mathbf{r}^{c}\left(\boldsymbol{\alpha}^{pred},\boldsymbol{\beta}^{pred}\right), \quad (3.84)$$

$$\mathbf{q}^{pred} = \mathbf{q}^{c}\left(\boldsymbol{\alpha}^{0},\boldsymbol{\beta}^{0}\right) + \mathbf{S}_{q\alpha}\left(\boldsymbol{\alpha}^{pred}-\boldsymbol{\alpha}^{0}\right) + \mathbf{S}_{q\beta}\left(\boldsymbol{\beta}^{pred}-\boldsymbol{\beta}^{0}\right) = \mathbf{q}^{c}\left(\boldsymbol{\alpha}^{opt},\boldsymbol{\beta}^{opt}\right). \quad (3.85)$$

The values $\left(\boldsymbol{\alpha}^{pred},\boldsymbol{\beta}^{pred},\mathbf{r}^{pred},\mathbf{q}^{pred}\right)$ can be eliminated from the expression of $P^{\min}$ by using Eqs. (3.84) and (3.85) together with Eq. (3.32) in Eq. (3.83) to obtain the following expression:

$$P^{\min} \triangleq V = \left[\left(\mathbf{r}^{d}\right)^{\dagger},\left(\mathbf{q}^{d}\right)^{\dagger}\right]\begin{bmatrix}\mathbf{D}_{11} & \mathbf{D}_{12}\\ \mathbf{D}_{12}^{\dagger} & \mathbf{D}_{22}\end{bmatrix}\begin{bmatrix}\mathbf{r}^{d}\left(\boldsymbol{\alpha}^{0},\boldsymbol{\beta}^{0}\right)\\ \mathbf{q}^{d}\left(\boldsymbol{\alpha}^{0},\boldsymbol{\beta}^{0}\right)\end{bmatrix} = \chi^{2}. \tag{3.86}$$

Note that the quadratic form on the rightmost-side of Eq.(3.86) is distributed according to a χ^2 distribution with $\left(N_r + N_q\right)$ degrees of freedom. Note that the validation metric V can be evaluated directly from the originally given data (i.e., from given parameters and responses, together with their original uncertainties), once the response sensitivities have been computed by either forward or adjoint methods (see, e.g., Cacuci 1981a, 1981b, 2003). Recall that the χ^2 (chi-square) distribution with n degrees of freedom of the continuous variable x, $0 \le x < \infty$, is defined as

$$P\left(x<\chi^{2}<x+dx\right)dx = \frac{1}{2^{n/2}\Gamma\left(n/2\right)}x^{n/2-1}e^{-x/2}dx,\ x>0,\ \left(n=1,2,\ldots\right). \tag{3.87}$$

The χ^2-distribution is a measure of the deviation of a "true distribution" (in this case – the distribution of experimental responses) from the hypothetic one (in this case – a Gaussian). Recall that the mean and variance of x are $\langle x\rangle = n$ and $\operatorname{var}\left(x\right) = 2n$. The value of χ^2 is computed using Eq. (3.86) to obtain

$$\begin{aligned} V \triangleq \chi^{2} &= \left(\mathbf{r}^{c}-\mathbf{r}^{m}\right)^{\dagger}\mathbf{D}_{11}\left(\mathbf{r}^{c}-\mathbf{r}^{m}\right)+2\left(\mathbf{r}^{c}-\mathbf{r}^{m}\right)^{\dagger}\mathbf{D}_{12}\left(\mathbf{q}^{c}-\mathbf{q}^{m}\right)\\ &+\left(\mathbf{q}^{c}-\mathbf{q}^{m}\right)^{\dagger}\mathbf{D}_{22}\left(\mathbf{q}^{c}-\mathbf{q}^{m}\right). \end{aligned} \tag{3.88}$$

The value of χ^2 computed using Eq. (3.88) provides a very valuable quantitative indicator for investigating the agreement between the computed and experimental responses, measuring essentially the consistency of the experimental responses with the model parameters. The value of V can be used as a validation metric for measuring the consistency between the computed and experimentally measured responses.

3.2 DISCUSSION AND PARTICULAR CASES

The derivations in the previous section were carried out in the response-space because, for large-scale practical problems, the number of measured responses is usually smaller than the number of model parameters. The only matrix inversion required in the response space is the computation of $\mathbf{D}^{-1}$ in Eq.(3.53), a matrix of size $\left(N_r + N_q\right)^2$. If the matrix $\mathbf{D}$ is too large to be inverted directly, its inversion can be performed by partitioning it as shown in Eqs (3.54) through (3.57). The inversion of $\mathbf{D}$ by partitioning requires only the inversion of the matrix $\mathbf{D}_{rr}$ of size N_r, and the inversion of the matrix $\left(\mathbf{D}_{qq} - \mathbf{D}_{rq}^{\dagger}\mathbf{D}_{rr}^{-1}\mathbf{D}_{rq}\right)$, which is of size N_q.

The BERRU-CMS methodology presented in Section 3.1 can also be used if one starts with the data assimilation and model calibration for one of the Models (either Model A or Model B), and subsequently couples the second model to the first one. Without the BERRU-CMS methodology, when the second Model (e.g., Model B) is coupled to the first one (e.g., Model A), both models would have to be calibrated anew, simultaneously, and the work performed initially for calibrating Model A alone would become useless. Using the BERRU-CMS methodology, however, the work initially performed for calibrating Model A would not be lost, but would simply be augmented by the specific additional terms arising from Model B, thus performing predictive modeling of coupled multi-physics systems in a sequential and more efficient way.

Furthermore, the explicit separation, effected in Eqs.(3.85) through (3.88), of contributions from Model A and Model B to the overall validation metric V enables the explicit evaluation of adding or subtracting measured responses. As is well known, large contributions to V indicate that the respective responses may be inconsistent or discrepant, and such discrepancies warrant further investigations.

It often happens in practice that, after one has already performed a model calibration, e.g., using Model A (involving N_α model parameters α_n and N_r experimentally measured responses r_i), additional measurements may become available and/or additional parameters (which were not considered in the initial data assimilation/model calibration/predictive modeling procedure) may need to be taken into account (e.g., model parameters for which quantified uncertainties became available only after the initial data assimilation/model calibration/predictive modeling procedure was already performed), all for the same Model A. The predictive modeling methodology presented in Section 3.1 can also be used as a most efficient procedure for systematically adding or subtracting responses and/or parameters for performing a subsequent data assimilation/model calibration/predictive modeling procedure on the same model. In this interpretation/usage of the predictive modeling methodology presented in Section 3.1, Model B is considered to be identical to Model A (i.e., Model B and Model A represent the same physical phenomena, described by identical mathematical equations). In this context, "efficient" means "without wasting the information already obtained in previous predictive modeling computations involving a different (higher or lower) number of responses and/or model parameters." As will be shown in the next Sub-section, the mathematical methodology for performing data assimilation/model calibration/predictive modeling by adding and/or subtracting measurements (responses) and/or model parameters to the same model (without needing to discard previous predictive modeling computations) actually amounts to considering particular cases of the general BERRU-CMS methodology presented in Section 3.1.

3.2.1 "One-Model" Case: Reduction of BERRU-CMS to BERRU-SMS

The "One-Model" Case is defined as performing predictive modeling solely for Model A, involving N_α model parameters α_n and N_r experimentally measured responses r_i. In this case, Eq.(3.44) through (3.47) become

$$\mathbf{D}_{rq} = \mathbf{0},\ \mathbf{D}_{qr} = \mathbf{0},\ \mathbf{D}_{qq} = \mathbf{0},\ \mathbf{D}_{rr} = \mathbf{S}_{r\alpha}\mathbf{C}_{\alpha\alpha}\mathbf{S}_{r\alpha}^{\dagger} - \mathbf{S}_{r\alpha}\mathbf{C}_{\alpha r} - \mathbf{C}_{\alpha r}^{\dagger}\mathbf{S}_{r\alpha}^{\dagger} + \mathbf{C}_{rr}. \quad (3.89)$$

$$\mathbf{X}_{\alpha} \equiv \mathbf{C}_{\alpha\alpha}\mathbf{S}_{r\alpha}^{\dagger} - \mathbf{C}_{\alpha r},\ \mathbf{Y}_{\alpha} \equiv 0,\ \mathbf{X}_{r} \equiv \mathbf{C}_{\alpha r}^{\dagger}\mathbf{S}_{r\alpha}^{\dagger} - \mathbf{C}_{rr},\ \mathbf{Y}_{r} = \mathbf{0}. \quad (3.90)$$

Furthermore, the predictive modeling equations (3.58) through (3.79) will reduce in this case to the final results of the BERRU-SMS methodology presented in Chapter 1 (Cacuci and Ionescu-Bujor, 2010a), namely:

$$\boldsymbol{\alpha}^{pred} = \boldsymbol{\alpha}^{0} - \left(\mathbf{C}_{\alpha\alpha}\mathbf{S}_{r\alpha}^{\dagger} - \mathbf{C}_{\alpha r}\right)\left[\mathbf{D}_{rr}\right]^{-1}\mathbf{r}^{d}\left(\boldsymbol{\alpha}^{0}\right), \quad (3.91)$$

$$\mathbf{r}^{pred} = \mathbf{r}^{m} - \left(\mathbf{C}_{\alpha r}^{\dagger}\mathbf{S}_{r\alpha}^{\dagger} - \mathbf{C}_{rr}\right)\left[\mathbf{D}_{rr}\right]^{-1}\mathbf{r}^{d}\left(\boldsymbol{\alpha}^{0}\right), \quad (3.92)$$

$$\mathbf{C}_{\alpha\alpha}^{pred} = \mathbf{C}_{\alpha\alpha} - \left(\mathbf{C}_{\alpha\alpha}\mathbf{S}_{r\alpha}^{\dagger} - \mathbf{C}_{\alpha r}\right)\left[\mathbf{D}_{rr}\right]^{-1}\left(\mathbf{C}_{\alpha\alpha}\mathbf{S}_{r\alpha}^{\dagger} - \mathbf{C}_{\alpha r}\right)^{\dagger}, \quad (3.93)$$

$$\mathbf{C}_{rr}^{pred} = \mathbf{C}_{rr} - \left(\mathbf{C}_{\alpha r}^{\dagger}\mathbf{S}_{r\alpha}^{\dagger} - \mathbf{C}_{rr}\right)\left[\mathbf{D}_{rr}\right]^{-1}\left(\mathbf{C}_{\alpha r}^{\dagger}\mathbf{S}_{r\alpha}^{\dagger} - \mathbf{C}_{rr}\right)^{\dagger}, \quad (3.94)$$

$$\mathbf{C}_{\alpha r}^{pred} = \mathbf{C}_{\alpha r} - \left(\mathbf{C}_{\alpha\alpha}\mathbf{S}_{r\alpha}^{\dagger} - \mathbf{C}_{\alpha r}\right)\left[\mathbf{D}_{rr}\right]^{-1}\left(\mathbf{C}_{\alpha r}^{\dagger}\mathbf{S}_{r\alpha}^{\dagger} - \mathbf{C}_{rr}\right)^{\dagger}. \quad (3.95)$$

Note that if the model is perfect (i.e., $\mathbf{C}_{\alpha\alpha} \equiv \mathbf{0}$ and $\mathbf{C}_{\alpha r} \equiv \mathbf{0}$), then Eqs. (3.91) through (3.95) would yield $\boldsymbol{\alpha}^{pred} = \boldsymbol{\alpha}^{0}$ and $\mathbf{r}^{pred} = \mathbf{r}^{c}\left(\boldsymbol{\alpha}^{0}, \boldsymbol{\beta}^{0}\right)$, predicted "perfectly," without any accompanying uncertainties (i.e., $\mathbf{C}_{rr}^{pred} \equiv \mathbf{0}$, $\mathbf{C}_{\alpha\alpha}^{pred} \equiv \mathbf{0}$, $\mathbf{C}_{\alpha r}^{pred} \equiv \mathbf{0}$). In other words, for a perfect model, the BERRU-CMS methodology predicts values for the responses and the parameters that coincide with the model's values (assumed to be perfect), and the experimental measurements would have no effect on the predictions (as would be expected, since imperfect measurements could not possibly improve the "perfect" model's predictions).

On the other hand, if the measurements were perfect, (i.e., $\mathbf{C}_{rr} \equiv \mathbf{0}$ and $\mathbf{C}_{\alpha r} \equiv \mathbf{0}$), but the model were imperfect, then Eqs. (3.91) through (3.95) would yield $\boldsymbol{\alpha}^{pred} = \boldsymbol{\alpha}^0 - \mathbf{C}_{\alpha\alpha}\mathbf{S}_{r\alpha}^{\dagger}\left[\mathbf{S}_{r\alpha}\mathbf{C}_{\alpha\alpha}\mathbf{S}_{r\alpha}^{\dagger}\right]^{-1}\mathbf{r}^d\left(\boldsymbol{\alpha}^0\right)$, $\mathbf{r}^{pred} \equiv \mathbf{r}^m$, $\mathbf{C}_{rr}^{pred} \equiv \mathbf{0}$, $\mathbf{C}_{\alpha r}^{pred} \equiv \mathbf{0}$, and $\mathbf{C}_{\alpha\alpha}^{pred} = \mathbf{C}_{\alpha\alpha} - \mathbf{C}_{\alpha\alpha}\mathbf{S}_{r\alpha}^{\dagger}\left[\mathbf{S}_{r\alpha}\mathbf{C}_{\alpha\alpha}\mathbf{S}_{r\alpha}^{\dagger}\right]^{-1}\mathbf{S}_{r\alpha}\mathbf{C}_{\alpha\alpha}$. In other words, in the case of perfect measurements, the BERRU-CMS predicted values for the responses would coincide with the measured values (assumed to be perfect), but the model's uncertain parameters would be calibrated by taking the measurements into account to yield improved nominal values and reduced parameters uncertainties.

3.2.2 BERRU-CMS predictive modeling for Model A with β additional parameters, but no additional responses

In this case, Eq. (3.44) through (3.47) become

$$\mathbf{D}_{rq} \equiv \mathbf{0},\ \mathbf{D}_{qr} \equiv \mathbf{0},\ \mathbf{D}_{qq} \equiv \mathbf{0}, \tag{3.96}$$

$$\begin{aligned}\mathbf{D}_{rr} = {} & \mathbf{S}_{r\alpha}\left(\mathbf{C}_{\alpha\alpha}\mathbf{S}_{r\alpha}^{\dagger} + \mathbf{C}_{\alpha\beta}\mathbf{S}_{r\beta}^{\dagger} - \mathbf{C}_{\alpha r}\right) + \mathbf{S}_{r\beta}\left(\mathbf{C}_{\alpha\beta}^{\dagger}\mathbf{S}_{r\alpha}^{\dagger} + \mathbf{C}_{\beta\beta}\mathbf{S}_{r\beta}^{\dagger} - \mathbf{C}_{\beta r}\right) \\ & - \mathbf{C}_{\alpha r}^{\dagger}\mathbf{S}_{r\alpha}^{\dagger} - \mathbf{C}_{\beta r}^{\dagger}\mathbf{S}_{r\beta}^{\dagger} + \mathbf{C}_{rr}.\end{aligned} \tag{3.97}$$

$$\mathbf{X}_{\alpha} \equiv \mathbf{C}_{\alpha\alpha}\mathbf{S}_{r\alpha}^{\dagger} + \mathbf{C}_{\alpha\beta}\mathbf{S}_{r\beta}^{\dagger} - \mathbf{C}_{\alpha r}, \tag{3.98}$$

$$\mathbf{X}_{\beta} \equiv \mathbf{C}_{\alpha\beta}^{\dagger}\mathbf{S}_{r\alpha}^{\dagger} + \mathbf{C}_{\beta\beta}\mathbf{S}_{r\beta}^{\dagger} - \mathbf{C}_{\beta r}, \tag{3.99}$$

$$\mathbf{X}_{r} \equiv \mathbf{C}_{\alpha r}^{\dagger}\mathbf{S}_{r\alpha}^{\dagger} + \mathbf{C}_{\beta r}^{\dagger}\mathbf{S}_{r\beta}^{\dagger} - \mathbf{C}_{rr}, \tag{3.100}$$

$$\mathbf{X}_{q} \equiv \mathbf{0},\ \mathbf{Y}_{\alpha} \equiv \mathbf{0},\ \mathbf{Y}_{r} \equiv \mathbf{0},\ \mathbf{Y}_{\beta} \equiv \mathbf{0},\ \mathbf{Y}_{q} \equiv \mathbf{0}, \tag{3.101}$$

$$\mathbf{D}_{11} \equiv \mathbf{D}_{rr}^{-1},\ \mathbf{D}_{12} = \mathbf{0},\ \mathbf{D}_{12}^{\dagger} = \mathbf{0},\ \mathbf{D}_{12}^{\dagger} = \mathbf{0},\ \mathbf{D}_{22} = \mathbf{0}, \tag{3.102}$$

$$\boldsymbol{\alpha}^{pred} = \boldsymbol{\alpha}^0 - \mathbf{X}_{\alpha}\mathbf{D}_{11}\mathbf{r}^d\left(\boldsymbol{\alpha}^0, \boldsymbol{\beta}^0\right), \tag{3.103}$$

$$\boldsymbol{\beta}^{pred} = \boldsymbol{\beta}^0 - \mathbf{X}_{\beta}\mathbf{D}_{11}\mathbf{r}^d\left(\boldsymbol{\alpha}^0, \boldsymbol{\beta}^0\right), \tag{3.104}$$

$$\mathbf{r}^{pred} = \mathbf{r}^m - \mathbf{X}_{r}\mathbf{D}_{11}\mathbf{r}^d\left(\boldsymbol{\alpha}^0, \boldsymbol{\beta}^0\right), \tag{3.105}$$

$$\mathbf{C}_{\alpha\alpha}^{pred} = \mathbf{C}_{\alpha\alpha} - \mathbf{X}_{\alpha}\mathbf{D}_{11}\mathbf{X}_{\alpha}^{\dagger}, \tag{3.106}$$

$$\mathbf{C}_{rr}^{pred} = \mathbf{C}_{rr} - \mathbf{X}_{r}\mathbf{D}_{11}\mathbf{X}_{r}^{\dagger}, \tag{3.107}$$

$$\mathbf{C}_{\alpha r}^{pred} = \mathbf{C}_{\alpha r} - \mathbf{X}_{\alpha}\mathbf{D}_{11}\mathbf{X}_{r}^{\dagger}, \tag{3.108}$$

$$\mathbf{C}_{\beta\beta}^{opt} = \mathbf{C}_{\beta\beta} - \mathbf{X}_{\beta}\mathbf{D}_{11}\mathbf{X}_{\beta}^{\dagger}, \tag{3.109}$$

$$\mathbf{C}_{\alpha\beta}^{pred} = \mathbf{C}_{\alpha\beta} - \mathbf{X}_{\alpha}\mathbf{D}_{11}\mathbf{X}_{\beta}^{\dagger}, \tag{3.110}$$

$$\mathbf{C}_{\beta r}^{pred} = \mathbf{C}_{\beta r} - \mathbf{X}_{\beta}\mathbf{D}_{11}\mathbf{X}_{r}^{\dagger}. \tag{3.111}$$

As the above expressions indicate, the predictive modeling formulation in the "response space" (as has been developed in Section 2) allows the consideration of additional parameters for a model without increasing the size of the matrix $\mathbf{D}_{rr}$ to be inverted.

3.2.3 BERRU-CMS predictive modeling for Model A with q additional responses, but no additional parameters

In this case, Eq. (3.44) through (3.47) become

$$\mathbf{D}_{rr} = \mathbf{S}_{r\alpha}\mathbf{C}_{\alpha\alpha}\mathbf{S}_{r\alpha}^{\dagger} - \mathbf{S}_{r\alpha}\mathbf{C}_{\alpha r} - \mathbf{C}_{\alpha r}^{\dagger}\mathbf{S}_{r\alpha}^{\dagger} + \mathbf{C}_{rr} \equiv \mathbf{C}_{d},\ Dim(\mathbf{D}_{rr}) = (N_r \times N_r), \tag{3.112}$$

$$\mathbf{D}_{rq} = \mathbf{S}_{r\alpha}\mathbf{C}_{\alpha\alpha}\mathbf{S}_{q\alpha}^{\dagger} - \mathbf{S}_{r\alpha}\mathbf{C}_{\alpha q} - \mathbf{C}_{\alpha r}^{\dagger}\mathbf{S}_{q\alpha}^{\dagger} + \mathbf{C}_{rq},\ Dim(\mathbf{D}_{rq}) = (N_r \times N_q), \tag{3.113}$$

$$\mathbf{D}_{qr} = \mathbf{S}_{q\alpha}\mathbf{C}_{\alpha\alpha}\mathbf{S}_{r\alpha}^{\dagger} - \mathbf{C}_{\alpha q}^{\dagger}\mathbf{S}_{r\alpha}^{\dagger} - \mathbf{S}_{q\alpha}\mathbf{C}_{\alpha r} + \mathbf{C}_{rq}^{\dagger},\ Dim(\mathbf{D}_{qr}) = (N_q \times N_r), \tag{3.114}$$

$$\mathbf{D}_{qq} = \mathbf{S}_{q\alpha}\mathbf{C}_{\alpha\alpha}\mathbf{S}_{q\alpha}^{\dagger} - \mathbf{S}_{q\alpha}\mathbf{C}_{\alpha q} - \mathbf{C}_{\alpha q}^{\dagger}\mathbf{S}_{q\alpha}^{\dagger} + \mathbf{C}_{qq},\ Dim(\mathbf{D}_{qq}) = (N_q \times N_q). \tag{3.115}$$

$$\mathbf{X}_{\alpha} \equiv \mathbf{C}_{\alpha\alpha}\mathbf{S}_{r\alpha}^{\dagger} - \mathbf{C}_{\alpha r}, \tag{3.116}$$

$$\mathbf{Y}_{\alpha} \equiv \mathbf{C}_{\alpha\alpha}\mathbf{S}_{q\alpha}^{\dagger} - \mathbf{C}_{\alpha q}, \tag{3.117}$$

$$\mathbf{X}_{\beta} \equiv \mathbf{0},\ \mathbf{Y}_{\beta} \equiv \mathbf{0}, \tag{3.118}$$

$$\mathbf{X}_{r} \equiv \mathbf{C}_{\alpha r}^{\dagger}\mathbf{S}_{r\alpha}^{\dagger} - \mathbf{C}_{rr}, \tag{3.119}$$

$$\mathbf{Y}_{r} \equiv \mathbf{C}_{\alpha r}^{\dagger}\mathbf{S}_{q\alpha}^{\dagger} - \mathbf{C}_{rq}^{\dagger} \tag{3.120}$$

$$\mathbf{X}_q \equiv \mathbf{C}_{\alpha q}^{\dagger}\mathbf{S}_{r\alpha}^{\dagger} - \mathbf{C}_{rq}^{\dagger}, \tag{3.121}$$

$$\mathbf{Y}_q \equiv \mathbf{C}_{\alpha q}^{\dagger}\mathbf{S}_{q\alpha}^{\dagger} - \mathbf{C}_{qq}, \tag{3.122}$$

$$\boldsymbol{\alpha}^{pred} = \boldsymbol{\alpha}^0 - \left[\mathbf{X}_\alpha\mathbf{D}_{11} + \mathbf{Y}_\alpha\mathbf{D}_{12}^{\dagger}\right]\mathbf{r}^d\left(\boldsymbol{\alpha}^0,\boldsymbol{\beta}^0\right) - \left[\mathbf{X}_\alpha\mathbf{D}_{12} + \mathbf{Y}_\alpha\mathbf{D}_{22}\right]\mathbf{q}^d\left(\boldsymbol{\alpha}^0,\boldsymbol{\beta}^0\right), \tag{3.123}$$

$$\mathbf{r}^{pred} = \mathbf{r}^m - \left[\mathbf{X}_r\mathbf{D}_{11} + \mathbf{Y}_r\mathbf{D}_{12}^{\dagger}\right]\mathbf{r}^d\left(\boldsymbol{\alpha}^0,\boldsymbol{\beta}^0\right) - \left[\mathbf{X}_r\mathbf{D}_{12} + \mathbf{Y}_r\mathbf{D}_{22}\right]\mathbf{q}^d\left(\boldsymbol{\alpha}^0,\boldsymbol{\beta}^0\right), \tag{3.124}$$

$$\mathbf{q}^{pred} = \mathbf{q}^m - \left[\mathbf{X}_q\mathbf{D}_{11} + \mathbf{Y}_q\mathbf{D}_{12}^{\dagger}\right]\mathbf{r}^d\left(\boldsymbol{\alpha}^0,\boldsymbol{\beta}^0\right) - \left[\mathbf{X}_q\mathbf{D}_{12} + \mathbf{Y}_q\mathbf{D}_{22}\right]\mathbf{q}^d\left(\boldsymbol{\alpha}^0,\boldsymbol{\beta}^0\right), \tag{3.125}$$

$$\mathbf{C}_{\alpha\alpha}^{pred} = \mathbf{C}_{\alpha\alpha} - \left[\mathbf{X}_\alpha\left(\mathbf{D}_{11}\mathbf{X}_\alpha^{\dagger} + \mathbf{D}_{12}\mathbf{Y}_\alpha^{\dagger}\right) + \mathbf{Y}_\alpha\left(\mathbf{D}_{21}\mathbf{X}_\alpha^{\dagger} + \mathbf{D}_{22}\mathbf{Y}_\alpha^{\dagger}\right)\right], \tag{3.126}$$

$$\mathbf{C}_{rr}^{pred} = \mathbf{C}_{rr} - \left[\mathbf{X}_r\left(\mathbf{D}_{11}\mathbf{X}_r^{\dagger} + \mathbf{D}_{12}\mathbf{Y}_r^{\dagger}\right) + \mathbf{Y}_r\left(\mathbf{D}_{21}\mathbf{X}_r^{\dagger} + \mathbf{D}_{22}\mathbf{Y}_r^{\dagger}\right)\right], \tag{3.127}$$

$$\mathbf{C}_{\alpha r}^{pred} = \mathbf{C}_{\alpha r} - \left[\mathbf{X}_\alpha\left(\mathbf{D}_{11}\mathbf{X}_r^{\dagger} + \mathbf{D}_{12}\mathbf{Y}_r^{\dagger}\right) + \mathbf{Y}_\alpha\left(\mathbf{D}_{21}\mathbf{X}_r^{\dagger} + \mathbf{D}_{22}\mathbf{Y}_r^{\dagger}\right)\right], \tag{3.128}$$

$$\mathbf{C}_{qq}^{pred} = \mathbf{C}_{qq} - \left[\mathbf{X}_q\left(\mathbf{D}_{11}\mathbf{X}_q^{\dagger} + \mathbf{D}_{12}\mathbf{Y}_q^{\dagger}\right) + \mathbf{Y}_q\left(\mathbf{D}_{21}\mathbf{X}_q^{\dagger} + \mathbf{D}_{22}\mathbf{Y}_q^{\dagger}\right)\right], \tag{3.129}$$

$$\mathbf{C}_{\alpha q}^{pred} = \mathbf{C}_{\alpha q} - \left[\mathbf{X}_\alpha\left(\mathbf{D}_{11}\mathbf{X}_q^{\dagger} + \mathbf{D}_{12}\mathbf{Y}_q^{\dagger}\right) + \mathbf{Y}_\alpha\left(\mathbf{D}_{21}\mathbf{X}_q^{\dagger} + \mathbf{D}_{22}\mathbf{Y}_q^{\dagger}\right)\right], \tag{3.130}$$

$$\mathbf{C}_{rq}^{pred} = \mathbf{C}_{rq} - \left[\mathbf{X}_r\left(\mathbf{D}_{11}\mathbf{X}_q^{\dagger} + \mathbf{D}_{12}\mathbf{Y}_q^{\dagger}\right) + \mathbf{Y}_r\left(\mathbf{D}_{21}\mathbf{X}_q^{\dagger} + \mathbf{D}_{22}\mathbf{Y}_q^{\dagger}\right)\right], \tag{3.131}$$

$$\mathbf{C}_{\alpha\beta}^{pred} = \mathbf{0},\; \mathbf{C}_{\beta\beta}^{opt} = \mathbf{0},\; \mathbf{C}_{\beta r}^{pred} = \mathbf{0},\; \mathbf{C}_{\beta q}^{opt} = \mathbf{0}. \tag{3.132}$$

Note also that (to first-order in response sensitivities) the covariance matrices of the computed responses arising from the uncertainties in the model parameters become:

$$\mathbf{C}_{rr}^{comp} \equiv \left\langle\left[\mathbf{r} - \mathbf{r}^c\left(\boldsymbol{\alpha}^0,\boldsymbol{\beta}^0\right)\right]\left[\mathbf{r} - \mathbf{r}^c\left(\boldsymbol{\alpha}^0,\boldsymbol{\beta}^0\right)\right]^{\dagger}\right\rangle = \mathbf{S}_{r\alpha}\mathbf{C}_{\alpha\alpha}\mathbf{S}_{r\alpha}^{\dagger}, \tag{3.133}$$

$$\mathbf{C}_{qq}^{comp} \equiv \left\langle\left[\mathbf{q} - \mathbf{q}^c\left(\boldsymbol{\alpha}^0,\boldsymbol{\beta}^0\right)\right]\left[\mathbf{q} - \mathbf{q}^c\left(\boldsymbol{\alpha}^0,\boldsymbol{\beta}^0\right)\right]^{\dagger}\right\rangle = \mathbf{S}_{q\alpha}\mathbf{C}_{\alpha\alpha}\mathbf{S}_{q\alpha}^{\dagger}, \tag{3.134}$$

$$\mathbf{C}_{rq}^{comp} \equiv \left\langle\left[\mathbf{r} - \mathbf{r}^c\left(\boldsymbol{\alpha}^0,\boldsymbol{\beta}^0\right)\right]\left[\mathbf{q} - \mathbf{q}^c\left(\boldsymbol{\alpha}^0,\boldsymbol{\beta}^0\right)\right]^{\dagger}\right\rangle = \mathbf{S}_{r\alpha}\mathbf{C}_{\alpha\alpha}\mathbf{S}_{q\alpha}^{\dagger}. \tag{3.135}$$

APPENDIX 3.A: Computation of the Covariance and Correlation Matrices of $\mathbf{D}_{rr}$, $\mathbf{D}_{rq}$ *and* $\mathbf{D}_{qq}$ *of the "Response Differences"* $\mathbf{r}^d$ *and* $\mathbf{q}^d$

Recall the definitions of $\mathbf{r}^d\left(\boldsymbol{\alpha}^0,\boldsymbol{\beta}^0\right)$ and $\mathbf{q}^d\left(\boldsymbol{\alpha}^0,\boldsymbol{\beta}^0\right)$ given in Eq. (3.48):

$$\mathbf{r}^d\left(\boldsymbol{\alpha}^0,\boldsymbol{\beta}^0\right)\triangleq\mathbf{r}^c\left(\boldsymbol{\alpha}^0,\boldsymbol{\beta}^0\right)-\mathbf{r}^m;\quad \mathbf{q}^d\left(\boldsymbol{\alpha}^0,\boldsymbol{\beta}^0\right)\triangleq\mathbf{q}^c\left(\boldsymbol{\alpha}^0,\boldsymbol{\beta}^0\right)-\mathbf{q}^m. \tag{3.136}$$

Recall also the linearized model responses $\mathbf{r}$ (for Model A) and $\mathbf{q}$ (for Model B), namely:

$$\mathbf{r}=\mathbf{r}^c\left(\boldsymbol{\alpha}^0,\boldsymbol{\beta}^0\right)+\mathbf{S}_{r\alpha}\left(\boldsymbol{\alpha}-\boldsymbol{\alpha}^0\right)+\mathbf{S}_{r\beta}\left(\boldsymbol{\beta}-\boldsymbol{\beta}^0\right)+\textit{higher order terms}, \tag{3.137}$$

$$\mathbf{q}=\mathbf{q}^c\left(\boldsymbol{\alpha}^0,\boldsymbol{\beta}^0\right)+\mathbf{S}_{q\alpha}\left(\boldsymbol{\alpha}-\boldsymbol{\alpha}^0\right)+\mathbf{S}_{q\beta}\left(\boldsymbol{\beta}-\boldsymbol{\beta}^0\right)+\textit{higher order terms}. \tag{3.138}$$

Replacing Eq. (3.39) in Eq.(3.44) and carrying out the respective algebraic computations shows that

$$\begin{aligned}
&\left\langle \mathbf{r}^d\left(\boldsymbol{\alpha}^0,\boldsymbol{\beta}^0\right)\left[\mathbf{r}^d\left(\boldsymbol{\alpha}^0,\boldsymbol{\beta}^0\right)\right]^\dagger\right\rangle \triangleq\left\langle\left(\mathbf{r}^c-\mathbf{r}^m\right)\left(\mathbf{r}^c-\mathbf{r}^m\right)^\dagger\right\rangle\\
&=\left\langle\left[\mathbf{r}-\mathbf{r}^m-\mathbf{S}_{r\alpha}\left(\boldsymbol{\alpha}-\boldsymbol{\alpha}^0\right)-\mathbf{S}_{r\beta}\left(\boldsymbol{\beta}-\boldsymbol{\beta}^0\right)\right]\left[\mathbf{r}-\mathbf{r}^m-\mathbf{S}_{r\alpha}\left(\boldsymbol{\alpha}-\boldsymbol{\alpha}^0\right)-\mathbf{S}_{r\beta}\left(\boldsymbol{\beta}-\boldsymbol{\beta}^0\right)\right]^\dagger\right\rangle\\
&=\left\langle\left(\mathbf{r}-\mathbf{r}^m\right)\left(\mathbf{r}-\mathbf{r}^m\right)^\dagger\right\rangle-\left\langle\left(\mathbf{r}-\mathbf{r}^m\right)\left(\boldsymbol{\alpha}-\boldsymbol{\alpha}^0\right)^\dagger\right\rangle\mathbf{S}_{r\alpha}^\dagger-\left\langle\left(\mathbf{r}-\mathbf{r}^m\right)\left(\boldsymbol{\beta}-\boldsymbol{\beta}^0\right)^\dagger\right\rangle\mathbf{S}_{r\beta}^\dagger-\\
&\quad\mathbf{S}_{r\alpha}\left\langle\left(\boldsymbol{\alpha}-\boldsymbol{\alpha}^0\right)\left(\mathbf{r}-\mathbf{r}^m\right)^\dagger\right\rangle+\mathbf{S}_{r\alpha}\left\langle\left(\boldsymbol{\alpha}-\boldsymbol{\alpha}^0\right)\left(\boldsymbol{\alpha}-\boldsymbol{\alpha}^0\right)^\dagger\right\rangle\mathbf{S}_{r\alpha}^\dagger+\\
&\quad\mathbf{S}_{r\alpha}\left\langle\left(\boldsymbol{\alpha}-\boldsymbol{\alpha}^0\right)\left(\boldsymbol{\beta}-\boldsymbol{\beta}^0\right)^\dagger\right\rangle\mathbf{S}_{r\beta}^\dagger-\mathbf{S}_{r\beta}\left\langle\left(\boldsymbol{\beta}-\boldsymbol{\beta}^0\right)\left(\mathbf{r}-\mathbf{r}^m\right)^\dagger\right\rangle+\\
&\quad\mathbf{S}_{r\beta}\left\langle\left(\boldsymbol{\beta}-\boldsymbol{\beta}^0\right)\left(\boldsymbol{\alpha}-\boldsymbol{\alpha}^0\right)^\dagger\right\rangle\mathbf{S}_{r\alpha}^\dagger+\mathbf{S}_{r\beta}\left\langle\left(\boldsymbol{\beta}-\boldsymbol{\beta}^0\right)\left(\boldsymbol{\beta}-\boldsymbol{\beta}^0\right)^\dagger\right\rangle\mathbf{S}_{r\beta}^\dagger\\
&=\mathbf{C}_{rr}-\mathbf{C}_{r\alpha}\mathbf{S}_{r\alpha}^\dagger-\mathbf{C}_{r\beta}\mathbf{S}_{r\beta}^\dagger-\mathbf{S}_{r\alpha}\mathbf{C}_{\alpha r}+\mathbf{S}_{r\alpha}\mathbf{C}_{\alpha\alpha}\mathbf{S}_{r\alpha}^\dagger+\mathbf{S}_{r\alpha}\mathbf{C}_{\alpha\beta}\mathbf{S}_{r\beta}^\dagger\\
&\quad-\mathbf{S}_{r\beta}\mathbf{C}_{\beta r}+\mathbf{S}_{r\beta}\mathbf{C}_{\beta\alpha}\mathbf{S}_{r\alpha}^\dagger+\mathbf{S}_{r\beta}\mathbf{C}_{\beta\beta}\mathbf{S}_{r\beta}^\dagger\\
&=\mathbf{S}_{r\alpha}\left(-\mathbf{C}_{\alpha r}+\mathbf{C}_{\alpha\alpha}\mathbf{S}_{r\alpha}^\dagger+\mathbf{C}_{\alpha\beta}\mathbf{S}_{r\beta}^\dagger\right)+\mathbf{S}_{r\beta}\left(-\mathbf{C}_{\beta r}+\mathbf{C}_{\beta\alpha}\mathbf{S}_{r\alpha}^\dagger+\mathbf{C}_{\beta\beta}\mathbf{S}_{r\beta}^\dagger\right)+\\
&\quad\mathbf{C}_{rr}-\mathbf{C}_{r\alpha}\mathbf{S}_{r\alpha}^\dagger-\mathbf{C}_{r\beta}\mathbf{S}_{r\beta}^\dagger\\
&\equiv\mathbf{D}_{rr}.
\end{aligned} \tag{3.139}$$

Replacing Eq. (3.40) in Eq (3.44) and carrying out the respective algebraic computations shows that

$$
\begin{aligned}
&\left\langle \mathbf{q}^{d}\left(\boldsymbol{\alpha}^{0},\boldsymbol{\beta}^{0}\right)\left[\mathbf{q}^{d}\left(\boldsymbol{\alpha}^{0},\boldsymbol{\beta}^{0}\right)\right]^{\dagger}\right\rangle \triangleq \left\langle \left(\mathbf{q}^{c}-\mathbf{q}^{m}\right)\left(\mathbf{q}^{c}-\mathbf{q}^{m}\right)^{\dagger}\right\rangle \\
&=\left\langle \left[\mathbf{q}-\mathbf{q}^{m}-\mathbf{S}_{q\alpha}\left(\boldsymbol{\alpha}-\boldsymbol{\alpha}^{0}\right)-\mathbf{S}_{q\beta}\left(\boldsymbol{\beta}-\boldsymbol{\beta}^{0}\right)\right]\times\left[\mathbf{q}-\mathbf{q}^{m}-\mathbf{S}_{q\alpha}\left(\boldsymbol{\alpha}-\boldsymbol{\alpha}^{0}\right)-\mathbf{S}_{q\beta}\left(\boldsymbol{\beta}-\boldsymbol{\beta}^{0}\right)\right]^{\dagger}\right\rangle \\
&=\left\langle \left(\mathbf{q}-\mathbf{q}^{m}\right)\left(\mathbf{q}-\mathbf{q}^{m}\right)^{\dagger}\right\rangle -\left\langle \left(\mathbf{q}-\mathbf{q}^{m}\right)\left(\boldsymbol{\alpha}-\boldsymbol{\alpha}^{0}\right)^{\dagger}\right\rangle \mathbf{S}_{q\alpha}^{\dagger}- \\
&\left\langle \left(\mathbf{q}-\mathbf{q}^{m}\right)\left(\boldsymbol{\beta}-\boldsymbol{\beta}^{0}\right)^{\dagger}\right\rangle \mathbf{S}_{q\beta}^{\dagger}-\mathbf{S}_{q\alpha}\left\langle \left(\boldsymbol{\alpha}-\boldsymbol{\alpha}^{0}\right)\left(\mathbf{q}-\mathbf{q}^{m}\right)^{\dagger}\right\rangle +\mathbf{S}_{q\alpha}\left\langle \left(\boldsymbol{\alpha}-\boldsymbol{\alpha}^{0}\right)\left(\boldsymbol{\alpha}-\boldsymbol{\alpha}^{0}\right)^{\dagger}\right\rangle \mathbf{S}_{q\alpha}^{\dagger} \\
&+\mathbf{S}_{q\alpha}\left\langle \left(\boldsymbol{\alpha}-\boldsymbol{\alpha}^{0}\right)\left(\boldsymbol{\beta}-\boldsymbol{\beta}^{0}\right)^{\dagger}\right\rangle \mathbf{S}_{q\beta}^{\dagger}-\mathbf{S}_{q\beta}\left\langle \left(\boldsymbol{\beta}-\boldsymbol{\beta}^{0}\right)\left(\mathbf{q}-\mathbf{q}^{m}\right)^{\dagger}\right\rangle + \\
&\mathbf{S}_{q\beta}\left\langle \left(\boldsymbol{\beta}-\boldsymbol{\beta}^{0}\right)\left(\boldsymbol{\alpha}-\boldsymbol{\alpha}^{0}\right)^{\dagger}\right\rangle \mathbf{S}_{q\alpha}^{\dagger}+\mathbf{S}_{q\beta}\left\langle \left(\boldsymbol{\beta}-\boldsymbol{\beta}^{0}\right)\left(\boldsymbol{\beta}-\boldsymbol{\beta}^{0}\right)^{\dagger}\right\rangle \mathbf{S}_{q\beta}^{\dagger} \\
&=\mathbf{C}_{qq}-\mathbf{C}_{q\alpha}\mathbf{S}_{q\alpha}^{\dagger}-\mathbf{C}_{q\beta}\mathbf{S}_{q\beta}^{\dagger}-\mathbf{S}_{q\alpha}\mathbf{C}_{\alpha q}+\mathbf{S}_{q\alpha}\mathbf{C}_{\alpha\alpha}\mathbf{S}_{q\alpha}^{\dagger}+\mathbf{S}_{q\alpha}\mathbf{C}_{\alpha\beta}\mathbf{S}_{q\beta}^{\dagger}-\mathbf{S}_{q\beta}\mathbf{C}_{\beta q}+ \\
&\mathbf{S}_{q\beta}\mathbf{C}_{\beta\alpha}\mathbf{S}_{q\alpha}^{\dagger}+\mathbf{S}_{q\beta}\mathbf{C}_{\beta\beta}\mathbf{S}_{q\beta}^{\dagger}=\mathbf{S}_{q\alpha}\left(-\mathbf{C}_{\alpha q}+\mathbf{C}_{\alpha\alpha}\mathbf{S}_{q\alpha}^{\dagger}+\mathbf{C}_{\alpha\beta}\mathbf{S}_{q\beta}^{\dagger}\right)+ \\
&\mathbf{S}_{q\beta}\left(-\mathbf{C}_{\beta q}+\mathbf{C}_{\beta\alpha}\mathbf{S}_{q\alpha}^{\dagger}+\mathbf{C}_{\beta\beta}\mathbf{S}_{q\beta}^{\dagger}\right)+\mathbf{C}_{qq}-\mathbf{C}_{q\alpha}\mathbf{S}_{q\alpha}^{\dagger}-\mathbf{C}_{q\beta}\mathbf{S}_{q\beta}^{\dagger}\equiv\mathbf{D}_{qq}
\end{aligned}
\tag{3.140}
$$

A similar sequence of computations shows that

$$
\begin{aligned}
&\left\langle \mathbf{r}^{d}\left(\boldsymbol{\alpha}^{0},\boldsymbol{\beta}^{0}\right)\left[\mathbf{q}^{d}\left(\boldsymbol{\alpha}^{0},\boldsymbol{\beta}^{0}\right)\right]^{\dagger}\right\rangle \triangleq \left\langle \left(\mathbf{r}^{c}-\mathbf{r}^{m}\right)\left(\mathbf{q}^{c}-\mathbf{q}^{m}\right)^{\dagger}\right\rangle \\
&=\left\langle \left[\mathbf{r}-\mathbf{r}^{m}-\mathbf{S}_{r\alpha}\left(\boldsymbol{\alpha}-\boldsymbol{\alpha}^{0}\right)-\mathbf{S}_{r\beta}\left(\boldsymbol{\beta}-\boldsymbol{\beta}^{0}\right)\right]\times\left[\mathbf{q}-\mathbf{q}^{m}-\mathbf{S}_{q\alpha}\left(\boldsymbol{\alpha}-\boldsymbol{\alpha}^{0}\right)-\mathbf{S}_{q\beta}\left(\boldsymbol{\beta}-\boldsymbol{\beta}^{0}\right)\right]^{\dagger}\right\rangle \\
&=\left\langle \left(\mathbf{r}-\mathbf{r}^{m}\right)\left(\mathbf{q}-\mathbf{q}^{m}\right)^{\dagger}\right\rangle -\left\langle \left(\mathbf{r}-\mathbf{r}^{m}\right)\left(\boldsymbol{\alpha}-\boldsymbol{\alpha}^{0}\right)^{\dagger}\right\rangle \mathbf{S}_{q\alpha}^{\dagger}-\left\langle \left(\mathbf{r}-\mathbf{r}^{m}\right)\left(\boldsymbol{\beta}-\boldsymbol{\beta}^{0}\right)^{\dagger}\right\rangle \mathbf{S}_{q\beta}^{\dagger}- \\
&\mathbf{S}_{r\alpha}\left\langle \left(\boldsymbol{\alpha}-\boldsymbol{\alpha}^{0}\right)\left(\mathbf{q}-\mathbf{q}^{m}\right)^{\dagger}\right\rangle +\mathbf{S}_{r\alpha}\left\langle \left(\boldsymbol{\alpha}-\boldsymbol{\alpha}^{0}\right)\left(\boldsymbol{\alpha}-\boldsymbol{\alpha}^{0}\right)^{\dagger}\right\rangle \mathbf{S}_{q\alpha}^{\dagger}+\mathbf{S}_{r\alpha}\left\langle \left(\boldsymbol{\alpha}-\boldsymbol{\alpha}^{0}\right)\left(\boldsymbol{\beta}-\boldsymbol{\beta}^{0}\right)^{\dagger}\right\rangle \mathbf{S}_{q\beta}^{\dagger} \\
&-\mathbf{S}_{r\beta}\left\langle \left(\boldsymbol{\beta}-\boldsymbol{\beta}^{0}\right)\left(\mathbf{q}-\mathbf{q}^{m}\right)^{\dagger}\right\rangle +\mathbf{S}_{r\beta}\left\langle \left(\boldsymbol{\beta}-\boldsymbol{\beta}^{0}\right)\left(\boldsymbol{\alpha}-\boldsymbol{\alpha}^{0}\right)^{\dagger}\right\rangle \mathbf{S}_{q\alpha}^{\dagger}+\mathbf{S}_{r\beta}\left\langle \left(\boldsymbol{\beta}-\boldsymbol{\beta}^{0}\right)\left(\boldsymbol{\beta}-\boldsymbol{\beta}^{0}\right)^{\dagger}\right\rangle \mathbf{S}_{q\beta}^{\dagger} \\
&=\mathbf{C}_{rq}-\mathbf{C}_{r\alpha}\mathbf{S}_{q\alpha}^{\dagger}-\mathbf{C}_{r\beta}\mathbf{S}_{q\beta}^{\dagger}-\mathbf{S}_{r\alpha}\mathbf{C}_{\alpha q}+\mathbf{S}_{qr}\mathbf{C}_{\alpha\alpha}\mathbf{S}_{q\alpha}^{\dagger}+\mathbf{S}_{qr}\mathbf{C}_{\alpha\beta}\mathbf{S}_{q\beta}^{\dagger}-\mathbf{S}_{r\beta}\mathbf{C}_{\beta q}+\mathbf{S}_{r\beta}\mathbf{C}_{\beta\alpha}\mathbf{S}_{q\alpha}^{\dagger} \\
&+\mathbf{S}_{r\beta}\mathbf{C}_{\beta\beta}\mathbf{S}_{q\beta}^{\dagger}=\mathbf{S}_{r\alpha}\left(-\mathbf{C}_{\alpha q}+\mathbf{C}_{\alpha\alpha}\mathbf{S}_{q\alpha}^{\dagger}+\mathbf{C}_{\alpha\beta}\mathbf{S}_{q\beta}^{\dagger}\right)+\mathbf{S}_{r\beta}\left(-\mathbf{C}_{\beta q}+\mathbf{C}_{\beta\alpha}\mathbf{S}_{q\alpha}^{\dagger}+\mathbf{C}_{\beta\beta}\mathbf{S}_{q\beta}^{\dagger}\right)+ \\
&\mathbf{C}_{rq}-\mathbf{C}_{r\alpha}\mathbf{S}_{q\alpha}^{\dagger}-\mathbf{C}_{r\beta}\mathbf{S}_{q\beta}^{\dagger}\equiv\mathbf{D}_{rq}=\mathbf{D}_{qr}^{\dagger}
\end{aligned}
\tag{3.141}
$$

APPENDIX 3.B: Computation of Predicted Optimal Covariance Matrices for Parameters and Responses

3.B.a. Computation of the predicted optimal covariance matrix $\mathbf{C}_{\alpha\alpha}^{pred}$ *for the parameters* $\boldsymbol{\alpha}$ *of Model A*

The predicted optimal covariance matrix for the parameters $\boldsymbol{\alpha}$ of Model A is defined as

$$\mathbf{C}_{\alpha\alpha}^{pred} \triangleq \left\langle \left(\boldsymbol{\alpha} - \boldsymbol{\alpha}^{pred}\right)\left(\boldsymbol{\alpha} - \boldsymbol{\alpha}^{pred}\right)^{\dagger} \right\rangle. \tag{3.142}$$

Writing Eq. (3.58) in the following compact form

$$\boldsymbol{\alpha}^{pred} = \boldsymbol{\alpha}^{0} - \left(\mathbf{X}_{\alpha}, \mathbf{Y}_{\alpha}\right)\mathbf{D}^{-1}\begin{bmatrix}\mathbf{r}^{d} \\ \mathbf{q}^{d}\end{bmatrix}, \tag{3.143}$$

replacing $\boldsymbol{\alpha}^{pred}$ from the above equation in Eq.(3.142), and carrying out the respective multiplications yields:

$$\begin{aligned}\mathbf{C}_{\alpha\alpha}^{pred} &= \left\langle \left(\boldsymbol{\alpha} - \boldsymbol{\alpha}^{0}\right)\left(\boldsymbol{\alpha} - \boldsymbol{\alpha}^{0}\right)^{\dagger} \right\rangle + \left\langle \left(\boldsymbol{\alpha} - \boldsymbol{\alpha}^{0}\right)\begin{pmatrix}\mathbf{r}^{d} \\ \mathbf{q}^{d}\end{pmatrix}^{\dagger} \right\rangle \mathbf{D}^{-1}\left(\mathbf{X}_{\alpha}, \mathbf{Y}_{\alpha}\right)^{\dagger} + \\ &+ \left(\mathbf{X}_{\alpha}, \mathbf{Y}_{\alpha}\right)\mathbf{D}^{-1}\left\langle \begin{pmatrix}\mathbf{r}^{d} \\ \mathbf{q}^{d}\end{pmatrix}\left(\boldsymbol{\alpha} - \boldsymbol{\alpha}^{0}\right)^{\dagger} \right\rangle \\ &+ \left(\mathbf{X}_{\alpha}, \mathbf{Y}_{\alpha}\right)\mathbf{D}^{-1}\left\langle \begin{pmatrix}\mathbf{r}^{d} \\ \mathbf{q}^{d}\end{pmatrix}\begin{pmatrix}\mathbf{r}^{d} \\ \mathbf{q}^{d}\end{pmatrix}^{\dagger} \right\rangle \mathbf{D}^{-1}\left(\mathbf{X}_{\alpha}, \mathbf{Y}_{\alpha}\right)^{\dagger}.\end{aligned} \tag{3.144}$$

The first-order model equations, i.e.,

$$\mathbf{r} = \mathbf{r}^{c}\left(\boldsymbol{\alpha}^{0}, \boldsymbol{\beta}^{0}\right) + \mathbf{S}_{r\alpha}\left(\boldsymbol{\alpha} - \boldsymbol{\alpha}^{0}\right) + \mathbf{S}_{r\beta}\left(\boldsymbol{\beta} - \boldsymbol{\beta}^{0}\right) + \textit{higher order terms}, \tag{3.145}$$

$$\mathbf{q} = \mathbf{q}^{c}\left(\boldsymbol{\alpha}^{0}, \boldsymbol{\beta}^{0}\right) + \mathbf{S}_{q\alpha}\left(\boldsymbol{\alpha} - \boldsymbol{\alpha}^{0}\right) + \mathbf{S}_{q\beta}\left(\boldsymbol{\beta} - \boldsymbol{\beta}^{0}\right) + \textit{higher order terms}, \tag{3.146}$$

can be used to recast the expressions involving the vectors of response deviations $\mathbf{r}^{d}$ and $\mathbf{q}^{d}$ in Eq. (3.141), as follows:

$$\left\langle \begin{pmatrix} \mathbf{r}^d \\ \mathbf{q}^d \end{pmatrix} (\boldsymbol{\alpha}-\boldsymbol{\alpha}^0)^\dagger \right\rangle = \left\langle \begin{pmatrix} (\mathbf{r}-\mathbf{r}^m) - \left[\mathbf{S}_{r\alpha}(\boldsymbol{\alpha}-\boldsymbol{\alpha}^0) + \mathbf{S}_{r\beta}(\boldsymbol{\beta}-\boldsymbol{\beta}^0)\right] \\ (\mathbf{q}-\mathbf{q}^m) - \left[\mathbf{S}_{q\alpha}(\boldsymbol{\alpha}-\boldsymbol{\alpha}^0) + \mathbf{S}_{q\beta}(\boldsymbol{\beta}-\boldsymbol{\beta}^0)\right] \end{pmatrix} (\boldsymbol{\alpha}-\boldsymbol{\alpha}^0)^\dagger \right\rangle$$

$$= \begin{pmatrix} \left\langle (\mathbf{r}-\mathbf{r}^m)(\boldsymbol{\alpha}-\boldsymbol{\alpha}^0)^\dagger \right\rangle - \mathbf{S}_{r\alpha} \left\langle (\boldsymbol{\alpha}-\boldsymbol{\alpha}^0)(\boldsymbol{\alpha}-\boldsymbol{\alpha}^0)^\dagger \right\rangle - \mathbf{S}_{r\beta} \left\langle (\boldsymbol{\beta}-\boldsymbol{\beta}^0)(\boldsymbol{\alpha}-\boldsymbol{\alpha}^0)^\dagger \right\rangle \\ \left\langle (\mathbf{q}-\mathbf{q}^m)(\boldsymbol{\alpha}-\boldsymbol{\alpha}^0)^\dagger \right\rangle - \mathbf{S}_{q\alpha} \left\langle (\boldsymbol{\alpha}-\boldsymbol{\alpha}^0)(\boldsymbol{\alpha}-\boldsymbol{\alpha}^0)^\dagger \right\rangle - \mathbf{S}_{q\beta} \left\langle (\boldsymbol{\beta}-\boldsymbol{\beta}^0)(\boldsymbol{\alpha}-\boldsymbol{\alpha}^0)^\dagger \right\rangle \end{pmatrix} \tag{3.147}$$

$$= \begin{pmatrix} \mathbf{C}_{r\alpha} - \mathbf{S}_{r\alpha}\mathbf{C}_{\alpha\alpha} - \mathbf{S}_{r\beta}\mathbf{C}_{\beta\alpha} \\ \mathbf{C}_{q\alpha} - \mathbf{S}_{q\alpha}\mathbf{C}_{\alpha\alpha} - \mathbf{S}_{q\beta}\mathbf{C}_{\beta\alpha} \end{pmatrix} = \begin{pmatrix} \mathbf{C}_{\alpha r}^\dagger - \mathbf{S}_{r\alpha}\mathbf{C}_{\alpha\alpha} - \mathbf{S}_{r\beta}\mathbf{C}_{\alpha\beta}^\dagger \\ \mathbf{C}_{\alpha q}^\dagger - \mathbf{S}_{q\alpha}\mathbf{C}_{\alpha\alpha} - \mathbf{S}_{q\beta}\mathbf{C}_{\alpha\beta}^\dagger \end{pmatrix} = \begin{pmatrix} -\mathbf{X}_\alpha^\dagger \\ -\mathbf{Y}_\alpha^\dagger \end{pmatrix}.$$

Replacing the above result into Eq.(3.144) and simplifying the resulting expression yields

$$\begin{aligned} \mathbf{C}_{\alpha\alpha}^{opt} &= \mathbf{C}_{\alpha\alpha} - (\mathbf{X}_\alpha, \mathbf{Y}_\alpha)\mathbf{D}^{-1} \begin{pmatrix} \mathbf{X}_\alpha^\dagger \\ \mathbf{Y}_\alpha^\dagger \end{pmatrix} \\ &= \mathbf{C}_{\alpha\alpha} - \left[\mathbf{X}_\alpha \left(\mathbf{D}_{11}\mathbf{X}_\alpha^\dagger + \mathbf{D}_{12}\mathbf{Y}_\alpha^\dagger\right) + \mathbf{Y}_\alpha \left(\mathbf{D}_{21}\mathbf{X}_\alpha^\dagger + \mathbf{D}_{22}\mathbf{Y}_\alpha^\dagger\right)\right]. \end{aligned} \tag{3.148}$$

3.B.b. Computation of the predicted optimal covariance matrix $\mathbf{C}_{rr}^{pred}$ *for the responses* $\mathbf{r}$ *of Model A*

The predicted optimal covariance matrix of the responses $\mathbf{r}$ of Model A is defined as

$$\mathbf{C}_{rr}^{pred} \triangleq \left\langle (\mathbf{r}-\mathbf{r}^{pred})(\mathbf{r}-\mathbf{r}^{pred})^\dagger \right\rangle. \tag{3.149}$$

Writing Eq.(3.60) in the form

$$\mathbf{r}^{pred} = \mathbf{r}^0 - (\mathbf{X}_r, \mathbf{Y}_r)\mathbf{D}^{-1} \begin{bmatrix} \mathbf{r}^d \\ \mathbf{q}^d \end{bmatrix}, \tag{3.150}$$

replacing $\mathbf{r}^{pred}$ from the above equation in Eq. (3.149), and carrying out the respective multiplications yields:

$$\begin{aligned} \mathbf{C}_{rr}^{pred} &= \left\langle (\mathbf{r}-\mathbf{r}^m)(\mathbf{r}-\mathbf{r}^m)^\dagger \right\rangle + (\mathbf{X}_r, \mathbf{Y}_r)\mathbf{D}^{-1} \left\langle \begin{pmatrix} \mathbf{r}^d \\ \mathbf{q}^d \end{pmatrix} (\mathbf{r}-\mathbf{r}^m)^\dagger \right\rangle \\ &+ \left\langle (\mathbf{r}-\mathbf{r}^m) \begin{pmatrix} \mathbf{r}^d \\ \mathbf{q}^d \end{pmatrix}^\dagger \right\rangle \mathbf{D}^{-1} \begin{pmatrix} \mathbf{X}_r^\dagger \\ \mathbf{Y}_r^\dagger \end{pmatrix} + (\mathbf{X}_r, \mathbf{Y}_r)\mathbf{D}^{-1} \left\langle \begin{pmatrix} \mathbf{r}^d \\ \mathbf{q}^d \end{pmatrix} \begin{pmatrix} \mathbf{r}^d \\ \mathbf{q}^d \end{pmatrix}^\dagger \right\rangle \mathbf{D}^{-1} \begin{pmatrix} \mathbf{X}_r^\dagger \\ \mathbf{Y}_r^\dagger \end{pmatrix}. \end{aligned} \tag{3.151}$$

Proceeding along the same conceptual path as in Eq. (3.147), it can be shown that

$$\left\langle \begin{pmatrix} \mathbf{r}^d \\ \mathbf{q}^d \end{pmatrix} \left(\mathbf{r}-\mathbf{r}^m\right)^\dagger \right\rangle = \begin{pmatrix} \mathbf{C}_{rr} - \mathbf{S}_{r\alpha}\mathbf{C}_{\alpha r} - \mathbf{S}_{r\beta}\mathbf{C}_{\beta r} \\ \mathbf{C}_{qr} - \mathbf{S}_{q\alpha}\mathbf{C}_{\alpha r} - \mathbf{S}_{q\beta}\mathbf{C}_{\beta r} \end{pmatrix} = \begin{pmatrix} -\mathbf{X}_r^\dagger \\ -\mathbf{Y}_r^\dagger \end{pmatrix} \tag{3.152}$$

Replacing Eq. (3.152) in Eq. (3.151) leads to

$$\begin{aligned} \mathbf{C}_{rr}^{pred} &= \mathbf{C}_{rr} - \left(\mathbf{X}_r,\ \mathbf{Y}_r\right)\mathbf{D}^{-1}\begin{pmatrix} \mathbf{X}_r^\dagger \\ \mathbf{Y}_r^\dagger \end{pmatrix} \\ &= \mathbf{C}_{rr} - \left[\mathbf{X}_r\left(\mathbf{D}_{11}\mathbf{X}_r^\dagger + \mathbf{D}_{12}\mathbf{Y}_r^\dagger\right) + \mathbf{Y}_r\left(\mathbf{D}_{21}\mathbf{X}_r^\dagger + \mathbf{D}_{22}\mathbf{Y}_r^\dagger\right)\right]. \end{aligned} \tag{3.153}$$

3.B.c. Computation of the predicted optimal correlation matrix $\mathbf{C}_{\alpha r}^{pred}$ for the parameters $\boldsymbol{\alpha}$ and responses $\mathbf{r}$ of Model A

The predicted optimal covariance matrix of the parameters $\boldsymbol{\alpha}$ and responses $\mathbf{r}$ of Model A is defined as

$$\mathbf{C}_{\alpha r}^{pred} \triangleq \left\langle \left(\boldsymbol{\alpha}-\boldsymbol{\alpha}^{pred}\right)\left(\mathbf{r}-\mathbf{r}^{pred}\right)^\dagger \right\rangle. \tag{3.154}$$

Replacing Eqs. (3.143) and (3.150) in Eq.(3.154) yields

$$\begin{aligned} \mathbf{C}_{\alpha r}^{pred} &= \left\langle \left(\boldsymbol{\alpha}-\boldsymbol{\alpha}^0\right)\left(\mathbf{r}-\mathbf{r}^m\right)^\dagger \right\rangle + \left[\mathbf{X}_\alpha,\ \mathbf{Y}_\alpha\right]\mathbf{D}^{-1}\left\langle \begin{pmatrix} \mathbf{r}^d \\ \mathbf{q}^d \end{pmatrix}\left(\mathbf{r}-\mathbf{r}^m\right)^\dagger \right\rangle \\ &+ \left\langle \left(\boldsymbol{\alpha}-\boldsymbol{\alpha}^0\right)\begin{pmatrix} \mathbf{r}^d \\ \mathbf{q}^d \end{pmatrix}^\dagger \right\rangle \mathbf{D}^{-1}\begin{bmatrix} \mathbf{X}_r^\dagger \\ \mathbf{Y}_r^\dagger \end{bmatrix} + \left[\mathbf{X}_\alpha,\ \mathbf{Y}_\alpha\right]\mathbf{D}^{-1}\left\langle \begin{pmatrix} \mathbf{r}^d \\ \mathbf{q}^d \end{pmatrix}\begin{pmatrix} \mathbf{r}^d \\ \mathbf{q}^d \end{pmatrix}^\dagger \right\rangle \mathbf{D}^{-1}\begin{bmatrix} \mathbf{X}_r^\dagger \\ \mathbf{Y}_r^\dagger \end{bmatrix}. \end{aligned} \tag{3.155}$$

Using the results in Eqs. (3.147) and (3.152) in Eq. (3.155) and carrying out the ensuing matrix algebra leads to

$$\begin{aligned} \mathbf{C}_{\alpha r}^{pred} &= \mathbf{C}_{\alpha r} - \left(\mathbf{X}_\alpha,\ \mathbf{Y}_\alpha\right)\mathbf{D}^{-1}\begin{pmatrix} \mathbf{X}_r^\dagger \\ \mathbf{Y}_r^\dagger \end{pmatrix} \\ &= \mathbf{C}_{rr} - \left[\mathbf{X}_\alpha\left(\mathbf{D}_{11}\mathbf{X}_r^\dagger + \mathbf{D}_{12}\mathbf{Y}_r^\dagger\right) + \mathbf{Y}_\alpha\left(\mathbf{D}_{21}\mathbf{X}_r^\dagger + \mathbf{D}_{22}\mathbf{Y}_r^\dagger\right)\right]. \end{aligned} \tag{3.156}$$

3.B.d. Computation of the predicted optimal covariance matrix for the parameters $\boldsymbol{\beta}$ of Model B

The predicted optimal covariance matrix for the parameters $\boldsymbol{\beta}$ of Model B is defined as

$$\mathbf{C}_{\beta\beta}^{pred} \triangleq \left\langle \left(\boldsymbol{\beta} - \boldsymbol{\beta}^{pred}\right)\left(\boldsymbol{\beta} - \boldsymbol{\beta}^{pred}\right)^{\dagger} \right\rangle . \tag{3.157}$$

Writing Eq. (3.59) in the form

$$\boldsymbol{\beta}^{pred} = \boldsymbol{\beta}^{0} - \left(\mathbf{X}_{\beta}, \mathbf{Y}_{\beta}\right)\mathbf{D}^{-1}\begin{bmatrix} \mathbf{r}^{d} \\ \mathbf{q}^{d} \end{bmatrix}, \tag{3.158}$$

replacing $\boldsymbol{\beta}^{pred}$ from the above equation in Eq. (3.157), and carrying out the respective multiplications yields:

$$\begin{aligned} \mathbf{C}_{\beta\beta}^{pred} &= \left\langle \left(\boldsymbol{\beta} - \boldsymbol{\beta}^{0}\right)\left(\boldsymbol{\beta} - \boldsymbol{\beta}^{0}\right)^{\dagger} \right\rangle + \left\langle \left(\boldsymbol{\beta} - \boldsymbol{\beta}^{0}\right)\begin{pmatrix} \mathbf{r}^{d} \\ \mathbf{q}^{d} \end{pmatrix}^{\dagger} \right\rangle \mathbf{D}^{-1}\left(\mathbf{X}_{\beta}, \mathbf{Y}_{\beta}\right)^{\dagger} + \\ &+ \left(\mathbf{X}_{\beta}, \mathbf{Y}_{\beta}\right)\mathbf{D}^{-1}\left\langle \begin{pmatrix} \mathbf{r}^{d} \\ \mathbf{q}^{d} \end{pmatrix}\left(\boldsymbol{\beta} - \boldsymbol{\beta}^{0}\right)^{\dagger} \right\rangle \\ &+ \left(\mathbf{X}_{\beta}, \mathbf{Y}_{\beta}\right)\mathbf{D}^{-1}\left\langle \begin{pmatrix} \mathbf{r}^{d} \\ \mathbf{q}^{d} \end{pmatrix}\begin{pmatrix} \mathbf{r}^{d} \\ \mathbf{q}^{d} \end{pmatrix}^{\dagger} \right\rangle \mathbf{D}^{-1}\left(\mathbf{X}_{\beta}, \mathbf{Y}_{\beta}\right)^{\dagger} . \end{aligned} \tag{3.159}$$

Following the same sequence of operations as for Model A leads to the result below:

$$\left\langle \left(\boldsymbol{\beta} - \boldsymbol{\beta}^{0}\right)\begin{pmatrix} \mathbf{r}^{d} \\ \mathbf{q}^{d} \end{pmatrix}^{\dagger} \right\rangle = \begin{pmatrix} \mathbf{C}_{\beta r} - \mathbf{C}_{\beta\alpha}\mathbf{S}_{r\alpha}^{\dagger} - \mathbf{C}_{\beta\beta}\mathbf{S}_{r\beta}^{\dagger} \\ \mathbf{C}_{\beta q} - \mathbf{C}_{\beta\alpha}\mathbf{S}_{q\alpha}^{\dagger} - \mathbf{C}_{\beta\beta}\mathbf{S}_{q\beta}^{\dagger} \end{pmatrix}^{\dagger} = \left(-\mathbf{X}_{\beta}, -\mathbf{Y}_{\beta}\right). \tag{3.160}$$

Replacing the result obtained in Eq. (3.160) into Eq. (3.159) and simplifying the resulting expression yields

$$\begin{aligned} \mathbf{C}_{\beta\beta}^{opt} &= \mathbf{C}_{\beta\beta} - \left(\mathbf{X}_{\beta}, \mathbf{Y}_{\beta}\right)\mathbf{D}^{-1}\begin{pmatrix} \mathbf{X}_{\beta}^{\dagger} \\ \mathbf{Y}_{\beta}^{\dagger} \end{pmatrix} \\ &= \mathbf{C}_{\beta\beta} - \left[\mathbf{X}_{\beta}\left(\mathbf{D}_{11}\mathbf{X}_{\beta}^{\dagger} + \mathbf{D}_{12}\mathbf{Y}_{\beta}^{\dagger}\right) + \mathbf{Y}_{\beta}\left(\mathbf{D}_{21}\mathbf{X}_{\beta}^{\dagger} + \mathbf{D}_{22}\mathbf{Y}_{\beta}^{\dagger}\right)\right]. \end{aligned} \tag{3.161}$$

3.B.e. Computation of the predicted optimal covariance matrix $\mathbf{C}_{qq}^{pred}$ *for the responses* $\mathbf{q}$ *of Model B*

The predicted optimal covariance matrix of the responses $\mathbf{q}$ of Model B is defined as

$$\mathbf{C}_{qq}^{pred} \triangleq \left\langle \left(\mathbf{q} - \mathbf{q}^{pred}\right)\left(\mathbf{q} - \mathbf{q}^{pred}\right)^{\dagger} \right\rangle. \tag{3.162}$$

Writing Eq. (3.61) in the form

$$\mathbf{q}^{pred} = \mathbf{q}^{0} - \left(\mathbf{X}_{q}, \mathbf{Y}_{q}\right)\mathbf{D}^{-1}\begin{bmatrix} \mathbf{r}^{d} \\ \mathbf{q}^{d} \end{bmatrix}, \tag{3.163}$$

replacing $\mathbf{q}^{pred}$ from the above equation in Eq. (3.162), and carrying out the respective multiplications yields:

$$\begin{aligned} \mathbf{C}_{qq}^{pred} &= \left\langle \left(\mathbf{q} - \mathbf{q}^{m}\right)\left(\mathbf{q} - \mathbf{q}^{m}\right)^{\dagger} \right\rangle + \left(\mathbf{X}_{q}, \mathbf{Y}_{q}\right)\mathbf{D}^{-1}\left\langle \begin{pmatrix} \mathbf{r}^{d} \\ \mathbf{q}^{d} \end{pmatrix}\left(\mathbf{q} - \mathbf{q}^{m}\right)^{\dagger} \right\rangle \\ &+ \left\langle \left(\mathbf{q} - \mathbf{q}^{m}\right)\begin{pmatrix} \mathbf{r}^{d} \\ \mathbf{q}^{d} \end{pmatrix}^{\dagger} \right\rangle \mathbf{D}^{-1}\begin{pmatrix} \mathbf{X}_{q}^{\dagger} \\ \mathbf{Y}_{q}^{\dagger} \end{pmatrix} + \left(\mathbf{X}_{q}, \mathbf{Y}_{q}\right)\mathbf{D}^{-1}\left\langle \begin{pmatrix} \mathbf{r}^{d} \\ \mathbf{q}^{d} \end{pmatrix}\begin{pmatrix} \mathbf{r}^{d} \\ \mathbf{q}^{d} \end{pmatrix}^{\dagger} \right\rangle \mathbf{D}^{-1}\begin{pmatrix} \mathbf{X}_{q}^{\dagger} \\ \mathbf{Y}_{q}^{\dagger} \end{pmatrix}. \end{aligned} \tag{3.164}$$

Proceeding along the same conceptual path as in Eq.(3.147), it can be shown that

$$\left\langle \left(\mathbf{q} - \mathbf{q}^{m}\right)\begin{pmatrix} \mathbf{r}^{d} \\ \mathbf{q}^{d} \end{pmatrix}^{\dagger} \right\rangle = \begin{pmatrix} \mathbf{C}_{rq}^{\dagger} - \mathbf{C}_{\alpha q}^{\dagger}\mathbf{S}_{r\alpha}^{\dagger} - \mathbf{C}_{\beta q}^{\dagger}\mathbf{S}_{r\beta}^{\dagger} \\ \mathbf{C}_{qq} - \mathbf{C}_{\alpha q}^{\dagger}\mathbf{S}_{q\alpha}^{\dagger} - \mathbf{C}_{\beta q}^{\dagger}\mathbf{S}_{q\beta}^{\dagger} \end{pmatrix}^{\dagger} = \left(-\mathbf{X}_{q}, -\mathbf{Y}_{q}\right) \tag{3.165}$$

Replacing Eq. (3.165) in Eq. (3.164) leads to

$$\begin{aligned} \mathbf{C}_{qq}^{pred} &= \mathbf{C}_{qq} - \left(\mathbf{X}_{q}, \mathbf{Y}_{q}\right)\mathbf{D}^{-1}\begin{pmatrix} \mathbf{X}_{q}^{\dagger} \\ \mathbf{Y}_{q}^{\dagger} \end{pmatrix} \\ &= \mathbf{C}_{qq} - \left[\mathbf{X}_{q}\left(\mathbf{D}_{11}\mathbf{X}_{q}^{\dagger} + \mathbf{D}_{12}\mathbf{Y}_{q}^{\dagger}\right) + \mathbf{Y}_{q}\left(\mathbf{D}_{21}\mathbf{X}_{q}^{\dagger} + \mathbf{D}_{22}\mathbf{Y}_{q}^{\dagger}\right)\right]. \end{aligned} \tag{3.166}$$

3.B.f. Computation of the predicted optimal correlation matrix $\mathbf{C}_{\beta q}^{pred}$ for the parameters $\boldsymbol{\beta}$ and responses q of Model B

The predicted optimal correlation matrix of the parameters $\boldsymbol{\beta}$ and responses $\mathbf{q}$ of Model B is defined as

$$\mathbf{C}_{\beta q}^{pred} \triangleq \left\langle \left(\boldsymbol{\beta}-\boldsymbol{\beta}^{pred}\right)\left(\mathbf{q}-\mathbf{q}^{pred}\right)^{\dagger} \right\rangle \tag{3.167}$$

Replacing Eqs.(3.158) and (3.163) in Eq. (3.167) yields

$$\begin{aligned}\mathbf{C}_{\beta q}^{pred} &= \left\langle \left(\boldsymbol{\beta}-\boldsymbol{\beta}^{0}\right)\left(\mathbf{q}-\mathbf{q}^{m}\right)^{\dagger} \right\rangle + \left[\mathbf{X}_{\beta},\ \mathbf{Y}_{\beta}\right]\mathbf{D}^{-1}\left\langle \begin{pmatrix}\mathbf{r}^{d}\\ \mathbf{q}^{d}\end{pmatrix}\left(\mathbf{q}-\mathbf{q}^{m}\right)^{\dagger} \right\rangle \\ &+ \left\langle \left(\boldsymbol{\beta}-\boldsymbol{\beta}^{0}\right)\begin{pmatrix}\mathbf{r}^{d}\\ \mathbf{q}^{d}\end{pmatrix}^{\dagger} \right\rangle \mathbf{D}^{-1}\begin{bmatrix}\mathbf{X}_{q}^{\dagger}\\ \mathbf{Y}_{q}^{\dagger}\end{bmatrix} + \left[\mathbf{X}_{\beta},\ \mathbf{Y}_{\beta}\right]\mathbf{D}^{-1}\left\langle \begin{pmatrix}\mathbf{r}^{d}\\ \mathbf{q}^{d}\end{pmatrix}\begin{pmatrix}\mathbf{r}^{d}\\ \mathbf{q}^{d}\end{pmatrix}^{\dagger} \right\rangle \mathbf{D}^{-1}\begin{bmatrix}\mathbf{X}_{q}^{\dagger}\\ \mathbf{Y}_{q}^{\dagger}\end{bmatrix}.\end{aligned} \tag{3.168}$$

Using the results from Eqs. (3.160) and (3.165) in Eq. (3.168) and carrying out the ensuing matrix algebra leads to

$$\begin{aligned}\mathbf{C}_{\beta q}^{pred} &= \mathbf{C}_{\beta q} - \left(\mathbf{X}_{\beta},\ \mathbf{Y}_{\beta}\right)\mathbf{D}^{-1}\begin{pmatrix}\mathbf{X}_{q}^{\dagger}\\ \mathbf{Y}_{q}^{\dagger}\end{pmatrix} \\ &= \mathbf{C}_{\beta q} - \left[\mathbf{X}_{\beta}\left(\mathbf{D}_{11}\mathbf{X}_{q}^{\dagger}+\mathbf{D}_{12}\mathbf{Y}_{q}^{\dagger}\right)+\mathbf{Y}_{\beta}\left(\mathbf{D}_{21}\mathbf{X}_{q}^{\dagger}+\mathbf{D}_{22}\mathbf{Y}_{q}^{\dagger}\right)\right].\end{aligned} \tag{3.169}$$

3.B.g. Computation of the predicted optimal correlation matrix $\mathbf{C}_{\alpha\beta}^{pred}$ for the parameters $\boldsymbol{\alpha}$ of Model A and the parameters $\boldsymbol{\beta}$ of Model B

The predicted optimal correlation matrix $\mathbf{C}_{\alpha\beta}^{pred}$ of the parameters $\boldsymbol{\alpha}$ of Model A and the parameters $\boldsymbol{\beta}$ of Model B is defined as

$$\mathbf{C}_{\alpha\beta}^{pred} \triangleq \left\langle \left(\boldsymbol{\alpha}-\boldsymbol{\alpha}^{pred}\right)\left(\boldsymbol{\beta}-\boldsymbol{\beta}^{pred}\right)^{\dagger} \right\rangle \tag{3.170}$$

Replacing (3.143) and (3.158) in Eq.(3.27) yields

$$\mathbf{C}_{\alpha\beta}^{pred} = \left\langle (\boldsymbol{\alpha}-\boldsymbol{\alpha}^0)(\boldsymbol{\beta}-\boldsymbol{\beta}^0)^\dagger \right\rangle + [\mathbf{X}_\alpha,\ \mathbf{Y}_\alpha]\mathbf{D}^{-1}\left\langle \begin{pmatrix} \mathbf{r}^d \\ \mathbf{q}^d \end{pmatrix} (\boldsymbol{\beta}-\boldsymbol{\beta}^0)^\dagger \right\rangle$$
$$+\left\langle (\boldsymbol{\alpha}-\boldsymbol{\alpha}^0)\begin{pmatrix} \mathbf{r}^d \\ \mathbf{q}^d \end{pmatrix}^\dagger \right\rangle \mathbf{D}^{-1}\begin{bmatrix} \mathbf{X}_\beta^\dagger \\ \mathbf{Y}_\beta^\dagger \end{bmatrix} + [\mathbf{X}_\alpha,\ \mathbf{Y}_\alpha]\mathbf{D}^{-1}\left\langle \begin{pmatrix} \mathbf{r}^d \\ \mathbf{q}^d \end{pmatrix}\begin{pmatrix} \mathbf{r}^d \\ \mathbf{q}^d \end{pmatrix}^\dagger \right\rangle \mathbf{D}^{-1}\begin{bmatrix} \mathbf{X}_\beta^\dagger \\ \mathbf{Y}_\beta^\dagger \end{bmatrix}. \tag{3.171}$$

Using Eqs. (3.147) and(3.160) in Eq. (3.171), and simplifying the resulting expressions gives

$$\mathbf{C}_{\alpha\beta}^{pred} = \mathbf{C}_{\alpha\beta} - (\mathbf{X}_\alpha,\ \mathbf{Y}_\alpha)\mathbf{D}^{-1}\begin{pmatrix} \mathbf{X}_\beta^\dagger \\ \mathbf{Y}_\beta^\dagger \end{pmatrix}$$
$$= \mathbf{C}_{\alpha\beta} - \left[\mathbf{X}_\alpha \left(\mathbf{D}_{11}\mathbf{X}_\beta^\dagger + \mathbf{D}_{12}\mathbf{Y}_\beta^\dagger \right) + \mathbf{Y}_\alpha \left(\mathbf{D}_{21}\mathbf{X}_\beta^\dagger + \mathbf{D}_{22}\mathbf{Y}_\beta^\dagger \right) \right]. \tag{3.172}$$

3.B.h. Computation of the predicted optimal correlation matrix $\mathbf{C}_{\alpha q}^{pred}$ for the parameters α of Model A and the and responses q of Model B

The predicted optimal correlation matrix $\mathbf{C}_{\alpha q}^{pred}$ of the parameters $\boldsymbol{\alpha}$ of Model A and responses $\mathbf{q}$ of Model B is defined as

$$\mathbf{C}_{\alpha q}^{pred} \triangleq \left\langle (\boldsymbol{\alpha}-\boldsymbol{\alpha}^{pred})(\mathbf{q}-\mathbf{q}^{pred})^\dagger \right\rangle \tag{3.173}$$

Replacing (3.144) and (3.163) in Eq. (3.173) yields

$$\mathbf{C}_{\alpha q}^{pred} = \left\langle (\boldsymbol{\alpha}-\boldsymbol{\alpha}^0)(\mathbf{q}-\mathbf{q}^0)^\dagger \right\rangle + [\mathbf{X}_\alpha,\ \mathbf{Y}_\alpha]\mathbf{D}^{-1}\left\langle \begin{pmatrix} \mathbf{r}^d \\ \mathbf{q}^d \end{pmatrix} (\mathbf{q}-\mathbf{q}^0)^\dagger \right\rangle$$
$$+\left\langle (\boldsymbol{\alpha}-\boldsymbol{\alpha}^0)\begin{pmatrix} \mathbf{r}^d \\ \mathbf{q}^d \end{pmatrix}^\dagger \right\rangle \mathbf{D}^{-1}\begin{bmatrix} \mathbf{X}_q^\dagger \\ \mathbf{Y}_q^\dagger \end{bmatrix} + [\mathbf{X}_\alpha,\ \mathbf{Y}_\alpha]\mathbf{D}^{-1}\left\langle \begin{pmatrix} \mathbf{r}^d \\ \mathbf{q}^d \end{pmatrix}\begin{pmatrix} \mathbf{r}^d \\ \mathbf{q}^d \end{pmatrix}^\dagger \right\rangle \mathbf{D}^{-1}\begin{bmatrix} \mathbf{X}_q^\dagger \\ \mathbf{Y}_q^\dagger \end{bmatrix} \tag{3.174}$$

Using Eqs.(3.147) and (3.152) in Eq. (3.174), and simplifying the resulting expressions gives

$$\mathbf{C}_{\alpha q}^{pred} = \mathbf{C}_{\alpha q} - (\mathbf{X}_\alpha,\ \mathbf{Y}_\alpha)\mathbf{D}^{-1}\begin{pmatrix} \mathbf{X}_q^\dagger \\ \mathbf{Y}_q^\dagger \end{pmatrix}$$
$$= \mathbf{C}_{\alpha q} - \left[\mathbf{X}_\alpha \left(\mathbf{D}_{11}\mathbf{X}_q^\dagger + \mathbf{D}_{12}\mathbf{Y}_q^\dagger \right) + \mathbf{Y}_\alpha \left(\mathbf{D}_{21}\mathbf{X}_q^\dagger + \mathbf{D}_{22}\mathbf{Y}_q^\dagger \right) \right]. \tag{3.175}$$

3.B.i. Computation of the predicted optimal correlation matrix $\mathbf{C}_{\beta r}^{pred}$ for the parameters $\boldsymbol{\beta}$ of Model B and the and responses $\mathbf{r}$ of Model A

The predicted optimal correlation matrix $\mathbf{C}_{\alpha q}^{pred}$ of the parameters $\boldsymbol{\beta}$ of Model B and responses $\mathbf{r}$ of Model A is defined as

$$\mathbf{C}_{\beta r}^{pred} \triangleq \left\langle \left(\boldsymbol{\beta}-\boldsymbol{\beta}^{pred}\right)\left(\mathbf{r}-\mathbf{r}^{pred}\right)^{\dagger} \right\rangle \tag{3.176}$$

Replacing (3.150) and (3.158) in Eq. (3.176) yields

$$\begin{aligned}\mathbf{C}_{\beta r}^{pred} &= \left\langle \left(\boldsymbol{\beta}-\boldsymbol{\beta}^{0}\right)\left(\mathbf{r}-\mathbf{r}^{0}\right)^{\dagger} \right\rangle + \left[\mathbf{X}_{\beta}, \mathbf{Y}_{\beta}\right]\mathbf{D}^{-1}\left\langle \begin{pmatrix}\mathbf{r}^{d}\\ \mathbf{q}^{d}\end{pmatrix}\left(\mathbf{r}-\mathbf{r}^{0}\right)^{\dagger} \right\rangle \\ &+\left\langle \left(\boldsymbol{\beta}-\boldsymbol{\beta}^{0}\right)\begin{pmatrix}\mathbf{r}^{d}\\ \mathbf{q}^{d}\end{pmatrix}^{\dagger} \right\rangle \mathbf{D}^{-1}\begin{bmatrix}\mathbf{X}_{r}^{\dagger}\\ \mathbf{Y}_{r}^{\dagger}\end{bmatrix} + \left[\mathbf{X}_{\beta}, \mathbf{Y}_{\beta}\right]\mathbf{D}^{-1}\left\langle \begin{pmatrix}\mathbf{r}^{d}\\ \mathbf{q}^{d}\end{pmatrix}\begin{pmatrix}\mathbf{r}^{d}\\ \mathbf{q}^{d}\end{pmatrix}^{\dagger} \right\rangle \mathbf{D}^{-1}\begin{bmatrix}\mathbf{X}_{r}^{\dagger}\\ \mathbf{Y}_{r}^{\dagger}\end{bmatrix}.\end{aligned} \tag{3.177}$$

Using Eqs. (3.152) and (3.160) in Eq.(3.177), and simplifying the resulting expressions gives

$$\begin{aligned}\mathbf{C}_{\beta r}^{pred} &= \mathbf{C}_{\beta r} - \left(\mathbf{X}_{\beta}, \mathbf{Y}_{\beta}\right)\mathbf{D}^{-1}\begin{pmatrix}\mathbf{X}_{r}^{\dagger}\\ \mathbf{Y}_{r}^{\dagger}\end{pmatrix} \\ &= \mathbf{C}_{\beta r} - \left[\mathbf{X}_{\beta}\left(\mathbf{D}_{11}\mathbf{X}_{r}^{\dagger}+\mathbf{D}_{12}\mathbf{Y}_{r}^{\dagger}\right)+\mathbf{Y}_{\beta}\left(\mathbf{D}_{21}\mathbf{X}_{r}^{\dagger}+\mathbf{D}_{22}\mathbf{Y}_{r}^{\dagger}\right)\right].\end{aligned} \tag{3.178}$$

3.B.j. Computation of the predicted optimal correlation matrix $\mathbf{C}_{rq}^{pred}$ for the responses $\mathbf{r}$ of Model A and the responses $\mathbf{q}$ of Model B

The predicted optimal correlation matrix $\mathbf{C}_{rq}^{pred}$ of the responses $\mathbf{r}$ of Model A and the responses $\mathbf{q}$ of Model B is defined as

$$\mathbf{C}_{rq}^{pred} \triangleq \left\langle \left(\mathbf{r}-\mathbf{r}^{pred}\right)\left(\mathbf{q}-\mathbf{q}^{pred}\right)^{\dagger} \right\rangle. \tag{3.179}$$

Replacing (3.150) and (3.163) in Eq. (3.179) yields

$$\mathbf{C}_{rq}^{pred} = \left\langle (\mathbf{r}-\mathbf{r}^m)(\mathbf{q}-\mathbf{q}^m)^\dagger \right\rangle + [\mathbf{X}_r, \mathbf{Y}_r]\mathbf{D}^{-1}\left\langle \begin{pmatrix} \mathbf{r}^d \\ \mathbf{q}^d \end{pmatrix} (\mathbf{q}-\mathbf{q}^m)^\dagger \right\rangle$$
$$+\left\langle (\mathbf{r}-\mathbf{r}^m)\begin{pmatrix} \mathbf{r}^d \\ \mathbf{q}^d \end{pmatrix}^\dagger \right\rangle \mathbf{D}^{-1}\begin{bmatrix} \mathbf{X}_q^\dagger \\ \mathbf{Y}_q^\dagger \end{bmatrix} + [\mathbf{X}_r, \mathbf{Y}_r]\mathbf{D}^{-1}\left\langle \begin{pmatrix} \mathbf{r}^d \\ \mathbf{q}^d \end{pmatrix}\begin{pmatrix} \mathbf{r}^d \\ \mathbf{q}^d \end{pmatrix}^\dagger \right\rangle \mathbf{D}^{-1}\begin{bmatrix} \mathbf{X}_q^\dagger \\ \mathbf{Y}_q^\dagger \end{bmatrix} \tag{3.180}$$

Using Eqs. (3.152) and (3.165) in Eq. (3.180), and simplifying the resulting expressions gives

$$\begin{aligned} \mathbf{C}_{rq}^{pred} &= \mathbf{C}_{rq} - (\mathbf{X}_r, \mathbf{Y}_r)\mathbf{D}^{-1}\begin{pmatrix} \mathbf{X}_q^\dagger \\ \mathbf{Y}_q^\dagger \end{pmatrix} \\ &= \mathbf{C}_{rq} - \left[\mathbf{X}_r\left(\mathbf{D}_{11}\mathbf{X}_q^\dagger + \mathbf{D}_{12}\mathbf{Y}_q^\dagger\right) + \mathbf{Y}_r\left(\mathbf{D}_{21}\mathbf{X}_q^\dagger + \mathbf{D}_{22}\mathbf{Y}_q^\dagger\right)\right]. \end{aligned} \tag{3.181}$$

3.3 MULTI-PRED CODE MODULE

The equations expressing the results of the BERRU-CMS methodology formulated by Cacuci (2014a) and presented in the previous Sections of this Chapter have been programmed in the computational software module MULTI-PRED (Cacuci et al, 2018). All routines in MULTI-PRED are written in FORTRAN 90 and are compatible with most Linux and window Systems. A C++ version will also become available. The following four cases can be selected for performing predictive modeling computations:

CASE 1: "*One Multi-Physics Model*": predictive modeling solely for Model A with N_a model parameters and N_r measured responses.

CASE 2: "*One Multi-Physics Model with Additional Model Parameters*": predictive modeling for Model A with N_b additional model parameters, but no additional responses.

CASE 3: "*One Multi-Physics Model with Additional Model Responses*": predictive modeling for Model A with N_q additional responses, but no additional parameters.

CASE 4: "*Two Coupled Multi-Physics Models*": predictive modelling for Model A coupled with Model B.

Three demonstration problems are provided to illustrate the application of the BERRU-CMS methodology. The first problem presents the application of the BERRU-CMS methodology to a simple particle diffusion problem which admits a closed-form analytical solution, thus facilitating a rapid understanding of this methodology and its predicted results; this problem will be discussed in detail in Chapter 4. The second demonstration problem presents the application of the BERRU-CMS methodology to the problem of inverse prediction, from detector responses in the presence of counting uncertainties, of the thickness of a homogeneous slab of material containing uniformly distributed gamma-emitting sources, for optically thin and thick slabs. This problem highlights the essential role played by the relative uncertainties (or, conversely, accuracies) of measured and computed responses, and will be discussed in detail in Chapter 5. The third demonstration problem presents the application of the BERRU-CMS methodology to the F-area cooling towers at the Savannah River National Lab. This problem demonstrates that the BERRU-CMS methodology reduces the predicted response uncertainties not only at locations where measurements are available, but also at locations where measurements are not available, and will be discussed in detail in Chapter 6. The source code of MULTI-PRED is provided on the accompanying CD.

3.3.1 MULTI-PRED Structure

The program Multi-Pred includes the following routines and modules:

- Main program: multi-pred.f90
- Module: ModuleGlobalParameters.f90
- Module: ModuleIO.f90
- Module: ModuleErrors.f90
- Subroutine: Files.f90
- Module: ModuleFiles.f90
- Subroutine: ReadInput.f90
- Module: ModuleReadWrite.f90
- Subroutine: MultiPredSolver.f90
- Module: ModuleMultiPred.f90
- Module: ModuleLapack.f90

The structure of the MULTI-PRED software module is organized as shown in Figure 3.1.

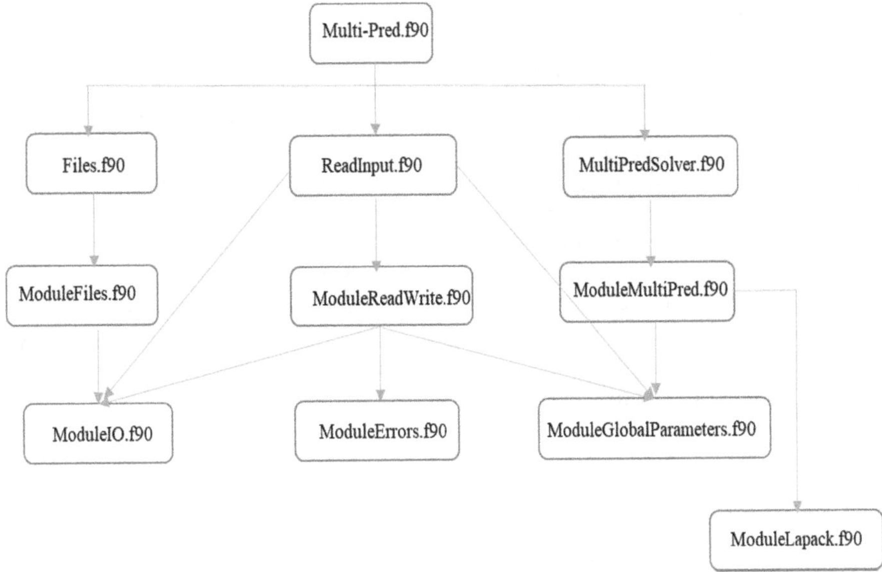

Figure 3.1. MULTI-PRED Code Structure

3.3.2 Directories

The computational software module **MULTI-PRED** comprises the following directories:

(1) multi-pred/source/
This folder contains the source codes.

(2) multi-pred/examples/
This folder contains 5 examples specified in the following subfolders.

(i) **../Neutron_Diffusion_Model_Case_1/**
This folder contains the input/output files for Multi-Pred Case 1 for a neutron diffusion model.

(ii) **../Cooling_Tower_Model_Case_1/**
This folder contains the input/output files for Multi-Pred Case 1 for the SRNL cooling tower model.

(iii) **../Cooling_Tower_Model_Case_2/**
This folder contains the input/output files for Multi-Pred Case 2 for the SRNL cooling tower model.

(iv) **../Cooling_Tower_Model_Case_3/**
This folder contains the input/output files for Multi-Pred Case 3 for the SRNL cooling tower model.

(v) **../Cooling_Tower_Model_Case_4/**
This folder contains the input/output files for Multi-Pred Case 4 for the SRNL cooling tower model.

(3) multi-pred/matrix_positive_definite_test/

This folder contains the source code for a stand-alone program used to test if a *symmetric* matrix is *positive definite* (SPD). Note that the covariance matrices $\mathbf{C}_{aa}(N_a \times N_a), \mathbf{C}_{rr}(N_r \times N_r)$, $\mathbf{C}_{bb}(N_b \times N_b)$ and $\mathbf{C}_{qq}(N_q \times N_q)$ must be SPD matrices. This program computes the Cholesky factorization of the matrix being tested. If it can be factorized, the program returns a flag indicating that the tested matrix is SPD. Running this test stand-alone program is optional, since the Cholesky factorization has also been implemented in **MULTI-PRED**. Also included in this folder is a large-scale matrix used for the SPD test. This matrix is a large symmetric positive definite matrix, with seemingly random sparsity pattern. It has a dimension of 60,000 by 60,000 with 410077 nonzero elements. Refer to the following website http://www.cise.ufl.edu/research/sparse/matrices/Andrews/Andrews for detailed information about this matrix.

3.3.3 Code Compilation and Execution

(1) Compile the software program in Linux

Enter the ***multi-pred/source/*** directory, and use the command *make*, an executable named *multi-pred* will be generated under the source directory.

The compiler used in the *makefile* is ifort (version 12.1.6 and above). It can also be compiled with gfortran (version 4.47 and above). An example makefile with the gfortran compiler, named *makefile.gfortran,* is also included in the *source* directory.

(2) Run the program

To run the program, copy the executable *multi-pred* into the example directories, then use the command:

./multi-pred superfile.inp

where the argument *superfile.inp* contains all the input/output files names. Output files will be generated in the respective example folders.

3.3.4 Input and Output File Organization

This Section describes the input and output files within the **MULTI-PRED** module.

3.3.4.1 Super File

The **MULTI-PRED** super-file is a text file that contains the names of input/output files and organizes the individual files for input and output operations. This super-file is read from the command line (UNIT=5) as an argument. The first line of the super-file is reserved for an identifier card, “MultiPredSup”. After the identifier line, each subsequent line is preceded by a category code and a filename. The category code and filename have to be enclosed in single quotes. The filenames can be changed by the user. The second line of the super-file is also reserved for the “dims” category; the corresponding input file defines the dimensions of the matrices and vectors used in **MULTI-PRED**. The lines after the second line are for data files. There are no restrictions regarding the order of the data files and their corresponding categories. Tables 3.1 through 3.4 show the format and complete list of super files for the **MULTI-PRED** Case 1, Case 2, Case 3 and Case 4, respectively.

Table 3.1. Super File Format for Multi-Pred Case 1

Category	**File Name**
MultiPredSup	
'dims'	'dimensions.inp'
'a_nom'	'a.inp'
'r_mea'	'rm.inp'
'r_com'	'rc.inp'
'C_aa'	'Caa.inp'
'C_ar'	'Car.inp'
'C_rr'	'Crr.inp'
'S_ra'	'Sra.inp'
'a_BE'	'aBE.out'
'r_BE'	'rBE.out'
'C_aaBE'	'CaaBE.out'
'C_rrBE'	'CrrBE.out'
'C_arBE'	'CarBE.out'
'Crr_comp'	"Crrcomp.out'
'chi2'	'chi2.out'

Table 3.2. Super File Format for Multi-Pred Case 2

Category	**File Name**
MultiPredSup	
'dims'	'dimensions.inp'
'a_nom'	'a.inp'
'r_mea'	'rm.inp'

'r_com'	'rc.inp'
'C_aa'	'Caa.inp'
'C_ar'	'Car.inp'
'C_rr'	'Crr.inp'
'S_ra'	'Sra.inp'
'b_nom'	'b.inp'
'C_bb'	'Cbb.inp'
'C_ab'	'Cab.inp'
'C_br'	'Cbr.inp'
'S_rb'	'Srb.inp'
'a_BE'	'aBE.out'
'r_BE'	'rBE.out'
'C_aaBE'	'CaaBE.out'
'C_rrBE'	'CrrBE.out'
'C_arBE'	'CarBE.out'
'Crr_comp'	"Crrcomp.out'
'b_BE'	'bBE.out'
'C_bbBE'	'CbbBE.out'
'C_abBE'	'CabBE.out'
'C_brBE'	'CbrBE.out'
'chi2'	'chi2.out'

Table 3.3 Super File Format for Multi-Pred Case 3

Category	**File Name**
MultiPredSup	
'dims'	'dimensions.inp'
'a_nom'	'a.inp'
'r_mea'	'rm.inp'
'r_com'	'rc.inp'
'C_aa'	'Caa.inp'

'C_ar'	'Car.inp'
'C_rr'	'Crr.inp'
'S_ra'	'Sra.inp'
'q_mea'	'qm.inp'
'q_com'	'qc.inp'
'C_qq'	'Cqq.inp'
'C_aq'	'Caq.inp'
'S_qa'	'Sqa.inp'
'a_BE'	'aBE.out'
'r_BE'	'rBE.out'
'C_aaBE'	'CaaBE.out'
'C_rrBE'	'CrrBE.out'
'C_arBE'	'CarBE.out'
'Crr_comp'	"Crrcomp.out'
'q_BE'	'qBE.out'
'C_qqBE'	'CqqBE.out'
'Cqq_comp'	"Cqqcomp.out'
'C_aqBE'	'CaqBE.out'
'C_rqBE'	'CrqBE.out'
'Crq_comp'	"Crqcomp.out'
'chi2'	'chi2.out'

Table 3.4 Super File Format for Multi-Pred Case 4

Category	**File Name**
MultiPredSup	
'dims'	'dimensions.inp'
'a_nom'	'a.inp'
'r_mea'	'rm.inp'
'r_com'	'rc.inp'
'C_aa'	'Caa.inp'

'C_ar'	'Car.inp'
'C_rr'	'Crr.inp'
'S_ra'	'Sra.inp'
'b_nom'	'b.inp'
'q_mea'	'qm.inp'
'q_com'	'qc.inp'
'C_bb'	'Cbb.inp'
'C_bq'	'Cbq.inp'
'C_qq'	'Cqq.inp'
'S_qb'	'Sqb.inp'
'C_ab'	'Cab.inp'
'C_aq'	'Caq.inp'
'C_br'	'Cbr.inp'
'C_rq'	'Crq.inp'
'S_rb'	'Srb.inp'
'S_qa'	'Sqa.inp'
'a_BE'	'aBE.out'
'r_BE'	'rBE.out'
'C_aaBE'	'CaaBE.out'
'C_rrBE'	'CrrBE.out'
'C_arBE'	'CarBE.out'
'Crr_comp'	''Crrcomp.out'
'b_BE'	'bBE.out'
'q_BE'	'qBE.out'
'C_bbBE'	'CbbBE.out'
'C_qqBE'	'CqqBE.out'
'C_bqBE'	'CbqBE.out'
'Cqq_comp'	''Cqqcomp.out'
'C_abBE'	'CabBE.out'
'C_aqBE'	'CaqBE.out'

'C_brBE'	'CbrBE.out'
'C_rqBE'	'CrqBE.out'
'Crq_comp'	"Crqcomp.out'
'chi2'	'chi2.out'

3.3.4.2. File "dimensions.inp"

The file *dimensions.inp* defines the following important control variables:

CaseNumber – Multi-Pred Case selection;

N_a: number of parameters for Model A;

N_r: number of responses for Model A;

N_b: number of additional parameters for Model A (Case 2) or the number of parameters of Model B (Case 4);

N_q: number of additional responses for Model A (Case 3) or the number of responses of Model B (Case 4);

The following is an example of *dimensions.inp* for the Cooling Tower Model Case 4. For this test case, Cooling Tower Model is separated into Model A and Model B. Model A comprises the first 42 parameters (of the total 52 model parameters) and the first 2 responses (of the total 3 model responses). Thus: for Model A, Na = 42 and Nr = 2. Model B comprises the last 10 parameters (of the total 52 model parameters) and the 3rd response (of the total 3 model responses of the Cooling Tower Model). Thus: for Model B, Nb = 10 and Nq =1.

```
/ Case options:
/   = 1 "One-Model" Case: predictive modeling solely for Model A with Na model parameters and Nr measured responses;
/   = 2 "One-Model" Case: predictive modeling for Model A with Nb additional parameters, but no additional responses;
/   = 3 "One-Model" Case: predictive modeling for Model A with Nq additionalresponses, but no additional parameters;
/   = 4 "Two-Model" Case: predictive modeling for Model A coupled with Model B.
/ Case selection (CaseNumber):
4
/Na -- number of parameters for model A
42
/Nr -- number of responses for model A
2
/Nb -- number of additional parameters:
/      -- for case 1: not used
```

```
  /        -- for case 2: number of parameters added to the Na
parameters for model A
  /       -- for case 3: not used
  /       -- for case 4: number of parameters of model B
  10
  /Nq -- number of additional responses:
  /       -- for case 1: not used
  /       -- for case 2: not used
  /        -- for case 3: number of responses added to the Nr
responses for model A
  /       -- for case 4: number of responses for model B
  1
```

The format of *dimensions.inp* is fixed as shown above. The user can change the numbers corresponding to the control variables, namely: CaseNumber, Na, Nr, Nb and Nq, respectively.

3.3.4.3 Contents and Organization of Input and Output Files

Tables 3.5 through 3.8 describe the contents of the input and output (I/O) files specified within the **MULTI-PRED** super-files listed in Tables 3.1 through 3.4, respectively. The vectors / matrices corresponding to each data file are also listed in Tables 3.5 through 3.8.

Table 3.5. Summary of Input and Output Files for MULTI-PRED Case 1

File	Unit	I/O	Corresponding vector/matrix	Descriptions
superfile.inp	5	input		File organization
dimensions.inp	20	input		Defines the Case selection and dimensions control
a.inp	21	input	$\boldsymbol{\alpha}(N_a)$	Nominal values of N_a parameters of model A
rm.inp	22	input	$\mathbf{r}_m(N_r)$	Nominal values of N_r measured responses of model A
rc.inp	23	input	$\mathbf{r}_c(N_r)$	Nominal values of N_r computed responses of model A

File	Unit	I/O	Corresponding vector/matrix	Descriptions
Caa.inp	24	input	$\mathbf{C}_{aa}(N_a \times N_a)$	Covariance matrix of N_a parameters of model A
Car.inp	25	input	$\mathbf{C}_{ar}(N_a \times N_r)$	Correlations between N_a parameters and N_r responses of Model A
Crr.inp	26	input	$\mathbf{C}_{rr}(N_r \times N_r)$	Covariance matrix of N_r responses of model A
Sra.inp	27	input	$\mathbf{S}_{ra}(N_r \times N_a)$	Absolute sensitivities of N_r responses of Model A w.r.t N_a parameters of Model A
aBE.out	51	output	$\boldsymbol{\alpha}^{be}(N_a)$	Best-estimate nominal values of parameters of Model A
rBE.out	52	output	$\mathbf{r}^{be}(N_r)$	Best-estimate nominal values of responses of Model A
CaaBE.out	53	output	$\mathbf{C}_{aa}^{be}(N_a \times N_a)$	Predicted covariance matrix of N_a parameters of Model A
CrrBE.out	54	output	$\mathbf{C}_{rr}^{be}(N_r \times N_r)$	Predicted covariance matrix of N_r responses of Model A
CarBE.out	55	output	$\mathbf{C}_{ar}^{be}(N_a \times N_r)$	Predicted correlation matrix between the N_a parameters and N_r responses of model A
Crrcomp.out	56	output	$\mathbf{C}_{rr}^{comp}(N_r \times N_r)$	Covariance matrix of N_r computed responses of model A
chi2.out	76	output	χ^2, scalar	Value of the consistency indicator

Table 3.6. Summary of Input and Output Files for MULTI-PRED Case 2

File	Unit	I/O	Corresponding vector/matrix	Descriptions
superfile.inp	5	input		File organization
dimensions.inp	20	input		Defines the Case selection and dimensions control
a.inp	21	Input	$\boldsymbol{\alpha}(N_a)$	Nominal values of N_a parameters of model A
rm.inp	22	Input	$\mathbf{r}_m(N_r)$	Nominal values of N_r measured responses of model A
rc.inp	23	Input	$\mathbf{r}_c(N_r)$	Nominal values of N_r computed responses of model A
Caa.inp	24	Input	$\mathbf{C}_{aa}(N_a \times N_a)$	Covariance matrix of N_a parameters of model A
Car.inp	25	Input	$\mathbf{C}_{ar}(N_a \times N_r)$	Correlations between N_a parameters and N_r responses of Model A
Crr.inp	26	Input	$\mathbf{C}_{rr}(N_r \times N_r)$	Covariance matrix of N_r responses of model A
Sra.inp	27	Input	$\mathbf{S}_{ra}(N_r \times N_a)$	Absolute sensitivities of N_r responses of Model A w.r.t N_a parameters of Model A
b.inp	31	Input	$\mathbf{b}(N_b)$	Nominal values of N_b additional parameters for model A
Cbb.inp	34	Input	$\mathbf{C}_{bb}(N_b \times N_b)$	Covariance matrix of N_b additional parameters
Cab.inp	41	Input	$\mathbf{C}_{ab}(N_a \times N_b)$	Correlations between N_a parameters of Model A and N_b additional parameters for Model A

File	Unit	I/O	Corresponding vector/matrix	Descriptions
Cbr.inp	43	In-put	$\mathbf{C}_{br}(N_b \times N_r)$	Correlations between N_b additional parameters for Model A and N_r responses of Model A
Srb.inp	45	in-put	$\mathbf{S}_{rb}(N_r \times N_b)$	Absolute sensitivities of N_r responses of Model A w.r.t N_b additional parameters for model A
aBE.out	51	out-put	$\boldsymbol{\alpha}^{be}(N_a)$	Best-estimate nominal values of N_a parameters of Model A
rBE.out	52	out-put	$\mathbf{r}^{be}(N_r)$	Best-estimate nominal values of N_r responses of Model A
CaaBE.out	53	out-put	$\mathbf{C}_{aa}^{be}(N_a \times N_a)$	Predicted covariance matrix of N_a parameters of Model A
CrrBE.out	54	out-put	$\mathbf{C}_{rr}^{be}(N_r \times N_r)$	Predicted covariance matrix of N_r responses of Model A
CarBE.out	55	out-put	$\mathbf{C}_{ar}^{be}(N_a \times N_r)$	Predicted correlation matrix between the N_a parameters and N_r responses of model A
Crrcomp.out	56	out-put	$\mathbf{C}_{rr}^{comp}(N_r \times N_r)$	Covariance matrix of N_r computed responses of model A
bBE.out	61	out-put	$\mathbf{b}^{be}(N_b)$	Best-estimate nominal values of N_b additional parameters
CbbBE.out	63	out-put	$\mathbf{C}_{bb}^{be}(N_b \times N_b)$	Predicted covariance matrix of N_b parameters of Model A
CabBE.out	71	out-put	$\mathbf{C}_{ab}^{be}(N_a \times N_b)$	Predicted correlation matrix between the N_a parameters

File	Unit	I/O	Corresponding vector/matrix	Descriptions
				of Model A and the N_b additional parameters for Model A
CbrBE.out	73	output	$\mathbf{C}_{br}^{be}(N_b \times N_r)$	Predicted correlation matrix between the N_b additional parameters for Model A and N_r responses of model A
chi2.out	76	output	χ^2, scalar	Value of the consistency indicator

Table 3.7. Summary of Input and Output Files for MULTI-PRED Case 3

File	Unit	I/O	Corresponding vector/matrix	Descriptions
superfile.inp	5	input		File organization
dimensions.inp	20	input		Defines the Case selection and dimensions control
a.inp	21	input	$\boldsymbol{\alpha}(N_a)$	Nominal values of N_a parameters of model A
rm.inp	22	input	$\mathbf{r}_m(N_r)$	Nominal values of N_r measured responses of model A
rc.inp	23	input	$\mathbf{r}_c(N_r)$	Nominal values of N_r computed responses of model A
Caa.inp	24	input	$\mathbf{C}_{aa}(N_a \times N_a)$	Covariance matrix of Na parameters of model A
Car.inp	25	input	$\mathbf{C}_{ar}(N_a \times N_r)$	Correlations between N_a parameters and N_r responses of Model A
Crr.inp	26	input	$\mathbf{C}_{rr}(N_r \times N_r)$	Covariance matrix of N_r responses of model A

File	Unit	I/O	Corresponding vector/matrix	Descriptions
Sra.inp	27	input	$\mathbf{S}_{ra}(N_r \times N_a)$	Absolute sensitivities of N_r responses of Model A w.r.t N_a parameters of Model A
qm.inp	32	input	$\mathbf{q}_m(N_q)$	Nominal values of N_q additional measured responses for model A
qc.inp	33	input	$\mathbf{q}_c(N_q)$	Nominal values of N_q additional computed responses for model A
Cqq.inp	36	input	$\mathbf{C}_{qq}(N_q \times N_q)$	Covariance matrix of N_q additional responses for Model A
Caq.inp	42	input	$\mathbf{C}_{aq}(N_a \times N_q)$	Correlations between N_a parameters of Model A and N_q additional responses for Model A
Crq.inp	44	input	$\mathbf{C}_{rq}(N_r \times N_q)$	Correlations between N_r responses of Model A and N_q additional responses for Model A
Sqa.inp	46	input	$\mathbf{S}_{qa}(N_q \times N_a)$	Absolute sensitivities of N_q additional responses for Model A w.r.t N_a parameters of Model A
aBE.out	51	output	$\boldsymbol{\alpha}^{be}(N_a)$	Best-estimate nominal values of parameters of Model A
rBE.out	52	output	$\mathbf{r}^{be}(N_r)$	Best-estimate nominal values of responses of Model A
CaaBE.out	53	output	$\mathbf{C}_{aa}^{be}(N_a \times N_a)$	Predicted covariance matrix of N_a parameters of Model A

File	Unit	I/O	Corresponding vector/matrix	Descriptions
CrrBE.out	54	output	$\mathbf{C}_{rr}^{be}(N_r \times N_r)$	Predicted covariance matrix of N_r responses of Model A
CarBE.out	55	output	$\mathbf{C}_{ar}^{be}(N_a \times N_r)$	Predicted correlation matrix between the N_a parameters and N_r esponses of model A
Crrcomp.out	56	output	$\mathbf{C}_{rr}^{comp}(N_r \times N_r)$	Covariance matrix of N_r computed responses of model A
qBE.out	62	output	$\mathbf{q}^{be}(N_q)$	Best-estimate nominal values of N_q additional responses for model A
CqqBE.out	64	output	$\mathbf{C}_{qq}^{be}(N_q \times N_q)$	Predicted covariance matrix of N_q additional responses for model A
Cqqcomp.out	66	output	$\mathbf{C}_{qq}^{comp}(N_q \times N_q)$	Covariance matrix of N_q additional computed responses for model A
CaqBE.out	72	output	$\mathbf{C}_{aq}^{comp}(N_a \times N_q)$	Predicted correlation matrix between the N_a parameters and of Model A and N_q additional responses for model A
CrqBE.out	74	output	$\mathbf{C}_{rq}^{be}(N_r \times N_q)$	Predicted correlation matrix of between N_r responses of Model A and N_q additional responses for model A
Crqcomp.out	75	output	$\mathbf{C}_{rq}^{comp}(N_r \times N_q)$	Correlation matrix of N_r computed responses of Model A and N_q additional computed responses for model A
chi2.out	76	output	χ^2, scalar	Value of the consistency indicator

Table 3.8. Summary of Input and Output Files for MULTI-PRED Case 4

File	Unit	I/O	Corresponding vector/matrix	Descriptions
superfile.inp	5	input		File organization
dimensions.inp	20	input		Defines the Case selection and dimensions control
a.inp	21	input	$\boldsymbol{\alpha}(N_a)$	Nominal values of N_a parameters of model A
rm.inp	22	input	$\mathbf{r}_m(N_r)$	Nominal values of N_r measured responses of model A
rc.inp	23	input	$\mathbf{r}_c(N_r)$	Nominal values of N_r computed responses of model A
Caa.inp	24	input	$\mathbf{C}_{aa}(N_a \times N_a)$	Covariance matrix of N_a parameters of model A
Car.inp	25	input	$\mathbf{C}_{ar}(N_a \times N_r)$	Correlations between N_a parameters and N_r responses of Model A
Crr.inp	26	input	$\mathbf{C}_{rr}(N_r \times N_r)$	Covariance matrix of N_r responses of model A
Sra.inp	27	input	$\mathbf{S}_{ra}(N_r \times N_a)$	Absolute sensitivities of N_r responses of Model A w.r.t N_a parameters of Model A
b.inp	31	input	$\mathbf{b}(N_b)$	Nominal values of N_b parameters of model B
qm.inp	32	input	$\mathbf{q}_m(N_q)$	Nominal values of N_q measured responses of model B
qc.inp	33	input	$\mathbf{q}_c(N_q)$	Nominal values of N_q computed responses of model B

File	Unit	I/O	Corresponding vector/matrix	Descriptions
Cbb.inp	34	input	$\mathbf{C}_{bb}(N_b \times N_b)$	Covariance matrix of N_b parameters of model B
Cbq.inp	35	input	$\mathbf{C}_{bq}(N_b \times N_q)$	Correlations between N_b parameters and N_q responses of Model B
Cqq.inp	36	input	$\mathbf{C}_{qq}(N_q \times N_q)$	Covariance matrix of N_q responses of model B
Sqb.inp	37	input	$\mathbf{S}_{qb}(N_q \times N_b)$	Absolute sensitivities of N_q responses of Model B w.r.t N_b parameters of Model B
Cab.inp	41	input	$\mathbf{C}_{ab}(N_a \times N_b)$	Correlation matrix between the N_a parameters of Model A and the N_b parameters of Model B
Caq.inp	42	input	$\mathbf{C}_{aq}(N_a \times N_q)$	Correlation matrix between the N_a parameters and of Model A and N_q responses of model B
Cbr.inp	43	input	$\mathbf{C}_{br}(N_b \times N_r)$	Correlation matrix between the N_b parameters of Model B and N_r responses of model A
Crq.inp	44	input	$\mathbf{C}_{rq}(N_r \times N_q)$	Correlation matrix of between N_r responses of Model A and N_q responses of model B
Srb.inp	45	input	$\mathbf{S}_{rb}(N_r \times N_b)$	Absolute sensitivities of N_r responses of Model A w.r.t N_b parameters of model B

File	Unit	I/O	Corresponding vector/matrix	Descriptions
Sqa.inp	46	input	$\mathbf{S}_{qa}(N_q \times N_a)$	Absolute sensitivities of N_q responses of Model B w.r.t N_a parameters of Model A
aBE.out	51	output	$\boldsymbol{\alpha}^{be}(N_a)$	Best-estimate nominal values of parameters of Model A
rBE.out	52	output	$\mathbf{r}^{be}(N_r)$	Best-estimate nominal values of responses of Model A
CaaBE.out	53	output	$\mathbf{C}_{aa}^{be}(N_a \times N_a)$	Predicted covariance matrix of N_a parameters of Model A
CrrBE.out	54	output	$\mathbf{C}_{rr}^{be}(N_r \times N_r)$	Predicted covariance matrix of N_r responses of Model A
CarBE.out	55	output	$\mathbf{C}_{ar}^{be}(N_a \times N_r)$	Predicted correlation matrix between the N_a parameters and N_r responses of model A
Crrcomp.out	56	output	$\mathbf{C}_{rr}^{comp}(N_r \times N_r)$	Covariance matrix of N_r computed responses of model A
bBE.out	61	output	$\mathbf{b}^{be}(N_b)$	Best-estimate nominal values of parameters of Model B
qBE.out	62	output	$\mathbf{q}^{be}(N_q)$	Best-estimate nominal values of responses of Model B
CbbBE.out	63	output	$\mathbf{C}_{bb}^{be}(N_b \times N_b)$	Predicted covariance matrix of N_b parameters of Model B
CqqBE.out	64	output	$\mathbf{C}_{qq}^{be}(N_q \times N_q)$	Predicted covariance matrix of N_q responses of Model B
CbqBE.out	65	output	$\mathbf{C}_{bq}^{be}(N_b \times N_q)$	Predicted correlation matrix between the N_b parameters

File	Unit	I/O	Corresponding vector/matrix	Descriptions
				and N_q responses of model B
Cqqcomp.out	66	output	$\mathbf{C}_{qq}^{comp}(N_q \times N_q)$	Covariance matrix of N_q computed responses of model B
CabBE.out	71	output	$\mathbf{C}_{ab}^{be}(N_a \times N_b)$	Predicted correlation matrix between the N_a parameters of Model A and the N_b parameters for Model B
CaqBE.out	72	output	$\mathbf{C}_{aq}^{be}(N_a \times N_q)$	Predicted correlation matrix between the N_a parameters and of Model A and N_q responses of model B
CbrBE.out	73	output	$\mathbf{C}_{br}^{be}(N_b \times N_r)$	Predicted correlation matrix between the N_b parameters of Model B and N_r responses of model A
CrqBE.out	74	output	$\mathbf{C}_{rq}^{be}(N_r \times N_q)$	Predicted correlation matrix of between N_r responses of Model A and N_q responses of model B
Crqcomp.out	75	output	$\mathbf{C}_{rq}^{comp}(N_r \times N_q)$	Correlation matrix of N_r computed responses of Model A and N_q computed responses of model B
chi2.out	76	output	χ^2, scalar	Value of the consistency indicator

3.3.5 *Input Data Files*

This Section describes in detail the *input* files (and their contents) that were listed in Table 3.11. All the data files are in the "sparse triplet matrix" file format, which is a commonly used ASCII file format for storing sparse matrices and compatible with most files in the Matrix Market format.
The *sparse triplet data structure* simply records, for each nonzero entry of the matrix, the row, column and value. The general format is as follows:

Line 1:	**M**	**N**	**Nz**
Line 2:	**Row_index**	**Col_index**	**Val**
Line 3:	**Row_index**	**Col_index**	**Val**
...	...	...	...
Line Nz+1:	**Row_index**	**Col_index**	**Val**

In the above format, the quantities **M** and **N** denote, respectively, the number of rows and columns in the original full matrix; **Nz** denotes total the number of nonzero elements in the matrix; **Row_index** and **Col_index** denote the row and column indices of each nonzero element; and **VAL** denotes the value of the nonzero element.

3.3.5.1 Input Data Files for MULTI-PRED Case 1

MULTI-PRED Case 1 requires the following 7 input data files as listed in Table 3.9, as well as in Table 3.5.

Table 3.9. Input Data Files for MULTI-PRED Case 1

Input Data File for Model A
a.inp
rm.inp
rc.inp
Caa.inp
Car.inp
Crr.inp
Sra.inp

The file structures for the inputs shown in Table 3.9 are described in detail below.

(1) a.inp

The input file *a.inp* contains the nominal values of all N_a parameters of Model A. For example, for the neutron diffusion model, the nominal values of the $N_a = 4$ parameters are given as follows:

$$\boldsymbol{\alpha} = \begin{pmatrix} 0.0197 \\ 0.16 \\ 1.0E+07 \\ 7.438 \end{pmatrix}. \tag{3.182}$$

The corresponding *a.inp* is as follows.

```
4   1    4
1   1    0.0197
2   1    0.16
3   1    1.0E+07
4   1    7.438
```

(2) rm.inp

The input file *rm.inp* contains the nominal values of N_r measured responses for Model A. For the neutron diffusion model, for example, the nominal values of the $N_r = 4$ measured responses are as follows:

$$\mathbf{r}_m = \begin{pmatrix} 3.40\text{E+09} \\ 3.59\text{E+09} \\ 3.77\text{E+09} \\ 3.74\text{E+09} \end{pmatrix}. \tag{3.183}$$

The corresponding *rm.inp* is as follows.

```
4   1    4
1   1    3.398068337E+09
2   1    3.586849912E+09
3   1    3.772511377E+09
4   1    3.735885053E+09
```

(3) rc.inp

The input file *rc.inp* contains the nominal values of N_r computed responses of Model A. For the neutron diffusion model, for example, the nominal values of the $N_r = 4$ computed responses are as follows:

$$\mathbf{r}_c = \begin{pmatrix} 3.77E+09 \\ 3.77E+09 \\ 3.66E+09 \\ 3.66E+09 \end{pmatrix}. \tag{3.184}$$

The corresponding *rc.inp* is as follows.

```
4    1     4
1    1     3.775631486E+09
2    1     3.775631486E+09
3    1     3.662632405E+09
4    1     3.662632405E+09
```

(4) Caa.inp

The input file *Caa.inp* contains the nonzero elements of the covariance matrix $\mathbf{C}_{\alpha\alpha}(N_\alpha \times N_\alpha)$ of model parameters of Model A. For the neutron diffusion model, for example, $\mathbf{C}_{\alpha\alpha}$ is:

$$C_{\alpha\alpha} = \begin{pmatrix} \left(9.85\times10^{-4}\right)^2 & 0 & 0 & 0 \\ 0 & \left(8.0\times10^{-3}\right)^2 & 0 & 0 \\ 0 & 0 & \left(1.5\times10^{6}\right)^2 & 0 \\ 0 & 0 & 0 & \left(7.44\times10^{-1}\right)^2 \end{pmatrix} \tag{3.185}$$

The corresponding *Caa.inp* is as follows.

```
4   4    4
1   1    9.70225E-07
2   2    6.40000E-05
3   3    2.25000E+12
4   4    5.5323844E-01
```

(5) Car.inp

The input file *Car.inp* contains the nonzero elements of the correlation matrix $\mathbf{C}_{ar}(N_a \times N_r)$ between the model parameters and measured responses of Model A. For the neutron diffusion model, for examples, the parameters and measured responses are not correlated; therefore, $\mathbf{C}_{ar}$ has the following structure:

$$\mathbf{C}_{\alpha r} = \begin{pmatrix} 0 & 0 & 0 & 0 \\ 0 & 0 & 0 & 0 \\ 0 & 0 & 0 & 0 \\ 0 & 0 & 0 & 0 \end{pmatrix}. \tag{3.186}$$

The corresponding *Car.inp* is as follows:

```
4    4       0
```

In other applications, the parameters and measured responses are correlated, i.e., $\mathbf{C}_{\alpha r} \neq \mathbf{0}$. An example of a non-zero correlation matrix is provided by the cooling tower model, for which $N_a = 52$, $N_r = 3$, and for which the correlation matrix $\mathbf{C}_{\alpha r}$ comprises 12 nonzero elements. Hence, for this example, *Car.inp* is as follows:

```
52   3      12
1    1      12.957508300000001
1    2      3.3548676099999999
1    3      -54.158679370000002
2    1      3.5102394000000001
2    2      3.0452589900000002
2    3      1.73334787
3    1      2.3294612799999999
3    2      1.8856921
3    3      -2.26657529
4    1      -447.08545706000001
4    2      -93.577718820000001
4    3      1831.03340159
```

(6) Crr.inp

The input file *Crr.inp* contains the nonzero elements of the covariance matrix $\mathbf{C}_{rr}(N_r \times N_r)$ between the model responses of Model A. For the neutron diffusion model, for example, $\mathbf{C}_{rr}$ is

$$C_{rr} = \begin{pmatrix} (1.7\times10^8)^2 & 0 & 0 & 0 \\ 0 & (2.15\times10^8)^2 & 0 & 0 \\ 0 & 0 & (1.89\times10^8)^2 & 0 \\ 0 & 0 & 0 & (1.87\times10^8)^2 \end{pmatrix}. \quad (3.187)$$

The corresponding *Crr.inp* is as follows:

```
4    4    4
1    1    2.886717104E+16
2    2    4.631577224E+16
3    3    3.557960521E+16
4    4    3.489209280E+16
```

(7) Sra.inp

The input file *Sra.inp* contains the nonzero elements of the absolute sensitivities matrix $\mathbf{S}_{ra}(N_r \times N_a)$. For the neutron diffusion model, for example, $\mathbf{S}_{ra}(N_r \times N_a)$ is

$$S \triangleq \left(\frac{\partial R_i}{\partial \alpha_j}\right) = \begin{pmatrix} -1.92\times10^{11} & -1.33\times10^5 & 3.78\times10^2 & 5.08\times10^8 \\ -1.92\times10^{11} & -1.33\times10^5 & 3.78\times10^2 & 5.08\times10^8 \\ -1.76\times10^{11} & -1.24\times10^9 & 3.66\times10^2 & 4.92\times10^8 \\ -1.76\times10^{11} & -1.24\times10^9 & 3.66\times10^2 & 4.92\times10^8 \end{pmatrix}. \quad (3.188)$$

The corresponding *Sra.inp* is as follows:

```
4    4    16
1    1    -1.916553399E+11
1    2    -1.330585230E+5
1    3    3.775631486E+2
1    4    5.076138055E+8
2    1    -1.916553399E+11
2    2    -1.330585230E+5
2    3    3.775631486E+2
2    4    5.076138055E+8
3    1    -1.758565925E+11
3    2    -1.239109567E+9
```

```
3   3   3.662632405E+2
3   4   4.924216731E+8
4   1   -1.758565925E+11
4   2   -1.239109567E+9
4   3   3.662632405E+2
4   4   4.924216731E+8
```

3.3.5.2 Input Data Files for MULTI-PRED Case 2

Table 3.10 presents the 12 input files required for MULTI-PRED Case 2; these files are also listed in Table 3.6. Of the 12 files listed in Table 3.10, 7 input data files have been previously described and the additional 5 input data files have the same structure as their counterparts for Model A.

Table 3.10. Input Data Files for MULTI-PRED Case 2

Input Data File for Model A	Inputs for the Coupled Matrices	Inputs for the N_b additional parameters for Model A
a.inp		b.inp
rm.inp		
rc.inp		
Caa.inp	Cab.inp	Cbb.inp
Car.inp	Cbr.inp	
Crr.inp		
Sra.inp	Srb.inp	

3.3.5.3 Input Data Files for MULTI-PRED Case 3

Table 3.11 presents the 13 input files required for MULTI-PRED Case 3; these files are also listed in Table 3.7.

Table 3.11. Input Data Files for MULTI-PRED Case 3

Input Data File for Model A	Inputs for the coupled matrices	Inputs for the N_q additional responses for Model A
a.inp		
rm.inp		qm.inp

Input Data File for Model A	Inputs for the coupled matrices	Inputs for the N_q additional responses for Model A
rc.inp		qc.inp
Caa.inp		
Car.inp	Caq.inp	
Crr.inp	Crq.inp	Cqq.inp
Sra.inp	Sqa.inp	

3.3.5.4 Input Data Files for MULTI-PRED Case 4

Table 3.12 presents the 20 input files required for MULTI-PRED Case 3; these files are also listed in Table 3.8.

Table 3.12. Input Data Files for MULTI-PRED Case 4

Input Data File for Model A	Inputs Data Files for the Coupled Matrices between Model A and Model B	Inputs Data Files for Model B
a.inp		b.inp
rm.inp		qm.inp
rc.inp		qc.inp
Caa.inp	Cab.inp	Cbb.inp
Car.inp	Caq.inp, Cbr.inp	Cbq.inp
Crr.inp	Crq.inp	Cqq.inp
ra.inp	Sqa.inp, Srb.inp	Sqb.inp

3.3.6 Output Data Files

The model output files are specified in the categories of the super files. All the output files are in the "sparse triplet matrix" file format. In addition, a data file for the consistency indicator, χ^2, is also generated.

3.3.6.1 Output Data Files for MULTI-PRED Case 1

Table 3.13 lists the output data files generated by MULTI-PRED Case 1; these output files are also listed in Table 3.5.

Table 3.13. Output Data Files for MULTI-PRED Case 1

Output Data File for Model A	
aBE.out	CarBE.out
rBE.out	Crrcomp.out
CaaBE.out	chi2.out
CrrBE.out	

(1) aBE.out

The output data file *aBE.out* contains the nonzero components of the resulting vector $\boldsymbol{\alpha}^{be}(N_a)$, which provide the best-estimate parameter values for Model A. This file has the same structure as the file *a.inp.* For the neutron diffusion model, for example, the best-estimate parameter values are:

$$\boldsymbol{\alpha}^{be} = \begin{pmatrix} 0.0198 \\ 0.1591 \\ 9.85\times10^{6} \\ 7.388 \end{pmatrix}. \tag{3.189}$$

The corresponding output data file *aBE.out* is as follows:

```
4   1   4
1   1   1.98418101E-02
2   1   1.59118840E-01
3   1   9.84778916E+06
4   1   7.38768248E+00
```

(2) rBE.out

The output file *rBE.out* contains the nonzero components of the vector $\mathbf{r}^{be}(N_r)$, which provide the best-estimate response values for Model A. This output file has the same structure as the file *rc.inp.* For the neutron diffusion model, for example, the best-estimate response values are:

$$\mathbf{r}^{be} = \begin{pmatrix} 3.66\times10^{9} \\ 3.66\times10^{9} \\ 3.56\times10^{9} \\ 3.56\times10^{9} \end{pmatrix}. \tag{3.190}$$

The corresponding output data file r*BE.out* is as follows:

```
4    1    4
1    1    3.66544187E+09
2    1    3.66544187E+09
3    1    3.55825935E+09
4    1    3.55825935E+09
```

(3) CaaBE.out

The output file C*aaBE.out* contains the nonzero components of the predicted optimal covariance matrix $\mathbf{C}_{aa}^{be}(N_a \times N_a)$ of parameters for Model A. This output file has the same structure as the file C*aa.inp*. For the neutron diffusion model, for example, the best-estimate covariance matrix $\mathbf{C}_{aa}^{be}(N_a \times N_a)$ has the following form:

$$\boldsymbol{C}_{aa}^{be} = \begin{pmatrix} 9.03\times10^{-7} & 6.75\times10^{-9} & 3.03\times10^{2} & 1.00\times10^{-4} \\ 6.75\times10^{-9} & 6.38\times10^{-5} & 7.37\times10^{1} & 2.44\times10^{-5} \\ 3.03\times10^{2} & 7.37\times10^{1} & 8.24\times10^{11} & -4.71\times10^{5} \\ 1.00\times10^{-4} & 2.44\times10^{-5} & -4.71\times10^{5} & 3.97\times10^{-1} \end{pmatrix} \tag{3.191}$$

The corresponding output data file Caa*BE.out* is as follows.

```
4    4    16
1    1    9.02992937E-07
1    2    6.48311998E-09
1    3    3.03370351E+02
1    4    1.00295979E-04
2    1    6.48311998E-09
2    2    6.38139054E-05
2    3    7.32951010E+01
2    4    2.43980122E-05
3    1    3.03370351E+02
3    2    7.32951010E+01
3    3    8.24405218E+11
```

```
3    4    -4.71401727E+05
4    1     1.00295979E-04
4    2     2.43980122E-05
4    3    -4.71401727E+05
4    4     3.97657348E-01
```

(4) CarBE.out

The output file C*arBE.out* contains the nonzero components of the predicted parameter-response correlation matrix $\mathbf{C}_{ar}^{be}(N_a \times N_r)$ for Model A. This output file has the same structure as the file C*ar.inp*. For the neutron diffusion model, for example, the correlation matrix $\mathbf{C}_{ar}^{be}(N_a \times N_r)$ has the following structure:

$$\boldsymbol{C}_{\alpha r}^{be} = \begin{pmatrix} -7.81\times10^{3} & -7.81\times10^{3} & 1.50\times10^{3} & 1.50\times10^{3} \\ 3.89\times10^{4} & 3.89\times10^{4} & -4.13\times10^{4} & -4.13\times10^{4} \\ 1.38\times10^{13} & 1.38\times10^{13} & 1.64\times10^{13} & 1.64\times10^{13} \\ 4.57\times10^{6} & 4.57\times10^{6} & 5.41\times10^{6} & 5.41\times10^{6} \end{pmatrix}. \tag{3.192}$$

The corresponding output data file CarB*E.out* is as follows:

```
4    4     16
1    1    -7.81261058E+03
1    2    -7.81261058E+03
1    3     1.50018202E+03
1    4     1.50018202E+03
2    1     3.88811594E+04
2    2     3.88811594E+04
2    3    -4.12791159E+04
2    4    -4.12791159E+04
3    1     1.38214773E+13
3    2     1.38214773E+13
3    3     1.63658795E+13
3    4     1.63658795E+13
4    1     4.56907323E+06
4    2     4.56907323E+06
4    3     5.41019607E+06
4    4     5.41019607E+06
```

(5) CrrBE.out

The output file *CrrBE.out* contains the nonzero components of the predicted covariance matrix $\mathbf{C}_{rr}^{be}(N_r \times N_r)$ of responses for Model A. This output file has the same file structure as the file *Crr.inp*. For the neutron diffusion model, for example, the correlation matrix $\mathbf{C}_{rr}^{be}(N_r \times N_r)$ is as follows:

$$\boldsymbol{C}_{rr}^{be} = \begin{pmatrix} 9.04\times10^{15} & 9.04\times10^{15} & 8.64\times10^{15} & 8.64\times10^{15} \\ 9.04\times10^{15} & 9.04\times10^{15} & 8.64\times10^{15} & 8.64\times10^{15} \\ 8.64\times10^{15} & 8.64\times10^{15} & 8.45\times10^{15} & 8.45\times10^{15} \\ 8.64\times10^{15} & 8.64\times10^{15} & 8.45\times10^{15} & 8.45\times10^{15} \end{pmatrix} \quad (3.193)$$

The corresponding output data file *CrrBE.out* is as follows:

```
4    4    16
1    1    9.03512848E+15
1    2    9.03512848E+15
1    3    8.63793079E+15
1    4    8.63793079E+15
2    1    9.03512848E+15
2    2    9.03512848E+15
2    3    8.63793079E+15
2    4    8.63793079E+15
3    1    8.63793079E+15
3    2    8.63793079E+15
3    3    8.44565029E+15
3    4    8.44565029E+15
4    1    8.63793079E+15
4    2    8.63793079E+15
4    3    8.44565029E+15
4    4    8.44565029E+15
```

(6) Crrcomp.out

The output file *Crrcomp.out* contains the nonzero components of the covariance matrix $\mathbf{C}_{rr}^{comp}(N_r \times N_r)$ of computed responses for Model A. This output file has the same structure as the file Crr.*inp*. For the neutron diffusion model, for example, the covariance matrix $\mathbf{C}_{rr}^{comp}(N_r \times N_r)$ is as follows:

$$\boldsymbol{C}_{rr}^{comp} = \begin{pmatrix} 4.99\times10^{17} & 4.99\times10^{17} & 4.82\times10^{17} & 4.82\times10^{17} \\ 4.99\times10^{17} & 4.99\times10^{17} & 4.82\times10^{17} & 4.82\times10^{17} \\ 4.82\times10^{17} & 4.82\times10^{17} & 4.66\times10^{17} & 4.66\times10^{17} \\ 4.82\times10^{17} & 4.82\times10^{17} & 4.66\times10^{17} & 4.66\times10^{17} \end{pmatrix} \tag{3.194}$$

The corresponding output data file *Crrcomp.out* is as follows:

```
4    4    16
1    1    4.98938357E+17
1    2    4.98938357E+17
1    3    4.82134716E+17
1    4    4.82134716E+17
2    1    4.98938357E+17
2    2    4.98938357E+17
2    3    4.82134716E+17
2    4    4.82134716E+17
3    1    4.82134716E+17
3    2    4.82134716E+17
3    3    4.66086473E+17
3    4    4.66086473E+17
4    1    4.82134716E+17
4    2    4.82134716E+17
4    3    4.66086473E+17
4    4    4.66086473E+17
```

(7) Chi2.out

The output file chi2.*out* contains the values for the consistency indicators χ^2 and $\frac{\chi^2}{N_r}$. For the neutron diffusion model, for example, MULTI-PRED outputs the following values for chi2.*out*:

```
chi^2 = 4.852
chi^2_d = (chi^2)/(number of responses)= 1.213
```

3.3.6.2 *Output Data Files for MULTI-PRED Case 2*

Table 3.14 presents the 11 output files generated for MULTI-PRED Case 2; these files are also listed in Table 3.6.

Table 3.14. Output Data Files for MULTI-PRED Case 2

Output Data File for Model A	**Outputs for the Coupled Matrices**	**Outputs for the N_b additional parameters for Model A**
aBE.out		bBE.out
rBE.out		
CaaBE.out	CabBE.out	CbbBE.out
CrrBE.out		
CarBE.out	CbrBE.out	
Crrcomp.out		
	chi2.out	

3.3.6.3 *Output Data Files for MULTI-PRED Case 3*

Table 3.15 presents the 13 output files generated for MULTI-PRED Case 3; these files are also listed in Table 3.7.

Table 3.15. Output Data Files for MULTI-PRED Case 3

Output Data File for Model A	**Outputs for the coupled matrices**	**Outputs for the N_q additional responses for Model A**
aBE.out		
rBE.out		qBE.out
CaaBE.out		
CrrBE.out	CrqBE.out	CqqBE.out
CarBE.out	CaqBE.out	
Crrcomp.out	Crqcomp.out	Cqqcomp.out
	chi2.out	

3.3.6.4 Output Data Files for MULTI-PRED Case 4

Table 3.16 presents the 18 output files generated for MULTI-PRED Case 4; these files are also listed in Table 3.8.

Table 3.16. Output Data Files for MULTI-PRED Case 4

Output Data File for Model A	**Outputs Data Files for the Coupled Matrices between Model A and Model B**	**Outputs Data Files for Model B**
aBE.out		bBE.out
rBE.out		qBE.out
CaaBE.out	CabBE.out	CbbBE.out
CrrBE.out	CrqBE.out	CqqBE.out
CarBE.out	CaqBE.out, CbrBE.out	CbqBE.out
Crrcomp.out	Crqcomp.out	Cqqcomp.out
chi2.out		

4. BERRU-CMS PREDICTIVE MODELING OF NUCLEAR REACTOR PHYSICS SYSTEMS

Abstract
This Chapter illustrates applications of the BERRU-CMS methodology to paradigm reactor physics systems, starting with the predictive modeling of a paradigm neutron diffusion model of a one-dimensional pool of water containing distributed neutron sources (e.g., spent fuel rods), and continuing with the OECD/NEA (ICSBEP 2010) benchmarks Godiva (a bare uranium sphere), Jezebel-239 and Jezebel-240 (bare plutonium spheres). The paradigm neutron diffusion model characterizes the main physical processes that would occur, for example, in a spent-fuel storage pool and was chosen because it is amenable to closed-form analytical solutions which highlight the salient features of the BERRU-CMS methodology. In addition, this example will also illustrate the basic concepts of computing response sensitivities using the adjoint sensitivity analysis procedure (Cacuci, 1981a, 1981b, 2003). In contradistinction to this didactical example involving the simple neutron diffusion equation, the application of the BERRU-CMS methodology to the benchmarks Godiva, Jezebel-239, and Jezebel-240 involves large-scale neutron transport computations characterized by many imprecisely-known model parameters, as follows: there are 2241 imprecisely known model parameters for Jezebel-239, 1458 imprecisely known parameters for Jezebel-240, and 2916 imprecisely known parameters for Godiva. Eight responses have been considered for Jezebel-239 (the effective multiplication factor, the center core fission rates for ^{233}U, ^{238}U, ^{237}Np, and ^{239}Pu, and the center core radiative capture rates for ^{55}Mn, ^{93}Nb and ^{63}Cu). Three responses (the effective multiplication factor, the center core fission rates for ^{233}U and ^{237}Np) were selected for Jezebel-240, and eleven responses were selected for Godiva (the reaction rate types listed for Jezebel-239, along with the radiative capture rates for ^{107}Ag, ^{127}I and ^{81}Br). The advantages of applying the BERRU-CMS methodology in the response-space (as opposed to the parameter-space) for these benchmark systems will become evident, as this methodology will be shown to reduce significantly the computational memory requirements for predictive modeling of systems involving many model parameters. The illustrative examples presented in this Chapter will underscore the fact that the BERRU-CMS-methodology ensures that increasing the amount of consistent information will increase the accuracy of the predicted results, while reducing their accompanying predicted uncertainties (standard deviations).

4.1 INTRODUCTION

This Chapter illustrates the application of the BERRU-CMS methodology presented in Chapter 3 to reactor physics systems. Section 4.2 presents the application of the BERRU-CMS methodology to a relatively simple neutron diffusion paradigm model of a one-dimensional pool of water containing distributed neutron sources (e.g., a long and deep spent-fuel storage pool). The material presented in Section

D. G. Cacuci, *BERRU Predictive Modeling*,
https://doi.org/10.1007/978-3-662-58395-1_4

4.2 is based on the work by Cacuci and Badea (2014). The diffusion equation was chosen because it is amenable to simple closed-form analytical solutions which highlight the salient features of the BERRU-CMS methodology. In addition, this example will also illustrate the basic concepts of computing response sensitivities using the adjoint sensitivity analysis methodology (ASAM) conceived by Cacuci (1981a, b). The diffusion model is described in Section 4.2, together with the computation of response sensitivities to the model's parameters (diffusion parameter, source, and cross sections) by means of the ASAM. The application of the BERRU-CMS methodology proceeds from the assimilation of a single experiment to the assimilation of four experiments in order to highlight the impact of assimilating additional information on reducing the uncertainties in the predicted results. The initial experimental measurement has deliberately been considered to be relatively imprecise, having a relatively large standard deviation. The impact of such a measurement is subsequently contrasted to the impact of assimilating a more precise measurement, having a relatively small standard deviation. It is shown that, in both cases, the assimilation of consistent experimental information leads to reduced predicted response uncertainties; the more precise the measurement, the larger the uncertainty reduction, as would be intuitively expected. Assimilating two, three and four consistent measurements yields even greater respective reductions in the predicted response uncertainties.

Section 4.3 presents the application of the BERRU-CMS methodology to various couplings of three reactor physics benchmarks, namely the bare uranium sphere Godiva, along with the bare plutonium spheres Jezebel-239 and Jezebel-240, which are actually coupled through shared nuclear data. The material presented in Section 4.3 is based on the work by Latten and Cacuci (2014). Initially, the predictive modeling formulas are applied to each benchmark separately, each considered as a stand-alone physical system, yielding best-estimate values with reduced predicted uncertainties for each of the individual benchmarks. Next, the predictive modeling formulas are applied to both Jezebel benchmarks considered as coupled systems. In this case, the BERRU-CMS formulas yield a different set of best-estimate predicted values and uncertainties. Finally, Godiva is added to both Jezebel configurations, and the predictive modeling formulas are applied to all three benchmarks considered as systems coupled through specific nuclear data (e.g., cross section covariances). The results obtained demonstrate that the BERRU-CMS formulas can be used for adding successively and efficiently new data (e.g., Godiva data) to extant data (e.g., Jezebel-239 combined with Jezebel-240) without loss or re-computation of quantities already computed. In all cases, the BERRU-CMS formulas yielded best-estimate response and parameter values with reduced predicted uncertainties for each individual benchmark. The results obtained indicate that the consideration of complete information, including couplings, provided jointly by all three benchmarks (as opposed to consideration of the benchmarks as separate systems) leads to more accurate predictions of nominal values for responses and model parameters, yielding larger reductions in the predicted uncertainties that accompany the predicted mean values of responses and model parameters.

4.2 BERRU-CMS PREDICTIVE MODELING OF NEUTRON DIFFUSION IN A SLAB

This section illustrates the main features of the BERRU-CMS methodology presented in Chapter 3 by means of a paradigm application to a simple, yet representative, reactor physics problem, which is amenable to a closed-form analytical solution. In addition, this example will highlight the basic concepts of computing response sensitivities using the ASAM conceived by Cacuci (1981a, b), which are essential for applying the BERRU methodology efficiently to large-scale systems.

Consider the diffusion of monoenergetic neutrons due to distributed sources of strength S $\text{neutrons}/\text{cm}^3 \cdot \text{s}$ within a slab of material of extrapolated thickness $2a$. The linear neutron diffusion equation that models mathematically this problem is

$$D\frac{d^2\varphi}{dx^2} - \Sigma_a\varphi + S = 0, \quad x \in (-a, a), \tag{4.1}$$

where $\varphi(x)$ denotes the neutron flux, D denotes the diffusion coefficient, Σ_a denotes the macroscopic absorption cross section, and S denotes a distributed source term within the slab. In view of the problem's symmetry, the origin $x = 0$ can be conveniently chosen at the middle (center) of the slab. The neutron flux is required to vanish on the slab's extrapolated outer surface, which is achieved by imposing the following boundary conditions for Eq. (4.1):

$$\varphi(\pm a) = 0. \tag{4.2}$$

A typical "system response" for the neutron diffusion problem modeled by Eqs. (4.1) and (4.2) is the reading of a detector placed within the slab at a distance b from the slab's midline (taken to be at $x = 0$). Such a response is characterized by the reaction rate

$$R(\mathbf{e}) \triangleq \Sigma_d\varphi(b), \tag{4.3}$$

where Σ_d represents the detector's equivalent reaction cross section. In practice, the model parameters Σ_a, D, S, Σ_d are derived from experimental data that resides in so-called "neutron cross section files", maintained by national or international institutions. Typically, these files provide only the evaluated ("recommended") nominal values and covariances for the various cross sections. Therefore. the imprecisely known model parameters for this problem are the positive constants

Σ_a, D, S, and Σ_d, which will be considered to be the components of the vector $\boldsymbol{\alpha}$ of system parameters, defined as $\boldsymbol{\alpha} \triangleq (\Sigma_a, D, S, \Sigma_d)$. The vector $\mathbf{e}(x)$ appearing in the functional dependence of R in Eq.(4.3) denotes the concatenation of $\varphi(x)$ with $\boldsymbol{\alpha}$, defined as

$$\mathbf{e} \triangleq (\varphi, \boldsymbol{\alpha}). \tag{4.4}$$

Consistent with the foregoing information for this paradigm problem, the mean nominal values $\boldsymbol{\alpha}^0 \triangleq (\Sigma_a^0, D^0, S^0, \Sigma_d^0)$ and the standard deviations $\mathbf{h}_\alpha \triangleq (\delta\Sigma_a, \delta D, \delta S, \delta\Sigma_d)$ for the model parameters are considered to be known. Specifically, consider that the slab of extrapolated thickness $a = 50\ cm$ consists of water with material properties having the following nominal values: $\Sigma_a^0 = 0.0197\, cm^{-1}$, $D^0 = 0{,}16\ cm$, containing distributed neutron sources emitting nominally $S^0 = 10^7 neutrons \cdot cm^{-3} \cdot s^{-1}$, which could represent a long and deep water pool containing spent nuclear fuel elements. These parameters are uncorrelated and have the following relative standard deviations: $\Delta\Sigma_a^0 / \Sigma_a^0 = 5\%$, $\Delta D^0 / D^0 = 5\%$, $\Delta S^0 / S^0 = 15\%$. Furthermore, consider that measurements are performed with an idealized, infinitely thin, detector immersed at locations $x = b$ in the water slab, having an indium-like nominal detector cross section $\Sigma_d^0 = 7.438\, cm^{-1}$, uncorrelated to the other parameters, with a standard deviation $\Delta\Sigma_d^0 / \Sigma_d^0 = 10\%$. Collecting this information and omitting, for simplicity, the respective units, it follows that the covariance matrix for the model parameters is

$$\mathbf{C}_{\alpha\alpha} = \begin{pmatrix} (9.85\times10^{-4})^2 & 0 & 0 & 0 \\ 0 & (8.0\times10^{-3})^2 & 0 & 0 \\ 0 & 0 & (1.5\times10^{6})^2 & 0 \\ 0 & 0 & 0 & (7.44\times10^{-1})^2 \end{pmatrix}. \tag{4.5}$$

The nominal value $\varphi^0(x)$ of the flux is determined by solving Eqs. (4.1).and (4.2) for the nominal parameter values $\boldsymbol{\alpha}^0 \triangleq (\Sigma_a^0, D^0, S^0, \Sigma_d^0)$, to obtain

$$\varphi^0(x) = \frac{S^0}{\Sigma_a^0}\left(1 - \frac{\cosh xk}{\cosh ak}\right), \quad k \triangleq \sqrt{\Sigma_a^0 / D^0}, \tag{4.6}$$

where $k \triangleq \sqrt{\Sigma_a^0 / D^0}$ denotes the nominal value of the reciprocal diffusion length. Inserting Eq. (4.6) together with the nominal value Σ_d^0 into Eq. (4.3) gives the nominal value of the detector response:

$$R\left(\mathbf{e}^0\right) = \frac{S^0 \Sigma_d^0}{\Sigma_a^0}\left(1 - \frac{\cosh bk}{\cosh ak}\right), \quad \mathbf{e}^0 \triangleq \left(\varphi^0, \boldsymbol{\alpha}^0\right). \tag{4.7}$$

Note that even though Eq. (4.1) is linear in φ, the solution $\varphi(x)$ depends nonlinearly on $\boldsymbol{\alpha}$ as evidenced by Eq. (4.6). The same is true of the response $R(\mathbf{e})$. Even though $R(\mathbf{e})$ is linear separately in φ and in $\boldsymbol{\alpha}$, as shown in Eq. (4.3), R is not simultaneously linear in φ and $\boldsymbol{\alpha}$, which leads to a nonlinear dependence of $R(\mathbf{e})$ on $\boldsymbol{\alpha}$. This fact is confirmed by the explicit expression of $R(\mathbf{e})$ given in Eq. (4.7).

4.2.1 Adjoint Sensitivity Analysis

The sensitivities of the response $R(\mathbf{e})$ to the variations

$$\mathbf{h}_\alpha \triangleq \left(\delta\Sigma_a, \delta D, \delta S, \delta\Sigma_d\right), \tag{4.8}$$

will be computed using the generally applicable ASAM for non-linear systems (Cacuci, 1980a, 1980b), who showed that the most general quantity that expresses the sensitivity of a response $R(\mathbf{e})$ to parameter variations $\mathbf{h}_\alpha$ is the Gateaux-differential $\delta R\left(\mathbf{e}^0;\mathbf{h}\right)$ of $R(e)$ at e^0 to variations

$$\mathbf{h} \triangleq \left(h_\varphi, \mathbf{h}_\alpha\right). \tag{4.9}$$

By definition, the G-differential $\delta R\left(\mathbf{e}^0;\mathbf{h}\right)$ of $R(\mathbf{e})$ at $\mathbf{e}^0$ is

$$\delta R\left(\mathbf{e}^0;\mathbf{h}\right) \triangleq \frac{d}{d\varepsilon}\left\{R\left(\mathbf{e}^0 + \varepsilon\mathbf{h}\right)\right\}_{\varepsilon=0}. \tag{4.10}$$

Hence, the explicit form of the G-differential of $R(\mathbf{e})$ defined in Eq.(4.3) becomes

$$\delta R\left(\mathbf{e}^0;\mathbf{h}\right) \triangleq \frac{d}{d\varepsilon}\left\{\left(\Sigma_d^0 + \varepsilon\,\delta\Sigma_d\right)\left[\varphi^0(b) + \varepsilon\, h_\varphi(b)\right]\right\}_{\varepsilon=0} = R'_\alpha\left(\mathbf{e}^0\right)\mathbf{h}_\alpha + R'_\varphi\left(\mathbf{e}^0\right)h_\varphi\,, \tag{4.11}$$

where the so-called "direct-effect" term $R'_\alpha\,\mathbf{h}_\alpha$ is defined as

$$R'_\alpha\left(\mathbf{e}^0\right)\mathbf{h}_\alpha \triangleq \delta\Sigma_d\varphi^0(b), \tag{4.12}$$

while the so-called "indirect-effect" term $R'_\varphi\, h_\varphi$ is defined as

$$R'_\varphi\left(\mathbf{e}^0\right)h_\varphi \triangleq \Sigma_d^0\, h_\varphi(b). \tag{4.13}$$

As indicated by Eq. (4.11), the operator $\delta R\left(\mathbf{e}^0;\mathbf{h}\right)$ is linear in $\mathbf{h}$; in particular, $R'_\varphi\, h_\varphi$ is a linear operator on h_φ. This linear dependence of $\delta R\left(\mathbf{e}^0;\mathbf{h}\right)$ on $\mathbf{h}$ is underscored by writing it henceforth as $DR\left(\mathbf{e}^0;\mathbf{h}\right)$ to denote the sensitivity of $R(\mathbf{e})$ at $\mathbf{e}^0$ to variations $\mathbf{h}$. The "direct-effect" term $R'_\alpha\,\mathbf{h}_\alpha$ can be evaluated at this stage by replacing Eq. (4.6) into Eq. (4.12), to obtain

$$R'_\alpha\left(\mathbf{e}^0\right)\mathbf{h}_\alpha = \delta\Sigma_d\frac{S^0}{\Sigma_a^0}\left(1 - \frac{\cosh bk}{\cosh ak}\right). \tag{4.14}$$

The "indirect-effect" term $R'_\varphi\, h_\varphi$, though, cannot be evaluated at this stage, since $h_\varphi(x)$ is not yet available. The first-order in $\left\|\mathbf{h}_\alpha\right\|$, the expression of $h_\varphi(x)$ is obtained by calculating the G-differentials of Eqs. (4.12).and (4.13), and then solving the resulting equations. The G-differentials of Eqs. (4.12) and (4.13) yield the following "forward sensitivity equations:"

$$L\left(\boldsymbol{\alpha}^0\right)h_\varphi + \left[L'_\alpha\left(\boldsymbol{\alpha}^0\right)\varphi^0\right]\mathbf{h}_\alpha = O\left(\left\|\mathbf{h}_\alpha\right\|^2\right), \tag{4.15}$$

together with the boundary conditions

$$h_\varphi(\pm a) = 0. \tag{4.16}$$

In Eq. (4.15), the operator $L\left(\boldsymbol{\alpha}^0\right)$ is defined as

$$L(\boldsymbol{\alpha}^0) \triangleq D^0 \frac{d^2}{dx^2} - \Sigma_a^0, \tag{4.17}$$

while the quantity

$$\left[L'_\alpha(\boldsymbol{\alpha}^0)\varphi^0\right]\mathbf{h}_\alpha \triangleq \delta D \frac{d^2\varphi^0}{dx^2} - \delta\Sigma_a\varphi^0 + \delta S \ , \tag{4.18}$$

which is the partial G-differential of $L\varphi$ at $\boldsymbol{\alpha}^0$ with respect to $\boldsymbol{\alpha}$, contains all of the first-order parameter variations $\mathbf{h}_\alpha$.

The Forward Sensitivity System, namely Eqs .(4.15) and (4.16), could be solved to obtain $h_\varphi(x)$, which could, in turn be used in Eq.(4.13) to obtain the indirect effect term $R'_\varphi h_\varphi$. For this simple paradigm diffusion problem, $h_\varphi(x)$ has the following closed-form expression:

$$\begin{aligned} h_\varphi(x) = {} & C_1(\cosh xk - \cosh ak) \\ & + C_2(x \sinh xk \cosh ak - a \sinh ak \cosh xk), \end{aligned} \tag{4.19}$$

where the constants C_1 and C_2 are defined as

$$C_1 \triangleq \frac{\left(\delta\Sigma_a S^0/\Sigma_a^0 - \delta S\right)}{\Sigma_a^0(\cosh ak)} \ , \tag{4.20}$$

and, respectively,

$$C_2 \triangleq \frac{\left(\delta D/D^0 - \delta\Sigma_a/\Sigma_a^0\right)S^0}{2\sqrt{D^0\Sigma_a^0}\,(\cosh ak)^2} \ . \tag{4.21}$$

Evaluating Eq. (4.19) at $x = b$ and replacing the resulting expression in Eq.(4.13) gives the “indirect-effect” term as

$$\begin{aligned} R'_\varphi(e^0)h_\varphi = {} & \Sigma_d^0 C_1(\cosh bk - \cosh ak) \\ & + \Sigma_d^0 C_2(b \sinh bk \cosh ak - a \sinh ak \cosh bk). \end{aligned} \tag{4.22}$$

In practice, however, Eqs. (4.15) and (4.16) would need to be solved repeatedly, to obtain $h_\varphi(x)$ for every parameter variation, which would be impractically expensive computationally for large-scale systems with many parameters. As generally shown by Cacuci (1981a, 1981b), the need to solve repeatedly the forward sensitivity equations for each component of $\mathbf{h}_\alpha$ can be circumvented by using the ASAM, if the "indirect-effect" term $R'_\varphi\left(e^0\right)h_\varphi$ can be expressed as a linear functional of h_φ. Since $R'_\varphi\left(\mathbf{e}^0\right)h_\varphi$ is indeed a linear functional of h_φ, as the examination of Eq. (4.13) readily reveals, it is possible to apply the ASAM to this paradigm problem. For this purpose, the indirect effect term $R'_\varphi\left(\mathbf{e}^0\right)h_\varphi$ will be represented as an inner product in an appropriately defined Hilbert space $\mathcal{H}_u$. For this illustrative problem, the appropriate space is the real Hilbert space $\mathcal{H}_u \triangleq \mathcal{L}_2(\Omega)$, with $\Omega \triangleq (-a,a)$, equipped with the inner product

$$\langle f(x), g(x)\rangle \triangleq \int_{-a}^{a} f(x)g(x)dx, \quad for\ f,g \in \mathcal{H}_u \triangleq \mathcal{L}_2(\Omega). \tag{4.23}$$

In $\mathcal{H}_u \triangleq \mathcal{L}_2(\Omega)$, the linear functional $R'_\varphi\left(e^0\right)h_\varphi$ defined in Eq. (4.13) can be represented as the inner product

$$R'_\varphi\left(e^0\right)h_\varphi \triangleq \Sigma_d^0 h_\varphi(b) = \int_{-a}^{a} \Sigma_d^0 h_\varphi(x)\,\delta(x-b)dx = \left\langle \Sigma_d^0 \delta(x-b), h_\varphi \right\rangle. \tag{4.24}$$

The next step underlying the ASAP is the construction of the operator $L^+\left(\boldsymbol{\alpha}^0\right)$ that is formally adjoint to $L\left(\boldsymbol{\alpha}^0\right)$. In view of Eq. (4.17), the formal adjoint of $L\left(\boldsymbol{\alpha}^0\right)$ is the operator

$$L^+\left(\boldsymbol{\alpha}^0\right) \triangleq D^0 \frac{d^2}{dx^2} - \Sigma_a^0 . \tag{4.25}$$

Note that $L^+\left(\boldsymbol{\alpha}^0\right)$ and $L\left(\boldsymbol{\alpha}^0\right)$ are formally self-adjoint. The qualifier "formally" must still be kept at this stage, since the boundary conditions for $L^+\left(\boldsymbol{\alpha}^0\right)$ have not been determined yet. The boundary conditions for $L^+\left(\boldsymbol{\alpha}^0\right)$ are obtained by forming the inner product of $\psi(x)$ with the left-side of Eq.(4.15) and integrating twice by parts to obtain

$$\int_{-a}^{a} \psi(x)\left[D^0 \frac{d^2 h_\varphi}{dx^2} - \Sigma_a^0 h_\varphi(x)\right] dx = \int_{-a}^{a}\left[D^0 \frac{d^2\psi(x)}{dx^2} - \Sigma_a^0 \psi(x)\right] h_\varphi(x)\, dx + \left\{ P\left[h_\varphi, \psi\right]\right\}_{x=-a}^{x=a}, \tag{4.26}$$

where the bilinear concomitant $\left\{ P\left[h_\varphi, \psi\right]\right\}_{x=-a}^{x=a}$ is defined as follows:

$$\left\{ P\left[h_\varphi, \psi\right]\right\}_{-a}^{a} \triangleq D^0 \left[\psi \frac{dh_\varphi}{dx} - h_\varphi \frac{d\psi}{dx}\right]_{-a}^{a}. \tag{4.27}$$

Note that the function $\psi(x)$ is still arbitrary at this stage, except for the requirement that $\psi \in \mathcal{H}_Q = \mathcal{L}_2(\Omega)$. Note also that the Hilbert spaces $\mathcal{H}_u$ and $\mathcal{H}_Q$ have now both become the same space, i.e., $\mathcal{H}_u = \mathcal{H}_Q = \mathcal{L}_2(\Omega)$. Since h_φ is known at $x = \pm a$ from Eq. (4.16) the boundary conditions for $L^+(\boldsymbol{\alpha}^0)$ can now be selected as

$$\psi(\pm a) = 0, \tag{4.28}$$

to ensure that unknown values of h_φ, such as the derivatives $\left\{dh_\varphi / dx\right\}_{-a}^{a}$, would be eliminated from the bilinear form $\left\{ P\left[h_\varphi, \psi\right]\right\}_{-a}^{a}$ in Eq. (4.27). Note that the implementation of both Eqs. (4.28) and (4.16) into Eq. (4.27) actually causes $\left\{ P\left[h_\varphi, \psi\right]\right\}_{-a}^{a}$ to vanish. Since the boundary conditions selected in Eq. (4.28) for the adjoint function $\psi(x)$ are the same as the boundary conditions for $h_\varphi(x)$ in Eq. (4.16), and since the operators $L^+(\boldsymbol{\alpha}^0)$ and $L(\boldsymbol{\alpha}^0)$ are formally self-adjoint, it follows that the operators $L^+(\boldsymbol{\alpha}^0)$ and $L(\boldsymbol{\alpha}^0)$ are indeed self-adjoint.

The last step in the construction of the adjoint system is the identification of the source term by requiring that the right-side of Eqs. (4.24) and (4.26) represent the same functional. This requirement leads to the relation

$$\nabla_\varphi R(\mathbf{e}^0) = \Sigma_d^0\, \delta(x - b), \tag{4.29}$$

so that the complete adjoint system becomes

$$L^+(\boldsymbol{\alpha}^0)\psi \triangleq D^0 \frac{d^2\psi}{dx^2} - \Sigma_a^0 \psi(x) = \Sigma_d^0\, \delta(x - b), \tag{4.30}$$

where the adjoint function $\psi(x)$ is subject to the boundary conditions $\psi(\pm a)=0$, as shown in Eq. (4.28). Using now Eq. (4.18) gives the following expression for the "indirect-effect" term $R'_{\varphi}\left(e^{0}\right)h_{\varphi}$:

$$R'_{\varphi}\left(e^{0}\right)h_{\varphi}=-\int_{-a}^{a}\psi(x)\left[\delta D\frac{d^{2}\varphi^{0}}{dx^{2}}-\delta\Sigma_{a}\varphi^{0}(x)+\delta S\right]dx\,, \tag{4.31}$$

where $\psi(x)$ is the solution of the adjoint sensitivity system defined by Eqs. (4.30) and (4.28).

As expected, the adjoint sensitivity system is independent of parameter variations $\mathbf{h}_{\alpha}$, so it needs to be solved only once to obtain the adjoint function $\psi(x)$. Very important, too, is the fact (characteristic of linear systems) that the adjoint system is independent of the original solution $\varphi^{0}(x)$, and can therefore be solved directly, without any knowledge of the neutron flux $\varphi^{0}(x)$. Of course, the adjoint system depends on the response, which provides the source term for the adjoint system. Solving the adjoint sensitivity system, consisting of Eqs. (4.30) and (4.28) yields the following expression for the adjoint function $\psi(x)$:

$$\psi(x)=\frac{\Sigma_{d}^{0}}{\sqrt{\Sigma_{a}^{0}D^{0}}}\left[\frac{\sinh(b-a)k}{\sinh 2ak}\sinh(x+a)k+H(x-b)\sinh(x-b)k\right], \tag{4.32}$$

where $H(x-b)$ is the Heaviside-step functional defined as

$$H(x)=\begin{cases}0, & for\ x<0\\ 1, & for\ x\geq 0\end{cases}. \tag{4.33}$$

Using Eq. (4.32) in Eq. (4.31) and carrying out the respective integrations over x yields, as expected, the same expression for the "indirect-effect" term $R'_{\varphi}\left(\mathbf{e}^{0}\right)h_{\varphi}$ as obtained in Eq. (4.22). Finally, the local sensitivity $DR\left(\mathbf{e}^{0};\mathbf{h}\right)$ of $R(\mathbf{e})$ at $\mathbf{e}^{0}$ to variations $\mathbf{h}_{\alpha}$ in the system parameters is computed by using Eqs. (4.11), (4.14), and the adjoint function in Eq.(4.31) to obtain the following expression:

$$
\begin{aligned}
DR\left(\mathbf{e}^{0};\mathbf{h}\right) = \delta\Sigma_{d}\frac{S^{0}}{\Sigma_{a}^{0}}\left(1-\frac{\cosh bk}{\cosh ak}\right) \\
+\Sigma_{d}^{0}\left[C_{1}\left(\cosh bk-\cosh ak\right)\right. \\
\left.+C_{2}\left(b\sinh bk\cosh ak-a\sinh ak\cosh bk\right)\right].
\end{aligned}
\tag{4.34}
$$

It is instructive to compare the expression of the local sensitivity $DR\left(\mathbf{e}^{0};\mathbf{h}\right)$ with the expression of the exact variation

$$
\left(\Delta R\right)_{exact} \triangleq R\left(\mathbf{e}^{0}+\mathbf{h}\right)-R\left(\mathbf{e}^{0}\right), \tag{4.35}
$$

which would be induced in the response $R(\mathbf{e})$ by parameter variations $\mathbf{h}_{\alpha}$. The exact variation $\left(\Delta R\right)_{exact}$ is readily obtained from Eq. (4.7) as

$$
\left(\Delta R\right)_{exact} = \frac{S^{0}+\delta S}{\Sigma_{a}^{0}+\delta\Sigma_{a}}\left(\Sigma_{d}^{0}+\delta\Sigma_{d}\right)\left(1-\frac{\cosh bk_{p}}{\cosh ak_{p}}\right)-R\left(\mathbf{e}^{0}\right), \tag{4.36}
$$

where

$$
k_{p} \triangleq \sqrt{\left(\Sigma_{a}^{0}+\delta\Sigma_{a}\right)/\left(D^{0}+\delta D\right)}. \tag{4.37}
$$

On the other hand, we can solve exactly the perturbed equation

$$
L\left(\boldsymbol{\alpha}^{0}+\boldsymbol{h}_{\alpha}\right)\left[\varphi^{0}+h_{\varphi}^{exact}(x)\right]+\left(S^{0}+\delta S\right)=0, \tag{4.38}
$$

subject to the boundary conditions given by Eq. (4.16), to obtain

$$
\begin{aligned}
h_{\varphi}^{exact}(x) = \frac{S^{0}\,\delta\Sigma_{a}/\Sigma_{a}^{0}-\delta S}{\left(\Sigma_{a}^{0}+\delta\Sigma_{a}\right)\cosh ak_{p}}\left(\cosh xk_{p}-\cosh ak_{p}\right) \\
+S^{0}\frac{\cosh\left(ak_{p}\right)\cosh\left(xk\right)-\cosh\left(ak\right)\cosh\left(xk_{p}\right)}{\Sigma_{a}^{0}\cosh ak_{p}\cosh ak}.
\end{aligned}
\tag{4.39}
$$

Comparing Eq. (4.39) to Eq. (4.19) readily confirms that the solution $h_{\varphi}(x)$ of the forward sensitivity equations is the first-order, in $\left\|\mathbf{h}_{\alpha}\right\|$, approximation of $h_{\varphi}^{exact}(x)$, i.e.,

$$h_{\varphi}^{exact}(x) = h_{\varphi}(x) + O\left(\|\mathbf{h}_{\alpha}\|^2\right). \tag{4.40}$$

Similarly, comparing Eq. (4.36) to Eq. (4.34) confirms, as expected, that the local sensitivity $DR(\mathbf{e}^0;\mathbf{h})$ is the first-order, in $\|\boldsymbol{h}_{\alpha}\|$, approximation of the exact response variation, namely:

$$R(\mathbf{e}^0 + \mathbf{h}) = R(\mathbf{e}^0) + DR(\mathbf{e}^0;\mathbf{h}) + O\left(\|\mathbf{h}_{\alpha}\|^2\right). \tag{4.41}$$

In view of Eq. (4.34), the expressions of the partial sensitivities of $R(\mathbf{e})$ to the various parameters are:

$$\frac{\partial R}{\partial S} = \frac{\Sigma_d^0}{\Sigma_a^0}\left(1 - \frac{\cosh bk}{\cosh ak}\right), \tag{4.42}$$

$$\frac{\partial R}{\partial \Sigma_d} = \frac{S^0}{\Sigma_a^0}\left(1 - \frac{\cosh bk}{\cosh ak}\right), \tag{4.43}$$

$$\frac{\partial R}{\partial \Sigma_a} = -\frac{S^0\Sigma_d^0}{\left(\Sigma_a^0\right)^2}\left(1 - \frac{\cosh bk}{\cosh ak}\right) + \frac{1}{2\sqrt{D^0\Sigma_a^0}}\frac{S^0\Sigma_d^0}{\Sigma_a^0}\frac{a\sinh ak\cosh bk - b\sinh bk\cosh ak}{\left(\cosh ak\right)^2}, \tag{4.44}$$

$$\frac{\partial R}{\partial D} = -\frac{1}{2}\sqrt{\frac{\Sigma_a^0}{D^0}}\frac{S^0\Sigma_d^0}{D^0\Sigma_a^0}\frac{a\sinh ak\cosh bk - b\sinh bk\cosh ak}{\left(\cosh ak\right)^2}. \tag{4.45}$$

Note that the relative sensitivities of R to Σ_d and S, respectively, are everywhere unity, i.e.,

$$\left(\partial R/\partial \Sigma_d\right)\left(\Sigma_d^0/R\right) = 1; \tag{4.46}$$

$$\left(\partial R/\partial S\right)\left(S^0/R\right) = 1. \tag{4.47}$$

4.2.2 BERRU-CMS Predictive Modeling

The predicted optimal values for parameters, responses, and their predicted reduced uncertainties are computed using the equations underlying the "One-Model Case" in Section 3.2.1 of Chapter 3. These equations are reproduced below for easy reference:

$$\boldsymbol{\alpha}^{pred} = \boldsymbol{\alpha}^0 - \left(\mathbf{C}_{\alpha\alpha}\mathbf{S}^{\dagger}_{r\alpha} - \mathbf{C}_{\alpha r}\right)\left[\mathbf{D}_{rr}\right]^{-1}\left(\mathbf{r}^c - \mathbf{r}^m\right), \tag{4.48}$$

$$\mathbf{r}^{pred} = \mathbf{r}^m - \left(\mathbf{C}^{\dagger}_{\alpha r}\mathbf{S}^{\dagger}_{r\alpha} - \mathbf{C}_{rr}\right)\left[\mathbf{D}_{rr}\right]^{-1}\left(\mathbf{r}^c - \mathbf{r}^m\right), \tag{4.49}$$

$$\mathbf{C}^{pred}_{\alpha\alpha} = \mathbf{C}_{\alpha\alpha} - \left(\mathbf{C}_{\alpha\alpha}\mathbf{S}^{\dagger}_{r\alpha} - \mathbf{C}_{\alpha r}\right)\left[\mathbf{D}_{rr}\right]^{-1}\left(\mathbf{C}_{\alpha\alpha}\mathbf{S}^{\dagger}_{r\alpha} - \mathbf{C}_{\alpha r}\right)^{\dagger}, \tag{4.50}$$

$$\mathbf{C}^{pred}_{rr} = \mathbf{C}_{rr} - \left(\mathbf{C}^{\dagger}_{\alpha r}\mathbf{S}^{\dagger}_{r\alpha} - \mathbf{C}_{rr}\right)\left[\mathbf{D}_{rr}\right]^{-1}\left(\mathbf{C}^{\dagger}_{\alpha r}\mathbf{S}^{\dagger}_{r\alpha} - \mathbf{C}_{rr}\right)^{\dagger}, \tag{4.51}$$

$$\mathbf{C}^{pred}_{\alpha r} = \mathbf{C}_{\alpha r} - \left(\mathbf{C}_{\alpha\alpha}\mathbf{S}^{\dagger}_{r\alpha} - \mathbf{C}_{\alpha r}\right)\left[\mathbf{D}_{rr}\right]^{-1}\left(\mathbf{C}^{\dagger}_{\alpha r}\mathbf{S}^{\dagger}_{r\alpha} - \mathbf{C}_{rr}\right)^{\dagger}. \tag{4.52}$$

$$\chi^2 = \left(\mathbf{r}^c - \mathbf{r}^m\right)^{\dagger}\left[\mathbf{D}_{rr}\right]^{-1}\left(\mathbf{r}^c - \mathbf{r}^m\right). \tag{4.53}$$

4.2.2.1 An Imprecise but Consistent Measurement

Consider now that a measurement is performed at the location $b = 10\,cm$ in the positive direction (i.e., to the right of the origin $x = 0$), yielding a nominal value of $r_1^m = 3.40 \times 10^9\, neutrons \cdot cm^{-3} \cdot s^{-1}$, with a relative standard deviation of 25%, i.e., $\Delta r_m = 8.50 \times 10^8 neutrons \cdot cm^{-3} \cdot s^{-1}$, which corresponds to a variance $C^m_{rr} = 7.22 \times 10^{17}\left(neutrons \cdot cm^{-3} \cdot s^{-1}\right)^2$. On the other hand, the nominal value of the computed response at $b = 10\,cm$ is obtained by using Eq. (4.7), which yields $R(10\,cm) \triangleq r_1^c = 3.77 \times 10^9 neutrons \cdot cm^{-3} \cdot s^{-1}$.

Omitting (for simplicity) the various units, the absolute sensitivities $\mathbf{S}^{1\times 4}_{r\alpha}$ of the response $R(10\,cm)$ to the model parameters are computed from Eqs. (4.42) through (4.45) to obtain the row vector:

$$\begin{aligned}\mathbf{S}^{1\times 4}_{r\alpha} &\triangleq \left[\partial R(10cm)/\partial\Sigma_a,\, \partial R(10cm)/\partial D,\, \partial R(10cm)/\partial S,\, \partial R(10cm)/\partial\Sigma_d\right] \\ &= \left[-1.92 \times 10^{11},\, -1.33 \times 10^5,\, 3.77 \times 10^2,\, 5.08 \times 10^8\right].\end{aligned} \tag{4.54}$$

In particular, the relative sensitivities of $R(10\,cm)$ to Σ_a and D are, respectively:

$$\left[\partial R(10\,cm)/\partial\Sigma_a\right]\left[\Sigma_a^0/R(10\,cm)\right] = -.99999 \tag{4.55}$$

$$\left[\partial R(10\,cm)/\partial D\right]\left[D^0/R(10\,cm)\right] = -5.64\times 10^{-6}. \tag{4.56}$$

Recall that the covariance matrix, $\mathbf{C}_{rr}^{comp}$, of the computed responses arises from the covariance matrix of the parameters and the response sensitivities; to first-order sensitivities, the response covariance has the following expression:

$$\mathbf{C}_{rr}^{comp} \triangleq \left\langle \left[\mathbf{r} - \mathbf{r}^c\left(\boldsymbol{\alpha}^0,\boldsymbol{\beta}^0\right)\right]\left[\mathbf{r} - \mathbf{r}^c\left(\boldsymbol{\alpha}^0,\boldsymbol{\beta}^0\right)\right]^\dagger \right\rangle = \mathbf{S}_{r\alpha}\mathbf{C}_{\alpha\alpha}\mathbf{S}_{r\alpha}^\dagger. \tag{4.57}$$

Since there is only one response in this example, Eq. (4.57) reduces to a scalar value for the covariance, C_{rr}^{comp}, of r_1^c. Thus, using the sensitivities from Eq. (4.54) together with the covariance matrix given in Eq. (4.5) yields the following value for the variance of $R(10\,cm)$:

$$C_{rr}^{comp}\left(r_1^c\right) = \left[\mathbf{S}_{r\alpha}^{1\times 4}\right]\mathbf{C}_{\alpha\alpha}\left[\mathbf{S}_{r\alpha}^{1\times 4}\right]^\dagger = 4.98\times 10^{17}\left(n\cdot cm^{-3}\cdot s^{-1}\right)^2 \triangleq \left[\Delta C_{rr}^{comp}\left(r_1^c\right)\right]^2, \tag{4.58}$$

where $\Delta C_{rr}^{comp}\left(r_1^c\right)$ denotes the absolute standard deviation of r_1^c; the relative standard deviation, $\Delta C_{rr}^{comp}\left(r_1^c\right)/r_1^c$, is readily obtained to be $\Delta C_{rr}^{comp}\left(r_1^c\right)/r_1^c = 18.72\%$. Note that this standard deviation is smaller than would be naively expected from the popular recipe of "taking the square root of the sum of the squared errors," which would in this case amount to 19.37%. The fact that the actual computed standard deviation is smaller than "the square root of the sum of the squared errors" stems from the fact that the sensitivity of $R(10\,cm)$ to D is vanishingly small, so that the error in D contributes a negligible amount to the computed error in $R(10\,cm)$. Note also that the measurement and computation can be considered as being consistent with one another, since
$(std)_{m,1} + (std)_{comp,1} = 1.56\times 10^9\, n\cdot cm^{-3}\cdot s^{-1} > |d_1| \equiv \left|r_1^c - r_1^m\right| = 3.7\times 10^8\, n\cdot cm^{-3}\cdot s^{-1}$;
Using the information from Eqs. (4.5), (4.54), (4.58) in Eqs. (4.48) through (4.52) leads to the following predicted best-estimate nominal values for the calibrated parameters and their accompanying predicted covariances:

$$\begin{aligned} &\Sigma_a^{pred} = 0.0197\,cm^{-1},\; D^{pred} = 0.16\,cm, \\ &S^{pred} = 9.7372\times 10^6\, neutrons\cdot cm^{-3}\cdot s^{-1},\; \Sigma_d^{pred} = 7.351\,cm^{-1}, \end{aligned} \tag{4.59}$$

$$C_{\alpha\alpha}^{pred} = \begin{pmatrix} 9.71\times10^{-4} & 0 & 0 & 0 \\ 0 & 0.80\times10^{-2} & 0 & 0 \\ 0 & 0 & 1.28\times10^{6} & 0 \\ 0 & 0 & 0 & 6.99\times10^{-1} \end{pmatrix}$$
$$\times \begin{pmatrix} 1.0 & -1.67\times10^{-7} & 1.03\times10^{-1} & 0.63\times10^{-1} \\ -1.67\times10^{-7} & 1.0 & 5.75\times10^{-7} & 3.50\times10^{-7} \\ 1.03\times10^{-1} & 5.75\times10^{-7} & 1.0 & -2.17\times10^{-1} \\ 0.63\times10^{-1} & 3.50\times10^{-7} & -2.17\times10^{-1} & 1.0 \end{pmatrix} \tag{4.60}$$
$$\times \begin{pmatrix} 9.71\times10^{-4} & 0 & 0 & 0 \\ 0 & 0.80\times10^{-2} & 0 & 0 \\ 0 & 0 & 1.28\times10^{6} & 0 \\ 0 & 0 & 0 & 6.99\times10^{-1} \end{pmatrix},$$

$$r_1^{pred} = 3.62\times10^{9}\, n\cdot cm^{-3}\cdot s^{-1},\ C_{rr}^{pred} = 2.95\times10^{17}\left(n\cdot cm^{-3}\cdot s^{-1}\right)^2, \tag{4.61}$$

$$C_{\alpha r}^{pred} = \left(5.43\times10^{8}\right)$$
$$\times\left(-2.08\times10^{-1} \quad -1.15\times10^{-6} \quad 7.18\times10^{-1} \quad 4.37\times10^{-1}\right) \tag{4.62}$$
$$\times \begin{pmatrix} 9.71\times10^{-4} & 0 & 0 & 0 \\ 0 & 0.80\times10^{-2} & 0 & 0 \\ 0 & 0 & 1.28\times10^{6} & 0 \\ 0 & 0 & 0 & 6.99\times10^{-1} \end{pmatrix}.$$

Note from Eq. (4.61) that the predicted absolute and, respectively, relative standard deviations are

$$\Delta C_{rr}^{pred} = 5.43\times10^{8}\, neutrons\cdot cm^{-3}\cdot s^{-1},\ \Delta C_{rr}^{pred} / r_1^{pred} = 14.99\%, \tag{4.63}$$

Figure 4.1 shows the spatial variation of the original nominal computed values and standard deviations (depicted using broken lines) together with the best estimate values and corresponding standard deviations (depicted using solid lines). The value of the consistency indicator is $\chi^2 = 0.12$, which is reasonable for a single measurement.

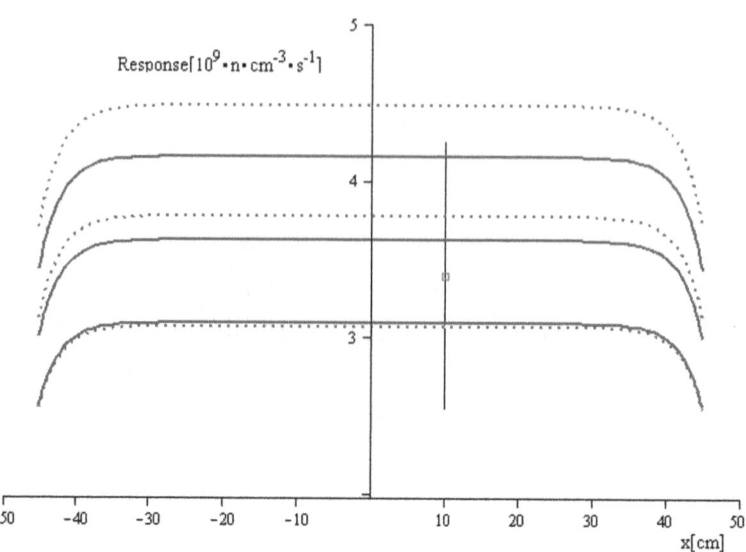

Fig. 4.1. An imprecise (relative standard deviation = 25%) but consistent measurement

Altogether, the information obtained in the foregoing leads to the following conclusions:

(i) As would be expected intuitively, the best-estimate response value, r_1^{pred}, is predicted by the BERRU-CMS methodology to fall between the experimentally measured and the originally computed values, somewhat closer to the computed values, because the relative standard deviation of the computed response, $\Delta C_{rr}^{comp} / R(10\,cm) = 18.72\%$, is smaller than the relative standard deviation of the measured response (25%), i.e.,

$$r_1^m = 3.40 < r_1^{pred} = 3.62 < r_1^c = 3.77\left[\times 10^9\, n \cdot cm^{-3} \cdot s^{-1}\right] \tag{4.64}$$

(ii) Again as intuitively expected, the assimilation of a consistent experiment reduces the best-estimate variance, C_{rr}^{pred}, of r_1^{pred} to a value that is smaller than the variances of both the computed and measured responses, i.e.,

$$C_{rr}^{pred} = 2.95 \times 10^{17} < C_{rr}^{comp} = 4.98 \times 10^{17} < C_{rr}^{m} = 7.22 \times 10^{17}\left[\left(neutrons \cdot cm^{-3} \cdot s^{-1}\right)^2\right]. \tag{4.65}$$

(iii) In the case under consideration, the absolute values of the response relative sensitivities to S, Σ_d and Σ_a are all equal to unity. When the relative sensitivities to several parameters are equal, the BERRU-calibration usually affects most markedly the parameters with the largest original standard deviation, while those with the smallest original standard deviation are adjusted the least. The above results underscore this general trend, in that the best estimate value of the source S (characterized by the largest original uncertainty) was adjusted the most, followed by the best-estimate value for Σ_d (the second largest original uncertainty), and Σ_a (which had the smallest original uncertainty). Since the relative sensitivity to D was vanishingly small, this parameter was not modified/calibrated following the assimilation of the experiment.

(iv) Comparing the matrices in Eqs. (4.5) and (4.60) indicates that the predicted best-estimate standard deviations for the parameters are also reduced by comparison to their original values. Due to the same reasons as discussed above (i.e., equal absolute values for the relative sensitivities of S, Σ_d, and Σ_a, and the values of the respective original standard deviations), the best-estimate standard deviations for S and Σ_d are reduced the most, the standard deviation for Σ_a is reduced the least, while that for D is not reduced (due to the extremely small accompanying sensitivity).

(v) Although the parameters were originally considered as uncorrelated, cf. Eq. (4.5), the assimilation of the experimental response induces correlations among the best-estimate parameters, as indicated by the correlation matrix in Eq. (4.60), as follows:

(a) Σ_a^{pred} becomes positively correlated with both S^{pred} and Σ_d^{pred};

(b) S^{pred} and Σ_d^{pred} become anti-correlated; and

(c) D^{pred} remains uncorrelated to the other best-estimate parameters, due to its vanishingly small relative sensitivity.

(vi) As indicated by the correlation matrix $C_{\alpha r}^{pred}$ in Eq. (4.62), the best-estimate response becomes anti-correlated to Σ_a^{pred}, remains uncorrelated to D^{pred}, and becomes positively correlated to both S^{pred} and Σ_d^{pred}, even though originally the response was considered to have been uncorrelated to any of the original model parameters.

(vii) Due to its vanishingly small relative sensitivity, the diffusion parameter is the sole quantity that has remained essentially unaffected by the assimilation of the experiment and subsequent calibration procedure, i.e., $D^0 = D^{pred}$ and $\Delta D^0 / D^0 = \Delta D^{pred} / D^{pred} = 5\%$.

4.2.2.2 A Precise and Consistent Measurement

Consider now that the measurement described in the previous section, with nominally measured value of $r_1^m = 3.40 \times 10^9 \, neutrons \cdot cm^{-3} \cdot s^{-1}$ is performed more precisely, with a relative standard deviation $\Delta r_1^m / r_1^m = 5\%$ (rather than 25%) or corresponding variance $C_{rr}^m = 2.89 \times 10^{16} \left(neutrons \cdot cm^{-3} \cdot s^{-1} \right)^2$. In this case, therefore, the experimental variance is considerably smaller than the computed response variance $C_{rr}^{comp} = 4.98 \times 10^{17} \left(neutrons \cdot cm^{-3} \cdot s^{-1} \right)^2$. The measurement and computation can be considered as being consistent with one another, since $(std)_{m,1} + (std)_{comp,1} = 1.41 \times 10^9 \, n \cdot cm^{-3} \cdot s^{-1} > |d_1| \triangleq |r_1^c - r_1^m| = 3.7 \times 10^8 \, n \cdot cm^{-3} \cdot s^{-1}$. Figure 4.2 shows the spatial variation of the original nominal computed values and standard deviations (depicted using broken lines) together with the best estimate response values and corresponding standard deviations (depicted using solid lines). The value of the consistency indicator is $\chi^2 = 0.27$, which quantitatively highlights the improvement over the imprecise measurement considered in the previous section.

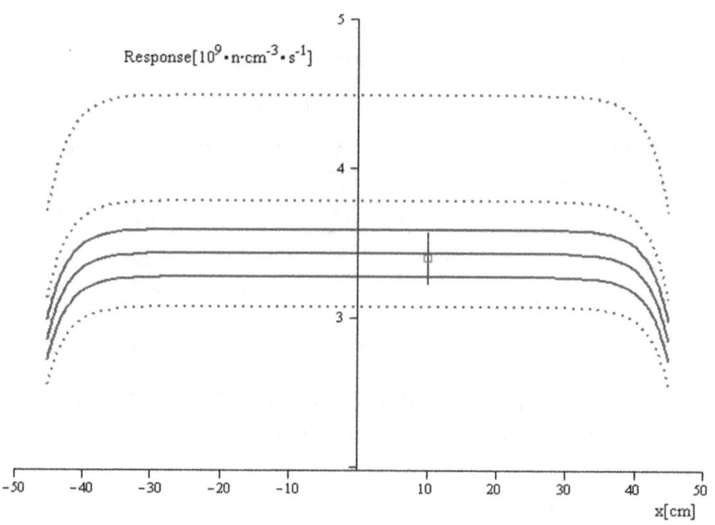

Fig. 4.2. A precise (relative standard deviation = 5%) and consistent measurement

After assimilating the experiment, the best-estimate predicted values are obtained as follows:

$$\Sigma_a^{pred} = 0.0198 \, cm^{-1}; \; D^{pred} = 0.16 \, cm;$$
$$S^{pred} = 9.393 \times 10^6 \, neutrons \cdot cm^{-3} \cdot s^{-1}; \; \Sigma_d^{pred} = 7.2371 \, cm^{-1}; \qquad (4.66)$$

$$C_{\alpha\alpha}^{pred} = \begin{pmatrix} 9.51\times10^{-4} & 0 & 0 & 0 \\ 0 & 0.80\times10^{-2} & 0 & 0 \\ 0 & 0 & 9.39\times10^{5} & 0 \\ 0 & 0 & 0 & 6.35\times10^{-1} \end{pmatrix}$$
$$\times\begin{pmatrix} 1.0 & -3.94\times10^{-7} & 3.34\times10^{-1} & 1.63\times10^{-1} \\ -3.94\times10^{-7} & 1.0 & 1.82\times10^{-6} & 8.91\times10^{-7} \\ 3.34\times10^{-1} & 1.82\times10^{-6} & 1.0 & -7.57\times10^{-1} \\ 1.63\times10^{-1} & 8.91\times10^{-7} & -7.57\times10^{-1} & 1.0 \end{pmatrix} \tag{4.67}$$
$$\times\begin{pmatrix} 9.51\times10^{-4} & 0 & 0 & 0 \\ 0 & 0.80\times10^{-2} & 0 & 0 \\ 0 & 0 & 9.39\times10^{5} & 0 \\ 0 & 0 & 0 & 6.35\times10^{-1} \end{pmatrix}.$$

The adjusted best-estimate response value, the corresponding best-estimate response variance, and the best-estimate response-parameter correlation matrix are obtained, respectively, as

$$r_1^{pred} = 3.42\times10^{9}\, n\cdot cm^{-3}\cdot s^{-1},\ C_{rr}^{pred} = 2.73\times10^{16}\left(n\cdot cm^{-3}\cdot s^{-1}\right)^2 \tag{4.68}$$

$$C_{\alpha r}^{pred} = \left(1.65\times10^{8}\right)$$
$$\times\left(-6.47\times10^{-2} \quad -3.52\times10^{-7} \quad 2.99\times10^{-1} \quad 1.46\times10^{-1}\right) \tag{4.69}$$
$$\times\begin{pmatrix} 9.51\times10^{-4} & 0 & 0 & 0 \\ 0 & 0.80\times10^{-2} & 0 & 0 \\ 0 & 0 & 9.39\times10^{5} & 0 \\ 0 & 0 & 0 & 6.35\times10^{-1} \end{pmatrix}.$$

The following conclusions can be drawn from the information shown in Eqs. (4.66) through (4.69):

(i) The more precise measurement considered in this section has a considerably stronger influence on the best-estimate predicted values than the less precise measurement considered in Subsection 4.4.2.1, pulling the predicted values strongly towards the experimentally measured value, as depicted in Figure 4.2. Thus, the predicted best-estimate response value $r_1^{pred} = 3.42\times10^{9}\, neutrons\cdot cm^{-3}\cdot s^{-1}$ is much closer to the experimentally measured value than to the originally computed value.

$$r_1^m = 3.40 < r_1^{pred} = 3.42 < r_1^c = 3.77\left[\times 10^9 \; n \cdot cm^{-3} \cdot s^{-1}\right]; \tag{4.70}$$

(ii) The best-estimate response variance C_{rr}^{pred} is also reduced more significantly in this case than in the case of the less precise measurement considered in Subsection 4.4.2.1; the best-estimate predicted relative standard deviation is 4.83%. As before, the predicted best-estimate response variance is smaller than both the measured and the computed response variances, namely:

$$C_{rr}^{pred} = 2.73 \times 10^{16} < C_{rr}^{m} = 2.89 \times 10^{16} < C_{rr}^{comp} = 4.98 \times 10^{17}\left[\left(neutrons \cdot cm^{-3} \cdot s^{-1}\right)^2\right]. \tag{4.71}$$

(iii) The response sensitivities to model parameters are obviously unaffected by measurements, so they have remained the same as in Subsection 4.4.2.1. Thus, the amounts of parameter calibration would also be expected to mirror the characteristics of the calibrations noticed in Subsection 4.4.2.1. This is indeed the case, as indicated by the results in Eq. (4.66): the largest adjustments occurred for the source S (characterized by the largest original uncertainty), followed by Σ_d (the second largest original uncertainty); Σ_a underwent a minute adjustment, while D was not adjusted (just as in Subsection 4.4.2.1). However, since the experimental response variance is much smaller than in the previous section, the corresponding parameter-adjustments are also larger than in Subsection 4.4.2.1.

(iv) Comparing the best-estimate parameter covariance matrix $C_{\alpha\alpha}^{pred}$ in Eq. (4.67) with the original parameter covariance matrix in Eq. (4.5) shows that the best-estimate standard deviations for the parameters are reduced more by the more precise experiment considered in this case (by comparison to the uncertainty reductions achieved in Subsection 4.4.2.1). The largest reductions of the standard deviations have occurred in the following order: for S (best-estimate relative standard deviation reduced to 10%), for Σ_d (best-estimate relative standard deviation reduced to 8.78%), and for Σ_a (best-estimate relative standard deviation reduced to 4.79%). Due to its small sensitivity, the uncertainty in D was hardly reduced (to 4.99%).

(v) Just as in Subsection 4.4.2.1, the originally uncorrelated parameters became correlated after calibration, as indicated by the correlation matrix $C_{\alpha\alpha}^{pred}$ in Eq. (4.67). The nature of the induced correlations has retained the same characteristics as in Subsection 4.4.2.1 (i.e., Σ_a^{pred} has become positively correlated with both S^{pred} and Σ_d^{pred}; S^{pred} and Σ_d^{pred} have become anti-correlated; and D^{pred} remained uncorrelated to the other best-estimate parameters), but the induced best-estimate correlations have become stronger (i.e., larger in absolute values) than in Subsection 4.4.2.1, by a factor of about three.

(vi) Just as in Subsection 4.4.2.1, the best-estimate response has become anti-correlated to Σ_a^{pred}, remained uncorrelated to D^{pred}, and became positively correlated to both S^{pred} and Σ_d^{pred}. However, as indicated by the correlation matrix $C_{\alpha r}^{pred}$ in Eq. (4.69), these correlations have become weaker, also by a factor of about three, by comparison to the assimilation of the imprecise measurement considered in Subsection 4.4.2.1.

(vii) As before, because of its vanishingly small relative sensitivity, the diffusion coefficient D^{pred} has remained essentially unaffected by the assimilation of the experiment and subsequent calibration procedure.

4.2.2.3 Two Consistent Measurements

Consider now that two measurements are performed at the locations $b = \pm 10\,cm$, symmetric with respect to the slab's mid-plane at $x = 0$. Consider that the measured nominal values of the detector response, and the corresponding absolute standard deviations (std) and relative standard deviations (rsd) are as follows:

$$\begin{aligned} &r_1^m \triangleq R(measured\; at\; 10\,cm) = 3.40 \times 10^9\, n \cdot cm^{-3} \cdot \sec^{-1}; \\ &(std)_{m,1} = 1.7 \times 10^8\, n \cdot cm^{-3} \cdot \sec^{-1};\; rsd\left(r_1^m\right) = 5\%; \end{aligned} \tag{4.72}$$

$$\begin{aligned} &r_2^m \triangleq R(measured\; at\; -10\,cm) = 3.59 \times 10^9\, n \cdot cm^{-3} \cdot \sec^{-1}; \\ &(std)_{m,2} = 2.15 \times 10^8\, n \cdot cm^{-3} \cdot \sec^{-1};\; rsd\left(r_2^m\right) = 6\%. \end{aligned} \tag{4.73}$$

Consider further that the above measurements are uncorrelated, so that the (2×2) covariance matrix of the measured responses is simply

$$\mathbf{C}_{rr}^{(2\times2)m} = \begin{pmatrix} \left(1.7 \times 10^8\right)^2 & 0 \\ 0 & \left(2.15 \times 10^8\right)^2 \end{pmatrix}. \tag{4.74}$$

The nominal computed response values at the above locations are obtained from Eq.(4.7); in view of the problem's symmetry, i.e.,

$$R(x) = R(-x) \tag{4.75}$$

a single computation of the response suffices to obtain

$$R(computed\;\; at\;\; 10\,cm) \triangleq r_1^c = 3.77 \times 10^9\, n \cdot cm^{-3} \cdot \sec^{-1}; \tag{4.76}$$

$$R(computed\ at\ -10\,cm) \triangleq r_2^c = 3.77\times10^9\, n\cdot cm^{-3}\cdot \sec^{-1}; \tag{4.77}$$

In general, since we deal with two responses, two distinct adjoint computations (one for each of the responses) would be needed to compute the matrix of response sensitivities to the four model parameters. Because of the symmetric responses considered in this illustrative example, however, a single computation of the adjoint function $\psi(x)$ suffices to obtain all eight sensitivities. Omitting the details of these straightforward computations, the results for the absolute values of the (2×4) matrix of sensitivities (omitting the obvious units, for simplicity) and also for the relative sensitivities, are as follows:

$$\mathbf{S}_{r\alpha}^{(2\times4)} \triangleq \left(\frac{\partial R_i}{\partial \alpha_j}\right)_{i=1,2;\,j=1,\dots,4} = \begin{pmatrix} -1.92\times10^{11} & -1.33\times10^{5} & 3.78\times10^{2} & 5.08\times10^{8} \\ -1.92\times10^{11} & -1.33\times10^{5} & 3.78\times10^{2} & 5.08\times10^{8} \end{pmatrix} \tag{4.78}$$

$$\mathbf{S}_{r\alpha}^{(2\times4)rel} \triangleq \left(\frac{\partial R_i}{\partial \alpha_j}\frac{\alpha_j}{R_i}\right)_{i=1,2;\,j=1,\dots,4} = \begin{pmatrix} -1.00 & -5.64\times10^{-6} & 1.00 & 1.00 \\ -1.00 & -5.64\times10^{-6} & 1.00 & 1.00 \end{pmatrix}. \tag{4.79}$$

As indicated by the above sensitivities, the responses are equally (and strongly) sensitive to the model parameters Σ_a, S, and Σ_d. The responses are insensitive to D, indicating that the neutron diffusion process in the region around the two measurement is practically irrelevant to the detectors' responses.

Omitting again (to simplify the notation) the explicit writing of the various units, replacing the absolute sensitivities from Eq. (4.78) and the parameter covariance matrix from Eq. (4.5) into Eq. (4.57) yields the following covariance matrix (i.e., uncertainties in computed responses due to the uncertainties in the model parameters) for the computed responses:

$$\mathbf{C}_{rr}^{(2\times2)comp} = \mathbf{S}_{r\alpha}^{(2\times4)}\mathbf{C}_{\alpha\alpha}\left(\mathbf{S}_{r\alpha}^{(2\times4)}\right)^{\dagger} = \begin{pmatrix} 4.99\times10^{17} & 4.99\times10^{17} \\ 4.99\times10^{17} & 4.99\times10^{17} \end{pmatrix}. \tag{4.80}$$

Note that $\mathbf{C}_{rr}^{(2\times2)comp}$ has rank of unity (rather than rank two), since the two computed (as opposed to measured!) responses are actually identical due to the problem's symmetry. As the result in Eq. (4.80) indicates, the absolute standard deviations of the two computed responses are identical, both having the value

$$(std)_{comp,1} = (std)_{comp,2} = 7.06\times10^{8}\, n\cdot cm^{-3}\cdot \sec^{-1}. \tag{4.81}$$

Note also that the sums of the respective measured and computed response standard deviations (in units of $n \cdot cm^{-3} \cdot \sec$) are larger than the "deviations" between the nominally measured and computed response values, namely:

$$(std)_{m,1} + (std)_{comp,1} = 8.76 \times 10^8 > d_1 \triangleq r_1^c - r_1^m = 3.78 \times 10^8; \tag{4.82}$$

$$(std)_{m,2} + (std)_{comp,2} = 9.22 \times 10^8 > d_2 = r_2^c - r_2^m = 1.89 \times 10^8. \tag{4.83}$$

The above relations indicate that both measurements are "consistent" with the computations. This fact is confirmed by computing the value of the "consistency indicator" χ^2, which in this case takes on the simple form

$$\chi^2 = \left(\mathbf{r}^c - \mathbf{r}^m\right)^\dagger \left[\mathbf{D}_{rr}\right]^{-1} \left(\mathbf{r}^c - \mathbf{r}^m\right), \; \mathbf{D}_{rr} = \mathbf{C}_{rr}^{(2\times 2)comp} + \mathbf{C}_{rr}^{(2\times 2)m}, \tag{4.84}$$

to obtain $\chi^2 = 0.327$, which indicates good consistency of the measured and computed data for the problem's two responses (i.e., two degrees of freedom).
The predicted optimal values for responses, parameters and their predicted reduced uncertainties are computed using the results from Eqs. (4.5) and (4.72) through (4.80) in Eqs. (4.48) through (4.52). Performing these computations leads to the following predicted value for the best-estimate parameters and (reduced) accompanying predicted standard deviations:

$$\Sigma_a^{pred} = 0.0198\,cm^{-1}, \; rsd\left(\Sigma_a^{pred}\right) = 4.8\%; \tag{4.85}$$

$$D^{pred} = 0{,}160\;cm, \; rsd\left(D^{pred}\right) = 5.00\%; \tag{4.86}$$

$$S^{pred} = 9.498 \times 10^6\, n \cdot cm^{-3} \cdot s^{-1}, \; rsd\left(S^{pred}\right) = 9.72\%; \tag{4.87}$$

$$\Sigma_d^{pred} = 7.272\,cm^{-1}, \; rsd\left(\Sigma_d^{pred}\right) = 8.7\%; \tag{4.88}$$

The predicted parameter covariance matrix is

$$\mathbf{C}_{\alpha\alpha}^{pred} = \begin{pmatrix} 9.03 \times 10^{-7} & -3.06 \times 10^{-12} & 3.06 \times 10^{2} & 1.01 \times 10^{-4} \\ -3.06 \times 10^{-12} & 6.40 \times 10^{-5} & 1.40 \times 10^{-2} & 4.63 \times 10^{-9} \\ 3.06 \times 10^{2} & 1.40 \times 10^{-2} & 8.53 \times 10^{11} & -4.62 \times 10^{5} \\ 1.01 \times 10^{-4} & 4.63 \times 10^{-9} & -4.62 \times 10^{5} & 4.01 \times 10^{-1} \end{pmatrix}, \tag{4.89}$$

indicating that the initially uncorrelated parameters have become correlated after assimilation of the experimental information.

The best-estimate predicted response values together with their predicted relative standard deviations are obtained as follows:

$$at\ (10\,cm):\ r_1^{pred} = 3.48\times10^9\, n\cdot cm^{-3}\cdot \sec^{-1};\ rsd\left(r_1^{pred}\right) = 3.8\%; \tag{4.90}$$

$$at\ (-10\,cm):\ r_2^{pred} = 3.48\times10^9\, n\cdot cm^{-3}\cdot \sec^{-1};\ rsd\left(r_2^{pred}\right) = 3.8\%; \tag{4.91}$$

The complete predicted response covariance matrix is

$$\mathbf{C}_{rr}^{(2\times2)pred} = \begin{pmatrix} 1.72\times10^{16} & 1.72\times10^{16} \\ 1.72\times10^{16} & 1.72\times10^{16} \end{pmatrix}, \tag{4.92}$$

indicating that the predicted best-estimate responses have become fully correlated after the assimilation of the experimental information even though the measurements were uncorrelated. This full correlation is actually introduced not by the measurements, but by the symmetry of the problem, via the respective response sensitivities [cf., Eq. (4.80)].

Finally, the predicted response-parameter correlation matrix is

$$\mathbf{C}_{\alpha r}^{pred} = \begin{pmatrix} -6.40\times10^3 & -2.93\times10^{-1} & 2.92\times10^{13} & 9.67\times10^6 \\ -6.40\times10^3 & -2.93\times10^{-1} & 2.92\times10^{13} & 9.67\times10^6 \end{pmatrix}, \tag{4.93}$$

indicating that, after assimilating experimental information, the predicted parameters and responses have become correlated. Note that $\mathbf{C}_{\alpha r}^{pred}$ is also of rank one (rather than rank two), just as $\mathbf{C}_{rr}^{pred}$, even though the two measurements differed from each other. These facts are consequences of the problem's symmetry (i.e., the fact the location of the measurements was symmetrical with respect to the problem's mid-plane, even though the nominal/mean values and standard deviations of the two measurements differed from each other).

Figure 4.3 depicts the computed responses with their computed standard deviations (depicted with dotted lines) and also the predicted best-estimate responses together with their predicted standard deviations (depicted with solid lines).

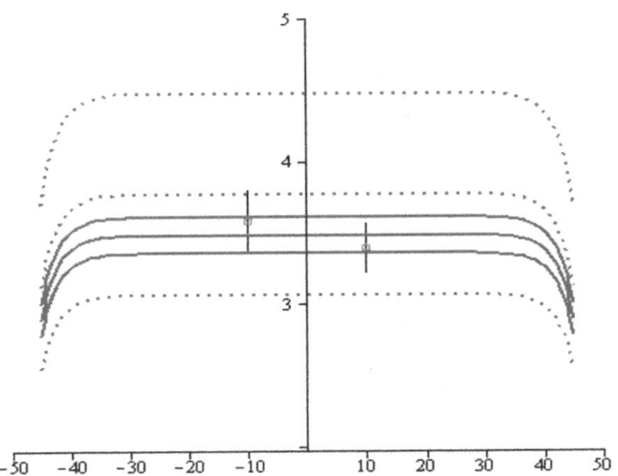

Fig. 4.3. Computed and predicted responses and their corresponding standard deviations when considering two consistent measurements for predictive modeling.

The following conclusions can be drawn from the results obtained in this sub-section:

(i) As depicted in Figure 4.3, the best-estimate response value $r^{pred} = 3.42 \times 10^9 \, neutrons \cdot cm^{-3} \cdot s^{-1}$ is much closer to the experimentally measured value than to the originally computed value.

$$r_1^m = 3.40 < r_1^{pred} = r_2^{pred} = 3.48 < r_2^m = 3.59 < r_1^c = r_2^c = 3.77 \left[\times 10^9 \; n \cdot cm^{-3} \cdot s^{-1} \right] \quad (4.94)$$

(ii) The predicted relative (and also the predicted absolute) response standard deviation is smaller than both the measured and the computed response variances:

$$\begin{aligned} rsd\left(r_1^{pred}\right) = rsd\left(r_2^{pred}\right) = 3.8\% < rsd\left(r_1^m\right) = 5\% < rsd\left(r_2^m\right) = 6\% \\ < rsd\left(r_1^c\right) = rsd\left(r_2^c\right) = 18.72\%; \end{aligned} \quad (4.95)$$

$$\begin{aligned} (std)_{pred,1} = (std)_{pred,2} = 1.32 < (std)_{m,1} = 1.7 < (std)_{m,2} = 2.15 \\ < (std)_{comp,1} = (std)_{comp,2} = 7.06 \left[\times 10^8 \, n \cdot cm^{-3} \cdot \sec \right] \end{aligned} \quad (4.96)$$

(iii) The largest adjustments occurred for the source S (characterized by the largest original uncertainty and largest sensitivity), followed by Σ_d (the second largest original uncertainty and also largest sensitivity); Σ_a underwent a minute adjustment (because its original uncertainty was small, although its sensitivity was comparable to those of S and Σ_d), while D was not adjusted (because of its vanishingly small sensitivity).

(iv) Comparing the best-estimate parameter covariance matrix $C_{\alpha\alpha}^{pred}$ in Eq. (4.89) with the original parameter covariance matrix in Eq. (4.5) shows that the best-estimate predicted standard deviations for the parameters are reduced. The largest reductions of the standard deviations occurred, in order, for S (best-estimate relative standard deviation reduced to 10%), Σ_d (best-estimate relative standard deviation reduced to 8.78%), and Σ_a (best-estimate relative standard deviation reduced to 4.79%), while that for D was hardly reduced (4.99%).

(v) The originally uncorrelated parameters became correlated after calibration, as indicated by the correlation matrix $C_{\alpha\alpha}^{pred}$ in Eq. (4.89): Σ_a^{pred} has become positively correlated with both S^{pred} and Σ_d^{pred}; S^{pred} and Σ_d^{pred} have become anti-correlated; and D^{pred} has remained uncorrelated to the other predicted best-estimate parameters.

(vi) As indicated by the correlation matrix $C_{\alpha r}^{pred}$ in Eq. (4.93), the best-estimate response has become anti-correlated to Σ_a^{pred}, remained essentially uncorrelated to D^{pred}, and became positively correlated to both S^{pred} and Σ_d^{pred}.

(vii) Because of its vanishingly small relative sensitivity, the diffusion parameter has remained essentially unaffected by the assimilation of the experiments and subsequent calibration within the predictive modeling procedure.

4.2.2.4 Three Consistent Measurements

Consider now that in addition to the two consistent measurements at the locations $b = \pm 10\,cm$ which were previously considered in Section 4.2.2.3, a third consistent measurement is performed at the location $b = -40\,cm$. Consider that the measured nominal values of the detector response, and the corresponding absolute standard deviations ("std") and relative standard deviations ("rsd") are as follows:

$$\begin{aligned} & r_1^m \triangleq R(measured\ at\ 10\,cm) = 3.40 \times 10^9\, n \cdot cm^{-3} \cdot \sec^{-1}; \\ & (std)_{m,1} = 1.7 \times 10^8\, n \cdot cm^{-3} \cdot \sec^{-1};\ rsd(r_1^m) = 5\%; \end{aligned} \tag{4.97}$$

$$\begin{aligned} & r_2^m \triangleq R(measured\ at\ -10\,cm) = 3.59 \times 10^9\, n \cdot cm^{-3} \cdot \sec^{-1}; \\ & (std)_{m,2} = 2.15 \times 10^8\, n \cdot cm^{-3} \cdot \sec^{-1};\ rsd(r_2^m) = 6\%. \end{aligned} \tag{4.98}$$

$$r_3^m \triangleq R(measured\, at\ -40\,cm) = 3.77\times 10^9 n\cdot cm^{-3}\cdot \sec^{-1};$$
$$(std)_{m,3} = 1.89\times 10^8 n\cdot cm^{-3}\cdot \sec^{-1};\ rsd(r_3^m) = 5\%. \tag{4.99}$$

Consider further that the above measurements are uncorrelated, so that the (3×3) covariance matrix of the measured responses is

$$\boldsymbol{C}_{rr}^{(3\times 3)m} = \begin{pmatrix} (1.7\times 10^8)^2 & 0 & 0 \\ 0 & (2.15\times 10^8)^2 & 0 \\ 0 & 0 & (1.89\times 10^8)^2 \end{pmatrix}. \tag{4.100}$$

The nominal values of the computed responses are as follows:

$$r_1^c \triangleq R(computed\ \ at\ \ 10\,cm) = 3.77\times 10^9 \left[n\cdot cm^{-3}\cdot \sec^{-1}\right]; \tag{4.101}$$

$$r_2^m \triangleq R(computed\, at\ -10\,cm) = 3.77\times 10^9 \left[n\cdot cm^{-3}\cdot \sec^{-1}\right]; \tag{4.102}$$

$$r_3^m \triangleq R(computed\, at\ -40\,cm) = 3.66\times 10^9 \left[n\cdot cm^{-3}\cdot \sec^{-1}\right]. \tag{4.103}$$

The (3×4) matrix comprising the absolute sensitivities (omitting the units, for simplicity) and, respectively, the relative sensitivities are as follows:

$$\boldsymbol{S}_{r\alpha}^{(3\times 4)} \triangleq \left(\frac{\partial R_i}{\partial \alpha_j}\right)_{i,j=1,\ldots,3} = \begin{pmatrix} -1.92\times 10^{11} & -1.33\times 10^5 & 3.78\times 10^2 & 5.08\times 10^8 \\ -1.92\times 10^{11} & -1.33\times 10^5 & 3.78\times 10^2 & 5.08\times 10^8 \\ -1.76\times 10^{11} & -1.24\times 10^9 & 3.66\times 10^2 & 4.92\times 10^8 \end{pmatrix} \tag{4.104}$$

$$\boldsymbol{S}_{r\alpha}^{(3\times 4)rel} \triangleq \left(\frac{\partial R_i}{\partial \alpha_j}\frac{\alpha_j}{R_i}\right) = \begin{pmatrix} -1.00 & -5.64\times 10^{-6} & 1.00 & 1.00 \\ -1.00 & -5.64\times 10^{-6} & 1.00 & 1.00 \\ -0.9458 & -5.41\times 10^{-2} & 1.00 & 1.00 \end{pmatrix}. \tag{4.105}$$

As indicated by the above sensitivities, the responses are equally (and strongly) sensitive to the model parameters Σ_a, S, and Σ_d. As in Section 4.2.2.3, the responses r_1^c and r_2^c are insensitive to D, indicating that the neutron diffusion process in the region around the two measurement is practically irrelevant to the detectors' responses. However, response r_3^c, which is located closer to the boundary, shows a small but non-negligible sensitivity to D, indicated that at $-40\,cm$, which is close to the boundary, the diffusion process begins to gain importance.

Omitting again (to simplify the notation) the explicit writing of the various units, replacing the absolute sensitivities from Eq. (4.104) and the parameter covariance matrix from Eq. (4.5) into Eq. (4.57) yields the following covariance matrix (i.e., uncertainties in computed responses due to the uncertainties in the model parameters) for the computed responses:

$$\boldsymbol{C}_{rr}^{(3\times3)comp} = \boldsymbol{S}_{r\alpha}^{3\times3}\boldsymbol{C}_{\alpha\alpha}\left(\boldsymbol{S}_{r\alpha}^{3\times3}\right)^{\dagger} = \begin{pmatrix} 4.99\times10^{17} & 4.99\times10^{17} & 4.82\times10^{17} \\ 4.99\times10^{17} & 4.99\times10^{17} & 4.82\times10^{17} \\ 4.82\times10^{17} & 4.82\times10^{17} & 4.66\times10^{17} \end{pmatrix}. \quad (4.106)$$

Note that rank of $\mathbf{C}_{rr}^{(3\times3)comp}$ is two (rather than rank three), since two of the computed (as opposed to measured!) responses are actually identical due to the problem's symmetry. The square roots of the elements of the diagonal of the matrix $\boldsymbol{C}_{rr}^{(3\times3)comp}$ yields the following values for the standard deviations of the respective computed response:

$$\begin{aligned} std\left(r_1^c\right) = std\left(r_2^c\right) &= 7.06\times10^8\left[n\cdot cm^{-3}\cdot \sec^{-1}\right], \\ std\left(r_3^c\right) &= 6.83\times10^8\left[n\cdot cm^{-3}\cdot \sec^{-1}\right]. \end{aligned} \quad (4.107)$$

Computing the sums of the respective measured and computed response standard deviations (in units of $n\cdot cm^{-3}\cdot\sec$) yields the relations

$$std\left(r_1^c\right)+std\left(r_1^m\right) = 8.76\times10^8 > \left|r_1^c - r_1^m\right| = 3.78\times10^8, \quad (4.108)$$

$$std\left(r_2^c\right)+std\left(r_2^m\right) = 9.22\times10^8 > \left|r_2^c - r_2^m\right| = 1.89\times10^8, \quad (4.109)$$

$$std\left(r_3^c\right)+std\left(r_3^m\right) = 8.71\times10^8 > \left|r_3^c - r_3^m\right| = 1.1\times10^8, \quad (4.110)$$

The above relations indicate that the sums of the respective measured and computed response standard deviations (in units of $n\cdot cm^{-3}\cdot\sec$) are larger than the "deviations" between the nominally measured and computed response values, which confirms that all three measurements are "consistent" with the computations. This fact is confirmed by computing the value of the "consistency indicator" χ^2, which is obtained as:

$$\chi^2 = 1.22 \tag{4.111}$$

The predicted optimal values for responses, parameters and their predicted reduced uncertainties are computed as in previous sections to obtain:

$$\Sigma_a^{pred} = 0.0198\, cm^{-1}, \;\; rsd\left(\Sigma_a^{pred}\right) = 4.78\%; \tag{4.112}$$

$$D^{pred} = 0{,}159\; cm, \;\; rsd\left(D^{pred}\right) = 5.00\%; \tag{4.113}$$

$$S^{pred} = 9.738 \times 10^6\, n \cdot cm^{-3} \cdot s^{-1}, \;\; rsd\left(S^{pred}\right) = 9.38\%; \tag{4.114}$$

$$\Sigma_d^{pred} = 7.351\, cm^{-1}, \;\; rsd\left(\Sigma_d^{pred}\right) = 8.59\%; \tag{4.115}$$

The predicted parameter covariance matrix is also obtained as in previous sections; the final result is:

$$\begin{aligned}
\mathbf{C}_{\alpha\alpha}^{pred} &= \begin{pmatrix} 9.03\times10^{-7} & 4.41\times10^{-9} & 3.04\times10^{2} & 1.00\times10^{-4} \\ 4.41\times10^{-9} & 6.39\times10^{-5} & 4.81\times10^{1} & 1.59\times10^{-5} \\ 3.04\times10^{2} & 4.81\times10^{1} & 8.34\times10^{11} & -4.68\times10^{5} \\ 1.00\times10^{-4} & 1.59\times10^{-5} & -4.68\times10^{5} & 3.98\times10^{-1} \end{pmatrix} \\
&= \begin{pmatrix} 9.50\times10^{-4} & 0 & 0 & 0 \\ 0 & 8.00\times10^{-3} & 0 & 0 \\ 0 & 0 & 9.13\times10^{5} & 0 \\ 0 & 0 & 0 & 6.31\times10^{-1} \end{pmatrix} \\
&\times \begin{pmatrix} 1.00 & 5.80\times10^{-4} & 3.50\times10^{-1} & 1.67\times10^{-1} \\ 5.80\times10^{-4} & 1.00 & 6.59\times10^{-3} & 3.15\times10^{-3} \\ 3.50\times10^{-1} & 6.59\times10^{-3} & 1.00 & -8.11\times10^{-1} \\ 1.67\times10^{-1} & 3.15\times10^{-3} & -8.11\times10^{-1} & 1.00 \end{pmatrix} \\
&\times \begin{pmatrix} 9.50\times10^{-4} & 0 & 0 & 0 \\ 0 & 8.00\times10^{-3} & 0 & 0 \\ 0 & 0 & 9.13\times10^{5} & 0 \\ 0 & 0 & 0 & 6.31\times10^{-1} \end{pmatrix}
\end{aligned} \tag{4.116}$$

indicating that the initially uncorrelated parameters have become correlated after assimilation of the experimental information. The best-estimate predicted response values together with their predicted relative standard deviations are obtained as follows:

$$at\ (10\,cm):\ r_1^{pred} = 3.61\times 10^9\, n\cdot cm^{-3}\cdot \sec^{-1};\ rsd\left(r_1^{pred}\right) = 3.0\%; \tag{4.117}$$

$$at\ (-10\,cm):\ r_2^{pred} = 3.61\times 10^9\, n\cdot cm^{-3}\cdot \sec^{-1};\ rsd\left(r_2^{pred}\right) = 3.0\%; \tag{4.118}$$

$$at\ (-40\,cm):\ r_3^{pred} = 3.50\times 10^9\, n\cdot cm^{-3}\cdot \sec^{-1};\ rsd\left(r_3^{pred}\right) = 3.0\%; \tag{4.119}$$

The covariance matrix of the predicted responses is obtained as:

$$\mathbf{C}_{rr}^{pred} = \begin{pmatrix} 1.19\times 10^{16} & 1.19\times 10^{16} & 1.14\times 10^{16} \\ 1.19\times 10^{16} & 1.19\times 10^{16} & 1.14\times 10^{16} \\ 1.14\times 10^{16} & 1.14\times 10^{16} & 1.11\times 10^{16} \end{pmatrix} \tag{4.120}$$

indicating that the predicted best-estimate responses have become fully correlated after the assimilation of the experimental information. The predicted response-parameter correlation matrix is obtained as

$$\begin{aligned} \mathbf{C}_{\alpha r}^{pred} &= \begin{pmatrix} -7.32\times 10^{3} & 2.54\times 10^{4} & 1.92\times 10^{13} & 6.34\times 10^{6} \\ -7.32\times 10^{3} & 2.54\times 10^{4} & 1.92\times 10^{13} & 6.34\times 10^{6} \\ 1.98\times 10^{3} & -5.44\times 10^{4} & 2.16\times 10^{13} & 7.14\times 10^{6} \end{pmatrix} \\ &= \begin{pmatrix} 1.09\times 10^{8} & 0 & 0 \\ 0 & 1.09\times 10^{8} & 0 \\ 0 & 0 & 1.06\times 10^{8} \end{pmatrix} \\ &\times \begin{pmatrix} -0.071 & 0.029 & 0.193 & 0.092 \\ -0.071 & 0.029 & 0.193 & 0.092 \\ 0.020 & -0.065 & 0.224 & 0.107 \end{pmatrix} \\ &\times \begin{pmatrix} 9.50\times 10^{-4} & 0 & 0 & 0 \\ 0 & 8.00\times 10^{-3} & 0 & 0 \\ 0 & 0 & 9.13\times 10^{5} & 0 \\ 0 & 0 & 0 & 6.31\times 10^{-1} \end{pmatrix} \end{aligned} \tag{4.121}$$

indicating that, after assimilating experimental information, the predicted parameters and responses have become correlated.

Figure 4.4 depicts the computed responses with their computed standard deviations (depicted with dotted lines) and also the predicted best-estimate responses together with their predicted standard deviations (depicted with solid lines).

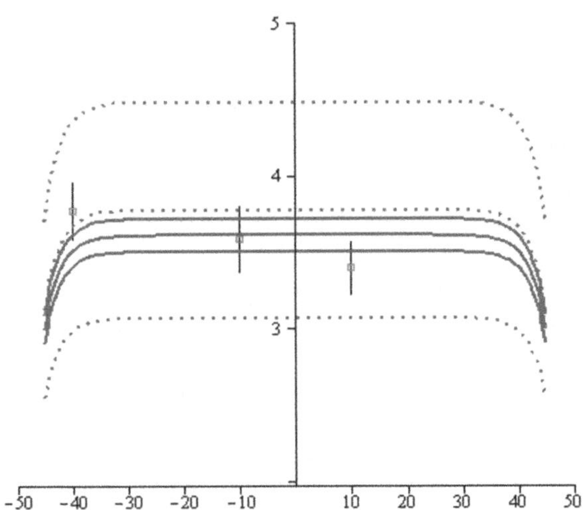

Fig. 4.4. ***Computed and predicted responses and their corresponding standard deviations when considering three consistent measurements for predictive modeling.***

The results obtained in this sub-section indicate that the conclusions drawn after having assimilated two consistent measurements have been reinforced by the assimilation of a third consistent experiment (as underscored by the value $\chi^2 = 1.22$ of the consistency indicator). Following the assimilation of the third measurement, the predicted standard deviations $rsd\left(r_1^{pred}\right) = 3.0\% < 3.8\%$ and $rsd\left(r_2^{pred}\right) = 3.0\% < 3.8\%$ of the these two responses (which were also considered in the previous section) were reduced more than when only two consistent experiments were assimilated. Furthermore, because the third measurement was taken close to the boundary, where diffusion effects become important, the diffusion coefficient D was affected, albeit just slightly.

4.2.2.5 *Four Consistent Measurements*

Consider now that in addition to the two measurements r_1^m and r_2^m considered in the previous sections at the locations $b = \pm 10\,cm$, as defined in Eqs. (4.72) and (4.73), two further measurements are performed at the locations $b = \pm 40\,cm$, also symmetrically with respect to the slab's mid-plane at $x = 0$. The two additional measurements at $b = \pm 40\,cm$ could be treated simultaneously with the two measurements r_1^m and r_2^m using Eqs. (4.48) through (4.53), i.e., using the formulas underlying the particular "one-model" predictive modeling procedure presented in Chapter 3, Section 3.2.1. However, this conceptually simple "one-model" predictive modeling procedure would require the inversion of (4×4) matrices, as will be shown in Sub-section 4.2.2.5(a), below. In this case, all of these previously obtained results in Section 4.2.2.3 (entitled "Two Consistent Measurements") would be lost if all four responses were considered ab initio. On the other hand, if the formulas underlying the particular case presented in Chapter 3, Section 3.2.3 ("Predictive modeling for Model A with q additional responses") of the BERRU-CMS methodology is used, then only two (2×2) matrices would need to be inverted, as will be shown in Sub-section 4.2.2.5(b). In this latter case, all of the results already obtained in in Section 4.2.2.3 can be used without needing any re-computations.

4.2.2.5.(a) *Application of the "One Model" BERRU-CMS Methodology*

In this sub-section, the two additional measurements at $b = \pm 40\,cm$ are treated simultaneously with the two measurements r_1^m and r_2^m using Eqs. (4.48) through (4.52) of the "one-model" predictive modeling procedure. For this purpose, the two measurements performed at the locations $b = \pm 40\,cm$ are denoted as r_3^m and r_4^m. All four measurements are considered to be uncorrelated, having the following mean values and standard deviations:

$$r_1^m \triangleq R(measured\ at\ 10\,cm) = 3.40\times 10^9 \left[n\cdot cm^{-3}\cdot \sec^{-1}\right]; \quad rsd\left(r_1^m\right) = 5\%; \tag{4.122}$$

$$r_2^m \triangleq R(measured\ at\ -10\,cm) = 3.59\times 10^9 \left[n\cdot cm^{-3}\cdot \sec^{-1}\right]; \quad rsd\left(r_2^m\right) = 6\%; \tag{4.123}$$

$$r_3^m \triangleq R(measured\ at\ -40\,cm) = 3.77\times 10^9 \left[n\cdot cm^{-3}\cdot \sec^{-1}\right]; \quad rsd\left(r_3^m\right) = 5\%; \tag{4.124}$$

$$r_4^m \triangleq R(measured\,at\;40\,cm) = 3.74\times10^9\left[n\cdot cm^{-3}\cdot\sec^{-1}\right];\quad rsd\left(r_4^m\right) = 5\%. \tag{4.125}$$

Since the above measurements are considered to be uncorrelated, it follows that the (4×4) covariance matrix of the measured responses is

$$C_{rr}^{(4\times4)m} = \begin{pmatrix} \left(1.7\times10^8\right)^2 & 0 & 0 & 0 \\ 0 & \left(2.15\times10^8\right)^2 & 0 & 0 \\ 0 & 0 & \left(1.89\times10^8\right)^2 & 0 \\ 0 & 0 & 0 & \left(1.87\times10^8\right)^2 \end{pmatrix}. \tag{4.126}$$

Using Eq. (4.7) yields the following nominal values for the computed responses at the four locations:

$$r_1^c \triangleq R(computed\;\;at\;\;10\,cm) = 3.77\times10^9\left[n\cdot cm^{-3}\cdot\sec^{-1}\right]; \tag{4.127}$$

$$r_2^c \triangleq R(computed\;\;at\;\;-10\,cm) = 3.77\times10^9\left[n\cdot cm^{-3}\cdot\sec^{-1}\right]; \tag{4.128}$$

$$r_3^c \triangleq R(computed\;\;at\;\;-40\,cm) = 3.66\times10^9\left[n\cdot cm^{-3}\cdot\sec^{-1}\right]; \tag{4.129}$$

$$r_4^c \triangleq R(computed\;\;at\;\;40\,cm) = 3.66\times10^9\left[n\cdot cm^{-3}\cdot\sec^{-1}\right]; \tag{4.130}$$

Because of the problems symmetry, i.e., $R(x) = R(-x)$, the sensitivities the responses and to the model parameters are computed using only two (rather than four) adjoint computations: one adjoint computation for obtaining the sensitivities of $R(at\pm10\,cm)$ and the second adjoint computation for obtaining the sensitivities of $R(at\pm40\,cm)$. Omitting again the units, the (4×4)-matrix of absolute sensitivities thus obtained is presented below:

$$S_{r\alpha}^{(4\times4)} \triangleq \left(\frac{\partial R_i}{\partial \alpha_j}\right)_{i,j=1,\dots,4} = \begin{pmatrix} -1.92\times10^{11} & -1.33\times10^5 & 3.78\times10^2 & 5.08\times10^8 \\ -1.92\times10^{11} & -1.33\times10^5 & 3.78\times10^2 & 5.08\times10^8 \\ -1.76\times10^{11} & -1.24\times10^9 & 3.66\times10^2 & 4.92\times10^8 \\ -1.76\times10^{11} & -1.24\times10^9 & 3.66\times10^2 & 4.92\times10^8 \end{pmatrix} \tag{4.131}$$

The (4×4)-matrix of relative sensitivities, which indicates the importance of model parameters in influencing the responses considered, can also be readily computed as:

$$\boldsymbol{S}_{r\alpha}^{(4\times4)rel} \triangleq \left(\frac{\partial R_i}{\partial \alpha_j}\frac{\alpha_j}{R_i}\right) = \begin{pmatrix} -1.00 & -5.64\times10^{-6} & 1.00 & 1.00 \\ -1.00 & -5.64\times10^{-6} & 1.00 & 1.00 \\ -9.46\times10^{-1} & -5.41\times10^{-2} & 1.00 & 1.00 \\ -9.46\times10^{-1} & -5.41\times10^{-2} & 1.00 & 1.00 \end{pmatrix}. \tag{4.132}$$

Omitting again the explicit writing of the various units, replacing the absolute sensitivities from Eq. (4.131) and the parameter covariance matrix from Eq. (4.5) into Eq. (4.57) yields the following covariance matrix (i.e., uncertainties in computed responses due to the uncertainties in the model parameters) for the computed responses:

$$\boldsymbol{C}_{rr}^{(4\times4)comp} = \boldsymbol{S}_{r\alpha}^{4\times4}\boldsymbol{C}_{\alpha\alpha}\left(\boldsymbol{S}_{r\alpha}^{4\times4}\right)^{\dagger} = \begin{pmatrix} 4.99\times10^{17} & 4.99\times10^{17} & 4.82\times10^{17} & 4.82\times10^{17} \\ 4.99\times10^{17} & 4.99\times10^{17} & 4.82\times10^{17} & 4.82\times10^{17} \\ 4.82\times10^{17} & 4.82\times10^{17} & 4.66\times10^{17} & 4.66\times10^{17} \\ 4.82\times10^{17} & 4.82\times10^{17} & 4.66\times10^{17} & 4.66\times10^{17} \end{pmatrix}. \tag{4.133}$$

Note that the rank of $\mathbf{C}_{rr}^{(4\times4)comp}$ is two (rather than four), since only two of the four computed responses are actually distinct from one another: the problem's symmetry imposes $r_1^c \equiv r_2^c$ and $r_3^c \equiv r_4^c$, so that only r_1^c and r_3^c have distinct computed values. This is in contrast to the measured responses: all four measured responses differ from one another.

Taking the square root of each of the diagonal elements of $\mathbf{C}_{rr}^{(4\times4)comp}$ in Eq. (4.133) yields the following values for the absolute standard deviations of the computed responses:

$$\begin{aligned} std\left(r_1^c\right) &= std\left(r_2^c\right) = 7.06\times10^{8}\left[n\cdot cm^{-3}\cdot\sec^{-1}\right], \\ std\left(r_3^c\right) &= std\left(r_4^c\right) = 6.83\times10^{8}\left[n\cdot cm^{-3}\cdot\sec^{-1}\right]. \end{aligned} \tag{4.134}$$

Computing the sums of the respective measured and computed response standard deviations (in units of $\left[n\cdot cm^{-3}\cdot\sec^{-1}\right]$) yields the following relations:

$$std\left(r_1^c\right) + std\left(r_1^m\right) = 8.76\times10^{8} > \left|r_1^c - r_1^m\right| = 3.78\times10^{8}, \tag{4.135}$$

$$std\left(r_2^c\right)+std\left(r_2^m\right)=9.22\times10^8>\left|r_2^c-r_2^m\right|=1.89\times10^8, \tag{4.136}$$

$$std\left(r_3^c\right)+std\left(r_3^m\right)=8.71\times10^8>\left|r_3^c-r_3^m\right|=1.1\times10^8, \tag{4.137}$$

$$std\left(r_4^c\right)+std\left(r_4^m\right)=8.69\times10^8>\left|r_4^c-r_4^m\right|=7.32\times10^7. \tag{4.138}$$

The above inequalities indicate that the sums of the respective measured and computed response standard deviations are larger than the corresponding "deviations" between the nominally measured and computed response values. In turn, these inequalities indicate that the measured and computed responses are indeed consistent with each other. The computation of the "consistency indicator" χ^2, defined as

$$\chi^2=\left(\mathbf{r}^c-\mathbf{r}^m\right)^\dagger\left[\mathbf{D}_{rr}\right]^{-1}\left(\mathbf{r}^c-\mathbf{r}^m\right),\ \mathbf{D}_{rr}=\mathbf{C}_{rr}^{(4\times4)comp}+\mathbf{C}_{rr}^{(4\times4)m}, \tag{4.139}$$

requires the inversion of the (4×4)-matrix $\mathbf{D}_{rr}$. Furthermore, the computation of the best-estimate predicted nominal values for the parameters and responses using Eqs. (4.48) through (4.52) also require the inversion of the (4×4)-matrix $\mathbf{D}_{rr}$. This path (namely, computing $\left[\mathbf{D}_{rr}\right]^{-1}$) will be followed in this sub-section. In the next sub-section, the results of this subsection will be reproduced by using the more efficient computational procedure presented in Chapter 3, Section 3.2.3 ("BERRU-CMS predictive modeling for Model A with q additional responses") which, for this particular case, would require the inversion of two (2×2)-matrices instead of the inverting the (4×4)-matrix $\mathbf{D}_{rr}$.

Computing the numerical values of the components of the matrix $\mathbf{D}_{rr}$ yields:

$$\boldsymbol{D}_{rr}=\begin{pmatrix}5.27\times10^{17} & 4.99\times10^{17} & 4.82\times10^{17} & 4.82\times10^{17}\\ 4.99\times10^{17} & 5.45\times10^{17} & 4.82\times10^{17} & 4.82\times10^{17}\\ 4.82\times10^{17} & 4.82\times10^{17} & 5.02\times10^{17} & 4.66\times10^{17}\\ 4.82\times10^{17} & 4.82\times10^{17} & 4.66\times10^{17} & 5.01\times10^{17}\end{pmatrix} \tag{4.140}$$

which subsequently gives

$$\left[\boldsymbol{D}_{rr}\right]^{-1}=\begin{pmatrix}2.38\times10^{-17} & -6.76\times10^{-18} & -8.41\times10^{-18} & -8.58\times10^{-18}\\ -6.76\times10^{-18} & 1.74\times10^{-17} & -5.24\times10^{-18} & -5.35\times10^{-18}\\ -8.41\times10^{-18} & -5.24\times10^{-18} & 2.14\times10^{-17} & -6.80\times10^{-18}\\ -8.58\times10^{-18} & -5.35\times10^{-18} & -6.80\times10^{-18} & 2.17\times10^{-17}\end{pmatrix}. \tag{4.141}$$

The numerical value of the "consistency indicator" obtained from Eq. (4.141) and (4.127) through (4.130) is $\chi^2\left(\text{per 4 degrees of freedom}\right)=1.21$, which indicates very good consistency of the measured and computed data for the problem's four responses.

The predicted optimal values for responses, parameters and their predicted reduced uncertainties are computed using the results from Eqs. (4.5) and (4.127)through (4.130) in Eqs. (4.48) through (4.52). Performing these computations leads to the following predicted value for the best-estimate parameters and (reduced) accompanying predicted standard deviations:

(i) Predicted nominal values for the model parameters:

$$\Sigma_a^{pred}=0.0198\ cm^{-1},\ rsd\left(\Sigma_a^{pred}\right)=4.79\%; \tag{4.142}$$

$$D^{pred}=0{,}1591\ cm,\ rsd\left(D^{pred}\right)=5.00\%; \tag{4.143}$$

$$S^{pred}=9.85\times 10^{6}\, n\cdot cm^{-3}\cdot s^{-1},\ rsd\left(S^{pred}\right)=9.21\%; \tag{4.144}$$

$$\Sigma_d^{pred}=7.388\, cm^{-1},\ \ rsd\left(\Sigma_d^{pred}\right)=8.53\%; \tag{4.145}$$

(ii) Predicted covariance matrices for the model parameters:

$$\mathbf{C}_{\alpha\alpha}^{pred} = \begin{pmatrix} 9.03\times10^{-7} & 6.75\times10^{-9} & 3.03\times10^{2} & 1.00\times10^{-4} \\ 6.75\times10^{-9} & 6.38\times10^{-5} & 7.37\times10^{1} & 2.44\times10^{-5} \\ 3.03\times10^{2} & 7.37\times10^{1} & 8.24\times10^{11} & -4.71\times10^{5} \\ 1.00\times10^{-4} & 2.44\times10^{-5} & -4.71\times10^{5} & 3.97\times10^{-1} \end{pmatrix}$$
$$= \begin{pmatrix} 9.50\times10^{-4} & 0 & 0 & 0 \\ 0 & 7.99\times10^{-3} & 0 & 0 \\ 0 & 0 & 9.08\times10^{5} & 0 \\ 0 & 0 & 0 & 6.30\times10^{-1} \end{pmatrix}$$
$$\times \begin{pmatrix} 1.0 & -8.89\times10^{-4} & 3.51\times10^{-1} & 1.67\times10^{-1} \\ -8.89\times10^{-4} & 1.0 & 1.02\times10^{-2} & 4.84\times10^{-3} \\ 3.51\times10^{-1} & 1.02\times10^{-2} & 1.0 & -8.24\times10^{-1} \\ 1.67\times10^{-1} & 4.84\times10^{-3} & -8.24\times10^{-1} & 1.0 \end{pmatrix} \tag{4.146}$$
$$\times \begin{pmatrix} 9.50\times10^{-4} & 0 & 0 & 0 \\ 0 & 7.99\times10^{-3} & 0 & 0 \\ 0 & 0 & 9.08\times10^{5} & 0 \\ 0 & 0 & 0 & 6.30\times10^{-1} \end{pmatrix}.$$

(iii) Predicted nominal values for the model responses:

$$at\ (10\,cm):\ r_1^{pred} = 3.66\times10^9\, n\cdot cm^{-3}\cdot \sec^{-1};\ rsd\left(r_1^{pred}\right) = 2.58\%; \tag{4.147}$$

$$at\ (-10\,cm):\ r_2^{pred} = 3.66\times10^9\, n\cdot cm^{-3}\cdot \sec^{-1};\ rsd\left(r_2^{pred}\right) = 2.59\%; \tag{4.148}$$

$$at\ (-40\,cm):\ r_3^{pred} = 3.56\times10^9\, n\cdot cm^{-3}\cdot \sec^{-1};\ rsd\left(r_3^{pred}\right) = 2.59\%; \tag{4.149}$$

$$at\ (40\,cm):\ r_4^{pred} = 3.56\times10^9\, n\cdot cm^{-3}\cdot \sec^{-1};\ rsd\left(r_4^{pred}\right) = 2.58\%; \tag{4.150}$$

(iv) Predicted covariance matrix for the model responses:

$$\mathbf{C}_{rr}^{pred} = \begin{pmatrix} 9.04\times10^{15} & 9.04\times10^{15} & 8.64\times10^{15} & 8.64\times10^{15} \\ 9.04\times10^{15} & 9.04\times10^{15} & 8.64\times10^{15} & 8.64\times10^{15} \\ 8.64\times10^{15} & 8.64\times10^{15} & 8.45\times10^{15} & 8.45\times10^{15} \\ 8.64\times10^{15} & 8.64\times10^{15} & 8.45\times10^{15} & 8.45\times10^{15} \end{pmatrix} \tag{4.151}$$

(v) Predicted parameter-response correlation matrix:

$$
\mathbf{C}_{\alpha r}^{pred} = \begin{pmatrix} 9.51\times 10^{7} & 0 & 0 & 0 \\ 0 & 9.51\times 10^{7} & 0 & 0 \\ 0 & 0 & 9.19\times 10^{7} & 0 \\ 0 & 0 & 0 & 9.19\times 10^{7} \end{pmatrix}
\times \begin{pmatrix} -8.65\times 10^{-2} & 5.12\times 10^{-2} & 1.60\times 10^{-1} & 7.62\times 10^{-2} \\ -8.65\times 10^{-2} & 5.12\times 10^{-2} & 1.60\times 10^{-1} & 7.62\times 10^{-2} \\ 1.72\times 10^{-2} & -5.62\times 10^{-2} & 1.96\times 10^{-1} & 9.34\times 10^{-2} \\ 1.72\times 10^{-2} & -5.62\times 10^{-2} & 1.96\times 10^{-1} & 9.34\times 10^{-2} \end{pmatrix}
\times \begin{pmatrix} 9.50\times 10^{-4} & 0 & 0 & 0 \\ 0 & 7.99\times 10^{-3} & 0 & 0 \\ 0 & 0 & 9.08\times 10^{5} & 0 \\ 0 & 0 & 0 & 6.30\times 10^{-1} \end{pmatrix}. \tag{4.152}
$$

As has already been mentioned in the paragraph following Eq. (4.139), the above results will be reproduced in the following Subsection by using the more efficient computational procedure presented in Chapter 3, Section 3.2.3 ("BERRU-CMS predictive modeling for Model A with q additional responses") which, for this particular case, would require the inversion of two (2×2)-matrices. The physical significance of the results obtained in the present sub-section will be discussed at the conclusion of the next sub-section, since the results will be identical, except that they will be obtained more efficiently by not considering all four responses simultaneously.

4.2.2.5.(b) *Application of the BERRU-CMS "One Model with Additional Responses" Methodology*

In this Subsection, the responses r_1^m and r_2^m remain as described by Eqs (4.122) and (4.123), respectively, but the responses r_3^m and r_4^m (measured at the locations $b=\pm 40\,cm$) will be considered as two additional responses, q_1^m and q_2^m, respectively, which will be treated using the block-matrix structure of the BERRU-CMS "One Model with Additional Responses" methodology.

$$
\begin{aligned}
& q_1^m \triangleq R(measured\ at\ \ -40\,cm) = 3.77\times 10^{9}\left[n\cdot cm^{-3}\cdot \sec^{-1}\right]; \\
& rsd\left(q_1^m\right) = 5\%;
\end{aligned} \tag{4.153}
$$

$$q_2^m \triangleq R(measured\,at\;40\,cm) = 3.74\times 10^9\left[n\cdot cm^{-3}\cdot \sec^{-1}\right]; \\ rsd\left(q_2^m\right) = 5\%. \tag{4.154}$$

For easy reference, it is useful to reproduce below the expressions underlying the special case of BERRU-CMS "One Model with Additional Responses" Methodology from Chapter 3, Section 3.2.3, for the specific values $N_r = N_q = 2$:

$$\mathbf{D}_{rr} = \mathbf{S}_{r\alpha}\mathbf{C}_{\alpha\alpha}\mathbf{S}_{r\alpha}^{\dagger} + \mathbf{C}_{rr},\; Dim\left(\mathbf{D}_{rr}\right) = (2\times 2), \tag{4.155}$$

$$\mathbf{D}_{rq} = \mathbf{S}_{r\alpha}\mathbf{C}_{\alpha\alpha}\mathbf{S}_{q\alpha}^{\dagger} + \mathbf{C}_{rq},\; Dim\left(\mathbf{D}_{rq}\right) = (2\times 2), \tag{4.156}$$

$$\mathbf{D}_{qr} = \mathbf{S}_{q\alpha}\mathbf{C}_{\alpha\alpha}\mathbf{S}_{r\alpha}^{\dagger} + \mathbf{C}_{rq}^{\dagger} = \mathbf{D}_{rq}^{\dagger},\; Dim\left(\mathbf{D}_{qr}\right) = (2\times 2), \tag{4.157}$$

$$\mathbf{D}_{qq} = \mathbf{S}_{q\alpha}\mathbf{C}_{\alpha\alpha}\mathbf{S}_{q\alpha}^{\dagger} + \mathbf{C}_{qq},\; Dim\left(\mathbf{D}_{qq}\right) = (2\times 2). \tag{4.158}$$

$$\mathbf{X}_{\alpha} \equiv \mathbf{C}_{\alpha\alpha}\mathbf{S}_{r\alpha}^{\dagger},\; \mathbf{Y}_{\alpha} \equiv \mathbf{C}_{\alpha\alpha}\mathbf{S}_{q\alpha}^{\dagger}, \tag{4.159}$$

$$\mathbf{X}_{r} \equiv -\mathbf{C}_{rr},\; \mathbf{Y}_{r} \equiv -\mathbf{C}_{rq}^{\dagger}, \tag{4.160}$$

$$\mathbf{X}_{q} \equiv -\mathbf{C}_{rq}^{\dagger},\; \mathbf{Y}_{q} \equiv -\mathbf{C}_{qq}, \tag{4.161}$$

$$\mathbf{D}_{11} = \mathbf{D}_{rr}^{-1} + \mathbf{D}_{rr}^{-1}\mathbf{D}_{rq}\mathbf{D}_{22}\mathbf{D}_{rq}^{\dagger}\mathbf{D}_{rr}^{-1},\; \mathbf{D}_{22} = \left(\mathbf{D}_{qq} - \mathbf{D}_{rq}^{\dagger}\mathbf{D}_{rr}^{-1}\mathbf{D}_{rq}\right)^{-1}, \tag{4.162}$$

$$\mathbf{D}_{12} = \mathbf{D}_{rr}^{-1}\mathbf{D}_{rq}\mathbf{D}_{22},\; \mathbf{D}_{12}^{\dagger} = -\mathbf{D}_{22}\mathbf{D}_{rq}^{\dagger}\mathbf{D}_{rr}^{-1}, \tag{4.163}$$

$$\boldsymbol{\alpha}^{pred} = \boldsymbol{\alpha}^{0} - \left[\mathbf{X}_{\alpha}\mathbf{D}_{11} + \mathbf{Y}_{\alpha}\mathbf{D}_{12}^{\dagger}\right]\left(\mathbf{r}^{c} - \mathbf{r}^{m}\right) - \left[\mathbf{X}_{\alpha}\mathbf{D}_{12} + \mathbf{Y}_{\alpha}\mathbf{D}_{22}\right]\left(\mathbf{q}^{c} - \mathbf{q}^{m}\right), \tag{4.164}$$

$$\mathbf{r}^{pred} = \mathbf{r}^{m} - \left[\mathbf{X}_{r}\mathbf{D}_{11} + \mathbf{Y}_{r}\mathbf{D}_{12}^{\dagger}\right]\left(\mathbf{r}^{c} - \mathbf{r}^{m}\right) - \left[\mathbf{X}_{r}\mathbf{D}_{12} + \mathbf{Y}_{r}\mathbf{D}_{22}\right]\left(\mathbf{q}^{c} - \mathbf{q}^{m}\right), \tag{4.165}$$

$$\mathbf{q}^{pred} = \mathbf{q}^{m} - \left[\mathbf{X}_{q}\mathbf{D}_{11} + \mathbf{Y}_{q}\mathbf{D}_{12}^{\dagger}\right]\left(\mathbf{r}^{c} - \mathbf{r}^{m}\right) - \left[\mathbf{X}_{q}\mathbf{D}_{12} + \mathbf{Y}_{q}\mathbf{D}_{22}\right]\left(\mathbf{q}^{c} - \mathbf{q}^{m}\right), \tag{4.166}$$

$$\mathbf{C}_{\alpha\alpha}^{pred} = \mathbf{C}_{\alpha\alpha} - \left[\mathbf{X}_{\alpha}\left(\mathbf{D}_{11}\mathbf{X}_{\alpha}^{\dagger} + \mathbf{D}_{12}\mathbf{Y}_{\alpha}^{\dagger}\right) + \mathbf{Y}_{\alpha}\left(\mathbf{D}_{21}\mathbf{X}_{\alpha}^{\dagger} + \mathbf{D}_{22}\mathbf{Y}_{\alpha}^{\dagger}\right)\right], \tag{4.167}$$

$$\mathbf{C}_{rr}^{pred} = \mathbf{C}_{rr} - \left[\mathbf{X}_{r}\left(\mathbf{D}_{11}\mathbf{X}_{r}^{\dagger} + \mathbf{D}_{12}\mathbf{Y}_{r}^{\dagger}\right) + \mathbf{Y}_{r}\left(\mathbf{D}_{21}\mathbf{X}_{r}^{\dagger} + \mathbf{D}_{22}\mathbf{Y}_{r}^{\dagger}\right)\right], \tag{4.168}$$

$$\mathbf{C}_{rq}^{pred} = \mathbf{C}_{rq} - \left[\mathbf{X}_r \left(\mathbf{D}_{11} \mathbf{X}_q^{\dagger} + \mathbf{D}_{12} \mathbf{Y}_q^{\dagger} \right) + \mathbf{Y}_r \left(\mathbf{D}_{21} \mathbf{X}_q^{\dagger} + \mathbf{D}_{22} \mathbf{Y}_q^{\dagger} \right) \right], \tag{4.169}$$

$$\mathbf{C}_{qq}^{pred} = \mathbf{C}_{qq} - \left[\mathbf{X}_q \left(\mathbf{D}_{11} \mathbf{X}_q^{\dagger} + \mathbf{D}_{12} \mathbf{Y}_q^{\dagger} \right) + \mathbf{Y}_q \left(\mathbf{D}_{21} \mathbf{X}_q^{\dagger} + \mathbf{D}_{22} \mathbf{Y}_q^{\dagger} \right) \right], \tag{4.170}$$

$$\mathbf{C}_{\alpha r}^{pred} = \mathbf{C}_{\alpha r} - \left[\mathbf{X}_\alpha \left(\mathbf{D}_{11} \mathbf{X}_r^{\dagger} + \mathbf{D}_{12} \mathbf{Y}_r^{\dagger} \right) + \mathbf{Y}_\alpha \left(\mathbf{D}_{21} \mathbf{X}_r^{\dagger} + \mathbf{D}_{22} \mathbf{Y}_r^{\dagger} \right) \right], \tag{4.171}$$

$$\mathbf{C}_{\alpha q}^{pred} = \mathbf{C}_{\alpha q} - \left[\mathbf{X}_\alpha \left(\mathbf{D}_{11} \mathbf{X}_q^{\dagger} + \mathbf{D}_{12} \mathbf{Y}_q^{\dagger} \right) + \mathbf{Y}_\alpha \left(\mathbf{D}_{21} \mathbf{X}_q^{\dagger} + \mathbf{D}_{22} \mathbf{Y}_q^{\dagger} \right) \right], \tag{4.172}$$

The measured responses r_1^m and r_2^m together with their standard deviations are as considered in Section 4.2.2.4, so that

$$\mathbf{r}^m \triangleq \left(r_1^m, r_2^m \right)^{\dagger}, \;\; \mathbf{C}_{rr}^m \triangleq \begin{pmatrix} \left(1.7 \times 10^8\right)^2 & 0 \\ 0 & \left(2.15 \times 10^8\right)^2 \end{pmatrix} \left[n \cdot cm^{-3} \cdot \sec^{-1} \right]^2. \tag{4.173}$$

The "additional" measured responses q_1^m and q_2^m, together with their corresponding standard deviations, are the components of the vector $\mathbf{q}^m$ and matrix $\mathbf{C}_{qq}^m$ defined below:

$$\mathbf{q}^m \triangleq \left(q_1^m, q_2^m \right)^{\dagger}, \;\; \mathbf{C}_{qq}^m \triangleq \begin{pmatrix} \left(1.89 \times 10^8\right)^2 & 0 \\ 0 & \left(1.87 \times 10^8\right)^2 \end{pmatrix} . \left[n \cdot cm^{-3} \cdot \sec^{-1} \right]^2 \tag{4.174}$$

Using Eq. (4.7) yields the same nominal values for the computed responses at the four locations as shown in Eqs. (4.127) through (4.130), noting that, in this case, the computed responses r_3^c and r_4^c are labeled q_1^c and q_2^c, respectively (i.e., $r_3^c \triangleq q_1^c$ and $r_4^c \triangleq q_2^c$). The sensitivities of the responses to the model parameters are computed using 2 adjoint computations [namely: one adjoint computation for obtaining the sensitivities of $R(at \pm 10\, cm)$ and the second adjoint computation for obtaining the sensitivities of $R(at \pm 40\, cm)$]; omitting again the units, the results thus obtained are presented below:

$$\mathbf{S}_{r\alpha} = \begin{pmatrix} -1.92 \times 10^{11} & -1.33 \times 10^5 & 3.78 \times 10^2 & 5.08 \times 10^8 \\ -1.92 \times 10^{11} & -1.33 \times 10^5 & 3.78 \times 10^2 & 5.08 \times 10^8 \end{pmatrix}, \tag{4.175}$$

$$\mathbf{S}_{q\alpha} = \begin{pmatrix} -1.76\times 10^{11} & -1.24\times 10^{9} & 3.66\times 10^{2} & 4.92\times 10^{8} \\ -1.76\times 10^{11} & -1.24\times 10^{9} & 3.66\times 10^{2} & 4.92\times 10^{8} \end{pmatrix}. \quad (4.176)$$

The relative sensitivities (which are used for ranking the importance of the various parameters to the various responses) corresponding to the absolute ones shown in Eq. (4.175) and (4.176) can also be are readily computed to obtain

$$\mathbf{S}_{r\alpha}^{relative} = \begin{pmatrix} 1.0 & -5.64\times 10^{-6} & 1.0 & 1.0 \\ 1.0 & -5.64\times 10^{-6} & 1.0 & 1.0 \end{pmatrix}, \quad (4.177)$$

$$\mathbf{S}_{q\alpha}^{relative} = \begin{pmatrix} -9.46\times 10^{-1} & -5.41\times 10^{-2} & 1.0 & 1.0 \\ -9.46\times 10^{-1} & -5.41\times 10^{-2} & 1.0 & 1.0 \end{pmatrix}. \quad (4.178)$$

As expected, the absolute sensitivities shown in Eqs. (4.175) correspond to the first two rows of the matrix $\boldsymbol{S}_{r\alpha}^{(4\times 4)}$ of sensitivities shown in Eq.(4.131) while the sensitivities shown in Eq. (4.176) correspond to the last two rows of the matrix $\boldsymbol{S}_{r\alpha}^{(4\times 4)}$. The same correspondence holds for the absolute sensitivities: the matrix shown in Eq. (4.177) corresponds to first two rows of the matrix $\boldsymbol{S}_{r\alpha}^{(4\times 4)rel}$ of sensitivities shown in Eq. (4.132), while the sensitivities shown in Eq. (4.178) correspond to the last two rows of the matrix $\boldsymbol{S}_{r\alpha}^{(4\times 4)rel}$.

The covariance matrices of the computed responses arising from the uncertainties in the model parameters are readily computed to obtain:

$$\mathbf{C}_{rr}^{comp} = \mathbf{S}_{r\alpha}\mathbf{C}_{\alpha\alpha}\mathbf{S}_{r\alpha}^{\dagger} = \begin{pmatrix} 4.99\times 10^{17} & 4.99\times 10^{17} \\ 4.99\times 10^{17} & 4.99\times 10^{17} \end{pmatrix}\left[n\cdot cm^{-3}\cdot \sec^{-1}\right]^{2}, \quad (4.179)$$

$$\mathbf{C}_{rq}^{comp} = \mathbf{S}_{r\alpha}\mathbf{C}_{\alpha\alpha}\mathbf{S}_{q\alpha}^{\dagger} = \begin{pmatrix} 4.82\times 10^{17} & 4.82\times 10^{17} \\ 4.82\times 10^{17} & 4.82\times 10^{17} \end{pmatrix}\left[n\cdot cm^{-3}\cdot \sec^{-1}\right]^{2} = \mathbf{C}_{qr}^{comp}, \quad (4.180)$$

$$\mathbf{C}_{qq}^{comp} = \mathbf{S}_{q\alpha}\mathbf{C}_{\alpha\alpha}\mathbf{S}_{q\alpha}^{\dagger} = \begin{pmatrix} 4.66\times 10^{17} & 4.66\times 10^{17} \\ 4.66\times 10^{17} & 4.66\times 10^{17} \end{pmatrix}\left[n\cdot cm^{-3}\cdot \sec^{-1}\right]^{2}. \quad (4.181)$$

It follows from Eq. (4.179) that the standard deviations of the computed responses are:

$$\begin{aligned} std\left(r_1^c\right) = std\left(r_2^c\right) = 7.06\times 10^{8}\left[n\cdot cm^{-3}\cdot \sec^{-1}\right], \\ std\left(q_1^c\right) = std\left(q_2^c\right) = 6.83\times 10^{8}\left[n\cdot cm^{-3}\cdot \sec^{-1}\right]. \end{aligned} \quad (4.182)$$

The consistency indicator is computed by using the corresponding formula from Chapter 3, Section 3.2.3, which is reproduced below:

$$\chi^2 = \left(\mathbf{r}^c - \mathbf{r}^m\right)^\dagger \mathbf{D}_{11}\left(\mathbf{r}^c - \mathbf{r}^m\right) + 2\left(\mathbf{r}^c - \mathbf{r}^m\right)^\dagger \mathbf{D}_{12}\left(\mathbf{q}^c - \mathbf{q}^m\right) + \left(\mathbf{q}^c - \mathbf{q}^m\right)^\dagger \mathbf{D}_{22}\left(\mathbf{q}^c - \mathbf{q}^m\right). \tag{4.183}$$

Introducing the required information into the above expression yields $\chi^2\left(\text{per degree of freedom} = 4\right) = 1.21$, which is the same value as was obtained using Eq. (4.139). Of course, the inequalities shown in Eqs. (4.135) through (4.138) also hold unchanged.

The matrices $\mathbf{D}_{rr}$, $\mathbf{D}_{rr}$, and $\mathbf{D}_{rr}$ can now be computed, in units of $\left[n \cdot cm^{-3} \cdot \sec^{-1}\right]^2$, to obtain:

$$\mathbf{D}_{rr} = \begin{pmatrix} 5.27\times10^{17} & 4.99\times10^{17} \\ 4.99\times10^{17} & 5.24\times10^{17} \end{pmatrix};\ \mathbf{D}_{rq} = \begin{pmatrix} 4.82\times10^{17} & 4.82\times10^{17} \\ 4.82\times10^{17} & 4.82\times10^{17} \end{pmatrix};$$
$$\mathbf{D}_{qr} = \begin{pmatrix} 4.82\times10^{17} & 4.82\times10^{17} \\ 4.82\times10^{17} & 4.82\times10^{17} \end{pmatrix};\ \mathbf{D}_{qq} = \begin{pmatrix} 5.21\times10^{17} & 4.66\times10^{17} \\ 4.66\times10^{17} & 5.01\times10^{17} \end{pmatrix}; \tag{4.184}$$

The matrices $\mathbf{D}_{11}$, $\mathbf{D}_{12}$, and $\mathbf{D}_{22}$ can be computed next (omitting the respective in units), to obtain:

$$\mathbf{D}_{11} = \begin{pmatrix} 2.46\times10^{-17} & -1.15\times10^{-17} \\ -1.15\times10^{-17} & 2.66\times10^{-17} \end{pmatrix};\ \mathbf{D}_{12} = \begin{pmatrix} -5.04\times10^{-18} & -7.94\times10^{-18} \\ -5.79\times10^{-18} & -9.11\times10^{-18} \end{pmatrix};$$
$$\mathbf{D}_{21} = \begin{pmatrix} -5.04\times10^{-18} & -5.79\times10^{-18} \\ -7.94\times10^{-18} & -9.11\times10^{-18} \end{pmatrix} = \mathbf{D}_{12}^\dagger;\ \mathbf{D}_{22} = \begin{pmatrix} 1.56\times10^{-17} & -4.09\times10^{-18} \\ -4.09\times10^{-18} & 2.21\times10^{-17} \end{pmatrix}. \tag{4.185}$$

It is important to note that the computation of the above matrices requires inversions of only two (2×2)-matrices instead of the (4×4)-matrix which needed to be inverted in Subsection 4.2.2.5 (a). The predicted nominal values for the model parameters are obtained from Eq. (4.164). The numerical values thus obtained coincide with the values previously obtained in Subsection 4.2.2.5 (a), in Eqs. (4.142) through (4.145) and will therefore not be reproduced here. The predicted covariance matrix for the model parameters is computed from Eq. (4.167); this computation yields the same numerical values as shown in Eq. (4.146). The predicted nominal values for the model responses are computed from Eq. (4.165). The values thus obtained are the same as obtained in Eqs. (4.147) through (4.150), and, for brevity, they will not be repeated here.

The predicted covariance matrices for the model responses are computed using Eqs. (4.168) through (4.172), to obtain:

$$\mathbf{C}_{rr}^{pred} = \begin{pmatrix} 9.04\times10^{15} & 9.04\times10^{15} \\ 9.04\times10^{15} & 9.04\times10^{15} \end{pmatrix}, \tag{4.186}$$

$$\mathbf{C}_{rq}^{pred} = \begin{pmatrix} 8.64\times10^{15} & 8.64\times10^{15} \\ 8.64\times10^{15} & 8.64\times10^{15} \end{pmatrix} = \mathbf{C}_{qr}^{pred}, \tag{4.187}$$

$$\mathbf{C}_{qq}^{pred} = \begin{pmatrix} 8.45\times10^{15} & 8.45\times10^{15} \\ 8.45\times10^{15} & 8.45\times10^{15} \end{pmatrix}. \tag{4.188}$$

As expected, the matrix shown in Eq. (4.186) corresponds to upper-left (2×2)-submatrix of the matrix shown in Eq. (4.151); the matrix shown in Eq. (4.187) corresponds to lower-right (2×2)-submatrix of the matrix shown in Eq. (4.151); and the matrix shown in Eq. (4.188) corresponds to upper-right as well as lower-left (due to symmetry) (2×2)-submatrices of the matrix shown in Eq. (4.151).

The predicted parameters-responses correlation matrices are computed using Eqs. (4.171) and (4.172) to obtain:

$$\mathbf{C}_{\alpha r}^{pred} = \begin{pmatrix} -7.81\times10^{3} & 3.89\times10^{4} & 1.38\times10^{13} & 4.57\times10^{6} \\ -7.81\times10^{3} & 3.89\times10^{4} & 1.38\times10^{13} & 4.57\times10^{6} \end{pmatrix}, \tag{4.189}$$

$$\mathbf{C}_{\alpha q}^{pred} = \begin{pmatrix} 1.50\times10^{3} & -4.13\times10^{4} & 1.64\times10^{13} & 5.41\times10^{6} \\ 1.50\times10^{3} & -4.13\times10^{4} & 1.64\times10^{13} & 5.41\times10^{6} \end{pmatrix}. \tag{4.190}$$

As expected, the matrix in shown Eq. (4.189) corresponds to the first two rows of the matrix shown in Eq. (4.152), while the matrix shown in Eq. (4.190) corresponds to the last two rows of the matrix shown in Eq. (4.152).

Figure 4.5 depicts the computed responses with their computed standard deviations (depicted using dotted lines) and also the predicted best-estimate responses together with their predicted standard deviations (depicted using solid lines).

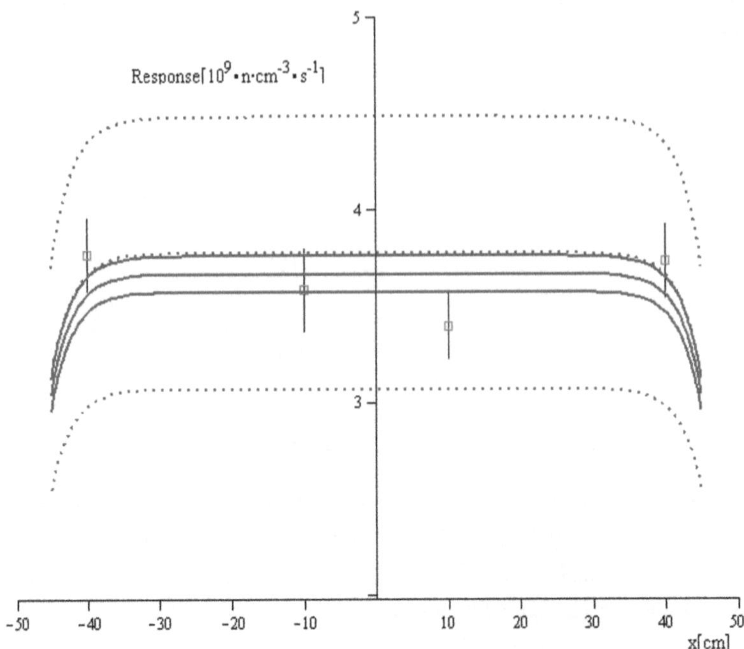

Figure 4.5. Computed and predicted responses and their corresponding standard deviations when considering four consistent measurements for predictive modeling.

The following important conclusions can be drawn from the results presented in this section:

(i) As the illustrative example presented in this section has demonstrated, using the BERRU-CMS predictive modeling methodology (Cacuci, 2014a) reduces the computational costs (by comparison to using the customary data assimilation procedures) when considering additional measurements, involving computations limited to the size of assimilating just the additional measurements, while preserving the same results as would have been obtained if all measurements had been processed simultaneously. In the illustrative example presented in this section, the BERRU-CMS predictive modeling methodology requires the inversion of matrices of size (2×2) rather than inversion of matrices of size (4×4), as would be required by methodologies in which all four responses must be considered simultaneously.

(ii) The BERRU-CMS predictive procedure preserves the problem's symmetry, so that $r_1^{pred} = r_2^{pred}$ and $q_1^{pred} = q_2^{pred}$, while all four predicted responses remain consistent with one another within their "one-standard deviation" ranges.

(iii) As in the cases of one and, respectively, two assimilated measurements, the predicted relative (and also the predicted absolute) response standard deviation after the assimilation of four consistent measurements is smaller than both the measured and the computed response variances. The results clearly highlight the fact that the assimilation of additional consistent measurements further reduces the uncertainty in the predicted response. In all cases, the assimilation of measurements has clearly reduced the uncertainty of the predicted responses well below the uncertainties in the computed responses due to uncertainties in the model parameters.

$$\begin{aligned} & rsd\left(r_i^{pred}\,\middle|\, with\ 4\ measurements\right) \cong 2.59\% < rsd\left(r_i^{pred}\,\middle|\, with\ 2\ measurements\right) \\ & \cong 3.8\% < rsd\left(r_i^{pred}\,\middle|\, with\ 1\ measurements\right) \cong 4.58\% < rsd\left(r_i^{m}\,\middle|\, all\ cases\right) \cong 5\% \\ & < rsd\left(r_i^{c}\,\middle|\, all\ cases\right) \cong 18.7\%; \end{aligned} \tag{4.191}$$

(iv) As the results in this Subsection showed, all of the conclusions reached in the previous sub-sections (i.e., after the assimilation of one, two, and three measurements, respectively) also remain valid when four consistent measurements are assimilated. The largest calibrations (adjustments) occurred for the source S, followed by Σ_d, and Σ_a, while D was hardly adjusted because of its vanishingly small sensitivity. The largest reductions of the standard deviations occurred, in order, for S, Σ_d, Σ_a, and D. The originally uncorrelated parameters became correlated after applying the predictive modeling procedure calibration: the predicted correlations have the same trend as already shown in the previous sub-sections, when assimilation one, two and/or three measurements.

4.3 BERRU-CMS PREDICTIVE MODELING OF MULTIPLE REACTOR PHYSICS BENCHMARKS

The material presented in this Section illustrates the application of the BERRU-CMS methodology to perform predictive modeling of three reactor physics benchmarks (Godiva, Jezebel-239, Jezebel-240). The most important model parameters (i.e., nuclei, cross sections, number densities) and responses (multiplication factors and reaction rate ratios) for the reactor physics benchmarks Jezebel-239, Jezebel-240, and Godiva are tabulated in Tables 4.1 and 4.2, below. The experimental values for each benchmark are taken from the International Handbook of Evaluated Criticality Safety Benchmark Experiments (ICSBEP, 2010), while the nuclear data for the model parameters stem from the ORNL code package SCALE6.1 (2011). The computational models of all three benchmarks are idealized spheres of varying radius. The corresponding experiments consisted of two or more pieces that could be assembled into a nearly spherical mass. In order to obtain experimentally measured reaction rates a thin foil was placed in the center of the assembly. Reaction rates are measured in two steps: first, a thin foil of the isotope of interest is placed in a critical assembly. This irradiates the foil for a specified length of time. Next, the foil is placed in a gamma-ray counting facility where the spectrum is measured at a standard detector distance.

Table 4.1: Atom densities for Jezebel-239, Jezebel-240 and Godiva

	Jezebel-239 Radius: 6.3849cm	Jezebel-240 Radius: 6.6595cm		Godiva Radius: 8.7407cm
Nuclide	Atom Density [atoms / barn-cm]	Atom Density, [atoms / barn-cm]	Nuclide	Atom Density, [atoms / barn-cm]
Ga	$1.3752 \cdot 10^{-3}$	$1.3722 \cdot 10^{-3}$	^{234}U	$4.9184 \cdot 10^{-4}$
^{239}Pu	$3.7047 \cdot 10^{-2}$	$2.9934 \cdot 10^{-2}$	^{235}U	$4.4994 \cdot 10^{-2}$
^{240}Pu	$1.7512 \cdot 10^{-3}$	$7.8754 \cdot 10^{-3}$	^{238}U	$2.4984 \cdot 10^{-3}$
^{241}Pu	$1.1674 \cdot 10^{-4}$	$1.2146 \cdot 10^{-3}$		
^{242}Pu	-	$1.5672 \cdot 10^{-4}$		

Table 4.2: Benchmark responses and parameters

	Jezebel-239	Jezebel-240	Godiva
N_r	8	3	11
N_α	2241	1458	2916
Nuclides in sphere	$^{69}Ga,\ ^{71}Ga,$ $^{239}Pu,\ ^{240}Pu,\ ^{241}$	$^{69}Ga,\ ^{71}Ga,$ $^{239}Pu,\ ^{240}Pu,\ ^{241}Pu,\ ^{242}Pu$	$^{234}U,\ ^{235}U,\ ^{238}U$
Nuclides in foils	$^{238}U,$ $^{233}U,$ $^{237}Np,$ $^{239}Pu,$ $^{55}Mn,$ $^{93}Nb,$ $^{63}Cu.$	$^{238}U,$ $^{237}Np.$	$^{238}U,$ $^{233}U,$ $^{237}Np,$ $^{239}Pu,$ $^{55}Mn,$ $^{93}Nb,$ $^{63}Cu,$ $^{81}Br,$ $^{107}Ag,$ $^{127}I.$

Eight responses were considered for Jezebel-239, as follows: the effective multiplication factor k_{eff}, the center core fission rates for ^{233}U, ^{238}U, ^{237}Np, and ^{239}Pu, as well as the center core radiative capture rates for ^{55}Mn, ^{93}Nb and ^{63}Cu. Three responses were selected for Jezebel-240: the effective multiplication factor k_{eff}, along with the center core fission rates for ^{233}U and ^{237}Np. The coupling between Jezebel-239 and Jezebel-240 is provided by the shared isotopes listed in Table 4.1. Eleven responses were selected for Godiva, including the effective multiplication factor k_{eff}, the eight reaction rate types already listed for Jezebel-239, and the radiative capture rates for ^{107}Ag, ^{127}I and ^{81}Br. All of the reaction rates reported in ICSBEP (2010) are values relative to the fission rate for ^{235}U.

The "model parameters" include individual cross sections for each material, nuclide, reaction type and energy-group. Thus, for each benchmark, the total number of parameters results from the following multiplication: $Nr.(materials)\times Nr.(nuclides)\times Nr.(reaction\,types)\times Nr.(energy\ groups)$. This total number of parameters is provided in Table 4.2, for each benchmark. Even though the Jezebel-240 benchmark contains more individual nuclides than Jezebel-239, the latter benchmark has a larger number of model parameters because it comprises more reaction rate measurements with external foils placed at the center of the sphere than does Jezebel-240.

As shown by the data presented in Table 4.1, both Jezebel-239 and Jezebel-240 contain ^{239}Pu, ^{240}Pu, ^{241}Pu, and Ga. These isotopes produce non-zero entries in the covariance matrix $\mathbf{C}_{\alpha\beta}$, which couples these benchmarks. Although Godiva does not have any isotopes in common with either of the Jezebel benchmarks, it is nevertheless coupled, albeit weakly, to the Jezebel benchmarks via the cross section covariance data for the fission cross section of uranium and plutonium.

In addition to the nominal values of the computed responses, the essential quantities needed prior to applying the BERRU-CMS predictive modeling formulas are the sensitivities of all model responses to all of the uncertain model parameters. The nominal values and response sensitivities were quantified using neutron transport computations with ORNL's deterministic neutron transport code DENOVO (Evans et al, 2010). Due to the symmetries inherent in the spherical geometries of each of the benchmarks, the DENOVO computations needed only an octant of each benchmark (sphere), with a 10x10x10 spatial discretization. This discretization was sufficiently accurate for the purposes of these illustrative examples; numerical discretization errors are not considered in the uncertainty analysis. Furthermore, the 27-energy-group structure of ENDF/B-VII using the nuclear data included in SCALE6.1 (2010) provided sufficient energy resolution for the purpose of this work. The energy-dependent sensitivity profiles were computed using the very efficient parallel Krylov-based adjoint sensitivity analysis procedure developed by Evans and Cacuci (2012), which was added to the fixed-source standard adjoint capabilities of the deterministic neutron transport solver DENOVO. The results for the most important sensitivities will be presented in Figures 4.6 through 4.10, in the following. As is well known, the *absolute sensitivity* of a response r to a parameter α_i^x is given by the (dimensional) derivative $\partial r/\partial \alpha_i^x$, while the *relative sensitivity* of response r to a parameter α_i^x is given by the (dimensionless) quantity $\left[\partial r/\partial \alpha_i^x\right]\left(\alpha_i^x/r\right)$. Since Figures 4.6 through 4.10 present relative sensitivities only, they will be denoted in the abbreviated form $\partial r/\partial \alpha_i^x$, where: (i) the response r represents either k_{eff} or a spectral index (i.e., reaction rate ratio); and (ii) the quantity α_i^x denotes a "model parameter", specified such that the subscript i will denote a specific isotope, while the superscript x indicates a specific neutron cross section, a benchmark-specific average prompt fission neutron multiplicity $\bar{\nu}$, or a benchmark-specific fission spectrum χ.

The energy-dependent profiles of the largest sensitivities of the benchmark-specific effective multiplication factor, k_{eff}, are depicted in Figures 4.6, 4.7 and 4.7, respectively, for the two Jezebel benchmarks and Godiva. Note that for all three assemblies, the respective sensitivity profiles of k_{eff} become vanishingly small for energies below about 10 keV. This common feature is expected in view of the fact that all three benchmarks are, by design, "fast assemblies", containing no neutron moderating materials.

As Figures 4.6 and 4.7 indicate, the largest sensitivities of the multiplication factors for the two Jezebel benchmarks are with respect to the number of neutrons emitted per fission, $\bar{\nu}$, and the fission cross section of ^{239}Pu. As depicted in Figure 4.6, the third and fourth largest sensitivities of the multiplication factor of Jezebel-239 are to the fission spectrum χ and the elastic scattering cross sections of ^{239}Pu. As Figure 4.6 also indicates, the sensitivity profile for the fission spectrum, χ, of ^{239}Pu changes sign over the respective energy range. It is therefore important to be aware that an energy-integrated value for this particular profile could yield a misleadingly low value/rank for the resulting energy-integrated sensitivity.

On the other hand, as shown in Figure 4.7, the third and fourth largest sensitivities of k_{eff} for Jezebel-240 are to the quantities describing the fission processes in ^{240}Pu, namely the average neutrons per fission in ^{240}Pu, $\bar{\nu}$, and the fission cross section of ^{240}Pu, which would be expected in view of the higher content of ^{240}Pu in Jezebel-240.

Figure 4.8 indicates that the largest sensitivities of Godiva's effective multiplication factor, k_{eff}, are, in the following order, to: (i) the average number of prompt neutrons emitted per fission, $\bar{\nu}$, in ^{235}U; (ii) the fission cross section of ^{235}U; (iii) the elastic scattering cross section; and (iv) the inelastic scattering cross section. Again, these results are expected since Godiva is a uranium system.

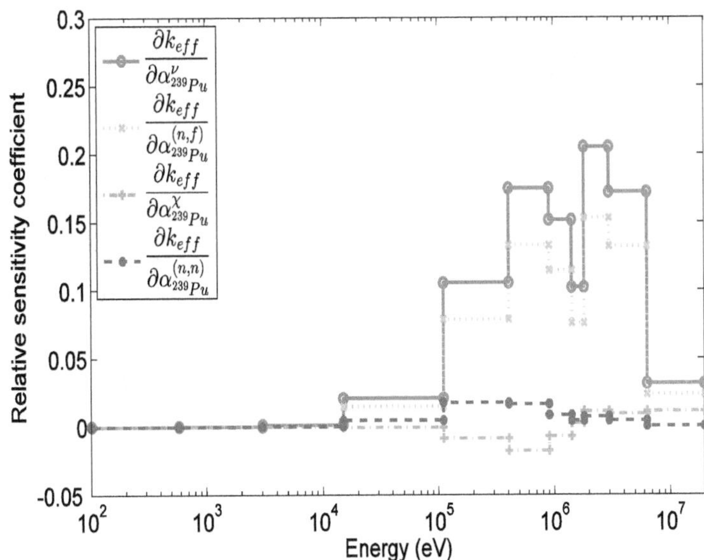

Figure 4.6: Sensitivity profile of k_{eff} to selected nuclides and reaction types for Jezebel-239.

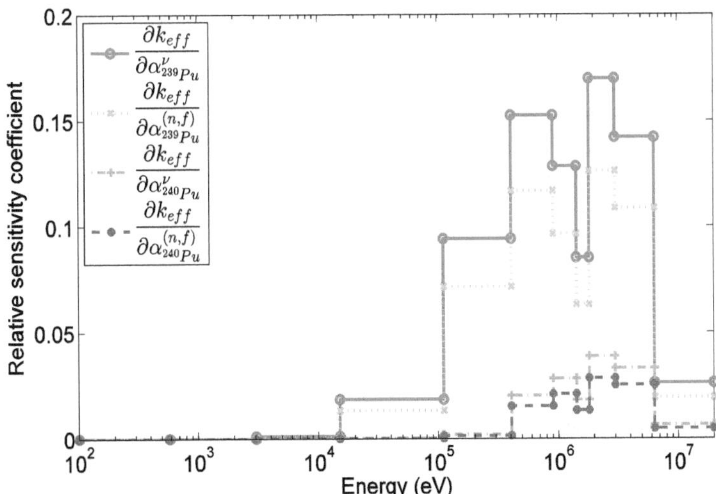

Figure 4.7: Sensitivity profile of k_{eff} to selected nuclides and reaction types for Jezebel-240

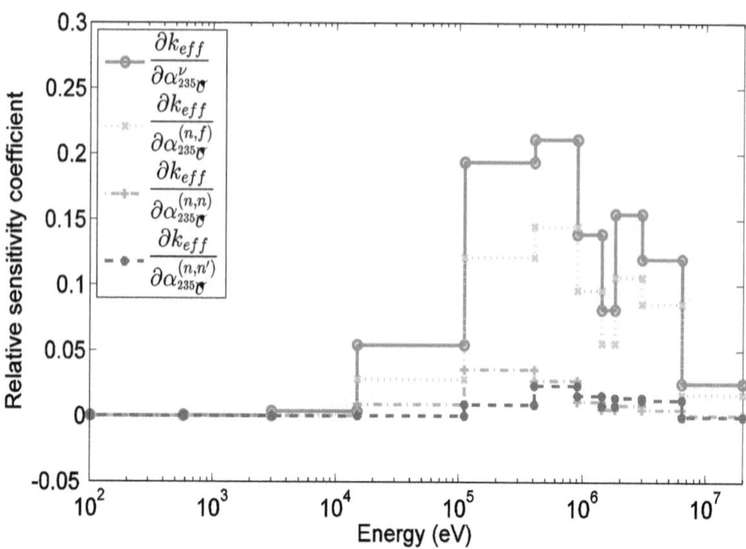

Figure 4.8: Sensitivity profile of k_{eff} to selected nuclides and reaction types for Godiva

Several thousands of energy-dependent profiles of sensitivities of reaction-rate-type responses to various cross sections and nuclides have been computed in preparation for applying the BERRU-CMS formulas. Figures 4.9 and 4.10 depict two typical energy-dependent sensitivity profiles of reaction rate responses (namely: a fission reaction rate, and a radiative capture reaction rate, respectively) associated with Jezebel-239. Specifically, Figure 4.9 depicts the energy-dependent profiles of sensitivities (to the most important cross sections of the respective nuclei) of the fission reaction rate of ^{237}Np normalized to the fission reaction rate of ^{235}U ; this response is denoted as $r_4 = \sigma_f\left(^{237}Np\right)/\sigma_f\left(^{235}U\right)$. Figure 4.10 depicts the energy-dependent profiles of sensitivities (to various cross sections of selected nuclei) of the radiative capture reaction rate of ^{63}Cu normalized to the fission reaction rate of ^{235}U ; this response is denoted as $r_8 = \sigma_\gamma\left(^{63}Cu\right)/\sigma_f\left(^{235}U\right)$. Since all reaction rate ratios considered in the present work share the same denominator, namely $\sigma_f\left(^{235}U\right)$, it follows that all responses associated with a specific benchmark have identical relative sensitivity profiles for the fission cross section of ^{235}U .

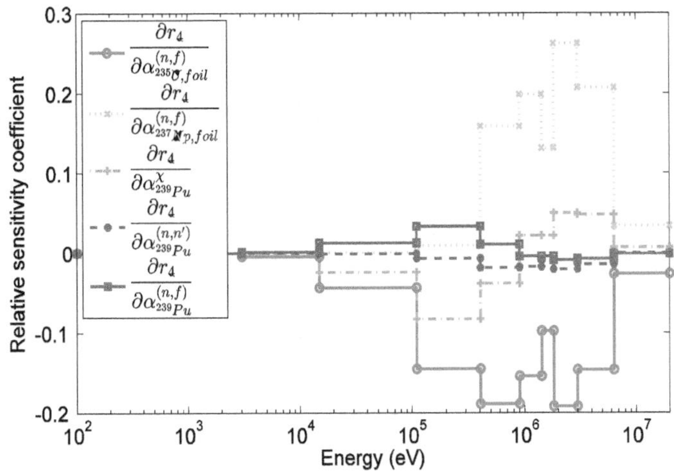

Figure 4.9: Sensitivity profile of $r_4 \triangleq \sigma_f\left({}^{237}Np\right)/\sigma_f\left({}^{235}U\right)$ to selected nuclides and reaction types for Jezebel-239.

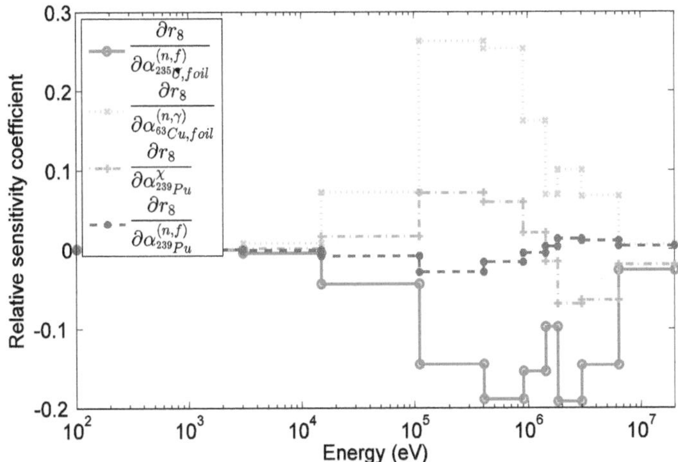

Figure 4.10: Sensitivity profiles of $r_8 \triangleq \sigma_\gamma\left({}^{63}Cu\right)/\sigma_f\left({}^{235}U\right)$ to selected nuclides and reaction types for Jezebel-239

The covariances ("uncertainties") of the computed responses are quantified by using Eqs. (4.57) to combine ("propagate") the covariance data for the model parameters (cross sections, average neutrons per fission, fission spectrum) with the sensitivity profiles presented in the foregoing. Table 4.3 presents the six largest contributors to the effective multiplication factor, k_{eff}, for Jezebel-239 and Jezebel-240, respectively. Comparing the rankings in Table 4.3 with the rankings depicted in Figures 4.6 and 4.7, respectively, readily indicates that the rankings of the largest sensitivities do *not* correspond with the rankings of the largest uncertainty contributors to the standard deviations in k_{eff}. This is because the contributions to the uncertainty in a response (in this case: k_{eff}) arise from matrix products of the form $\mathbf{S}_{r\alpha}\mathbf{C}_{\alpha\alpha}\mathbf{S}^{\dagger}_{r\alpha}$, in which an individual contribution arises (roughly speaking) from products of variances and squared-sensitivities. For both of the Jezebel benchmarks, the largest contribution to the overall standard deviation in the respective k_{eff} arises from the average prompt fission neutron multiplicity $\bar{\nu}$ of ^{239}Pu, but this is the only contributor for which the sensitivity-rankings also correspond to the "uncertainty contribution" ranking (both being the largest). As already mentioned, the two Jezebel configurations are coupled via the covariance block $\mathbf{C}_{\alpha\beta}$ of shared isotopes, so that the total uncertainty of a particular response is, in general, affected by these additional coupling contributions. For both of the Jezebel benchmarks, it was assumed that $\mathbf{S}_{r\beta}=\mathbf{0}$ and $\mathbf{S}_{q\alpha}=\mathbf{0}$, so that the largest uncertainty contributions were not affected by such cross-terms. In general, however, the coupling terms between two tightly-coupled multi-physics systems may produce the largest uncertainty contribution to certain responses.

Table 4.3: The six largest contributors to the uncertainty in k_{eff} for Jezebel-239 and Jezebel-240

Nuclide-Quantity Pair Jezebel-239	Relative Standard Deviation in %	Nuclide-Quantity Pair Jezebel-240	Relative Standard Deviation in %
$^{239}Pu\ \bar{\nu}$	1.21E-01	$^{239}Pu\ \bar{\nu}$	1.02E-01
$^{239}Pu(n,n')$	8.64E-02	$^{239}Pu(n,n')$	6.35E-02
$^{239}Pu(n,n)$	4.54E-02	$^{239}Pu(n,n)$	3.75E-02
$^{239}Pu(n,f)$	3.48E-02	$^{240}Pu\ \bar{\nu}$	3.31E-02
$^{239}Pu(n,\gamma)$	7.34E-03	$^{239}Pu(n,f)$	2.93E-02

Initially, each of the benchmarks was considered individually as a stand-alone system, and the BERRU-SMS expressions were used to obtain the best-estimate nominal values and reduced predicted uncertainties for each benchmark individually;

the quantities thus obtained are labeled with the superscript "be" in Figures 4.11 through 4.15. Subsequently, the BERRU-CMS methodology was applied to the two Jezebel benchmarks considered as coupled systems; the best-estimate quantities thus obtained are labeled with the superscript "opt,1" in Figures 4.11 through 4.15. Finally, the Godiva benchmark was coupled to the two Jezebel benchmarks, and the BERRU-CMS methodology was applied to all three benchmarks simultaneously; the best-estimate values thus obtained are labeled with the superscript "opt,2" in Figures 4.11 through 4.15.

Figure 4.11 depicts the measured, computed and best-estimate values for k_{eff}, along with $(\pm)$ one standard deviation, all normalized to the nominal measured value, for the combinations of benchmarks described above. The quantities appearing in the legend for Figure 4.11 (as well as in Figures 4.12 through 4.15), have the following meanings: (i) r_c and σ_c denote the *mean value* and corresponding *standard deviation* for a *computed response*; (ii) r_m and σ_m denote the mean value and corresponding standard deviation for a *measured response*; (iii) r^{be} and σ^{be} denote the mean value and corresponding standard deviation for a *best-estimate response* resulting from the application of the BERRU-CMS formulas to *each benchmark individually*; (iv) $r^{opt,1}$ and $\sigma^{opt,1}$ denote the mean value and corresponding standard deviation for a *predicted response* resulting from the application of the BERRU-CMS formulas to *the benchmarks Jezebel-239 and Jezebel-240 considered as coupled multi-physics systems*; (v) $r^{opt,2}$ and $\sigma^{opt,2}$ denote the mean value and corresponding standard deviation for a *predicted response* resulting from the application of the BERRU-CMS formulas to *all three benchmarks (Jezebel-239, Jezebel-240 and Godiva) considered as coupled multi-physics systems.*

As shown in Figure 4.11, the computed value for k_{eff} is slightly super critical for all three benchmarks, and their computed standard deviations are significantly larger than the measured standard deviations of the corresponding experiments. The relative values of the measured and computed standard deviations indicate that the experimental nominal values are more accurately known than the computed ones. Applying the BERRU-SMS formulas to each benchmark individually shifts the best-estimate ("be") predicted mean values of each of the respective multiplication factors, k_{eff}, closer to the respective experimentally measured nominal values (which are more accurately known than the computed ones), and reduces its accompanying predicted standard deviation to a value that is actually smaller than the corresponding measured standard deviation. When both Jezebel benchmarks are treated as a coupled multi-physics system, the BERRU-CMS methodology calibrates the predicted ("opt,1") mean value of the multiplication factor even closer to the experimentally measured mean value, and reduces the predicted standard deviation even more. Finally, as depicted in Figure 4.11, including Godiva and applying the BERRU-CMS formulas to all three benchmarks simultaneously has little effect additional effect on the predicted ("opt,2") mean value of the multiplication factor, as can be expected from the weak coupling between the two plutonium benchmarks to the uranium benchmark.

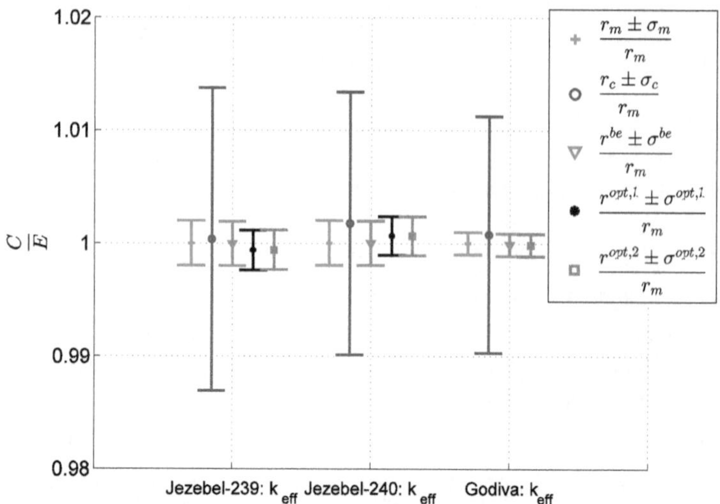

Figure 4.11: Measured, computed and best-estimate values for k_{eff} and one standard deviation (all normalized to the nominal measured value and denoted as Computation/Experiment) for the cases: (i) separate benchmarks ("be"); (ii) coupled Jezebel-239 and Jezebel-240 ("opt, 1"); (iii) all 3 benchmarks coupled ("opt, 2").

Figures 4.12 through 4.15 present the predictive modeling results for the reaction rate rations (also called "spectral indices") responses. These responses are grouped into two categories: (i) a spectral index that was measured in two or all three benchmarks will be called a "similar response"; and (ii) a spectral index that was measured just in a single benchmark will be called a "dissimilar response". An example of a "similar" response is the fission rate of ^{238}U relative to the fission rate of ^{235}U, which was measured in all three benchmarks. On the other hand, the ^{107}Ag radiative capture ratio in Godiva was measured only in Godiva, so it is a "dissimilar response". Furthermore, within each category of responses, fission rates and radiative capture rates are grouped together.

Figure 4.12 depicts the predictive modeling results obtained for the fission rates of ^{238}U and ^{237}Np when the BERRU-CMS formulas were applied to each benchmark individually ("be"), to the two Jezebel benchmarks considered as two coupled systems ("opt,1"), and all three benchmarks considered as coupled multi-physics systems ("opt,1"). Since these ratios were measured in all three benchmarks, they are labeled "similar" responses. Notably, for all three benchmarks, the computed fission rates of ^{238}U are all smaller than the experimentally measured ones, while the computed fission rates of ^{237}Np are larger (for Jezebel-240 and Godiva) than the experimentally measured ones. In all cases, though, the predictive modeling procedure calibrates the respective responses to predicted values which fall in between the measured and the computed ones, while reducing the respective predicted standard deviations.

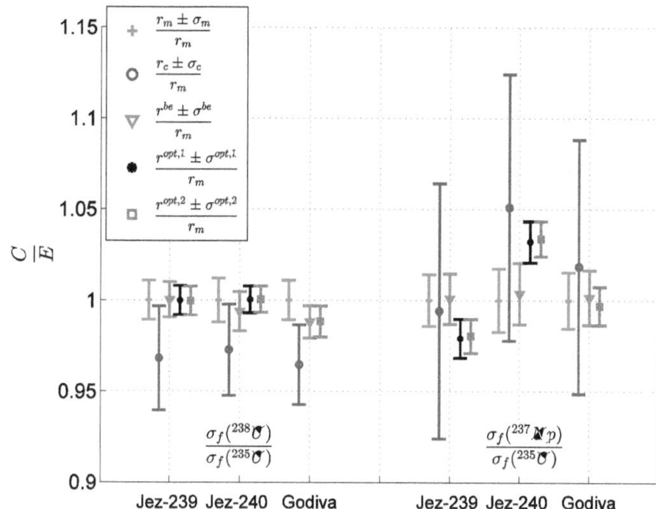

***Figure 4.12:** Measured, computed and best-estimate values for fission rate ratios of ^{238}U and ^{237}Np ("similar" responses) and one standard deviation (all normalized to the nominal measured value, denoted as Computation/Experiment) for the following cases: (i) separate benchmarks ("be"); (ii) coupled Jezebel-239 and Jezebel-240 ("opt, 1"); (iii) all 3 benchmarks coupled ("opt, 2").*

Figure 4.13 depicts the results obtained from the BERRU-CMS formulas for the fission rate ratios of ^{233}U and ^{239}Pu to each benchmark individually ("be"), to the two Jezebel benchmarks considered as two coupled systems ("opt,1"), and all three benchmarks considered as coupled multi-physics systems ("opt,2"). Although these ratios were measured only in two benchmarks, they are still labeled "similar" responses. As Figure 4.13 indicates, the computed the fission rates of both ^{233}U and ^{239}Pu are smaller than the experimentally measured ones, for both benchmarks, and, as expected, the predictive modeling procedure calibrates the respective responses to predicted values which fall in between the measured and the computed ones, while reducing the respective predicted standard deviations. Notably, even though the computed and experimentally measured values of the fission ration of ^{239}Pu are seemingly *discrepant* (since the *difference* between the computed and the experimentally measured nominal values is *larger than the sum* of one computed plus one experimentally measured standard deviations), the BERRU-CMS procedure yields a predicted nominal value and predicted standard deviation that reconcile the originally discrepant (computed vs. measured) values.

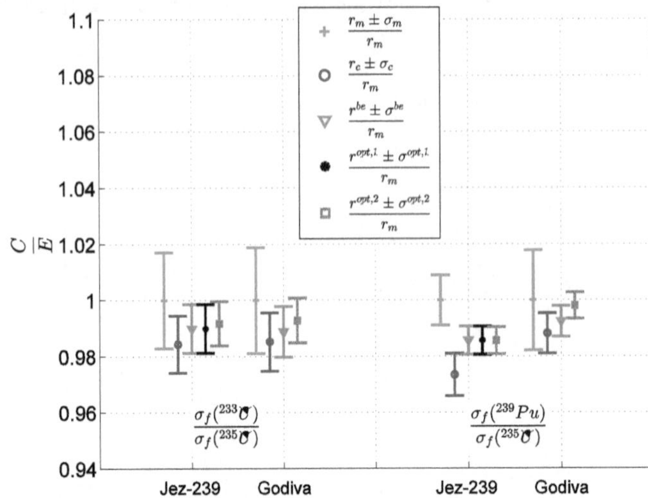

***Figure 4.13:** Measured, computed and best-estimate values for fission rate ratios of 233U and 239Pu ("similar" responses) and one standard deviation (all normalized to the nominal measured value and denoted as Computation/Experiment) for the cases: (i) separate benchmarks ("be"); (ii) coupled Jezebel-239 and Jezebel-240 ("opt, 1"); (iii) all 3 benchmarks coupled ("opt, 2").*

Figure 4.14 depicts the results obtained from the BERRU-CMS formulas for the radiative capture rate (ratios) in ^{55}Mn, ^{93}Nb and ^{63}Cu ("similar" responses) for the same combinations of benchmarks as described in the foregoing. The computed radiative capture rate (ratios) in ^{55}Mn, ^{93}Nb differ from (in this case, they are larger than) the experimentally measured ones, for both Godiva and Jezebel-239, and, as expected, the predictive modeling procedure calibrates the respective responses to predicted values that fall in between the measured and the computed ones, while reducing the respective predicted standard deviations. As in the case of the discrepant computed and experimentally measured values of the fission ratio of ^{239}Pu depicted in Figure 4.13, the BERRU-CMS procedure yields a predicted nominal value and predicted standard deviation that reconcile the originally computed and measured discrepant values of the radiative capture rate ratio in ^{93}Nb. In the case of ^{63}Cu, the computed radiative capture rate is smaller than the experimentally measured one, for both Godiva and Jezebel-239; the BERRU-CMS formulas calibrates the respective responses to predicted values which fall in between the measured and the computed ones, while reducing the respective predicted standard deviations.

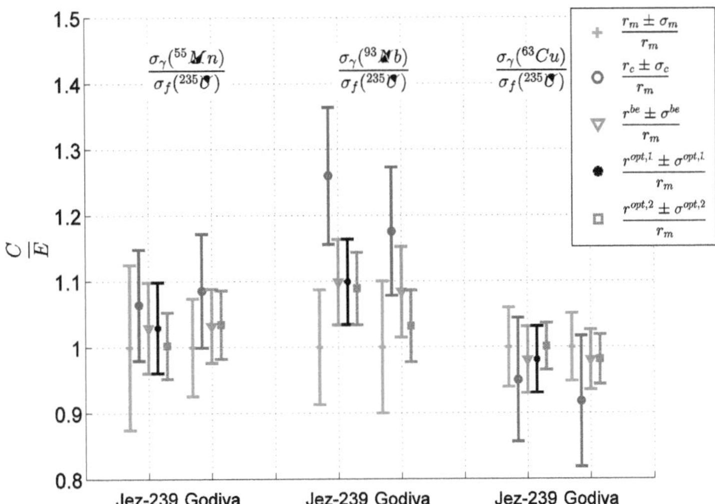

***Figure 4.14:** Measured, computed and best-estimate values for radiative capture rate ratios in ^{55}Mn, ^{93}Nb and ^{63}Cu ("similar" responses) and one standard deviation (all normalized to the nominal measured value and denoted as Computation/ Experiment) for the cases: (i) separate benchmarks ("be"); (ii) coupled Jezebel-239 and Jezebel-240 ("opt, 1"); (iii) all 3 benchmarks coupled ("opt, 2").*

On the other hand, Figure 4.15 depicts results for the "dissimilar" radiative capture rate (ratios) responses in ^{107}Ag, ^{127}I, and ^{81}Br, all for Godiva. As before, the BERRU-CMS formulas calibrate the respective responses to predicted values that fall in between the measured and the computed ones, while reducing the respective predicted standard deviations. However, the BERRU-CMS procedure cannot fully bridge the gap between the markedly discrepant computed and experimentally measured values of the radiative capture ratio of ^{81}Br.

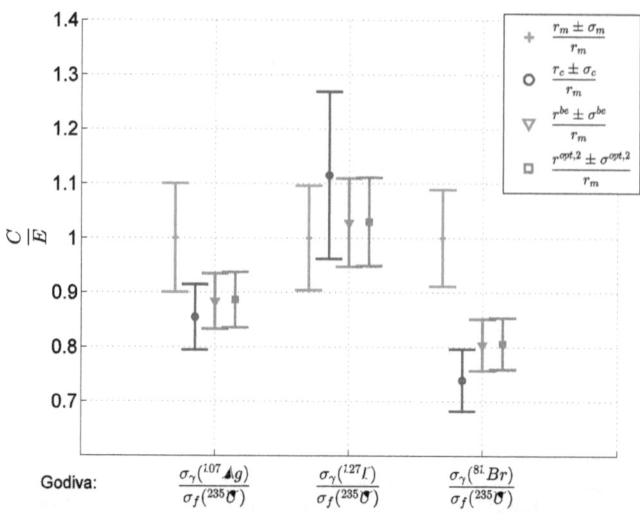

Figure 4.15: Measured, computed and best-estimate values for radiative capture rate ratios in ^{107}Ag, ^{127}I, and ^{81}Br in Godiva ("dissimilar" responses), along with the cor responding one standard deviation (all normalized to the nominal measured value and denoted as Computation/Experiment) for the cases: (i) separate benchmarks ("be"); (ii) all 3 benchmarks coupled ("opt, 2").

The BERRU-CMS methodology calibrates simultaneously not only the models' responses but also models' parameters, to obtain best-estimate values, accompanied by reduced uncertainties, for the predicted responses as well as for the predicted models' parameters. It is not feasible to present here the complete results of applying the BERRU-CMS formulas to the benchmarks' thousands of parameters, but, as an illustrative example, we present in Figures 4.16 through 4.19 the predicted results for the elastic and inelastic cross sections of ^{239}Pu for the Jezebel benchmarks. To facilitate the discussion of the results presented in these tables, the cross sections associated with Jezebel-239 will be generically denoted "α" while the cross sections associated with Jezebel-240 will be generically denoted "β". Thus, the nominal value of the elastic scattering cross section of ^{239}Pu in Jezebel-239 will be denoted as $\alpha_{^{239}Pu}^{(n,n),0}$, while the same elastic scattering cross section of ^{239}Pu in Jezebel-240 will be denoted as $\beta_{^{239}Pu}^{(n,n),0}$. The *individually calibrated* (i.e., by considering each benchmark as an independent system) ^{239}Pu-elastic cross section for Jezebel-239 is denoted as $\alpha_{^{239}Pu}^{(n,n),be}$ while the same calibrated cross section for Jezebel-240 is labeled $\beta_{^{239}Pu}^{(n,n),be}$. Furthermore, the *jointly calibrated* (i.e., by considering the two benchmarks as coupled) ^{239}Pu elastic cross section for Jezebel-239 and Jezebel-240 are denoted as $\alpha_{^{239}Pu}^{(n,n),opt}$ and $\beta_{^{239}Pu}^{(n,n),opt}$, respectively. Note that the nominal and the optimal cross sections are approximately the same for both Jezebel benchmarks, i.e. $\alpha_{^{239}Pu}^{(n,n),0} \cong \beta_{^{239}Pu}^{(n,n),0}$ and $\alpha_{^{239}Pu}^{(n,n),opt} \cong \beta_{^{239}Pu}^{(n,n),opt}$ respectively.

Figure 4.16 presents the predicted relative calibrations while Figure 4.17 depicts the absolute predicted nominal values for the ^{239}Pu elastic scattering cross section. Note that the change in the predicted nominal values is over 12% when Jezebel-239 is considered as an independent system and both Jezebel benchmarks are considered to be coupled, but is smaller (up to 8%) when the Jezebel-240 benchmarks is considered individually.

The predictive modeling results for the ^{239}Pu *inelastic* (n, n') scattering cross section are presented in Figures 4.18 and 4.19. Since the inelastic scattering cross section is anti-correlated to the elastic scattering cross section, an increase in the calibrated elastic scattering cross section is therefore counterbalanced by a decrease in the calibrated inelastic scattering cross section. As depicted in Figure 4.18, the (negative) relative change in the inelastic scattering cross section reaches from 5% to up to 37%.

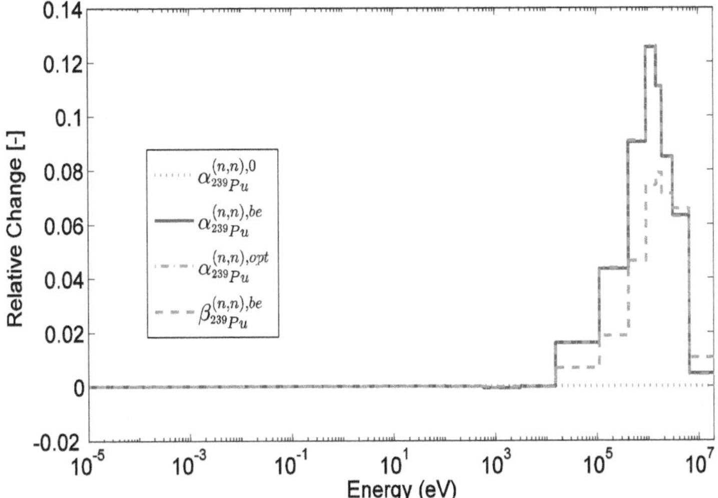

***Figure 4.16:** Relative change in individually (superscript "be") and jointly (superscript "opt") calibrated ^{239}Pu elastic scattering cross section for Jezebel-239 and Jezebel-240*

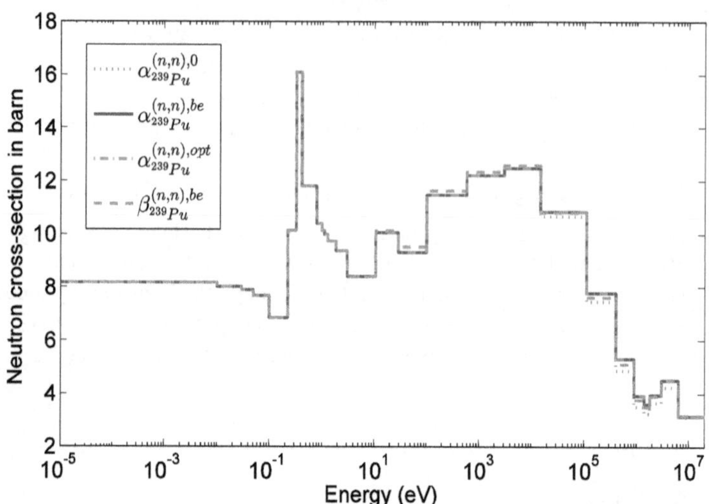

Figure 4.17: ***Nominal value (superscript "0"), individually (superscript "be") and jointly (superscript "opt") calibrated*** ^{239}Pu ***elastic scattering cross section for Jezebel-239 and Jezebel-240***

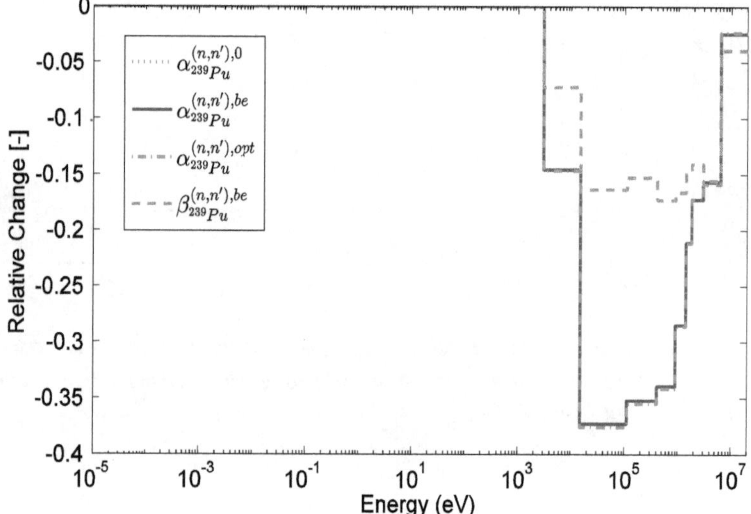

Figure 4.18: ***Relative changes in individually (superscript "be") and jointly (superscript "opt") calibrated*** ^{239}Pu ***inelastic scattering cross section for Jezebel-239 and Jezebel-240***

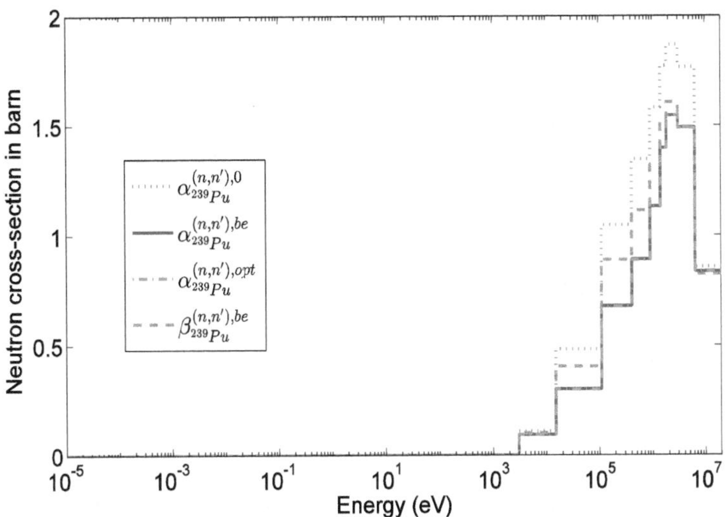

Figure 4.19: Nominal value (superscript "0"), individually and jointly (superscript "opt") adjusted ^{239}Pu ***inelastic scattering cross section for Jezebel-239 and Jezebel-240***

In summary, this Section has presented important applications of the BERRU-PM (predictive modeling) methodology to the predictive modeling of the reactor physics benchmarks Jezebel-239, Jezebel-240 and Godiva. The benchmarks' parameters included individual cross sections for each material, nuclide, reaction type and energy-group, fission spectra, and average number of prompt neutrons emitted per fission, while the responses included the respective benchmarks' effective multiplication factors, various center core fission rate ratios, and various center core radiative capture rate ratios. The BERRU-PM methodology was applied to the predictive modeling of each benchmark individually, to the two Jezebel benchmarks considered as coupled systems, and also to all three benchmarks coupled through various cross section parameters. In all cases, the BERRU-PM methodology yielded best-estimate response and parameter values with reduced predicted uncertainties for each individual benchmark. The results obtained indicate that the interdependence for similar responses (which were measured in more than a single benchmark) is stronger than for dissimilar responses (which were measured in just in a single benchmark); the latter are only marginally affected by the simultaneous predictive modeling. More generally, the consideration of the complete information, including couplings, provided jointly by all three benchmarks (as opposed to consideration of the benchmarks as separate systems) leads to more accurate predictions of nominal values for responses and model parameters, yielding larger reductions in the predicted uncertainties that accompany the predicted mean values of responses and model parameters. The results obtained are consistent with the principles of information theory, which underlies the BERRU-PM methodology: the more consistent information is taken into account, the larger the reduction in the predicted uncertainties, and hence the more precise the predictions. In other words, the more *a priori* consistent information is available, the larger is the uncertainty reduction in the predicted responses and model parameters.

Practical systems usually involve many more parameters than responses. For this reason, the BERRU-PM methodology was developed *ab initio* in the response-space, to reduce (to dimensions dictated by the number of responses) the size of matrices that would necessarily have to be inverted. To reduce even more the memory requirements, the BERRU-PM methodology was specifically developed to avoid inversions of matrices of the size of the total number of responses (i.e., $N_r + N_q$), while achieving identical results by inverting only matrices of dimension N_r and/or N_q (where N_r and N_q denote, respectively, the number of responses considered in the two coupled multi-physics systems). The computations underlying the results presented in this Section benefited by significant reductions (by a factor of about 2.5) in memory requirements when inverting matrices of sizes N_r and/or N_q rather than of size $N_r + N_q$. Notably, *matrix multiplications* and *data transfer* operations (rather than matrix inversions), particularly involving the parameter covariance matrices $\mathbf{C}_{\alpha\alpha}$ or $\mathbf{C}_{\beta\beta}$, may become the operations most expensive in terms of computational resources required for the implementation of the BERRU-PM methodology. Optimizing the trade-offs between the number of flops (floating point operations per second), memory and data transfer requirements for implementing the BERRU-PM methodology is a hardware-dependent issue that varies from application to application.

5. INVERSE BERRU PREDICTIVE MODELING OF RADIATION TRANSPORT IN THE PRESENCE OF COUNTING UNCERTAINTIES

Abstract

Using a paradigm problem of inverse prediction, from detector responses in the presence of counting uncertainties, of the thickness of a homogeneous slab of material containing uniformly distributed gamma-emitting sources, this Chapter presents an investigation of the possible reasons for the apparent failure of the traditional inverse-problem methods based on the minimization of chi-square-type functionals to predict accurate results for optically thick slabs. This Chapter also presents a comparison of the results produced by such traditional methods with the results produced by applying the BERRU methodology, for optically thin and thick slabs. For *optically thin* slabs, the results presented in this Chapter show that both the traditional chi-square-minimization method and the BERRU methodology predict the slab's thickness accurately. However, the BERRU methodology is considerably more efficient computationally, and a single application of the BERRU methodology predicts the thin slab's thickness at least as precisely as the traditional chi-square-minimization method, even though the measurements used in the BERRU methodology were ten times less accurate than the ones used for the traditional chi-square-minimization method. For *optically thick* slabs, however, the results presented in this Chapter show that:

(i) The traditional inverse-problem methods based on the minimization of chi-square-type functionals fail to predict the slab's thickness.

(ii) The BERRU methodology under-predicts the slab's actual physical thickness when imprecise experimental results are assimilated, even though the predicted responses agrees within the imposed error criterion with the experimental results; (iii) The BERRU methodology correctly predicts the slab's actual physical thickness when precise experimental results are assimilated, while also predicting the physically correct response within the selected precision criterion.

(iv) The BERRU methodology is vastly more efficient computationally, while yielding significantly more accurate results, than the traditional chi-square-minimization methodology.

(v) The accuracy of the results predicted by using the BERRU methodology in the "inverse predictive" mode is limited more by the precision of the measurements, rather than by the BERRU methodology's underlying computational algorithm.

D. G. Cacuci, *BERRU Predictive Modeling*,
https://doi.org/10.1007/978-3-662-58395-1_5

5.1 INTRODUCTION

This Chapter presents the application of the BERRU methodology in the inverse mode to predict the thickness of a homogeneous slab of material containing uniformly distributed gamma-emitting sources by using detector responses affected by counting uncertainties. The material presented in this Chapter is based on the work by Cacuci (2017). The Boltzmann particle and radiation transport equation describes all possible interactions of particles within the host medium while taking into account the medium's detailed material properties and geometry. Most often, the Boltzmann equation is used for solving direct problems, namely to determine the unknown distribution of particles in a medium with known composition and geometry and known locations and magnitudes of all sources of particles. The second, and far more difficult, type of problems are the "inverse problems," in which the Boltzmann equation is used to determine, often from an imperfectly-well known particle distribution, the characteristics of the host medium or characteristics of the sources that have generated the respective particles. In particular, "measurement problems" seek to determine from measurements the properties of the host medium (e.g., composition, geometry, including internal interfaces), or the properties of the source (e.g., strength, location, direction), and/or the size of the medium on its boundaries. Therefore, a "measurement problem" is "inverse" to the "direct problem." Such "inverse" particle transport problems are encountered in fields as diverse as astrophysics (in which one measures the intensity and spectral distribution of light in order to infer properties of stars), nuclear medicine (where radioisotopes are injected into patients and the radiation emitted is used in diagnostics to reconstruct body properties, e.g. tumors), non-destructive fault detection in materials, underground (oil, water) logging, and detection of sensitive materials. Some authors further group such inverse problems into "invasive", when the interior particle distribution is accessible for measurements, as opposed to "non-invasive" ones, in which measurements can be performed only on or outside of the medium's boundaries.

The existence of a solution for an inverse problem is in most cases secured by defining the data space to be the set of solutions to the direct problem. This approach may fail if the data is incomplete, perturbed or noisy. Furthermore, inverse problems involving differential operators are notoriously ill-posed, because the differentiation operator is not continuous with respect to any physically meaningful observation topology. If the uniqueness of a solution cannot be secured from the given data, additional data and/or a priori knowledge about the solution need to be used to restrict the set of admissible solutions. In particular, stability of the solution is the most difficult to ensure and verify. If an inverse problem fails to be stable, then small round-off errors or noise in the data will amplify to a degree that renders a computed solution useless.

As already discussed in Section 2.5 of Chapter 2, the traditional procedures employed to compute approximately the "solution" for an ill-posed problem are called regularization procedures or methods, and are customarily categorized as "explicit" or "implicit". In particle and radiation transport problems, the historically older explicit methods attempt to manipulate the direct transport equation in conjunction with measurements in order to estimate explicitly the unknown source and/or other unknown characteristics of the medium. On the other hand, implicit methods combine measurements with repeated solutions of the direct problem obtained with different values of the unknowns, iterating until an a priori selected functional, usually representing the user-defined "goodness of fit" between measurements and direct computations, is reduced to a value deemed to be "acceptable" by the user. Inverse problems are fundamentally ill-posed and/or ill-conditioned, unstable to uncertainties in the transport model parameters and/or the experimental measurements. These features have increasingly favored the development of implicit methods, which allow, to various degrees, the consideration of uncertainties in the "inverse problem" algorithms. Examples of inverse time-independent radiative transfer problem have been reviewed by McCormick (1992) while examples of inverse source problems for time-independent neutron transport have been provided by Sanchez and McCormick (2007). Neglecting scattering and uncertainties in the underlying cross sections and material properties, Bledsoe et al (2011a, 2011b) used the "differential evolution" and "Levenberg-Marquardt" methods to investigate inverse gamma-ray transport problems by minimizing an a priori chosen chi-square-type functional that estimates the "differences between measured and computed quantities of interest". As discussed by Bledsoe et al (2011a, 2011b), the current state-of-the-art methods are susceptible to being trapped in apparent local minima even for one-dimensional spherical systems.

In this Chapter, the results produced by traditional inverse-problem methods based on the minimization of chi-square-type functionals will be compared with the results produced by applying the BERRU methodology, which was presented in Chapter 2, in the "inverse mode." The paradigm inverse radiation transport problem that will be used for this purpose is the inverse prediction, from detector responses in the presence of counting uncertainties, of the thickness of a homogeneous slab of material containing uniformly distributed gamma-emitting sources. This paradigm problem is presented in Section 5.2, along with the exact results of interest in the absence of counting uncertainties, i.e., when the detector's response is deterministic and exact. Section 5.2 also presents typical results that are produced by traditional methods for solving "inverse problems," which are based on the minimization of a user-defined generalized least-squares-type (chi-square) functional that provides a measure of the discrepancies between the computed and the measured responses in the presence of counting uncertainties. Section 5.3 presents the results of applying the BERRU methodology to the paradigm problem of determining the slab's thickness from detector responses in the presence of counting uncertainties, for the following physical systems (i) optically very thin slabs; (ii) optically thin slab; (iii) optically thick slabs; (iv) optically very thick slabs; (v) optically extremely thick slab, and (vi) prediction limit for single-precision computations. These illustrative examples demonstrating that the BERRU methodology is considerably more efficient computationally and delivers results of vastly superior accuracy by comparison to the traditional chi-square minimization techniques. Section 5.4 summarizes the conclusions drawn from the results presented in this Chapter regarding the application of the BERRU methodology to inverse problems.

5.2 A PARADIGM INVERSE RADIATION TRANSPORT PROBLEM: PREDICTION OF SLAB THICKNESS FROM DETECTOR RESPONSES IN THE PRESENCE OF COUNTING UNCERTAINTIES

Consider a one-dimensional slab of homogeneous material extending from $z=0$ to $z=a\left[cm\right]$, placed in air and characterized by a total interaction coefficient $\mu\left[cm^{-1}\right]$. The slab contains a uniformly distributed source of strength $Q\left[photons/cm^3 \sec\right]$ emitting isotropically monoenergetic photons within the slab. It is assumed that there is no scattering into the energy lines. Under these conditions, the angular flux of photons within the slab is described by the Boltzmann transport equation without scattering and with "vacuum" incoming boundary condition, i.e.,

$$\omega\frac{d\psi(z,\omega)}{dz}+\mu\psi(z,\omega)=\frac{Q}{2},\quad 0<z\leq a,\ \omega>0, \tag{5.1}$$

$$\psi(0,\omega)=0. \tag{5.2}$$

where $\psi(z,\omega)$ denotes the neutron angular flux at position z and direction $\omega\triangleq\cos\theta$, where θ denotes the angle between the photon's direction and the z-axis. The solution of Eqs. (5.1) and (5.2) can be readily obtained as

$$\psi(z,\omega)=\frac{Q}{2\mu}\left[1-\exp(\mu z/\omega)\right]. \tag{5.3}$$

Consider further that the leakage flux of uncollided photons is measured by an "infinite plane" detector placed in air at some location $z>a$ external to the slab. The detector's response function, denoted as $\Sigma_d\left[cm^{-1}\right]$, is considered to be a perfectly well-known constant. If the detection process were a perfectly deterministic process, rather than a stochastic one, it would follow from Eq. (5.3) that the "exact detector response", denoted as $r(\mu a)$, would be given by the expression

$$r(\mu a)\triangleq\Sigma_d\int_0^1\psi(z,\omega)d\omega=\frac{Q\Sigma_d}{2\mu}\left[1-E_2(\mu a)\right], \tag{5.4}$$

where the exponential-integral function is defined as

$$E_n(x)=\int_0^1 u^{n-2}e^{-x/u}du,\quad n=0,1,2,... \tag{5.5}$$

5.2.1 Determination of Slab Thickness from Detector Response in the Absence of Uncertainties

Since the aim of this application is the determination of the slab's optical thickness from detector measurements, the quantities Σ_d, μ, and Q will be considered to be perfectly well known, for simplicity. If the detector were perfect and if its response $r(\mu a)$ were the consequence of an exactly-known deterministic counting process, Eq. (5.4) could be "inverted" to obtain the slab's optical thickness (μa) by solving deterministically the following nonlinear equation:

$$E_2(x) = 1 - \frac{2\mu r(x)}{Q\Sigma_d} \triangleq C, \ x \triangleq \mu a. \tag{5.6}$$

When $r(x)$ is known, the right-side of Eq. (5.6) is a known constant, denoted as C. Since the function $E_1(x)$ is everywhere positive, i.e., $E_1(x) > 0, \ for \ 0 < x < \infty$, it follows that

$$\frac{dE_2(x)}{dx} = -E_1(x) < 0, \ \ 0 < x < \infty. \tag{5.7}$$

The result in Eq. (5.7) indicates that $E_2(x)$ is a monotonically decreasing function of x as $x \geq 0$ increases, and the "amount of decrease" increases as x increases. In other words, the value of $E_2(x)$ decreases monotonically, at an increasingly slower rate, as x increases. Since $E_2(0) = 1$ and $E_2(x) \xrightarrow{x \to \infty} 0$, it follows that $E_2(x)$ will take on at most once each value in the interval $1 \geq E_2(x) = C > 0$ as x increases monotonically in the interval $0 \leq x < \infty$. Hence, despite the fact that the axis $x = 0$ is asymptotically tangent to $E_2(x)$ in the limit when $x \to \infty$, Eq. (5.6) admits just a single real-valued root. Consequently, for each value of $r(\mu a)$, which determines the value of C, there corresponds a single, well-defined, slab optical thickness $\mu a = x$. In other words, Eq. (5.6) does not admit degenerate roots, in the sense that more than one distinct value of the slab's optical thickness $(\mu a = x)$ might correspond to the same value $r(\mu a)$. The fact that Eq.(5.6) admits a single real-valued root is also underscored by recalling the asymptotic expansions for $E_2(x)$, i.e.,:

$$E_2(x) \sim \frac{e^{-x}}{x+2}\left[1+\frac{2}{(x+2)^2}+\frac{2(2-2x)}{(x+2)^4}+\frac{2(6x^2-16x+4)}{(x+2)^6}+\ldots\right] \triangleq A(x), \quad x \triangleq \mu a > 1, \tag{5.8}$$

$$E_2(x) \sim 1+x\left[\log(x)-0.422784\right]-\frac{x^2}{2}+\frac{x^3}{12}-\frac{x^4}{72}+\ldots \triangleq B(x), \quad x \triangleq \mu a < 1. \tag{5.9}$$

The asymptotic expansion in Eq. (5.8) can be used to compute the real-valued root of Eq. (5.6) for $C < 0.8$; (ii) both asymptotic expansions given in Eqs. (5.8) and (5.9) can be used to compute the real-valued root of Eq. (5.6) when $0.2 < C < 0.8$; (iii) the asymptotic expansion in Eq. (5.9) can be used to compute the real-valued root of Eq. (5.6) when $C > 0.2$. The left- and right-sides of the equations

$$A(x) = C, \quad B(x) = C, \tag{5.10}$$

where $A(x)$ and $B(x)$ are defined in Eqs. (5.8) and (5.9), respectively, are plotted in Figure 5.1, in which the quantities $Q = 1\left[photons / cm^3 \sec\right]$, $\Sigma_d = 1\left[cm^{-1}\right]$, $\mu = 1\left[cm^{-1}\right]$ were normalized to unity, for convenience. The intersection of the horizontal line with the decreasing curve depicting the function $E_2(x)$ provides the location of the real root of Eq. (5.6). It is also evident from Eqs. (5.6), (5.8) and (5.9) that in the limit of infinitely thin or infinitely thick slabs, respectively, the corresponding "readings" by perfect detectors would be

$$r(0) = 0, \quad r(\infty) = \frac{Q\Sigma_d}{2\mu}. \tag{5.11}$$

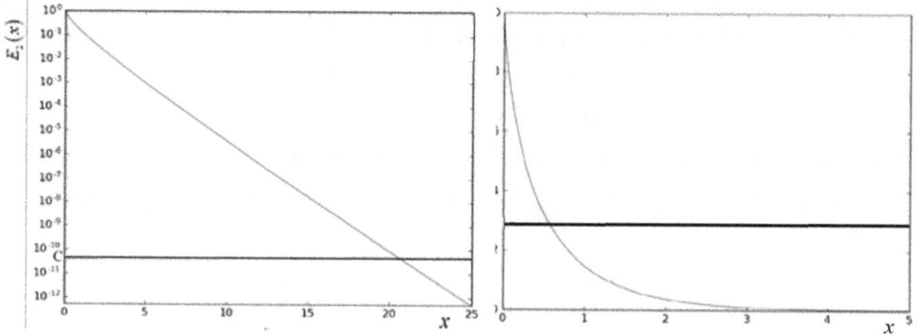

Figure 5.1: Location of the (unique) real root of Eq. (5.6): Left: linear-linear scale; Right: log-linear scale.

5.2.2 *Traditional Chi-Square Minimization Method for Determining the Slab's Thickness from Detector Responses in the Presence of Counting Uncertainties*

It is reasonable to expect that the slab's unknown optical thickness should be obtainable from detector measurements, since the detector measurements implicitly "know" the exact thickness of the slab, which is reflected in the respective number of photons reaching the detector. Also, in the limit of infinite experimental precision and accuracy, the detector response must indicate the exact thickness of the slab, as shown in the previous section. As is well known, the process of detecting photons (as well as other particles) can be described by a Poisson distribution. When a sufficiently large number of events are counted, as is usually the case with photon detection, the respective Poisson distribution can be approximated well by a normal (Gaussian) distribution. For this paradigm example, it suffices to consider that the kth-experimentally-measured response, which will be denoted as $r_{\exp}^{(k)}$, is obtained as a random event drawn from a normal distribution having the mean equal to the exact response, $r(\mu a)$, and the standard deviation equal to $\beta r(\mu a)$, where β is the relative standard deviation (in %), so that

$$r_{\exp}^{(k)} = random\ normal\left[r,\ \beta r\right],\quad k = 1,\dots,K. \tag{5.12}$$

A direct attempt to determine the slab's optical thickness would be by plotting the difference

$$\delta \triangleq \left(r_{\text{model}} - r_{\exp}\right), \tag{5.13}$$

between a random realization of a detector response, $r_{\exp}$, and the "model response," r_{model}, defined from Eq. (5.4) as

$$r_{\text{model}} \triangleq \frac{Q\Sigma_d}{2\mu}\left[1 - E_2\left(\mu a_{\text{model}}\right)\right]. \tag{5.14}$$

Since only counting uncertainties in the detector response will be considered in this work, the quantities Σ_d, μ, and Q will be considered, as in the previous Section, to be perfectly well known and be normalized to unity, i.e., $Q = 1\left[photons / cm^3 \sec\right]$, $\Sigma_d = 1\left[cm^{-1}\right]$, $\mu = 1\left[cm^{-1}\right]$. Figure 5.2 depicts the behavior of the quantity $\delta \triangleq \left(r_{\text{model}} - r_{\exp}\right)$, plotted as a function of μa_{model}, for four values of the actual optical thickness μa (namely: $\mu a = 0.1$, $\mu a = 1.0$, $\mu a = 3.0$ and $\mu a = 10.0$), using software provided by Mattingly (2015). The corresponding detector response, r_{exp}, is considered to be distributed normally with a mean equal to (the exact) r_{model}, and having a relative standard deviation of 1%, i.e., $std.dev\left(r_{\text{exp}}\right)=(0.01)r_{\text{exp}}$. As the plots in Figure 5.2 indicate, for measurements having a relative standard deviation of 1% (i.e., fairly accurate measurements), the "zero-crossings" of the respective differences $\delta \triangleq \left(r_{\text{model}}-r_{\text{exp}}\right)$ are clearly identified for optically thin slabs, as exemplified by the graphs for $\mu a=0.1$ and $\mu a=1$. These zeros also correctly correspond to the values $\mu a_{\text{model}}=0.1$ and $\mu a_{\text{model}}=1$, respectively. On the other hand, for measurements having a relative standard deviation of 1%, the plots corresponding to $\mu a=3.0$ and $\mu a=10.0$ in Figure 5.2 indicate that the "zero-crossings" of the corresponding differences $\delta \triangleq \left(r_{\text{model}}-r_{\text{exp}}\right)$ can no longer be identified beyond about three mean free paths (i.e., $\mu a>3$); the respective "zero-crossings" appear to be multiple-valued, perhaps even degenerate.

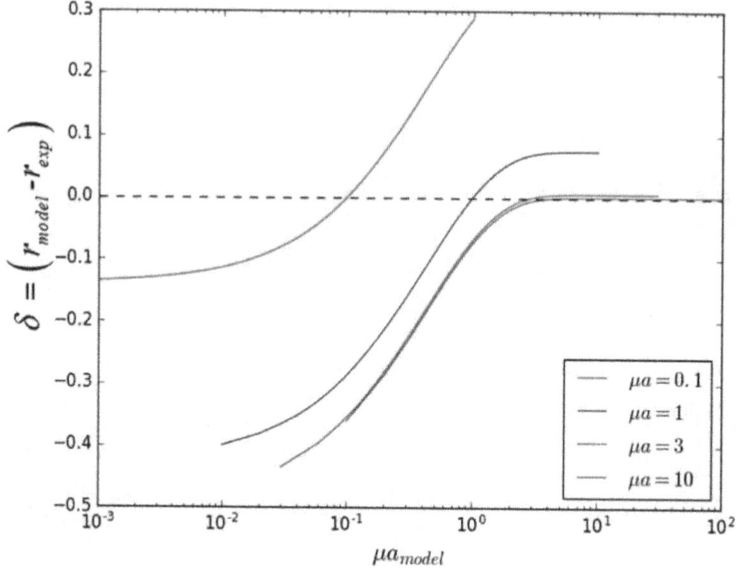

***Figure 5.2:** Variation of the difference between the computed detector response, r_{model}, and a measured (normally distributed, with a relative standard deviation of 1%) detector response, r_{exp}, as a function of the model's optical thickness $\left(\mu a_{\text{model}}\right)$.*

The software provided by Mattingly (2015) can also be used to plot the quantity $\delta^2 \triangleq \left(r_{\text{model}}-r_{\text{exp}}^{(k)}\right)^2$ as a function of the model's optical thickness $\left(\mu a_{\text{model}}\right)$, for various values of the actual optical thickness, μa, and by considering (as before) that the corresponding detector response, r_{exp}, is distributed normally with a mean equal to (the exact) r_{model}. Figures 5.3 through 5.6 present plots of $\delta^2 \triangleq \left(r_{\text{model}} - r_{\text{exp}}^{(k)}\right)^2$ for the same values (namely: $\mu a = 0.1$, $\mu a = 1.0$, $\mu a = 3.0$ and $\mu a = 10.0$) of the actual optical thickness, μa, as considered in Figure 5.2, for ten measurements $r_{\text{exp}}^{(k)}$, k=1,...,10, which are considered to be distributed normally with a mean equal to (the exact response) r_{model} and a relative standard deviation of 1% [i.e., $std.dev\left(r_{\text{exp}}\right) = (0.01)\, r_{\text{exp}}$].

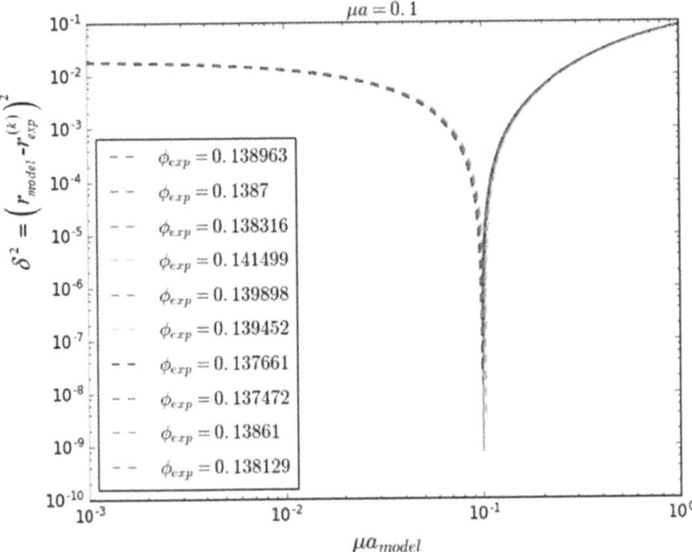

***Figure 5.3:** Variation of $\delta^2 \triangleq \left(r_{\text{model}} - r_{\text{exp}}^{(k)}\right)^2$ as a function of the model's optical thickness (μa_{model}) for a slab of actual optical thickness $\mu a = 0.1$, for measurements with a relative standard deviation of 1%.*

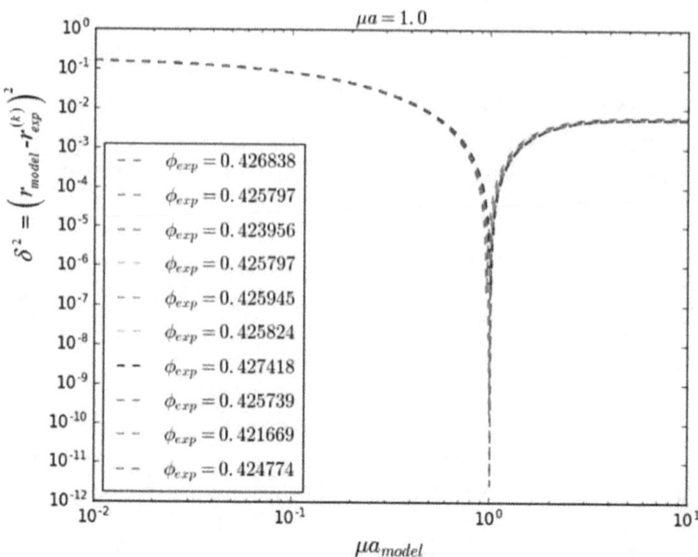

Figure 5.4: Variation of $\delta^2 \triangleq \left(r_{\text{model}} - r_{\text{exp}}^{(k)}\right)^2$ ***as a function of the model's optical thickness*** (μa_{model}) ***for a slab of actual optical thickness*** $\mu a = 1.0$***, for measurements with a relative standard deviation of 1%.***

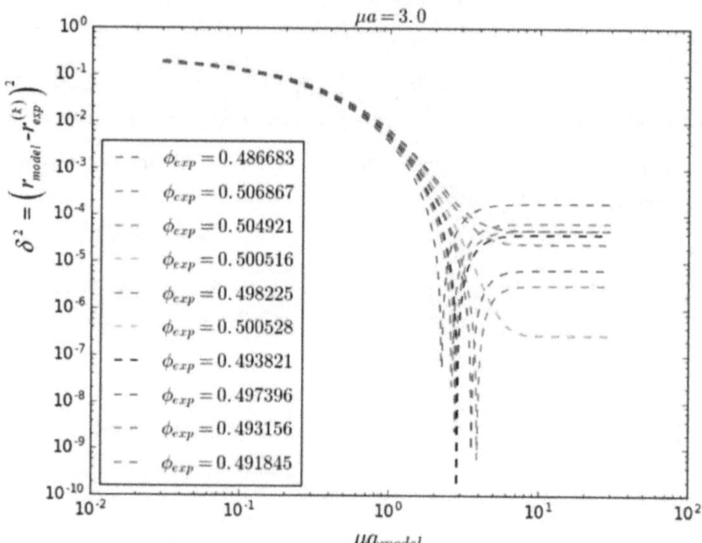

Figure 5.5: Variation of $\delta^2 \triangleq \left(r_{\text{model}} - r_{\text{exp}}^{(k)}\right)^2$ ***as a function of the model's optical thickness*** (μa_{model}) ***for a slab of actual optical thickness*** $\mu a = 3.0$***, for measurements with a relative standard deviation of 1%.***

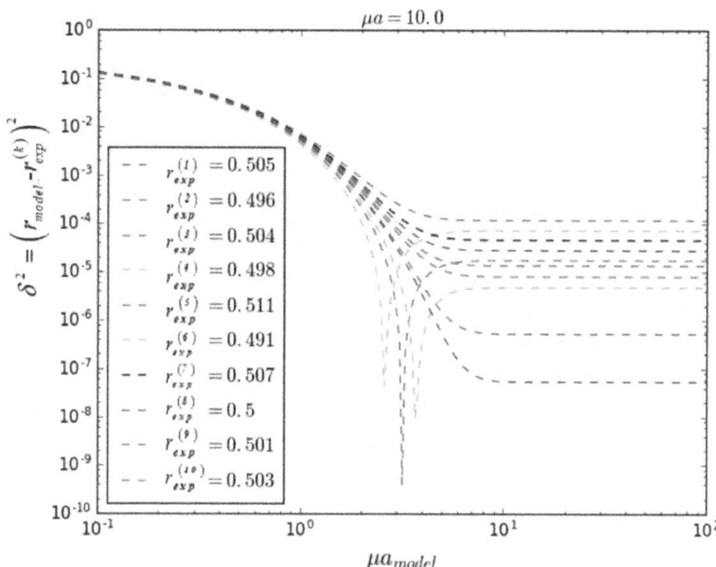

Figure 5.6: Variation of $\delta^2 \triangleq \left(r_{\text{model}} - r_{\text{exp}}^{(k)}\right)^2$ ***as a function of the model's optical thickness*** (μa_{model}) ***for a slab of actual optical thickness*** $\mu a = 10.0$ ***, for measurements with a relative standard deviation of 1%.***

Figures 5.3 and 5.4 indicate that the minimum of the quantity $\delta^2 \triangleq \left(r_{\text{model}} - r_{\text{exp}}^{(k)}\right)^2$ appears to be uniquely corresponding to the actual value of the slab's thickness, irrespective of the precise value of the measurements. In other words, for slabs that are optically thin, the minimum of the quantity $\delta^2 \triangleq \left(r_{\text{model}} - r_{\text{exp}}^{(k)}\right)^2$ is unique, insensitive to the precision of the respective measurements, and identifies the slab's actual optical thickness correctly and accurately.

A very different situation becomes evident in Figure 5.5 for a slab of optical thickness $\mu a = 3.0$: depending on the value of the respective measurement, the corresponding quantity $\delta^2 \triangleq \left(r_{\text{model}} - r_{\text{exp}}^{(k)}\right)^2$ displays a minimum at various locations within the interval $1.0 < \mu a < 4.0$, or may display no minimum at all. The various minima depicted in Figure 5.5 either under-predict or over-predict, in an apparent random fashion, the actual optical slab thickness of $\mu a = 3.0$. Similar conclusions can be drawn from the results depicted in Figure 5.6, for a (thick) slab of optical thickness $\mu a = 10.0$. The results in Figure 5.6 indicate that, depending on the value of the respective measurement, the corresponding quantity $\delta^2 \triangleq \left(r_{\text{model}} - r_{\text{exp}}^{(k)}\right)^2$ displays a minimum at various locations within the interval $1.0 < \mu a < 4.0$, or may display no minimum at all. In this case, however, there are no over-predictions of the slab's correct thickness: all of the minima under-predict, in an apparent random fashion, the actual optical slab thickness $\mu a = 10.0$.

Figures 5.3 and 5.4 have indicated that for optically thin slabs, the precision of measurements does not affect the location of the unique minimum of the quantity $\delta^2 \triangleq \left(r_{\text{model}} - r_{\text{exp}}^{(k)}\right)^2$, and the actual thickness of the respective slab is determined sufficiently accurately (for practical purposes) by the unique location of this minimum. As indicated by the results depicted in Figures 5.5 and 5.6, however, the precision of the measurements decisively affects the results for optically thick slabs. It would be intuitively expected that more precise measurements would yield results "more tightly grouped" around a "better defined" minimum, and hence lead to more accurate predictions of the actual thickness for optically thick slabs. This intuitive expectation is supported by the typical results presented in Figures 5.7 and 5.8 for a thick slab of actual optical thickness $\mu a = 10.0$. The results Figure 5.7 correspond to measurements following a normal distribution with a mean equal to (the exact response) r_{model} and a relative standard deviation of 10%. The results presented in Figure 5.8 are for extremely precise measurements assumed to be normally distributed with a mean equal to (the exact response) r_{model} and having a relative standard deviation of 0.001%.

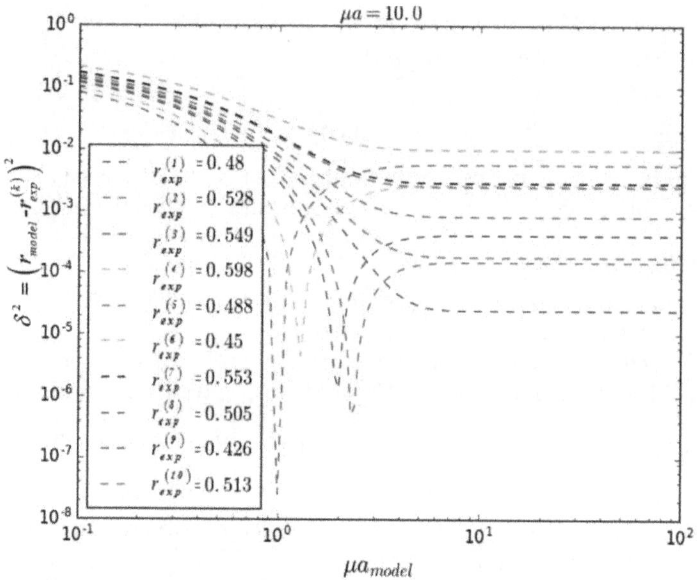

***Figure 5.7:** Variation of $\delta^2 \triangleq \left(r_{\text{model}} - r_{\text{exp}}^{(k)}\right)^2$ as a function of the model's optical thickness (μa_{model}) for a slab of actual optical thickness $\mu a = 10.0$, for measurements with a relative standard deviation of 10%.*

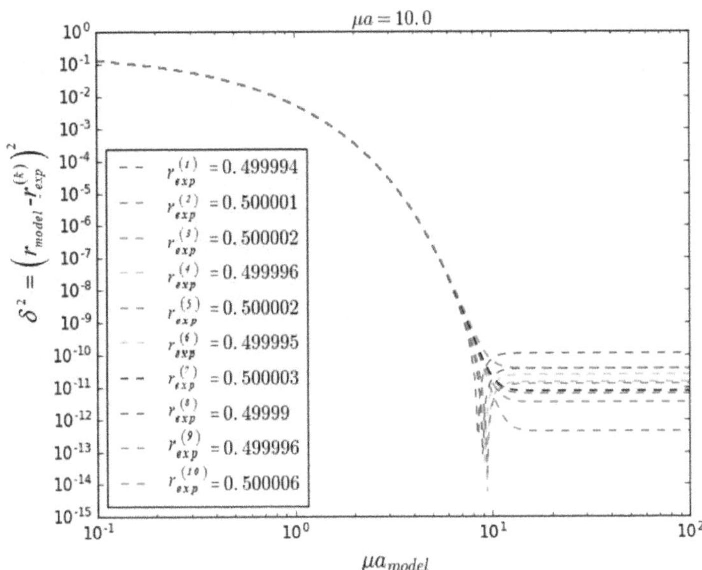

Figure 5.8: Variation of $\delta^2 \triangleq \left(r_{\text{model}} - r^{(k)}_{\text{exp}}\right)^2$ ***as a function of the model's optical thickness*** (μa_{model}) ***for a slab of actual optical thickness*** $\mu a = 10.0$***, for extremely precise measurements with a relative standard deviation of 0.001%.***

Comparing the results depicted in Figure 5.7 with those depicted in Figure 5.6 shows that the quantity $\delta^2 \triangleq \left(r_{\text{model}} - r^{(k)}_{\text{exp}}\right)^2$ corresponding to the less precise measurements (relative standard deviation of 10% for Figure 5.6) displays either no minima, or minima that depend sensitively on the individual measurements, just as displayed by the results for the more precise measurements (relative standard deviation of 1%) presented in Figure 5.7. Furthermore, the minima displayed by the less precise measurements (in Figure 5.7) fall within the interval $1.0 < \mu a < 3.0$, thus being even less indicative of the correct slab thickness than the indication provided by the more precise measurements (in Figure 5.6). This conclusion is further strengthened by comparing all of the results presented in Figures 5.6, 5.7 and 5.8 for a slab of optical thickness $\mu a = 10.0$, namely that the quantity $\delta^2 \triangleq \left(r_{\text{model}} - r^{(k)}_{\text{exp}}\right)^2$ may display no minimum for some measurements, and when it does display a minimum, they respective minimum depends sensitively on the respective measurements. Furthermore, the more accurate the measurements (i.e., the smaller the respective standard deviations), the tighter together grouped are the measurement values; hence, the minima of the squared-differences δ^2 corresponding to the respective measurements are "grouped" more tightly together, and the respective "group of minima" is closer to the correct slab thickness.

As has been mentioned in Section 5.1, the current state-of-the-art methods for determining the optical dimension of a uniform homogeneous medium from K uncertain photon measurements external to the medium rely on minimizing a user-defined chi-square-type functional of the following form:

$$\chi^2 \triangleq \sum_{k=1}^{K} \left[\frac{r_{\text{model}}\left(\mu a_{\text{model}}^{(k)}\right) - r_{\text{exp}}^{(k)}}{std.dev\left(r_{\text{exp}}\right)} \right]^2 . \tag{5.15}$$

The value $(\mu a)_{\min}$ which yields the minimum value, $\chi^2_{\min}$, of χ^2 is considered to be the slab's optical thickness. Since, as shown in Figures 5.5 through 5.8, some of the summands in Eq. (5.15) may not admit any real-valued minimum, while the other summands have minima that do not coincide with one another, it is not surprising that a numerical algorithm for minimizing the χ^2-functional may yield some minimum value that has no physical meaning, in that the actual physical slab thickness would differ from the value $(\mu a)_{\min}$. On the other hand, in the absence of counting uncertainties, the detector's response yields a unique slab thickness, as demonstrated in Section 5.2.1. If the measurements are inaccurate, then any minimization of the expression in Eq. (5.15) will lead to erroneous physical results, in that the result delivered by any minimization procedure will not be physically correct. Furthermore, for equally precise measurements, the larger the optical thickness of the slab, the more unphysical will likely be the result of the minimization procedure. Altogether, therefore, the results presented in this Section indicate that the reason for the failure of the current state-of-the-art methods to predict accurately the actual thickness of optically thicker slabs stems not from the numerical method used to minimize the χ^2- functional, but stems from the very formulation of the χ^2-functional, which makes this functional to be extremely sensitive to the random value of each measurement. In the next Section, it will be shown that the BERRU methodology, which incorporates considerably more features of the model than the methods based on minimizing a user-defined χ^2-functional, alleviates the shortcomings of the latter methods, yielding results that are physically accurate up to machine precision.

5.3 APPLYING THE BERRU METHODOLOGY FOR THE INVERSE DETERMINATION OF SLAB THICKNESS IN THE PRESENCE OF COUNTING UNCERTAINTIES

For the paradigm system consisting of the slab and detector considered in this Chapter, "Model B" reduces to a point (i.e., the point detector), so the BERRU relations relevant to this system are (the BERRU-SMS results) as follows:

$$\boldsymbol{\alpha}^{pred} = \boldsymbol{\alpha}^0 - \left(\mathbf{C}_{\alpha\alpha}\mathbf{S}^{\dagger}_{r\alpha} - \mathbf{C}_{\alpha r}\right)\left[\mathbf{D}_{rr}\right]^{-1}\mathbf{r}^d\left(\boldsymbol{\alpha}^0\right), \tag{5.16}$$

$$\mathbf{r}^{pred} = \mathbf{r}^m - \left(\mathbf{C}^{\dagger}_{\alpha r}\mathbf{S}^{\dagger}_{r\alpha} - \mathbf{C}_{rr}\right)\left[\mathbf{D}_{rr}\right]^{-1}\mathbf{r}^d\left(\boldsymbol{\alpha}^0\right), \tag{5.17}$$

$$\mathbf{C}^{pred}_{\alpha\alpha} = \mathbf{C}_{\alpha\alpha} - \left(\mathbf{C}_{\alpha\alpha}\mathbf{S}^{\dagger}_{r\alpha} - \mathbf{C}_{\alpha r}\right)\left[\mathbf{D}_{rr}\right]^{-1}\left(\mathbf{C}_{\alpha\alpha}\mathbf{S}^{\dagger}_{r\alpha} - \mathbf{C}_{\alpha r}\right)^{\dagger}, \tag{5.18}$$

$$\mathbf{C}^{pred}_{rr} = \mathbf{C}_{rr} - \left(\mathbf{C}^{\dagger}_{\alpha r}\mathbf{S}^{\dagger}_{r\alpha} - \mathbf{C}_{rr}\right)\left[\mathbf{D}_{rr}\right]^{-1}\left(\mathbf{C}^{\dagger}_{\alpha r}\mathbf{S}^{\dagger}_{r\alpha} - \mathbf{C}_{rr}\right)^{\dagger}, \tag{5.19}$$

$$\mathbf{C}^{pred}_{\alpha r} = \mathbf{C}_{\alpha r} - \left(\mathbf{C}_{\alpha\alpha}\mathbf{S}^{\dagger}_{r\alpha} - \mathbf{C}_{\alpha r}\right)\left[\mathbf{D}_{rr}\right]^{-1}\left(\mathbf{C}^{\dagger}_{\alpha r}\mathbf{S}^{\dagger}_{r\alpha} - \mathbf{C}_{rr}\right)^{\dagger}. \tag{5.20}$$

$$\chi^2 = \left(\mathbf{r}^c - \mathbf{r}^m\right)^{\dagger}\left[\mathbf{D}_{rr}\right]^{-1}\left(\mathbf{r}^c - \mathbf{r}^m\right). \tag{5.21}$$

where

$$\mathbf{D}_{rr} \triangleq \mathbf{C}_{rc} - \mathbf{S}_{r\alpha}\mathbf{C}_{\alpha r} - \mathbf{C}^{\dagger}_{\alpha r}\mathbf{S}^{\dagger}_{r\alpha} + \mathbf{C}_{rr}, \tag{5.22}$$

and where the "computed response covariance matrix", $\mathbf{C}_{rc}$, is defined as

$$\mathbf{C}_{rc} \triangleq \mathbf{S}_{r\alpha}\mathbf{C}_{\alpha\alpha}\mathbf{S}^{\dagger}_{r\alpha}\,. \tag{5.23}$$

Note that if the model is perfect (i.e., $\mathbf{C}_{\alpha\alpha} = \mathbf{0}$ and $\mathbf{C}_{\alpha r} = \mathbf{0}$), then Eqs. (5.16) through (5.20) would yield $\boldsymbol{\alpha}^{pred} = \boldsymbol{\alpha}^0$ and $\mathbf{r}^{pred} = \mathbf{r}^c\left(\boldsymbol{\alpha}^0, \boldsymbol{\beta}^0\right)$, predicted "perfectly," without any accompanying uncertainties (i.e., $\mathbf{C}^{pred}_{rr} = \mathbf{0}$, $\mathbf{C}^{pred}_{\alpha\alpha} = \mathbf{0}$, $\mathbf{C}^{pred}_{\alpha r} = \mathbf{0}$). In other words, for a perfect model, the BERRU methodology predicts values for the responses and the parameters that coincide with the model's values (assumed to be perfect), and the experimental measurements would have no effect on the predictions (as would be expected, since imperfect measurements could not possibly improve the "perfect" model's predictions). On the other hand, if the measurements were perfect, (i.e., $\mathbf{C}_{rr} = \mathbf{0}$ and $\mathbf{C}_{\alpha r} = \mathbf{0}$), but the model were imperfect, then Eqs. (5.16) through (5.20) would yield $\boldsymbol{\alpha}^{pred} = \boldsymbol{\alpha}^{0} - \mathbf{C}_{\alpha\alpha}\mathbf{S}_{r\alpha}^{\dagger}\left[\mathbf{S}_{r\alpha}\mathbf{C}_{\alpha\alpha}\mathbf{S}_{r\alpha}^{\dagger}\right]^{-1}\mathbf{r}^{d}\left(\boldsymbol{\alpha}^{0}\right)$, $\mathbf{C}_{\alpha\alpha}^{pred} = \mathbf{C}_{\alpha\alpha} - \mathbf{C}_{\alpha\alpha}\mathbf{S}_{r\alpha}^{\dagger}\left[\mathbf{S}_{r\alpha}\mathbf{C}_{\alpha\alpha}\mathbf{S}_{r\alpha}^{\dagger}\right]^{-1}\mathbf{S}_{r\alpha}\mathbf{C}_{\alpha\alpha}$, $\mathbf{r}^{pred} = \mathbf{r}^{m}$, $\mathbf{C}_{rr}^{pred} = \mathbf{0}$, $\mathbf{C}_{\alpha r}^{pred} = \mathbf{0}$. In other words, in the case of perfect measurements, the BERRU predicted values for the responses would coincide with the measured values (assumed to be perfect), but the model's uncertain parameters would be calibrated by taking the measurements into account to yield improved nominal values and reduced parameters uncertainties.

When Eqs. (5.16) through (5.20) are employed for forward predictive modeling, all of the quantities on the right sides of these equations are known, and the best-estimate predicted quantities are those on the left-side of the respective equations. Note that the detector measures, albeit statistically, the exact response, which implicitly comprises the information about the exact optical thickness of the medium under investigation. Each measured response represents a "point" or "element" sampled from the counting statistical distribution characterizing the detected particles (photons). For simplicity and without loss of generality, the counting statistics are considered to be Gaussian, so that each measured detector response, $r_m^{(k)}$, has the value $r_m^{(k)} = random\ normal\left[r^{exact},\ sd\left(r_m^{(exact)}\right)\right]$, $k = 1, \ldots, K_n$, where K_n denotes the total number of experiments performed in the "batch n". On the other hand, when Eqs. (5.16) through (5.20) are employed for inverse predictive modeling, the set of parameters $\boldsymbol{\alpha}^{0}$ are unknown, so the first set of measurements is used to provide a preliminary (initial) estimate of these parameter values. Subsequent measurements are assimilated to improve the previous predictions of both the response and parameter values, until the last predicted values for the response and/or parameter satisfy some a priori imposed accuracy criteria. The detailed inverse predictive algorithm is as follows:

I. Perform the initial set of measurements

1. Perform the *initial* set of measurements, $r_{\exp}^{(k)}$, by drawing random results from the normal distribution $r_m^{(k)} = random\ normal\left[r^{exact},\ \beta r^{exact}\right]$, $k = 1, \ldots, K_0$.

2. Compute the *initial* "sample average": $r_{m,ave}^{(K_0)} = \dfrac{1}{K_0}\sum_{k=1}^{K_0} r_m^{(k)}$.

3. Compute the initial "measurement variance": $C_m^{(K_0)} = \dfrac{1}{K_0 - 1}\sum_{k=1}^{K_0}\left[r_m^{(k)} - r_{m,ave}^{(K_0)}\right]^2$.

4. Compute the *initial* "sample standard deviation" $SD_m^{(K_0)} = \sqrt{C_m^{(K_0)}}$.

5. Compute the *initial* estimated parameter value $\alpha^{(1)}$ by using the model, i.e., by solving the nonlinear equation $E_2\left[\alpha^{(1)}\right]=1-\frac{2\mu r_{m,ave}^{(K_0)}}{Q\Sigma_d}$.

6. Compute the *initial* sensitivities of the response to the uncertain (unknown) model parameter. In general, this is performed by using the adjoint sensitivity analysis methodology (Cacuci, 1981a, 1981b). For the paradigm problem under consideration, the only uncertain model parameter is the medium's optical thickness, so the detector response's sensitivity is readily obtained as: $S^{(1)}=\frac{Q\Sigma_d}{2\mu}E_1\left[\alpha^{(1)}\right]$.

7. Define the *initial* parameter standard deviation: $sd\left(\alpha^{(1)}\right)=\gamma\alpha^{(1)}$ and the variance $C_\alpha^{(1)}=\left[\gamma\alpha^{(1)}\right]^2$. The effects of this *initial* parameter standard deviation" can be assessed by considering various values for γ. In this study, however, the fixed value $\gamma=10^{-1}$ has been used throughout.

8. Use Eq. (5.23) to compute the *initial* computed response covariance: $C_{rc}^{(1)}=S^{(1)\dagger}C_\alpha^{(1)}S^{(1)}$.

9. Assuming, in the absence of information to the contrary, that the measured responses are uncorrelated to the model parameters (in this case: the slab's optical thickness), use Eq. (5.22) to compute the following initial value: $D_r^{(1)}=C_r^{(1)}+C_m^{(K_0)}$

10. Use Eq. (5.20) to compute the *initial* parameter response covariance: $C_{\alpha r}^{(1)}=C_\alpha^{(1)}S^{(1)\dagger}\left[D_r^{(1)}\right]^{-1}C_m^{(K_0)}$.

11. Since the *initial* parameter value was computed by solving the inverse problem using the "average measurement", set the *initial* computed response value to be the same as the *initial* measurement: $r_{comp}^{(1)}=r_{m,ave}^{(K_0)}$.

II. Commence performing experiments to be used for the "inverse predictive modeling" computations

1. Perform $n=1,\ldots,N$ sets of measurements, $r_m^{(k)}$, $k=1,\ldots,K_n$, , by sampling from the normal distribution $r_m^{(k)}=random\ normal\left[r^{exact},\ sd\left(r_m^{(exact)}\right)\right]$.

2. For each set of experiments, K_n, compute the following quantities:

   (a) the "sample average": $r_{m,ave}^{(K_n)}=\frac{1}{K_n}\sum_{k=1}^{K_n}r_m^{(k)}$;

(b) the "measurement variance": $C_m^{(K_n)} = \frac{1}{K_1 - 1}\sum_{k=1}^{K_1}\left[r_m^{(k)} - r_{m,ave}^{(K_n)}\right]^2$;

(c) the "sample standard deviation" $SD_m^{(K_n)} = \sqrt{C_m^{(K_n)}}$;

(d) the measured response, $r_{meas}^{(n)} \equiv r_{m,ave}^{(K_n)}$, and its covariance $C_{meas}^{(n)} \equiv C_m^{(K_n)}$.

3. Use Eq. (5.16) to compute the new "predicted response" values:

$$r_{pred}^{(n+1)} = r_{meas}^{(n)} + \left[C_{meas}^{(n)} - C_{\alpha r}^{(n)\dagger} S^{(n)\dagger}\right]\left[D_r^{(n)}\right]^{-1}\left[r_{comp}^{(n)} - r_{meas}^{(n)}\right];$$

4. Use Eq. (5.17) to compute the new "predicted parameter" values:

$$\alpha_{pred}^{(n+1)} = \alpha^{(n)} + \left[C_{\alpha r}^{(n)} - C_{\alpha}^{(n)} S^{(n)\dagger}\right]\left[D_r^{(n)}\right]^{-1}\left[r_{comp}^{(n)} - r_{meas}^{(n)}\right];$$

5. Use Eq.(5.18) to compute the new "predicted parameter covariances:

$$C_{\alpha}^{(n+1)} = C_{\alpha}^{(n)} - \left[C_{\alpha}^{(n)} S^{(n)\dagger} - C_{\alpha r}^{(n)}\right]\left[D_r^{(n)}\right]^{-1}\left[C_{\alpha}^{(n)} S^{(n)\dagger} - C_{\alpha r}^{(n)}\right]^{\dagger};$$

6. Use Eq. (5.19) to compute the new "predicted response covariances":

$$C_{r,pred}^{(n+1)} = C_{meas}^{(n)} - \left[C_{\alpha r}^{(n)\dagger} S^{(n)\dagger} - C_{meas}^{(n)}\right]\left[D_r^{(n)}\right]^{-1}\left[C_{\alpha r}^{(n)\dagger} S^{(n)\dagger} - C_{meas}^{(n)}\right]^{\dagger} \text{ with } C_{\alpha r}^{(n)} \neq 0$$

7. Use Eq. (5.20) to compute the new "predicted response-parameter covariances":

$$C_{\alpha r}^{(n+1)} = C_{\alpha r}^{(n)} - \left[C_{\alpha}^{(n)} S^{(n)\dagger} - C_{\alpha r}^{(n)}\right]\left[D_r^{(n)}\right]^{-1}\left[C_{\alpha r}^{(n)\dagger} S^{(n)\dagger} - C_{meas}^{(n)}\right]^{\dagger}$$

8. Use Eq. (5.21) to compute the "validation metric" or predicted "consistency indicator" (CI):

$$(CI)^{n+1} \triangleq \left(\chi^2\right)^{n+1} = \left[r_{comp}^{(n)} - r_{meas}^{(n)}\right]^{\dagger}\left[D_r^{(n)}\right]^{-1}\left[r_{comp}^{(n)} - r_{meas}^{(n)}\right]$$

9. Optionally: to quantify the possible effects of nonlinearities, perform the new $(n+1)^{th}$ computation with the "calibrated model parameters":

$$r_{comp}^{(n+1)} = \frac{Q\Sigma_d}{2\mu}\left[1 - E_2\left(\alpha_{pred}^{(n+1)}\right)\right];$$

$$S^{(n+1)} = \frac{Q\Sigma_d}{2\mu} E_1\left(\alpha_{pred}^{(n+1)}\right);$$

$$C_{rc}^{(n+1)} = S^{(n+1)\dagger} C_{\alpha}^{(n+1)} S^{(n+1)};$$

$$\alpha^{(n+1)} \equiv \alpha_{pred}^{(n+1)}$$

Note: the recomputed matrix $C_{rc}^{(n+1)}$ may differ from $C_{r,pred}^{(n+1)}$ because of model non-linearities; the latter matrix is used as the current best-estimate for the covariance matrix of the experimental measurements, to compute the matrix below.

III. Perform the next set of measurements

1. Prepare for the next batch of experiments by using computing the quantity

$$D_r^{(n+1)} = C_r^{(n+1)} - S^{(n+1)} C_{\alpha r}^{(n+1)} - C_{\alpha r}^{(n+1)\dagger} S^{(n+1)\dagger} + C_{r,pred}^{(n+1)};$$

2. Stop when $\left|\dfrac{r_{comp}^{(n+1)} - r_{pred}^{(n+1)}}{r_{comp}^{(n+1)}}\right| < \varepsilon$.

Recall that the experimentally measured detector results reflect the physics of the situations in that the experimental results represent random realizations of a distribution that has the exact response, r^{exact}, as its mean. Thus, the detector results embody (i.e., "know") the exact slab thickness, even though this thickness is unknown to the experimentalist who is attempting to determine it from the model and the experimental results, using the BERRU methodology described in the previous Section. Since the successively predicted responses contain directly the effects of all of the measured responses (which reflect the actual physics of the problem) while the successively computed responses contain indirectly the effects of the successively predicted slab thicknesses, the convergence stopping criterion for the BERRU iterations is imposed on the convergence between the predicted and computed responses, rather than on the convergence of the computationally predicted slab optical thickness. It is logical to strive towards attaining agreement between computational results and experimental measurements as directly as possible, whenever possible.

For the illustrative examples to follow, the distribution of response measurements is considered to be the normal distribution with mean equal to r^{exact}, and with relative standard deviation β, the value of which will be varied to study its influence on the accuracy of the prediction of the unknown optical thickness of the slab under consideration. Simulated experimental results drawn from a normal distribution with a relative standard deviation of 10% $(\beta = 10^{-1})$ will be considered to be "imprecise;" the experimental results drawn from a normal distribution with a relative standard deviation of 0.1% $(\beta = 10^{-3})$ will be considered as being "precise;" and the experimental results drawn from a normal distribution with a relative standard deviation of 0.001% $(\beta = 10^{-5})$ will be considered as being "very precise."

5.3.1 Prediction of Optically Very Thin Slab (Exact Optical Thickness: $\mu a = 0.1$)

a) Imprecise measurements $(\beta = 10^{-1})$

The exact detector response stemming from a slab of optical thickness $\mu a = 0.1$ is $r^{exact} = 1.387275 \times 10^{-1}\, photons\, /\, cm^2 sec$, as shown in the last row of Table 5.1. Consider a set $K_1 = 100$ of imprecise measurements, characterized by a relative standard deviation $\beta = 10\%$, drawn from a random normal distribution with the mean taken to be the exact response, r^{exact}. The results predicted by the BERRU methodology are: (i) the "predicted response value"; (ii) the "predicted response standard deviation"; (iii) the "predicted slab thickness (parameter)"; and (iv) the "predicted standard deviation of the slab thickness". These results are shown in columns 2 through 5 of Table 5.1. It is seen that the first $(n = 1)$ set of imprecise measurements predicts the exact response within a standard deviation of $0.01\, photons\, /\, cm^2 sec$, and the exact optical slab thickness within a standard deviation of 8.89×10^{-3}. Assimilating the second $(n = 2)$ set of 100 measurements, which are just as imprecise as the first set, nevertheless improves even further the prediction of the exact response and slab thickness while reducing even further the respective standard deviations. This reduction in the predicted standard deviations that accompany the predicted response and parameter (slab thickness), respectively, is a consequence of the properties of the BERRU methodology.

Table 5.1: Results predicted by BERRU methodology for a slab of exact thickness $\mu a = 0.1$ after successively assimilating 2 batches of 100 imprecise experiments

$\mu a = 0.1$; $\beta = 10^{-1}$; $\varepsilon = 10^{-3}$; $K_n = 100$;

$Measured\ response = normal\left(r^{exact}, \beta r^{exact}\right)$

N	Experimental Response Mean Value	Predicted Response	Predicted Response SD	Predicted Parameter	Predicted Parameter SD
1	1.405016 x10^{-1}	1.395441 x10^{-1}	1.002145 x10^{-2}	9.98860 x10^{-2}	8.896069 x10^{-3}
2	1.401943 x10^{-1}	1.389812 x10^{-1}	7.129884 x10^{-3}	1.00278 x10^{-1}	7.818043 x10^{-4}
		Exact Response	Exact Response SD	Exact Parameter	
		1.387275x 10^{-1}	1.387275 x10^{-2}	0.1	

b) *Very precise measurements* $\left(\beta = 10^{-5}\right)$

Consider a set $K_1 = 100$ of very precise measurements (relative standard deviation $\beta = 10^{-5}$) drawn from the same random normal distribution, i.e., with the distribution's mean taken to be $r^{exact} = 1.387275 \times 10^{-1}\, photons\, /\, cm^2 sec$. Using these very precise measurements, the BERRU methodology predicts the exact response value within a standard deviation of 1.3×10^{-6} and the slab thickness within a standard deviation of 2×10^{-6}, respectively, as shown in Table 5.2. These results clearly indicate the important consequences of precise measurements, which enable the BERRU methodology to produce considerably more precise predictions than when less precise experiments are assimilated.

Table 5.2: Results predicted by BERRU methodology for a slab of exact thickness $\mu a = 0.1$ after assimilating one batch of 100 very precise experiments ($\beta = 10^{-5}$)

$\mu a = 0.1$; $\beta = 10^{-5}$; $\varepsilon = 10^{-8}$; $K_n = 100$;

$Measured\ response = normal\left(r^{exact}, \beta r^{exact}\right)$

N	Experimental Response Mean Value	Predicted Response	Predicted Response SD	Predicted Parameter	Predicted Parameter SD
1	1.387274 x10^{-1}	1.387277 x10-1	1.303206 x10^{-6}	1.00002 x10^{-1}	2.003542 x10^{-6}
		Exact Response	Exact Response SD	Exact Parameter	
		1.387275x 10^{-1}	1.387275 x10^{-2}	0.1	

The results presented in Table 5.2 indicate that a single application of the BERRU methodology using very precise measurements predicts the slab thickness within 6 significant digits. The response is also predicted within 6 significant digits. The measurements' precision is the most important factor that affects the accuracy of the prediction of the slab's thickness using the BERRU methodology.

5.3.2 Prediction of Optically Thin Slab (Exact Optical Thickness $\mu a = 1.0$)

a) Measurements with 10% relative standard deviation $\left(\beta = 10^{-1}\right)$

Consider a set $K_1 = 100$ of rather imprecise measurements (relative standard deviation $\beta = 10^{-1}$) drawn from the random normal distribution with the mean taken to be the exact response $r^{exact} = 4.257522 \times 10^{-1}\, photons / cm^2 sec$. The results predicted by the BERRU methodology are presented in columns 2 through 5 of Table 5.3. It is seen that the first $(n = 1)$ set of imprecise measurements predicts the exact response within a standard deviation of 2.85×10^{-2}, and the exact optical slab thickness is predicted within a standard deviation of 9.65×10^{-2}. As expected from the properties of the BERRU methodology, the assimilation of the second $(n = 2)$ set of 100 measurements further improves the prediction of the exact response and slab thickness and reduces further the respective standard deviations, even though the second set of experiments is just as imprecise as the first set.

Table 5.3: Results predicted by BERRU methodology for a slab of exact thickness $\mu a = 1$ after assimilating two batches of 100 experiments with $\beta = 10^{-1}$.

$\mu a = 1$; $\beta = 10^{-1}$; $\varepsilon = 10^{-3}$; $K_n = 100$;

$Measured\ response = normal\left(r^{exact}, \beta r^{exact}\right)$

N	Experimental Response Mean Value	Predicted Response	Predicted Response SD	Predicted Parameter	Predicted Parameter SD
1	4.311969 x10^{-1}	4.276684 x10^{-1}	2.854158 x10^{-2}	9.867036 x10^{-1}	9.655633 x10^{-2}
2	4.302537 x10^{-1}	4.245931 x10^{-1}	1.054078 x10^{-2}	9.895185 x10^{-1}	9.397102 x10^{-2}
		Exact Response	Exact Response SD	Exact Parameter	
		4.257522 x10^{-1}	4.257522 x10^{-2}	1.00	

b) Measurements with 0.001% relative standard deviation $\left(\beta = 10^{-5}\right)$

Consider a set $K_1 = 100$ of precise measurements (relative standard deviation $\beta = 10^{-5}$) drawn from the same random normal distribution, with $r^{exact} = 4.257522 \times 10^{-1}\ photons / cm^2 sec$ as the distribution's mean. As shown in Table 5.4, using these precise measurements, the BERRU methodology predicts the response within 7 significant digits. These results indicate, as before, the important consequences of precise measurements, which enable the BERRU to produce considerably more precise predictions than when less precise experiments are assimilated.

Table 5.4: Results predicted by BERRU methodology for a slab of exact thickness $\mu a = 1$ ***after assimilating one batch of 100 experiments with*** $\beta = 10^{-5}$.

$$\mu a = 1; \beta = 10^{-5}; \ \varepsilon = 10^{-8} \ ; K_n = 100 ;$$

$$Measured\ response = normal\left(r^{exact}, \beta r^{exact}\right)$$

N	Experimental Response Mean Value	Predicted Response	Predicted Response SD	Predicted Parameter	Predicted Parameter SD
1	4.257528 x10^{-1}	4.257528 x10^{-1}	3.999515 x10^{-6}	1.000005	5.106484 x10^{-6}
		Exact Response	Exact Response SD	Exact Parameter	
		4.257522 x10^{-1}	4.257522 x10^{-6}	1.00	

The results presented in Table 5.4 indicate that a single application of the BERRU methodology using very precise measurements predicts the slab thickness within 6 significant digits. The response is also predicted within 6 significant digits. Once again, the measurements' precision is the most important factor that affects the accuracy of the prediction of the slab's thickness using the BERRU methodology.

5.3.3 *Prediction of Optically Thick Slab (Exact Optical Thickness=3.0)*

a) Measurements with 10% relative standard deviation $\left(\beta = 10^{-1}\right)$

Consider a set $K_1 = 100$ of rather imprecise measurements (relative standard deviation $\beta = 10^{-1}$) drawn from the random normal distribution with the mean taken to be the exact response $r^{exact} = 4.94679 \times 10^{-1}\, photons\,/\,cm^2 sec$. The results predicted by the BERRU methodology are shown in columns 2 through 5 of Table 5.5. It is seen that the first $(n = 1)$ set of imprecise measurements predicts the exact response within a standard deviation of 3.25×10^{-2}, and the exact optical slab thickness is predicted within a standard deviation of 0.273. Assimilating the second $(n = 2)$ set of 100 measurements, which are just as imprecise as the first set, improves only slightly the prediction of the exact response and of the slab thickness.

Table 5.5: Results predicted by BERRU methodology for a slab of exact thickness $\mu a = 3$ ***after assimilating batches of 100 experiments with*** $\beta = 10^{-1}$.

$$\mu a = 3\ ;\beta = 10^{-1};\ \varepsilon = 10^{-3}\ ;K_n = 100\,;$$
$$Measured\ response = normal\left(r^{exact}, \beta r^{exact}\right)$$

N	Experimental Response Mean Value	Predicted Response	Predicted Response SD	Predicted Parameter	Predicted Parameter SD
1	5.010051 x10^{-1}	4.967511 x10^{-1}	3.255519 x10^{-2}	2.739635	2.736269 x10^{-1}
2	4.999093 x10^{-1}	4.926791 x10^{-1}	2.490237 x10^{-3}	2.741372	2.733279 x10^{-1}
		Exact Response	Exact Response SD	Exact Parameter	
		4.94679 x10^{-1}	4.94679 x10^{-2}	3.00	

b) Measurements with 0.001% relative standard deviation $\left(\beta = 10^{-5}\right)$

Consider a set $K_1 = 100$ of precise measurements (relative standard deviation $\beta = 10^{-5}$) drawn from the same random normal distribution, with $r^{exact} = 4.94679 \times 10^{-1}\,photons\,/\,cm^2 sec$ as the distribution's mean. Using these precise measurements, the BERRU methodology predicts the response within a standard deviation of $4.65 \times 10^{-6}\,photons\,/\,cm^2 sec$, and predicts the slab thickness within six significant digits, respectively, as shown in Table 5.6. As before, these results again indicate that precise measurements enable the BERRU to produce considerably more precise predictions than when less precise experiments are assimilated.

Table 5.6: Results predicted by BERRU methodology for a slab of exact thickness $\mu a = 3$ ***after assimilating batches of 100 experiments with*** $\beta = 10^{-5}$.

$$\mu a = 3\,;\beta = 10^{-5};\ \varepsilon = 10^{-8}\ ;K_n = 100\,;$$
$$Measured\ response = normal\left(r^{exact}, \beta r^{exact}\right)$$

N	Experimental Response Mean Value	Predicted Response	Predicted Response SD	Predicted Parameter	Predicted Parameter SD
1	4.946797 x10^{-1}	4.946797 x10^{-1}	4.647000 x10^{-6}	3.000097	9.973716 x10^{-5}
		Exact Response	Exact Response SD	Exact Parameter	
		4.94679 x10^{-1}	4.94679 x10^{-6}	3.00	

5.3.4 Prediction of Optically Very Thick Slab (Exact Optical Thickness $\mu a = 7.0$)

a) Measurements with 10% relative standard deviation $(\beta = 10^{-1})$

Consider a set $K_1 = 100$ of rather imprecise measurements (relative standard deviation $\beta = 10^{-1}$) drawn from the random normal distribution with the mean taken to be the exact response $r^{exact} = 4.999482 \times 10^{-1}\, photons / cm^2 sec$. The results predicted by the BERRU methodology are shown in columns 2 through 5 of Table 5.7. It is seen that the first $(n = 1)$ set of imprecise measurements predicts the exact response within a standard deviation of 9.41×10^{-4}, but the exact optical slab thickness is severely under-predicted. Assimilating the second $(n = 2)$ set of 100 measurements, which are just as imprecise as the first set, improves significantly the prediction of the exact response, but improves just marginally the prediction of the slab thickness. Additional imprecise experiments would not improve significantly the prediction of the slab thickness.

Table 5.7: Results predicted by BERRU methodology for a slab of exact thickness $\mu a = 7$ ***after assimilating batches of 100 experiments with*** $\beta = 10^{-1}$.

$$\mu a = 7\ ;\ \beta = 10^{-1};\ \varepsilon = 10^{-3}\ ;\ K_n = 100\ ;$$

$$Measured\ response = normal\left(r^{exact}, \beta r^{exact}\right)$$

N	Experimental Response Mean Value	Predicted Response	Predicted Response SD	Predicted Parameter
1	5.063417x10^{-1}	5.020369x10^{-1}	3.288029x10^{-2}	3.770365
2	5.052342x10^{-1}	4.979026x10^{-1}	9.418037x10^{-4}	3.771262
		Exact Response	Exact Response SD	Exact Parameter
		4.999482x10^{-1}	4.999482x10^{-6}	7.00

b) Measurements with 0.001% relative standard deviation $(\beta = 10^{-5})$

Consider a set $K_1 = 100$ of precise measurements (relative standard deviation $\beta = 10^{-5}$) drawn from the same random normal distribution, with $r^{exact} = 4.999482 \times 10^{-1}\, photons / cm^2 sec$ as the distribution's mean. It is seen from the results presented in Table 5.8 that the first $(n = 1)$ set of precise measurements predicts the exact response within a standard deviation of 4.66×10^{-6}. In addition, the BERRU methodology predicts the slab's thickness within a standard deviation of 0.112. The second $(n=2)$ set of precise measurements further improve the predicted values of both the response and the slab's thickness. As before, these results again indicate that precise measurements enable the BERRU methodology to produce considerably more precise predictions than when less precise experiments are assimilated.

Table 5.8: Results predicted by BERRU methodology for a slab of exact thicknes $\mu a=7$ ***after assimilating batches of 100 experiments with*** $\beta=10^{-5}$.

$$\mu a=7\,;\,\beta=10^{-5}\,;\;\varepsilon=10^{-8}\,;\;K_n=100\,;$$
$$Measured\ response=normal\left(r^{exact},\beta r^{exact}\right)$$

N	Experimental Response Mean Value	Predicted Response	Predicted Response SD	Predicted Parameter	Predicted Parameter SD
1	4.99948 x10^{-1}	4.999489 x10^{-1}	4.665733 x10^{-6}	7.010649	1.11982 x10^{-2}
2	4.9994 x10^{-1}	4.999488 x10^{-1}	4.117474 x10^{-6}	7.009803	7.217122 x10^{-3}
		Exact Response	Exact Response SD	Exact Parameter	
		4.999482 x10^{-1}	4.999482 x10^{-6}	7.00	

The results presented in Tables 5.7 and 5.8 for the slab having the exact optical thickness $\mu a=7$ reinforce the conclusions drawn from Tables 5.5 and 5.6 for the slab having the exact optical thickness $\mu a=3$, namely that: (i) the BERRU methodology under-predicts the slab's actual physical thickness when imprecise experimental results are assimilated, even though the predicted responses agrees within the imposed error criterion with the experimental results; and (ii) the BERRU methodology correctly predicts the slab's actual physical thickness when precise experimental results are assimilated, while also predicting the physically correct response within the selected precision criterion.

5.3.5 *Prediction of Optically Extremely Thick Slab (Exact Optical Thickness* $\mu a = 10.0$*)*

a) Measurements with 10% relative standard deviation $\left(\beta = 10^{-1}\right)$

Table 5.9 presents results predicted by the BERRU methodology when sets comprising increasingly more experiments, all having relative standard deviations of 10%, are being assimilated. After assimilating a set of $K_n = 5$ experiments, the BERRU methodology predicts the correct value of the response with 2 digits of accuracy, but the slab's thickness is under-predicted by a factor of 5. Increasing the numbers of similarly imprecise measurements from $K_n = 5$ experiments to $K_n = 100$ experiments per set does not appreciably increase the precision of the predicted response, but increases the accuracy of the predicted value of the slab thickness by a factor of about two, although the exact value remains severely under-predicted, due to the relatively large standard deviation ($\beta = 10^{-1}$) considered for the experimental responses.

Table 5.9: Results predicted by BERRU methodology for a slab of exact thickness $\mu a = 10$ ***after assimilating batches of experiments with*** $\beta = 10^{-1}$

$$\mu a = 10\,;\ \beta = 10^{-1};\ \varepsilon = 10^{-3}\ ;\ K_n = 5\,;$$

$$Measured\ response = normal\left(r^{exact}, \beta r^{exact}\right)$$

N	Experimental Response Mean Value	Predicted Response Value	Predicted Response SD	Predicted Parameter-Value
1	4.959541 x10^{-1}	4.920234 x10^{-1}	1.644221 x10^{-2}	2.022075
$\mu a = 10\,;\ \beta = 10^{-1};\ \varepsilon = 10^{-3}\ ;\ K_n = 10\,;$				
1	5.079625 x10^{-1}	4.960152 x10^{-1}	2.033668 x10^{-2}	2.406645
$\mu a = 10\,;\ \beta = 10^{-1};\ \varepsilon = 10^{-3}\ ;\ K_n = 50$				
1	4.993500 x10^{-1}	4.977449 x10^{-1}	3.236887 x10^{-2}	3.355578
$\mu a = 10\,;\ \beta = 10^{-1};\ \varepsilon = 10^{-3}\ ;\ K_n = 100\,;$				
1	5.063922 x10^{-1}	5.020869 x10^{-1}	3.288343 x10^{-2}	3.790445
2	5.052846 x10^{-1}	4.979521 x10^{-1}	9.238754 x10^{-4}	3.791330
		Exact Response Value	Exact Response SD	Exact Parameter Value

4.999981x10^{-1}	4.999981x10^{-2}	10.0

b) Measurements with 1% relative standard deviation $\left(\beta = 10^{-2}\right)$

Table 5.10 presents results predicted by the BERRU methodology when sets comprising increasingly more experiments, all having relative standard deviations of 1%, are being assimilated.

Table 5.10: Results predicted by BERRU methodology for a slab of exact thickness $\mu a = 10$ ***after assimilating batches of experiments with*** $\beta = 10^{-2}$.

$\mu a = 10\,;\,\beta = 10^{-2}\,;\;\varepsilon = 10^{-5}\;;K_n = 5\,;$

$Measured\ response = normal\left(r^{exact}, \beta r^{exact}\right)$

N	Experimental Response Mean Value	Predicted Response Value	Predicted Response SD	Predicted Parameter Value
1	4.995937 x10-1	4.992142 x10-1	1.654840 x10-3	3.910644
	$\mu a = 10\,;\,\beta = 10^{-2}\,;\;\varepsilon = 10^{-5}\;;K_n = 10\,;$			
1	5.007945x10-^{1}	4.996136x10^{-1}	2.052874x10^{-3}	4.322908
	$\mu a = 10\,;\,\beta = 10^{-2}\,;\;\varepsilon = 10^{-5}\;;K_n = 50$			
1	4.999333x10^{-1}	4.997729x10^{-1}	3.238153x10^{-3}	5.315490
	$\mu a = 10\,;\,\beta = 10^{-2}\,;\;\varepsilon = 10^{-5}\;;K_n = 100\,;$			
1	5.006375x10^{-1}	5.020869x10^{-1}	3.288731x10^{-3}	5.769938
2	5.005267x10^{-1}	4.997939x10^{-1}	1.352363x10^{-4}	5.771909
		Exact Response Value	Exact Response SD	Exact Parameter Value
		4.999981x10^{-1}	4.999981x10^{-3}	10.0

After assimilating a set of $K_n = 5$ such experiments, the results presented in Table 5.10 indicate that the BERRU methodology predicts the correct value of the response with 3 digits of accuracy, but the slab's thickness is under-predicted by a factor of 2.5. Increasing the numbers of similar measurements from $K_n = 5$ experiments to $K_n = 100$ experiments per set does not increase significantly the precision of the predicted response, but increases the accuracy of the predicted value of the slab thickness, although the exact value remains under-predicted by about *40%*, which is the prediction limit for the experimental responses drawn from a normal distribution with a relative standard deviation of *1%*.

c) *Measurements with 0.1% relative standard deviation* $\left(\beta = 10^{-3}\right)$

Table 5.11 presents results predicted by the BERRU methodology when sets comprising increasingly more experiments, all having relative standard deviations of 0.1%, are being assimilated.

Table 5.11: Results predicted by BERRU methodology for a slab of exact thickness $\mu a = 10$ ***after assimilating batches of experiments with*** $\beta = 10^{-3}$.

$$\mu a = 10\,;\, \beta = 10^{-3};\ \varepsilon = 10^{-5}\,;\, K_n = 5\,;$$

$$Measured\ response = normal\left(r^{exact}, \beta r^{exact}\right)$$

N	Experimental Response Mean Value	Predicted Response Value	Predicted Response SD	Predicted Parameter Value
1	$4.999576\text{x}10^{-1}$	$4.999218\text{x}10^{-1}$	$1.67085\ \text{x}10^{-4}$	5.929063
	$\mu a = 10\,;\, \beta = 10^{-3};\ \varepsilon = 10^{-5}\ ;\, K_n = 10\,;$			
1	$5.000777\ \text{x}10^{-1}$	$4.999618\ \text{x}10^{-1}$	$2.082969\ \text{x}10^{-4}$	6.345481
	$\mu a = 10\,;\, \beta = 10^{-3};\ \varepsilon = 10^{-5}\ ;\, K_n = 50$			
1	$4.999916\text{x}10^{-1}$	$4.999756\text{x}10^{-1}$	$3.240331\text{x}10^{-4}$	7.316728
	$\mu a = 10\,;\, \beta = 10^{-3};\ \varepsilon = 10^{-5}\ ;\, K_n = 100\,;$			
1	$5.000620\text{x}10^{-1}$	$5.000190\ \text{x}10^{-1}$	$3.289465\ \text{x}10^{-4}$	7.756322
2	$5.000509\text{x}10^{-1}$	$4.999778\ \text{x}10^{-1}$	$1.913978\ \text{x}10^{-5}$	7.760068
		Exact Response Value	Exact Response SD	Exact Parameter Value
		$4.999981\text{x}10^{-1}$	$4.999981\text{x}10^{-4}$	10.0

After assimilating a set of $K_n = 5$ such experiments, the results presented in Table 5.11 indicate that the BERRU methodology predicts the correct value of the response with 4 digits of accuracy, but the slab's thickness is under-predicted by 40%. Increasing the numbers of measurements having the same standard deviation from $K_n = 5$ experiments to $K_n = 100$ experiments per set does not increase significantly the precision of the predicted response, but increases the accuracy of the predicted value of the slab thickness, although the exact value remains under-predicted by about *20%*, which is the prediction limit for the experimental responses drawn from a normal distribution with a relative standard deviation of *0.1%*.

d) *Measurements with 0.01% relative standard deviation* $\left(\beta = 10^{-4}\right)$

Table 5.12 presents results predicted by the BERRU methodology when sets comprising increasingly more experiments, all having relative standard deviations of *0.01%*, are being assimilated. After assimilating a set of $K_n = 5$ such experiments, the results presented in Table 5.12 indicate that the BERRU methodology predicts the correct value of the response with 5 digits of accuracy, but the slab's thickness is under-predicted by *20%*. Increasing the numbers of measurements from $K_n = 5$ experiments to $K_n = 100$ experiments per set increases the accuracy of the predicted value of the slab thickness, although the exact value remains under-predicted by about *7%*, which is the prediction limit for the experimental responses drawn from a normal distribution with a relative standard deviation of *0.01%*.

Table 5.12: Results predicted by BERRU methodology for a slab of exact thickness $\mu a = 10$ ***after assimilating batches of experiments with*** $\beta = 10^{-4}$.

$$\mu a = 10\,;\, \beta = 10^{-4}\,;\ \varepsilon = 10^{-5}\,;\, K_n = 5\,;$$

$$Measured\ response = normal\left(r^{exact}, \beta r^{exact}\right)$$

N	Experimental Response Mean Value	Predicted Response Value	Predicted Response SD	Predicted Parameter Value
1	4.999940x10^-1	4.999908x10^-1	1.697100x10^-5	7.959722
	$\mu a = 10\,;\, \beta = 10^{-4}\,;\ \varepsilon = 10^{-5}\ ;\, K_n = 10\,;$			
1	5.000060x10^-1	4.999949x10^-1	2.146285x10^-5	8.334934
	$\mu a = 10\,;\, \beta = 10^{-4}\,;\ \varepsilon = 10^{-5}\ ;\, K_n = 50$			
1	4.999974x10^-1	4.999958x10^-1	3.250068x10^-5	9.056938
	$\mu a = 10\,;\, \beta = 10^{-4}\,;\ \varepsilon = 10^{-5}\ ;\, K_n = 100\,;$			
	5.000045x10^-1	5.000002x10^-1	3.294413x10^-5	9.338401
		Exact Response Value	Exact Response SD	Exact Parameter Value
		4.999981x10^-1	4.999981x10^-5	10.0

e) *Measurements with 0.001% relative standard deviation* $\left(\beta = 10^{-5}\right)$

Table 5.13 presents results predicted by the BERRU methodology when sets comprising increasingly more experiments, all having relative standard deviations of *0.001%*, are being assimilated. After assimilating a set of $K_n = 5$ such experiments, the results presented in Table 5.13 indicate that the BERRU methodology predicts the correct value of the response with 5 digits of accuracy, while the slab's thickness is under-predicted by *5%*. Increasing the numbers of measurements from $K_n = 5$ experiments to $K_n = 100$ experiments per set enables the BERRU methodology to predict practically the exact value of the response, and also enables the prediction of the slab thickness within a (negative) difference of *0.02* (*2%*) of the exact value.

Table 5.13: Results predicted by BERRU methodology for a slab of exact thickness $\mu a = 10$ ***after assimilating batches of experiments with*** $\beta = 10^{-5}$.

$$\mu a = 10\,;\, \beta = 10^{-5}\,;\; \varepsilon = 10^{-8}\,;\, K_n = 5\,;$$

$$Measured\ response = normal\left(r^{exact}, \beta r^{exact}\right)$$

N	Experimental Response Mean Value	Predicted Response	Predicted Response SD	Predicted Parameter	Predicted Parameter SD
1	4.999977 x10-1	4.999975 x10-1	1.796069 x10-6	9.563645	6.465910 x10-1
	$\mu a = 10\,;\, \beta = 10^{-5}\,;\; \varepsilon = 10^{-8}\,;\, K_n = 10\,;$				
1	4.999989 x10-1	4.999981 x10-1	2.566375 x10-6	9.786590	7.628210 x10-1
	$\mu a = 10\,;\, \beta = 10^{-5}\,;\; \varepsilon = 10^{-8}\,;\, K_n = 50\,;$				
1	4.999980 x10-1	4.999979 x10-1	3.486978 x10-6	9.861132	9.23650 x10-1
2	4.999986 x10-1	4.999981 x10-1	1.680819 x10-6	9.987424	7.737391 x10-1
	$\mu a = 10\,;\, \beta = 10^{-5}\,;\; \varepsilon = 10^{-8}\,;\, K_n = 100\,;$				
1	4.999987 x10-1	4.999983 x10-1	3.466796 x10-6	9.945714	9.359559 x10-1
2	4.999986 x10-1	4.999981 x10-1	1.927486 x10-6	9.983171	8.739583 x10-1
		Exact Response	Exact Response SD	Exact Parameter	
		4.999981 x10-1	4.999981 x10-6	10.0	

The results presented in Tables 5.9 through 5.13 for the slab having the exact optical thickness $\mu a = 10$ reinforce the conclusions previously drawn from the analysis of the slabs of exact optical thickness $\mu a = 3$ and $\mu a = 7$, respectively, namely that:

(i) The BERRU methodology under-predicts the slab's actual physical thickness when imprecise experimental results are assimilated, even though the predicted responses agrees within the imposed error criterion with the experimental results.

(ii) The BERRU methodology correctly predicts the slab's actual physical thickness when precise experimental results are assimilated, while also predicting the physically correct response within the selected precision criterion.

5.3.6 Prediction Limit for Single-Precision Computations: Slab of Exact Optical Thickness $\mu a = 15$.

For single precision computations, the limits of prediction accuracy when applying the BERRU methodology are illustrated by the results presented in Table 5.14 for a slab of exact optical thickness $\mu a = 15$. Assimilating *169* extremely precise experiments, distributed normally with a relative standard deviation $\beta = 10^{-7}$ around the exact response value, the BERRU methodology predicts the exact response value with 10 significant digits and the exact thickness within *0.2%.* This is a remarkable achievement for such a "deep penetration" paradigm problem, in which exponentially fewer gamma rays originating deeply within the slab escape to its surface.

Table 5.14: Prediction limit for single-precision computations using the BERRU methodology

$$\mu a = 15\,;\ \beta = 10^{-7}\,;\ \varepsilon = 10^{-9}\,;\ K_n = 169\,;$$

$$Measured\ response = normal\left(r^{exact}, \beta r^{exact}\right)$$

N	Experimental Response Mean Value	Predicted Response	Predicted Response SD	Predicted Parameter Value
	$5.000000\text{x}10^{-1}$	$5.000000\text{x}10^{-1}$	$3.498662\text{x}10^{-8}$	15.41315
		Exact Response	Exact Response SD	Exact Parameter
		$4.999999909\text{x}10^{-1}$	$4.999999909\text{x}10^{-8}$	15.0

5.4 SUMMARY AND CONCLUSIONS

This Chapter has presented a paradigm problem that models the inverse prediction, from detector responses in the presence of counting uncertainties, of the thickness of a homogeneous slab of material containing uniformly distributed gamma-emitting sources. Using this paradigm inverse problem, this Chapter has investigated the possible reasons for the apparent failure of the traditional inverse-problem methods based on the minimization of chi-square-type functionals, and has also compared the results produced by such methods with the results produced by applying the BERRU methodology, for optically thin and thick slabs.

For optically thin slabs, the results presented in this Chapter have shown that both the traditional chi-square-minimization method and the BERRU methodology predict the slab's thickness accurately. For optically thick slabs, the results obtained in this Chapter have led to following conclusions:

(i) The traditional inverse-problem methods based on the minimization of chi-square-type functionals fail to predict the slab's thickness;

(ii) The BERRU methodology under-predicts the slab's actual physical thickness when imprecise experimental results are assimilated, even though the predicted responses agrees within the imposed error criterion with the experimental results;

(iii) The BERRU methodology correctly predicts the slab's actual physical thickness when precise experimental results are assimilated, while also predicting the physically correct response within the selected precision criterion.

For single precision computations, the limits of prediction accuracy when applying the BERRU methodology were illustrated by assimilating 169 extremely precise experiments, distributed normally with a relative standard deviation $\beta = 10^{-7}$ around the exact response value, and showing that the BERRU methodology predicts the exact response value with 10 significant digits and the exact thickness within *0.2%*, which is a remarkable achievement for such a "deep penetration" paradigm problem. The results presented in this Chapter correspond to realistic measured standard deviations, obtainable routinely in gamma-ray measurements, except for the "very precise" measurements, which were used for illustrative purposes, to highlight the fact that the accuracy of the results predicted by using the BERRU methodology in the "inverse predictive" mode is limited by the precision of the measurements, rather than by the BERRU methodology or by its underlying computational algorithm.

6 BERRU-CMS APPLICATION TO SAVANNAH RIVER NATIONAL LABORATORY'S F-AREA COOLING TOWERS

Abstract. The BERRU-CMS methodology is applied in this Chapter to the model of a mechanical draft cooling tower (MDCT) located at Savannah River National Laboratory (SRNL) in order to improve the predictions of this model by combining computational information with measurements of outlet air humidity, outlet air and outlet water temperatures. At the outlet of this cooling tower, where measurements of the quantities of interest are available, the BERRU-CMS methodology reduces the predicted uncertainties for these quantities to values that are smaller than either the computed or the measured uncertainties. The BERRU-CMS methodology has also been applied to reduce the uncertainties for quantities of interest inside the tower's fill section, where no direct measurements are available. The maximum reductions of uncertainties occur at the locations where direct measurements are available. At other locations, the predicted response uncertainties are reduced by the BERRU-CMS methodology to values that are smaller than the modeling uncertainties arising from the imprecisely known model parameters.

6.1 INTRODUCTION

The material presented in this Chapter is based on work presented in the following publications: Cacuci and Fang (2016), Di Rocco and Cacuci (2016), Fang et al (2016), Cacuci and Di Rocco (2017), Di Rocco et al (2017), Cacuci and Fang (2017), and Fang et al (2017). A mechanical draft cooling tower (MDCT) dissipates waste heat from an industrial process into the atmosphere. Cooling towers are essential for the functioning of large energy-producing plants, including nuclear reactors. The rate of thermal energy dissipation is computed using a numerical simulation model of the cooling tower together with measurements of outlet air relative humidity, outlet air and water temperatures. In addition to computing the temperature drop of the cooling water as it passes through the tower, a cooling tower model that derives heat dissipation rates from thermal imagery needs to convert the remotely measured cooling tower throat or area-weighted temperature to a cooling water inlet temperature. Therefore, a cooling tower model comprises two main components, as follows: (i) an inner model which computes the amount of cooling undergone by the water as it passes through the tower as a function of inlet cooling water temperature and ambient weather conditions (air temperature and humidity); and (ii) an outer model which uses a remotely measured throat or area-weighted temperature and iterates on the inlet water temperature to match the target temperature of interest. The cooling tower model produces an estimate of the rate at which energy is being discharged to the atmosphere by evaporation and sensible heat transfer. The sensible heat transfer is estimated using the computed change in air or water

D. G. Cacuci, *BERRU Predictive Modeling*,
https://doi.org/10.1007/978-3-662-58395-1_6

enthalpy as it passes through the MDCT. If the MDCT fans are on, a prescribed mass flow rate of air and water is used. If the MDCT fans are off, an additional mechanical energy equation is iteratively solved to determine the mass flow rate of air. The flow regime in the fill section of a cooling tower, which can be cross-flow or counter-flow, determines the type of the respective cooling tower. The air mass flow rate is specified when the cooling tower operates in the mechanical draft mode. When the fan is turned off, the cooling tower operates in the natural draft/wind-aided mode, in which case the air mass flow rate is calculated using the numerical model.

Section 6.2 presents the mathematical model of the counter-flow cooling tower considered in the Chapter, along with the numerical method for computing the steady state distributions of the following quantities: (i) the water mass flow rates at the exit of each control volume along the height of the fill section of the cooling tower; (ii) the water temperatures at the exit of each control volume along the height of the fill section of the cooling tower; (iii) the air temperatures at the exit of each control volume along the height of the fill section of the cooling tower; (iv) the humidity ratios at the exit of each control volume along the height of the fill section of the cooling tower; and (v) the air relative humidity at the exit of each control volume. Section 6.3 presents the development of the *adjoint cooling tower sensitivity model* for computing *efficiently and exactly* the sensitivities (i.e., functional derivatives) of the model responses (i.e., quantities of interest) to all 52 model parameters. The adjoint sensitivity model is developed by applying the general *adjoint sensitivity analysis methodology* (ASAM) *for nonlinear systems*, originally presented by Cacuci (1981a). Even though the forward cooling tower model is nonlinear in the state functions, the adjoint sensitivity model is *linear* in the adjoint state functions, which correspond one-to-one to the forward state functions mentioned in the foregoing. Using the adjoint state functions, the sensitivities of each model response to all of the 52 model parameters can be computed exactly using *a single adjoint model computation.* Section 6.4 presents the numerical results and the ranking of model parameters in the order of their importance for contributing to response uncertainties.

During the period from April, 2004 through August, 2004, a total of 8079 measured benchmark data sets for F-area cooling towers (fan-on case) were recorded every fifteen minutes at SRNL for F-Area Cooling Towers (Garrett et al, 2005). Each of these data sets contained measurements of the following (four) quantities: (i) outlet air temperature measured with the sensor called "Tidbit", which will be denoted as $T_{a,out(Tidbit)}$; (ii) outlet air temperature measured with the sensor called "Hobo", which will be denoted as $T_{a,out(Hobo)}$; (iii) outlet water temperature, which will be denoted as $T_{w,out}^{meas}$; (iv) outlet air relative humidity, which will be denoted as RH^{meas}. Section 6.5 presents the histogram plots of these 7668 measurement sets (each set containing measurements of $T_{a,out(Tidbit)}, T_{a,out(Hobo)}$, $T_{w,out}^{meas}$, and RH^{meas}) together with the corresponding statistical analyses. Section 6.6 presents the application of the BERRU-CMS methodology to reduce the uncertainties for quantities of interest inside the tower's fill section, where no direct measurements are available. The maximum reductions of uncertainties occur at the locations where direct measurements are available. At other locations, the predicted response uncertainties are reduced by the BERRU-CMS methodology to values that are smaller than the modeling uncertainties arising from the imprecisely known model parameters.

6.2 MATHEMATICAL MODEL OF THE COUNTER-FLOW COOLING TOWER

The counter-flow cooling tower considered in this Chapter was originally developed by Aleman and Garrett (2015) and is schematically presented in Figure 6.1. As this figure indicates, forced air flow enters the tower through the "rain section" above the water basin, flows upward through the fill section and the drift eliminator, and exits at the tower's top through an exhaust that encloses a fan. Hot water enters above the fill section and is sprayed onto the top of the fill section to create a uniform, downward falling, film flow through the fill's numerous meandering vertical passages. Film fills are designed to maximize the water free surface area and the residence time inside of the fill section. Heat and mass transfer occurs at the falling film's free surface between the water film and the upward air flow. The drift eliminator above the spray zone removes entrained water droplets from the upward flowing air. Below the fill section, the water droplets fall into a collection basin, placed at the bottom of the cooling tower. The heat and mass transfer processes occur overwhelmingly in the fill section. Modeling the heat and mass transfer processes between falling water film and rising air in the cooling tower's fill section is accomplished solving the following balance equations: (A) liquid continuity; (B) liquid energy balance; (C) water vapor continuity; (D) air/water vapor energy balance. The assumptions used in deriving these equations are as follows:

1. the air and/or water temperatures are uniform throughout each stream at any cross section;
2. the cooling tower has uniform cross-sectional area;
3. the heat and mass transfer occur solely in the direction normal to flows;
4. the heat and mass transfer through tower walls to the environment is negligible;
5. the heat transfer from the cooling tower fan and motor assembly to the air is negligible;
6. the air and water vapor mix as ideal gasses;
7. the flow between flat plates is unsaturated through the fill section;
8. the heat and mass transfer processes in the rain section is negligible.

The fill section is modeled by discretizing it in vertically stacked control volumes as depicted in Figure 6.2. The heat and mass transfer between the falling water film and the rising air in a typical control volume of the cooling tower's fill section is presented in Figure 6.3.

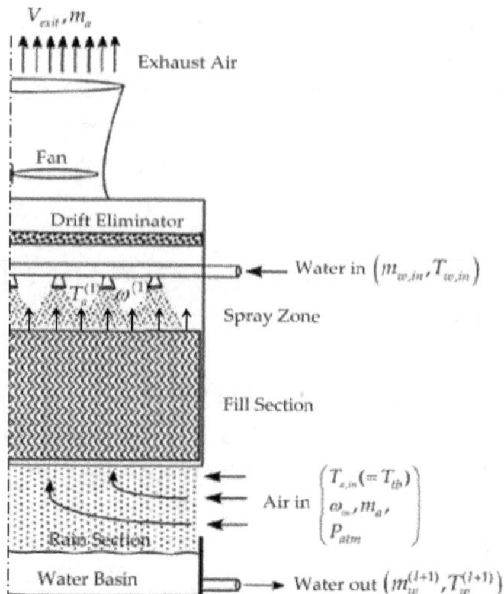

Figure 6.1: Flow through a counter-flow cooling tower.

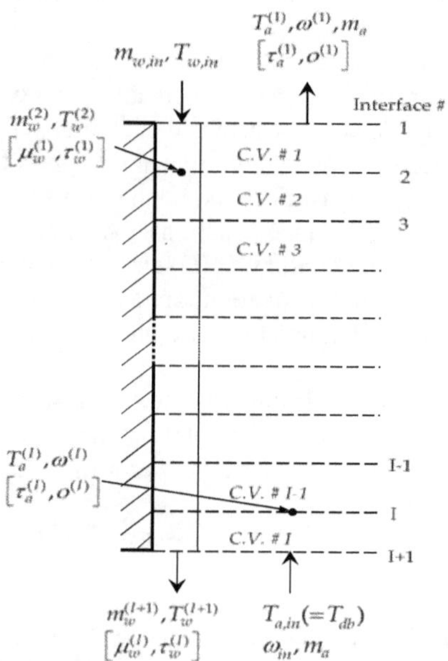

Figure 6.2: Control volumes $(i = 1,..,I)$ ***comprising the counter-flow cooling tower, together with the symbols denoting the forward state functions*** $\left(m_w^{(i)}, T_w^{(i)}, T_a^{(i)}, \omega^{(i)},\ i = 1,..,I\right)$ ***and the adjoint state functions*** $\left(\mu_w^{(i)}, \tau_w^{(i)}, \tau_a^{(i)}, o^{(i)};\ i = 1,..,I\right)$***, respectively.***

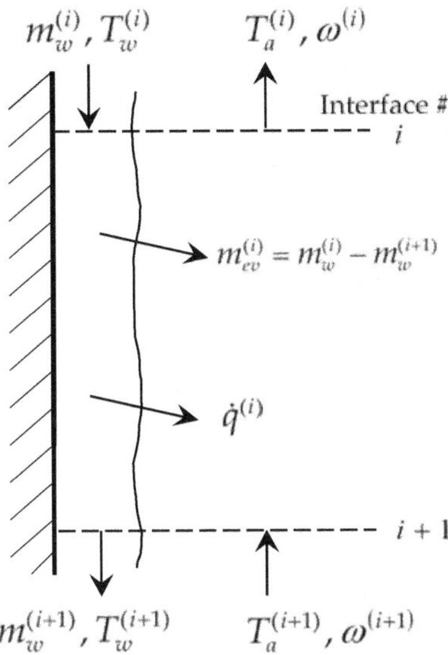

Figure 6.3: Heat and mass transfer between falling water film and rising air in a typical control volume of the cooling tower's fill section.

In the mechanical draft mode, the mass flow rate of dry air is specified. With the fan off and hot water flowing through the cooling tower, air will continue to flow through the tower due to buoyancy. Wind pressure at the air inlet to the cooling tower will also enhance air flow through the tower. The air flow rate is determined from the overall mechanical energy equation for the dry air flow. The state functions underlying the cooling tower model (cf., Figures 6.1 through 6.3) are as follows:

1. the water mass flow rates, denoted as $m_w^{(i)}$ $(i = 2,...,50)$, at the exit of each control volume, *i*, along the height of the fill section of the cooling tower;
2. the water temperatures, denoted as $T_w^{(i)}$ $(i = 2,...,50)$, at the exit of each control volume, *i*, along the height of the fill section of the cooling tower;
3. the air temperatures, denoted as $T_a^{(i)}$ $(i = 1,...,49)$, at the exit of each control volume, *i*, along the height of the fill section of the cooling tower; and
4. the humidity ratios, denoted as . $\omega^{(i)}$ $(i = 1,...,49)$, at the exit of each control volume, *i*, along the height of the fill section of the cooling tower.

It is convenient to consider the above state functions to be components of the following (column) vectors:

$$\mathbf{m}_w \triangleq \left[m_w^{(2)},...,m_w^{(I+1)} \right]^{\dagger}, \mathbf{T}_w \triangleq \left[T_w^{(2)},...,T_w^{(I+1)} \right]^{\dagger}, \quad \mathbf{T}_a \triangleq \left[T_a^{(1)},...,T_a^{(I)} \right]^{\dagger}, \boldsymbol{\omega} \triangleq \left[\omega^{(1)},...,\omega^{(I)} \right]^{\dagger} \qquad (6.1)$$

In this work, the dagger *(†)* will be used to denote "transposition", and all vectors will be considered to be column vectors. The governing conservation equations within the total of *I=49* control volumes represented in Figure 6.2 are as follows (Aleman and Garrett, 2015):

A. Liquid continuity equations:

(i) *Control Volume i=1:*

$$N_1^{(1)}(\mathbf{m}_w, \mathbf{T}_w, \mathbf{T}_a, \boldsymbol{\omega}; \boldsymbol{\alpha}) \triangleq m_w^{(2)} - m_{w,in} + \frac{M(m_a, \boldsymbol{\alpha})}{\overline{R}} \left[\frac{P_{vs}^{(2)}(T_w^{(2)}, \boldsymbol{\alpha})}{T_w^{(2)}} - \frac{\omega^{(1)} P_{atm}}{T_a^{(1)}(0.622 + \omega^{(1)})} \right] = 0; \tag{6.2}$$

(ii) *Control Volumes i=2,..., I-1:*

$$N_1^{(i)}(\mathbf{m}_w, \mathbf{T}_w, \mathbf{T}_a, \boldsymbol{\omega}; \boldsymbol{\alpha}) \triangleq m_w^{(i+1)} - m_w^{(i)} + \frac{M(m_a, \boldsymbol{\alpha})}{\overline{R}} \left[\frac{P_{vs}^{(i+1)}(T_w^{(i+1)}, \boldsymbol{\alpha})}{T_w^{(i+1)}} - \frac{\omega^{(i)} P_{atm}}{T_a^{(i)}(0.622 + \omega^{(i)})} \right] = 0; \tag{6.3}$$

(iii) *Control Volume i=I:*

$$N_1^{(I)}(\mathbf{m}_w, \mathbf{T}_w, \mathbf{T}_a, \boldsymbol{\omega}; \boldsymbol{\alpha}) \triangleq m_w^{(I+1)} - m_w^{(I)} + \frac{M(m_a, \boldsymbol{\alpha})}{\overline{R}} \left[\frac{P_{vs}^{(I+1)}(T_w^{(I+1)}, \boldsymbol{\alpha})}{T_w^{(I+1)}} - \frac{\omega^{(I)} P_{atm}}{T_a^{(I)}(0.622 + \omega^{(I)})} \right] = 0; \tag{6.4}$$

B. Liquid energy balance equations:

(i) *Control Volume i=1:*

$$\begin{aligned} N_2^{(1)}(\mathbf{m}_w, \mathbf{T}_w, \mathbf{T}_a, \boldsymbol{\omega}; \boldsymbol{\alpha}) &\triangleq m_{w,in} h_f(T_{w,in}, \boldsymbol{\alpha}) - (T_w^{(2)} - T_a^{(1)}) H(m_a, \boldsymbol{\alpha}) \\ &- m_w^{(2)} h_f^{(2)}(T_w^{(2)}, \boldsymbol{\alpha}) - (m_{w,in} - m_w^{(2)}) h_{g,w}^{(2)}(T_w^{(2)}, \boldsymbol{\alpha}) = 0; \end{aligned} \tag{6.5}$$

(ii) *Control Volumes i=2,..., I-1:*

$$\begin{aligned} N_2^{(i)}(\mathbf{m}_w, \mathbf{T}_w, \mathbf{T}_a, \boldsymbol{\omega}; \boldsymbol{\alpha}) &\triangleq m_w^{(i)} h_f^{(i)}(T_w^{(i)}, \boldsymbol{\alpha}) - (T_w^{(i+1)} - T_a^{(i)}) H(m_a, \boldsymbol{\alpha}) \\ &- m_w^{(i+1)} h_f^{(i+1)}(T_w^{(i+1)}, \boldsymbol{\alpha}) - (m_w^{(i)} - m_w^{(i+1)}) h_{g,w}^{(i+1)}(T_w^{(i+1)}, \boldsymbol{\alpha}) = 0; \end{aligned} \tag{6.6}$$

(iii) *Control Volume i=I:*

$$N_2^{(I)}\left(\mathbf{m}_w,\mathbf{T}_w,\mathbf{T}_a,\boldsymbol{\omega};\boldsymbol{\alpha}\right) \triangleq m_w^{(I)} h_f^{(I)}\left(T_w^{(I)},\boldsymbol{\alpha}\right) - \left(T_w^{(I+1)} - T_a^{(I)}\right) H\left(m_a,\boldsymbol{\alpha}\right)$$
$$- m_w^{(I+1)} h_f^{(I+1)}\left(T_w^{(I+1)},\boldsymbol{\alpha}\right) - \left(m_w^{(I)} - m_w^{(I+1)}\right) h_{g,w}^{(I+1)}\left(T_w^{(I+1)},\boldsymbol{\alpha}\right) = 0; \tag{6.7}$$

C. Water vapor continuity equations:

(i) Control Volume i=1:

$$N_3^{(1)}\left(\mathbf{m}_w,\mathbf{T}_w,\mathbf{T}_a,\boldsymbol{\omega};\boldsymbol{\alpha}\right) \triangleq \omega^{(2)} - \omega^{(1)} + \frac{m_{w,in} - m_w^{(2)}}{\left|m_a\right|} = 0; \tag{6.8}$$

(ii) Control Volumes i=2,..., I-1:

$$N_3^{(i)}\left(\mathbf{m}_w,\mathbf{T}_w,\mathbf{T}_a,\boldsymbol{\omega};\boldsymbol{\alpha}\right) \triangleq \omega^{(i+1)} - \omega^{(i)} + \frac{m_w^{(i)} - m_w^{(i+1)}}{\left|m_a\right|} = 0; \tag{6.9}$$

(iii) Control Volume i=I:

$$N_3^{(I)}\left(\mathbf{m}_w,\mathbf{T}_w,\mathbf{T}_a,\boldsymbol{\omega};\boldsymbol{\alpha}\right) \triangleq \omega_{in} - \omega^{(I)} + \frac{m_w^{(I)} - m_w^{(I+1)}}{\left|m_a\right|} = 0; \tag{6.10}$$

D. The air/water vapor energy balance equations:

(i) Control Volume i=1:

$$N_4^{(1)}\left(\mathbf{m}_w,\mathbf{T}_w,\mathbf{T}_a,\boldsymbol{\omega};\boldsymbol{\alpha}\right) \triangleq \left(T_a^{(2)} - T_a^{(1)}\right) C_p^{(1)}\left(\frac{T_a^{(1)} + 273.15}{2},\boldsymbol{\alpha}\right) - \omega^{(1)} h_{g,a}^{(1)}\left(T_a^{(1)},\boldsymbol{\alpha}\right)$$
$$+ \frac{\left(T_w^{(2)} - T_a^{(1)}\right) H\left(m_a,\boldsymbol{\alpha}\right)}{\left|m_a\right|} + \frac{\left(m_{w,in} - m_w^{(2)}\right) h_{g,w}^{(2)}\left(T_w^{(2)},\boldsymbol{\alpha}\right)}{\left|m_a\right|} + \omega^{(2)} h_{g,a}^{(2)}\left(T_a^{(2)},\boldsymbol{\alpha}\right) = 0; \tag{6.11}$$

(ii) Control Volumes i=2,..., I-1:

$$N_4^{(i)}\left(\mathbf{m}_w,\mathbf{T}_w,\mathbf{T}_a,\boldsymbol{\omega};\boldsymbol{\alpha}\right)\triangleq\left(T_a^{(i+1)}-T_a^{(i)}\right)C_p^{(i)}\left(\frac{T_a^{(i)}+273.15}{2},\boldsymbol{\alpha}\right)$$
$$-\omega^{(i)}h_{g,a}^{(i)}\left(T_a^{(i)},\boldsymbol{\alpha}\right)+\frac{\left(T_w^{(i+1)}-T_a^{(i)}\right)H\left(m_a,\boldsymbol{\alpha}\right)}{\left|m_a\right|} \tag{6.12}$$
$$+\frac{\left(m_w^{(i)}-m_w^{(i+1)}\right)h_{g,w}^{(i+1)}\left(T_w^{(i+1)},\boldsymbol{\alpha}\right)}{\left|m_a\right|}+\omega^{(i+1)}h_{g,a}^{(i+1)}\left(T_a^{(i+1)},\boldsymbol{\alpha}\right)=0;$$

(iii) *Control Volume i=I:*

$$N_4^{(I)}\left(\mathbf{m}_w,\mathbf{T}_w,\mathbf{T}_a,\boldsymbol{\omega};\boldsymbol{\alpha}\right)\triangleq\left(T_{a,in}-T_a^{(I)}\right)C_p^{(I)}\left(\frac{T_a^{(I)}+273.15}{2},\boldsymbol{\alpha}\right)$$
$$-\omega^{(I)}h_{g,a}^{(I)}\left(T_a^{(I)},\boldsymbol{\alpha}\right)+\frac{\left(T_w^{(I+1)}-T_a^{(I)}\right)H\left(m_a,\boldsymbol{\alpha}\right)}{\left|m_a\right|}+ \tag{6.13}$$
$$\frac{\left(m_w^{(I)}-m_w^{(I+1)}\right)h_{g,w}^{(I+1)}\left(T_w^{(I+1)},\boldsymbol{\alpha}\right)}{\left|m_a\right|}+\omega_{in}h_{g,a}\left(T_{a,in},\boldsymbol{\alpha}\right)=0.$$

The components of the vector $\boldsymbol{\alpha}$, which appears in Eqs. (6.2) through (6.13) comprise the model parameters which are generically denoted as α_i, i.e.,

$$\boldsymbol{\alpha}\triangleq\left(\alpha_1,\ldots,\alpha_{N_\alpha}\right), \tag{6.14}$$

where N_α denotes the total number of model parameters. These model parameters are experimentally derived quantities, and their complete distributions are not known; however, we have determined the first four moments (means, variance/covariance, skewness, and kurtosis) as detailed in Section 6.5.

In the original work by Aleman and Garrett (2015), the above equations were solved using a two stage-iterative method comprising an "inner-iteration" using Newton's method within each control volume, followed by an outer iteration aimed at achieving overall convergence. This procedure, though, did not converge at all of the points of interest. Therefore, the original solution method of Aleman and Garrett (2015) has been replaced by Newton's method together with the GMRES linear iterative solver for sparse matrices (Saad and Schultz, 1986) provided in the NSPCG package (Oppe et al, 1988), which turned out to be the most efficient among those we have tested. This GMRES method (Saad and Schultz, 1986) approximates the exact solution-vector of a linear system by using the Arnoldi iteration to find the approximate solution-vector by minimizing the norm of the residual vector over a Krylov subspace. The specific computational steps are as follows:

(a) Write Eqs. (6.1) through (6.13) in vector form as

$$\mathbf{N}(\mathbf{u}) = \mathbf{0}, \tag{6.15}$$

where the following definitions were used:

$$\mathbf{N} \triangleq \left(N_1^{(1)},..,N_1^{(I)},...,N_4^{(1)},...,N_4^{(I)}\right)^\dagger, \ \mathbf{u} \triangleq (\mathbf{m}_w, \mathbf{T}_w, \mathbf{T}_a, \boldsymbol{\omega})^\dagger; \tag{6.16}$$

(b) Set the initial guess, $\mathbf{u}_0$, to be the inlet boundary conditions;

(c) Steps d through g, below, constitute the outer iteration loop; for $n = 0,1,2,...$, iterate over the following steps until convergence:

(d) Inner iteration loop: for $m = 1,2,...$, use the iterative GMRES linear solver with the Modified Incomplete Cholesky (MIC) preconditioner, with restarts, to solve, until convergence, the following system to compute the vector $\delta\mathbf{u}$:

$$\mathbf{J}(\mathbf{u}_n)\delta\mathbf{u} = -\mathbf{N}(\mathbf{u}_n), \tag{6.17}$$

where n is the current outer loop iteration number, and the Jacobian matrix of derivatives of Eqs. (6.3) through (6.13) with respect to the state functions is the block-matrix

$$\mathbf{J}(\mathbf{u}_n) \triangleq \begin{pmatrix} \mathbf{A}_1 & \mathbf{B}_1 & \mathbf{C}_1 & \mathbf{D}_1 \\ \mathbf{A}_2 & \mathbf{B}_2 & \mathbf{C}_2 & \mathbf{D}_2 \\ \mathbf{A}_3 & \mathbf{B}_3 & \mathbf{C}_3 & \mathbf{D}_3 \\ \mathbf{A}_4 & \mathbf{B}_4 & \mathbf{C}_4 & \mathbf{D}_4 \end{pmatrix}, \tag{6.18}$$

where the block-matrix-components of $\mathbf{J}(\mathbf{u}_n)$ will be defined below.

The functional derivatives of Eqs. (6.2) through (6.13) with respect to the vector-valued state function $\mathbf{u} \triangleq (\mathbf{m}_w, \mathbf{T}_w, \mathbf{T}_a, \boldsymbol{\omega})^\dagger$, where $\mathbf{m}_w \triangleq \left[m_w^{(2)},...,m_w^{(I+1)}\right]^\dagger$; $\mathbf{T}_w \triangleq \left[T_w^{(2)},...,T_w^{(I+1)}\right]^\dagger$; $\mathbf{T}_a \triangleq \left[T_a^{(1)},...,T_a^{(I)}\right]^\dagger$; and $\boldsymbol{\omega} \triangleq \left[\omega^{(1)},...,\omega^{(I)}\right]^\dagger$, will be denoted as follows:

$$a_\ell^{i,j} \triangleq \frac{\partial N_\ell^{(i)}}{\partial m_w^{(j+1)}}; \ \ell = 1,2,3,4; \ i = 1,...,I; \ j = 1,...,I; \tag{6.19}$$

$$b_\ell^{i,j} \triangleq \frac{\partial N_\ell^{(i)}}{\partial T_w^{(j+1)}}; \ \ell = 1,2,3,4; \ i = 1,...,I; \ j = 1,...,I; \tag{6.20}$$

$$c_\ell^{i,j} \triangleq \frac{\partial N_\ell^{(i)}}{\partial T_a^{(j)}};\ \ell = 1,2,3,4;\ i = 1,\ldots,I;\ j = 1,\ldots,I; \tag{6.21}$$

$$d_\ell^{i,j} \triangleq \frac{\partial N_\ell^{(i)}}{\partial \omega^{(j)}};\ \ell = 1,2,3,4;\ i = 1,\ldots,I;\ j = 1,\ldots,I; \tag{6.22}$$

The specific expressions of each of the above functional derivatives are as follows:

1. The derivatives of the "liquid continuity equations" [cf., Eqs. (6.2) through (6.4)] with respect to $m_w^{(j)}$ are as follows:

$$\frac{\partial N_1^{(i)}}{\partial m_w^{(j+1)}} \triangleq a_1^{i,j} = 0;\ i = 1,\ldots,I;\ j = 1,\ldots,I;\ j \neq i-1,\ i; \tag{6.23}$$

$$\frac{\partial N_1^{(i)}}{\partial m_w^{(i)}} \triangleq a_1^{i,i-1} = -1;\ i = 2,\ldots,I;\ j = i-1; \tag{6.24}$$

$$\frac{\partial N_1^{(i)}}{\partial m_w^{(i+1)}} \triangleq a_1^{i,i} = 1;\ i = 1,\ldots,I;\ j = i. \tag{6.25}$$

For subsequent use, the above quantities are considered to be the components of the $I \times I$ matrix $\mathbf{A}_1$ defined as follows:

$$\mathbf{A}_1 \triangleq \left(a_1^{i,j}\right)_{I\times I} = \begin{pmatrix} 1 & 0 & . & 0 & 0 \\ -1 & 1 & . & 0 & 0 \\ . & . & . & . & . \\ 0 & 0 & . & 1 & 0 \\ 0 & 0 & . & -1 & 1 \end{pmatrix}. \tag{6.26}$$

2. The derivatives of the "liquid continuity equations" [cf. Eqs. (6.2) through (6.4)] with respect to $T_w^{(j)}$ are as follows:

$$\frac{\partial N_1^{(i)}}{\partial T_w^{(j+1)}} \triangleq b_1^{i,j} = 0;\ i = 1,\ldots,I;\ j = 1,\ldots,I;\ j \neq i; \tag{6.27}$$

$$\frac{\partial N_1^{(i)}}{\partial T_w^{(i+1)}} \triangleq b_1^{i,i} = -\frac{M(m_a,\boldsymbol{\alpha})}{\overline{R}} \frac{P_{vs}^{(i+1)}\left(T_w^{(i+1)},\boldsymbol{\alpha}\right)}{\left[T_w^{(i+1)}\right]^2}\left(\frac{a_1}{T_w^{(i+1)}}+1\right); \quad i=1,\ldots,I;\; j=i. \tag{6.28}$$

For subsequent use, the above quantities are considered to be the components of the $I \times I$ diagonal matrix $\mathbf{B}_1$ defined as follows:

$$\mathbf{B}_1 \triangleq \left(b_1^{i,j}\right)_{I\times I} = \begin{pmatrix} b_1^{1,1} & 0 & . & 0 & 0 \\ 0 & b_1^{2,2} & . & 0 & 0 \\ . & . & . & . & . \\ 0 & 0 & . & b_1^{I-1,I-1} & 0 \\ 0 & 0 & . & 0 & b_1^{I,I} \end{pmatrix}. \tag{6.29}$$

3. The derivatives of the "liquid continuity equations" [cf. Eqs. (6.2) through (6.4)] with respect to $T_a^{(j)}$ are as follows:

$$\frac{\partial N_1^{(i)}}{\partial T_a^{(j)}} \triangleq c_1^{i,j} = 0;\; i=1,\ldots,I;\; j=1,\ldots,I;\; j\neq i; \tag{6.30}$$

$$\frac{\partial N_1^{(i)}}{\partial T_a^{(i)}} \triangleq c_1^{i,i} = \frac{M(m_a,\boldsymbol{\alpha})}{\overline{R}} \frac{\omega^{(i)} P_{atm}}{\left[T_a^{(i)}\right]^2\left(0.622+\omega^{(i)}\right)}; \quad i=1,\ldots,I;\; j=i. \tag{6.31}$$

For subsequent use, the above quantities are considered to be the components of the $I \times I$ diagonal matrix $\mathbf{C}_1$ defined as follows:

$$\mathbf{C}_1 \triangleq \left(c_1^{i,j}\right)_{I\times I} = \begin{pmatrix} c_1^{1,1} & 0 & . & 0 & 0 \\ 0 & c_1^{2,2} & . & 0 & 0 \\ . & . & . & . & . \\ 0 & 0 & . & c_1^{I-1,I-1} & 0 \\ 0 & 0 & . & 0 & c_1^{I,I} \end{pmatrix}. \tag{6.32}$$

4. The derivatives of the "liquid continuity equations" [cf. Eqs. (6.2) through (6.4)] with respect to $\omega^{(i)}$ are as follows:

$$\frac{\partial N_1^{(i)}}{\partial \omega^{(j)}} \triangleq d_1^{i,j} = 0;\ i = 1,\dots,I;\ j = 1,\dots,I;\ j \neq i; \tag{6.33}$$

$$\frac{\partial N_1^{(i)}}{\partial \omega^{(i)}} \triangleq d_1^{i,i} = \frac{M(m_a,\boldsymbol{\alpha})}{\overline{R}} \frac{P_{atm}}{\left[0.622+\omega^{(i)}\right]T_a^{(i)}} \left\{ \frac{\omega^{(i)}}{\left[0.622+\omega^{(i)}\right]} - 1 \right\}; \tag{6.34}$$
$$i = 1,\dots,I;\ j = i.$$

For subsequent use, the above quantities are considered to be the components of the $I \times I$ diagonal matrix $\mathbf{D}_1$ defined as follows:

$$\mathbf{D}_1 \triangleq \left(d_1^{i,j}\right)_{I\times I} = \begin{pmatrix} d_1^{1,1} & 0 & . & 0 & 0 \\ 0 & d_1^{2,2} & . & 0 & 0 \\ . & . & . & . & . \\ 0 & 0 & . & d_1^{I-1,I-1} & 0 \\ 0 & 0 & . & 0 & d_1^{I,I} \end{pmatrix}. \tag{6.35}$$

5. The derivatives of the liquid energy balance equations [cf. Eqs. (6.5) through (6.7)] with respect to $m_w^{(j)}$ are as follows:

$$\frac{\partial N_2^{(i)}}{\partial m_w^{(j+1)}} \triangleq a_2^{i,j} = 0;\ i = 1,\dots,I;\ j = 1,\dots,I;\ j \neq i-1,\ i; \tag{6.36}$$

$$\frac{\partial N_2^{(i)}}{\partial m_w^{(i)}} \triangleq a_2^{i,i-1} = h_f^{(i)}\left(T_w^{(i)},\boldsymbol{\alpha}\right) - h_g^{(i+1)}\left(T_w^{(i+1)},\boldsymbol{\alpha}\right); \tag{6.37}$$
$$i = 2,\dots,I;\ j = i-1;$$

$$\frac{\partial N_2^{(i)}}{\partial m_w^{(i+1)}} \triangleq a_2^{i,i} = h_g^{(i+1)}\left(T_w^{(i+1)},\boldsymbol{\alpha}\right) - h_f^{(i+1)}\left(T_w^{(i+1)},\boldsymbol{\alpha}\right); \tag{6.38}$$
$$i = 1,\dots,I;\ j = i.$$

For subsequent use, the above quantities are considered to be the components of the $I \times I$ matrix $\mathbf{A}_2$ defined as follows:

$$\mathbf{A}_2 \triangleq \left(a_2^{i,j}\right)_{I\times I} = \begin{pmatrix} a_2^{1,1} & 0 & . & 0 & 0 \\ a_2^{2,1} & a_2^{2,2} & . & 0 & 0 \\ . & . & . & . & . \\ 0 & 0 & . & a_2^{I-1,I-1} & 0 \\ 0 & 0 & . & a_2^{I,I-1} & a_2^{I,I} \end{pmatrix}. \tag{6.39}$$

6. The derivatives of the liquid energy balance equations [cf. Eqs. (6.5) through (6.7)] with respect to $T_w^{(j)}$ are as follows:

$$\frac{\partial N_2^{(i)}}{\partial T_w^{(j+1)}} \triangleq b_2^{i,j} = 0;\ i = 1,\ldots,I;\ j = 1,\ldots,I;\ j \neq i-1, i; \tag{6.40}$$

$$\frac{\partial N_2^{(i)}}{\partial T_w^{(i)}} \triangleq b_2^{i,i-1} = m_w^{(i)} \frac{\partial h_f^{(i)}}{\partial T_w^{(i)}};\ i = 2,\ldots,I;\ j = i-1; \tag{6.41}$$

$$\begin{aligned} \frac{\partial N_2^{(i)}}{\partial T_w^{(i+1)}} \triangleq b_2^{i,i} &= -m_w^{(i+1)} \frac{\partial h_f^{(i+1)}}{\partial T_w^{(i+1)}} - \left(m_w^{(i)} - m_w^{(i+1)}\right) \frac{\partial h_{g,w}^{(i+1)}}{\partial T_w^{(i+1)}} \\ &\quad - H\left(m_a, \boldsymbol{\alpha}\right);\ i = 1,\ldots,I;\ j = i. \end{aligned} \tag{6.42}$$

For subsequent use, the above quantities are considered to be the components of the $I \times I$ diagonal matrix $\mathbf{B}_2$ defined as follows:

$$\mathbf{B}_2 \triangleq \left(b_2^{i,j}\right)_{I\times I} = \begin{pmatrix} b_2^{1,1} & 0 & . & 0 & 0 \\ b_2^{2,1} & b_2^{2,2} & . & 0 & 0 \\ . & . & . & . & . \\ 0 & 0 & . & b_2^{I-1,I-1} & 0 \\ 0 & 0 & . & b_2^{I,I-1} & b_2^{I,I} \end{pmatrix}. \tag{6.43}$$

7. The derivatives of the liquid energy balance equations [cf. Eqs. (6.5) through (6.7)] with respect to $T_a^{(j)}$ are as follows:

$$\frac{\partial N_2^{(i)}}{\partial T_a^{(j)}} \triangleq c_2^{i,j} = 0;\ i = 1,\ldots,I;\ i = 1,\ldots,I;\ j \neq i; \tag{6.44}$$

$$\frac{\partial N_2^{(i)}}{\partial T_a^{(i)}} \triangleq c_2^{i,i} = H\left(m_a, \boldsymbol{\alpha}\right);\ i = 1,\ldots,I;\ j = i. \tag{6.45}$$

For subsequent use, the above quantities are considered to be the components of the $I \times I$ diagonal matrix $\mathbf{C}_2$ defined as follows:

$$\mathbf{C}_2 \triangleq \left(c_2^{i,j}\right)_{I\times I} = \begin{pmatrix} c_2^{1,1} & 0 & . & 0 & 0 \\ 0 & c_2^{2,2} & . & 0 & 0 \\ . & . & . & . & . \\ 0 & 0 & . & c_2^{I-1,I-1} & 0 \\ 0 & 0 & . & 0 & c_2^{I,I} \end{pmatrix}. \tag{6.46}$$

9. The derivatives of the liquid energy balance equations [cf. Eqs. (6.5) – (6.7)] with respect to $\omega^{(j)}$ are as follows:

$$\frac{\partial N_2^{(i)}}{\partial \omega^{(j)}} \triangleq d_2^{i,j} = 0;\ i = 1,\ldots,I;\ j = 1,\ldots,I. \tag{6.47}$$

For subsequent use, the above quantities are considered to be the components of the $I \times I$ matrix

$$\mathbf{D}_2 \triangleq \left[d_2^{i,j}\right]_{I\times I} = \mathbf{0}. \tag{6.48}$$

9. The derivatives of the water vapor continuity equations [cf. Eqs. (6.8) through (6.10)] with respect to $m_w^{(j)}$ are as follows:

$$\frac{\partial N_3^{(i)}}{\partial m_w^{(j+1)}} \triangleq a_3^{i,j} = 0;\ i = 1,\ldots,I;\ j = 1,\ldots,I;\ j \neq i-1, i; \tag{6.49}$$

$$\frac{\partial N_3^{(i)}}{\partial m_w^{(i)}} \triangleq a_3^{i,i-1} = \frac{1}{m_a};\ i = 2,\ldots,I;\ j = i-1; \tag{6.50}$$

$$\frac{\partial N_3^{(i)}}{\partial m_w^{(i+1)}} \triangleq a_3^{i,i} = -\frac{1}{m_a};\ i = 1,\ldots,I;\ j = i. \tag{6.51}$$

For subsequent use, the above quantities are considered to be the components of the $I \times I$ matrix $\mathbf{A}_3$ defined as follows:

$$\mathbf{A}_3 \triangleq \left(a_3^{i,j}\right)_{I\times I} = \frac{1}{m_a}\begin{pmatrix} -1 & 0 & . & 0 & 0 \\ 1 & -1 & . & 0 & 0 \\ . & . & . & . & . \\ 0 & 0 & . & -1 & 0 \\ 0 & 0 & . & 1 & -1 \end{pmatrix}.. \tag{6.52}$$

10. The derivatives of the water vapor continuity equations [cf. Eqs. (6.8) through (6.10)] with respect to $T_w^{(j)}$ are as follows:

$$\frac{\partial N_3^{(i)}}{\partial T_w^{(j+1)}} \triangleq b_3^{i,j} = 0;\; i = 1,...,I;\; j = 1,...,I. \tag{6.53}$$

For subsequent use, the above quantities are considered to be the components of the $I \times I$ matrix

$$\mathbf{B}_3 \triangleq \left[b_3^{i,j}\right]_{I\times I} = \mathbf{0}. \tag{6.54}$$

11. The derivatives of the water vapor continuity equations [cf. Eqs. (6.8) through (6.10)] with respect to $T_a^{(j)}$ are as follows:

$$\frac{\partial N_3^{(i)}}{\partial T_a^{(j)}} \triangleq c_3^{i,j} = 0;\; i = 1,...,I;\; j = 1,...,I. \tag{6.55}$$

For subsequent use, the above quantities are considered to be the components of the $I \times I$ matrix

$$\mathbf{C}_3 \triangleq \left[c_3^{i,j}\right]_{I\times I} = \mathbf{0}. \tag{6.56}$$

12. The derivatives of the water vapor continuity equations [cf. Eqs. (6.8) through (6.10)] with respect to $\omega^{(j)}$ are as follows:

$$\frac{\partial N_3^{(i)}}{\partial \omega^{(j)}} \triangleq d_3^{i,j} = 0;\; i = 1,...,I;\; j = 1,...,I;\; j \neq i, i+1; \tag{6.57}$$

$$\frac{\partial N_3^{(i)}}{\partial \omega^{(i)}} \triangleq d_3^{i,i} = -1;\; i = 1,...,I;\; j = i. \tag{6.58}$$

$$\frac{\partial N_3^{(i)}}{\partial \omega^{(i+1)}} \triangleq d_3^{i,i+1} = 1;\ i = 1,...,I-1; j = i+1. \tag{6.59}$$

For subsequent use, the above quantities are considered to be the components of the $I \times I$ matrix $\mathbf{D}_3$ defined as follows:

$$\mathbf{D}_3 \triangleq \left(d_3^{i,j}\right)_{I\times I} = \begin{pmatrix} -1 & 1 & . & 0 & 0 \\ 0 & -1 & . & 0 & 0 \\ . & . & . & . & . \\ 0 & 0 & . & -1 & 1 \\ 0 & 0 & . & 0 & -1 \end{pmatrix}. \tag{6.60}$$

13. The derivatives of the air/water vapor energy balance equations [cf. Eqs. (6.11) through (6.13)] with respect to $m_w^{(j)}$ are as follows:

$$\frac{\partial N_4^{(i)}}{\partial m_w^{(j+1)}} \triangleq a_4^{i,j} = 0;\ i = 1,...,I;\ j = 1,...,I;\ j \neq i-1, i; \tag{6.61}$$

$$\frac{\partial N_4^{(i)}}{\partial m_w^{(i)}} \triangleq a_4^{i,i-1} = \frac{h_{g,w}^{(i+1)}\left(T_w^{(i+1)}, \boldsymbol{\alpha}\right)}{m_a};\ i = 2,...,I;\ j = i-1; \tag{6.62}$$

$$\frac{\partial N_4^{(i)}}{\partial m_w^{(i+1)}} \triangleq a_4^{i,i} = -\frac{h_{g,w}^{(i+1)}\left(T_w^{(i+1)}, \boldsymbol{\alpha}\right)}{m_a};\ i = 1,...,I;\ j = i. \tag{6.63}$$

For subsequent use, the above quantities are considered to be the components of the $I \times I$ matrix $\mathbf{A}_4$ defined as follows:

$$\mathbf{A}_4 \triangleq \left(a_4^{i,j}\right)_{I\times I} = \begin{pmatrix} a_4^{1,1} & 0 & . & 0 & 0 \\ a_4^{2,1} & a_4^{2,2} & . & 0 & 0 \\ . & . & . & . & . \\ 0 & 0 & . & a_4^{I-1,I-1} & 0 \\ 0 & 0 & . & a_4^{I,I-1} & a_4^{I,I} \end{pmatrix}. \tag{6.64}$$

14. The derivatives of the air/water vapor energy balance equations [cf. Eqs. (6.11) through (6.13)] with respect $T_w^{(j)}$ are as follows:

$$\frac{\partial N_4^{(i)}}{\partial T_w^{(j+1)}} \triangleq b_4^{i,j} = 0;\ i = 1,\ldots,I;\ j \neq i; \tag{6.65}$$

$$\frac{\partial N_4^{(i)}}{\partial T_w^{(i+1)}} \triangleq b_4^{i,i} = \frac{1}{m_a}\left[\left(m_w^{(i)} - m_w^{(i+1)}\right)\frac{\partial h_{g,w}^{(i+1)}}{\partial T_w^{(i+1)}} + H\left(m_a,\boldsymbol{\alpha}\right)\right]; \\ i = 1,\ldots,I;\ j = i. \tag{6.66}$$

For subsequent use, the above quantities are considered to be the components of the $I \times I$ diagonal matrix $\mathbf{B}_4$ defined as follows:

$$\mathbf{B}_4 \triangleq \left(b_4^{i,j}\right)_{I\times I} = \begin{pmatrix} b_4^{1,1} & 0 & . & 0 & 0 \\ 0 & b_4^{2,2} & . & 0 & 0 \\ . & . & . & . & . \\ 0 & 0 & . & b_4^{I-1,I-1} & 0 \\ 0 & 0 & . & 0 & b_4^{I,I} \end{pmatrix}. \tag{6.67}$$

15. The derivatives of the air/water vapor energy balance equations [cf. Eqs. (6.11) through (6.13)] with respect to $T_a^{(j)}$ are as follows:

$$c_4^{i,j} \triangleq \frac{\partial N_4^{(i)}}{\partial T_a^{(j)}} = 0;\ i = 1,\ldots,I;\ j \neq i,i+1; \tag{6.68}$$

$$\frac{\partial N_4^{(i)}}{\partial T_a^{(i)}} \triangleq c_4^{i,i} = \left(T_a^{(i+1)} - T_a^{(i)}\right)\frac{\partial C_p^{(i)}}{\partial T_a^{(i)}} - C_p^{(i)}\left(\frac{T_a^{(i)} + 273.15}{2},\boldsymbol{\alpha}\right) \\ - \omega^{(i)}\frac{\partial h_{g,a}^{(i)}}{\partial T_a^{(i)}} - \frac{H\left(m_a,\boldsymbol{\alpha}\right)}{m_a};\ i = 1,\ldots,I;\ j = i. \tag{6.69}$$

$$\frac{\partial N_4^{(i)}}{\partial T_a^{(i+1)}} \triangleq c_4^{i,i+1} = C_p^{(i)}\left(\frac{T_a^{(i)} + 273.15}{2},\boldsymbol{\alpha}\right) + \omega^{(i+1)}\frac{\partial h_{g,a}^{(i+1)}}{\partial T_a^{(i+1)}}; \\ i = 1,\ldots,I-1;\ j = i+1. \tag{6.70}$$

For subsequent use, the above quantities are considered to be the components of the $I \times I$ diagonal matrix defined as follows:

$$\mathbf{C}_4 \triangleq \left(c_4^{i,j}\right)_{I\times I} = \begin{pmatrix} c_4^{1,1} & c_4^{1,2} & . & 0 & 0 \\ 0 & c_4^{2,2} & . & 0 & 0 \\ . & . & . & . & . \\ 0 & 0 & . & c_4^{I-1,I-1} & c_4^{I-1,I} \\ 0 & 0 & . & 0 & c_4^{I,I} \end{pmatrix}. \tag{6.71}$$

16. The derivatives of the air/water vapor energy balance equations [cf. Eqs. (6.11) through (6.13)] with respect to $\omega^{(i)}$ are as follows:

$$\frac{\partial N_4^{(i)}}{\partial \omega^{(j)}} \triangleq d_4^{i,j} = 0;\; i = 1,\ldots,I;\; j \neq i, i+1; \tag{6.72}$$

$$\frac{\partial N_4^{(i)}}{\partial \omega^{(i)}} \triangleq d_4^{i,i} = -h_{g,a}^{(i)}\left(T_a^{(i)},\boldsymbol{\alpha}\right);\; i = 1,\ldots,I;\; j = i; \tag{6.73}$$

$$\frac{\partial N_4^{(i)}}{\partial \omega^{(i+1)}} \triangleq d_4^{i,i+1} = h_{g,a}^{(i+1)}\left(T_a^{(i+1)},\boldsymbol{\alpha}\right);\; i = 1,\ldots,I-1;\; j = i+1. \tag{6.74}$$

For subsequent use, the above quantities are considered to be the components of the $I \times I$ diagonal matrix $\mathbf{D}_4$ defined as follows:

$$\mathbf{D}_4 \triangleq \left(d_4^{i,j}\right)_{I\times I} = \begin{pmatrix} d_4^{1,1} & d_4^{1,2} & . & 0 & 0 \\ 0 & d_4^{2,2} & . & 0 & 0 \\ . & . & . & . & . \\ 0 & 0 & . & d_4^{I-1,I-1} & d_4^{I-1,I} \\ 0 & 0 & . & 0 & d_4^{I,I} \end{pmatrix}. \tag{6.75}$$

The Jacobian represented by Eq. (6.18) is a non-symmetric sparse matrix of order 196 by 196, with 14 nonzero diagonals. The non-symmetric diagonal storage format is used to store the respective 14 nonzero diagonals, so that the "condensed" Jacobian matrix has dimensions 196 by 14. Since the Jacobian is highly non-symmetric, the cost of the iterations of the GMRES solver grows as $O(m^2)$, where m is the iteration number within the GMRES solver. To reduce this computational cost, the GMRES solver is configured to run with the restart feature. The optimized value for the restart frequency is 10 for this specific application. The MIC preconditioner can speed up the convergence of the GMRES solver (Saad et al 1986) using the parameters OMEGA and LVFILL in the modified incomplete factorization methods for the MIC preconditioner; for this application the following values were found to be optimal: OMEGA = 0.000000001 and LVFILL = 1. The Jacobian is not updated inside the sparse GMRES solver. The default convergence of GMRES is tested with the following criterion:

$$\left[\frac{\langle \tilde{\mathbf{z}}^{(m)}, \tilde{\mathbf{z}}^{(m)} \rangle}{\langle \delta\mathbf{u}^{(m)}, \delta\mathbf{u}^{(m)} \rangle}\right]^{\frac{1}{2}} < \zeta \tag{6.76}$$

where $\tilde{\mathbf{z}}^{(m)}$ denotes the pseudo-residual at m^{th}-iteration of the GMRES solver, $\delta\mathbf{u}^{(m)}$ is the solution of Eq. (6.17) at m^{th}-iteration, and ζ denotes the stopping test value for the GMRES solver.

(e) Set

$$\mathbf{u}_{n+1} = \mathbf{u}_n + \delta\mathbf{u}, \tag{6.77}$$

where n is the current outer loop iteration number, and update the Jacobian.

(f) Test for convergence of the outer loop until the error in the solution is less than a specified maximum value. For solving Eqs. (6.2) through (6.13), the following error criterion has been used:

$$error = max\left(\frac{\left|\delta m_w^{(i)}\right|}{m_w^{(i)}}, \frac{\left|\delta T_w^{(i)}\right|}{T_w^{(i)}}, \frac{\left|\delta T_a^{(i)}\right|}{T_a^{(i)}}, \frac{\left|\delta \omega^{(i)}\right|}{\omega^{(i)}}\right) < 10^{-6}. \tag{6.78}$$

(g) Set $n = n + 1$ and go to step d.

The above solution strategy for solving Eqs. (6.2) through (6.13) converged successfully for all the 8079 benchmark data sets. As described previously, each of these data sets contained measurements of the following quantities: (i) outlet air temperature measured with the sensor called "Tidbit"; (ii) outlet air temperature measured with the sensor called "Hobo"; (iii) outlet water temperature; (iv) outlet air relative humidity. For each of these benchmark data sets, the outer loop iterations described above (i.e., steps c through g) converge in 4 iterations; for each outer loop iteration, the GMRES solver used for solving Eq. (6.17) converges in 12 iterations. The "zero-to-zero" verification of the solution's accuracy using Eqs. (6.2) through (6.13) gives an error of the order of 10^{-7}.

In view of the above-mentioned measurements, the responses of interest are as follows:

(a) the vector $\mathbf{m}_w \triangleq \left[m_w^{(2)},...,m_w^{(I+1)}\right]^\dagger$ of water mass flow rates at the exit of each control volume i, $(i = 1,...,49)$;

(b) the vector $\mathbf{T}_w \triangleq \left[T_w^{(2)},...,T_w^{(I+1)}\right]^\dagger$ of water temperatures at the exit of each control volume i, $(i = 1,...,49)$;

(c) the vector $\mathbf{T}_a \triangleq \left[T_a^{(1)},...,T_a^{(I)}\right]^\dagger$ of air temperatures at the exit of each control volume i, $(i=1,...,49)$;

(d) the vector $\mathbf{RH} \triangleq \left[RH^{(1)},...,RH^{(I)}\right]^\dagger$, having as components the air relative humidity at the exit of each control volume i, $(i=1,...,49)$.

While the water mass flow rates $m_w^{(i)}$, the water temperatures $T_w^{(i)}$, and the air temperatures $T_a^{(i)}$ are obtained directly as the solutions of Eqs.(6.2) through (6.13), the air relative humidity, $RH^{(i)}$, is computed for each control volume using the expression :

$$RH^{(i)} = \frac{P_v\left(\omega^{(i)},\boldsymbol{\alpha}\right)}{P_{vs}\left(T_a^{(i)},\boldsymbol{\alpha}\right)} \times 100 = \frac{\left(\dfrac{\omega^{(i)} P_{atm}}{\omega^{(i)}+0.622}\right)}{\left(e^{a_0+\frac{a_1}{T_a^{(i)}}}\right)} \times 100. \tag{6.79}$$

The bar plots, showing the respective values of the water mass flow rates $m_w^{(i)}$, the water temperatures $T_w^{(i)}$, the air temperatures $T_a^{(i)}$, and the air relative humidity, $RH^{(i)}$, at the exit of each control volume, are presented in Figures 6.4 through 6.7, below.

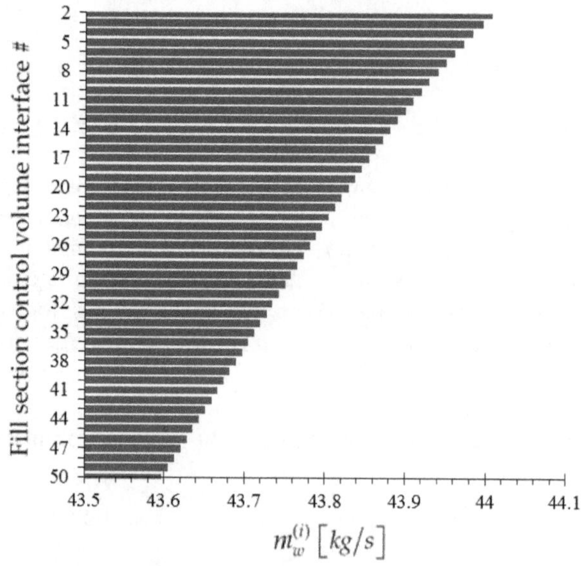

***Figure 6.4:** Bar plot of the water mass flow rates $m_w^{(i)}$, $(i=2,...,50)$, at the exit of each control volume along the height of the fill section of the cooling tower.*

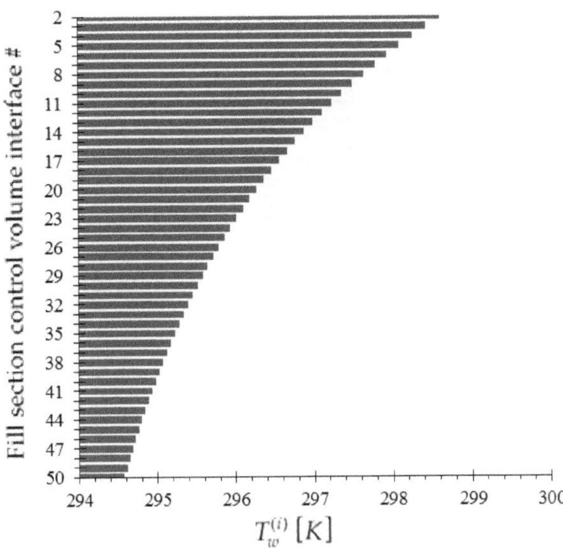

Figure 6.5: Bar plot of the water temperatures $T_w^{(i)}$, $(i = 2,...,50)$, at the exit of each control volume along the height of the fill section of the cooling tower.

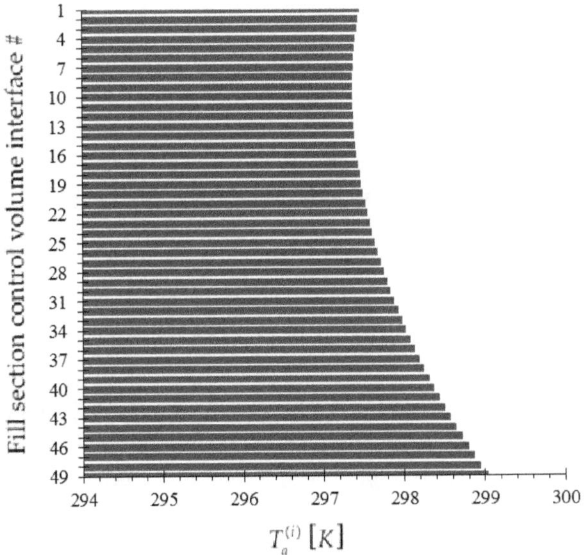

Figure 6.6: Bar plot of the air temperatures $T_a^{(i)}$, $(i = 1,...,49)$, at the exit of each control volume along the height of the fill section of the cooling tower.

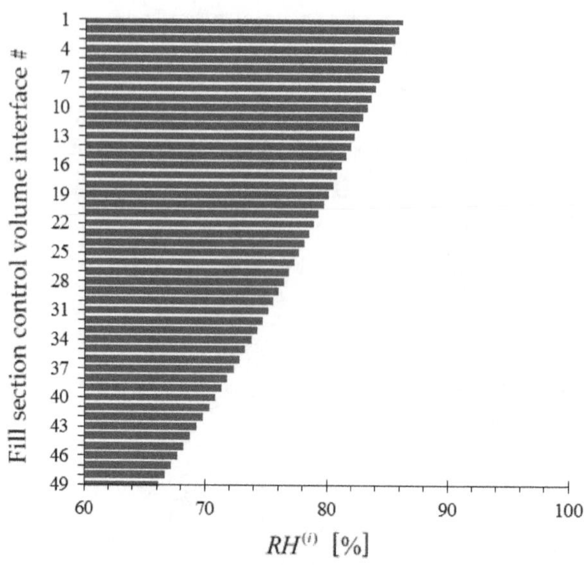

***Figure 6.7:** Bar plot of the air relative humidity $RH^{(i)}$, $(i = 1,...,49)$, at the exit of each control volume along the height of the fill section of the cooling tower.*

6.3 COOLING TOWER ADJOINT SENSITIVITY MODEL: DEVELOPMENT AND SOLUTION VERIFICATION

All of the responses of interest in this work, e.g., the experimentally measured and/or computed responses discussed in the previous Sections, can be generally represented in the functional form $R(\mathbf{m}_w, \mathbf{T}_w, \mathbf{T}_a, \boldsymbol{\omega}; \boldsymbol{\alpha})$, where R is a known functional of the model's state functions and parameters. As generally shown by Cacuci (1981a, 1981b), the sensitivity of such as this response to arbitrary variations in the model's parameters $\delta\boldsymbol{\alpha} \triangleq (\delta\alpha_1, ..., \delta\alpha_{N_\alpha})$ and state functions $\delta\mathbf{m}_w, \delta\mathbf{T}_w, \delta\mathbf{T}_a, \delta\boldsymbol{\omega}$ is provided by the response's Gateaux (G-) differential $DR(\mathbf{m}_w^0, \mathbf{T}_w^0, \mathbf{T}_a^0, \boldsymbol{\omega}^0; \boldsymbol{\alpha}^0; \delta\mathbf{m}_w, \delta\mathbf{T}_w, \delta\mathbf{T}_a, \delta\boldsymbol{\omega}; \delta\boldsymbol{\alpha})$, which is defined as follows:

$$
\begin{aligned}
&DR\left(\mathbf{m}_w^0,\mathbf{T}_w^0,\mathbf{T}_a^0,\boldsymbol{\omega}^0;\boldsymbol{\alpha}^0;\delta\mathbf{m}_w,\delta\mathbf{T}_w,\delta\mathbf{T}_a,\delta\boldsymbol{\omega};\delta\boldsymbol{\alpha}\right)\triangleq\\
&\frac{d}{d\varepsilon}\left[R\left(\mathbf{m}_w^0+\varepsilon\delta\mathbf{m}_w,\mathbf{T}_w^0+\varepsilon\delta\mathbf{T}_w,\mathbf{T}_a^0+\varepsilon\delta\mathbf{T}_a,\boldsymbol{\omega}^0+\varepsilon\delta\boldsymbol{\omega};\boldsymbol{\alpha}^0+\varepsilon\delta\boldsymbol{\alpha}\right)\right]_{\varepsilon=0}\\
&=DR_{direct}+DR_{indirect},
\end{aligned}
\tag{6.80}
$$

where the so-called "direct effect" term, DR_{direct}, and the so-called "indirect effect" term, $DR_{indirect}$, are defined, respectively, as follows:

$$
DR_{direct}\triangleq\sum_{i=1}^{N_\alpha}\left(\frac{\partial R}{\partial\alpha_i}\delta\alpha_i\right),
\tag{6.81}
$$

$$
\begin{aligned}
DR_{indirect}&\triangleq\sum_{i=1}^{I}\left(\frac{\partial R}{\partial m_w^{(i+1)}}\delta m_w^{(i+1)}+\frac{\partial R}{\partial T_w^{(i+1)}}\delta T_w^{(i+1)}+\frac{\partial R}{\partial T_a^{(i)}}\delta T_a^{(i)}+\frac{\partial R}{\partial\omega^{(i)}}\delta\omega^{(i)}\right)\\
&=\mathbf{R}_1\cdot\delta\mathbf{m}_w+\mathbf{R}_2\cdot\delta\mathbf{T}_w+\mathbf{R}_3\cdot\delta\mathbf{T}_a+\mathbf{R}_4\cdot\delta\boldsymbol{\omega},
\end{aligned}
\tag{6.82}
$$

where the components of the vectors $\mathbf{R}_\ell\triangleq\left(r_\ell^{(1)},\ldots,r_\ell^{(I)}\right)$, $\ell=1,2,3,4$ are defined as follows:

$$
r_1^{(i)}\triangleq\frac{\partial R}{\partial m_w^{(i+1)}};\quad r_2^{(i)}\triangleq\frac{\partial R}{\partial T_w^{(i+1)}};\quad r_3^{(i)}\triangleq\frac{\partial R}{\partial T_a^{(i)}};\quad r_4^{(i)}\triangleq\frac{\partial R}{\partial\omega^{(i)}};\qquad i=1,\ldots,I.
\tag{6.83}
$$

Since the model parameters are related to the model's state functions through Eqs. (6.2) through (6.13), it follows that variations in the model parameter will induce variations in the state variables. More precisely, it has been shown by Cacuci (1981a, 1981b) that to first-order in the parameter variations, the respective variations in the state variables can be computed by solving the G-differentiated model equations, namely:

$$
\frac{d}{d\varepsilon}\left[\mathbf{N}\left(\mathbf{u}^0+\varepsilon\delta\mathbf{u};\boldsymbol{\alpha}^0+\varepsilon\delta\boldsymbol{\alpha}\right)\right]_{\varepsilon=0}=\mathbf{0},
\tag{6.84}
$$

Performing the above differentiation on Eqs. (6.2) through (6.13) yields the following forward sensitivity system:

$$\begin{pmatrix} \mathbf{A}_1 & \mathbf{B}_1 & \mathbf{C}_1 & \mathbf{D}_1 \\ \mathbf{A}_2 & \mathbf{B}_2 & \mathbf{C}_2 & \mathbf{D}_2 \\ \mathbf{A}_3 & \mathbf{B}_3 & \mathbf{C}_3 & \mathbf{D}_3 \\ \mathbf{A}_4 & \mathbf{B}_4 & \mathbf{C}_4 & \mathbf{D}_4 \end{pmatrix} \begin{pmatrix} \delta \mathbf{m}_w \\ \delta \mathbf{T}_w \\ \delta \mathbf{T}_a \\ \delta \boldsymbol{\omega} \end{pmatrix} = \begin{pmatrix} \mathbf{Q}_1 \\ \mathbf{Q}_2 \\ \mathbf{Q}_3 \\ \mathbf{Q}_4 \end{pmatrix}, . \tag{6.85}$$

where the components of the vectors $\mathbf{Q}_\ell \triangleq \left(q_\ell^{(1)},...,q_\ell^{(I)}\right)$, $\ell = 1,2,3,4$ are defined as follows:

$$q_\ell^{(i)} \triangleq \sum_{j=1}^{N_\alpha} \left(\frac{\partial N_\ell^{(i)}}{\partial \alpha_j} \delta \alpha_j \right); \; i = 1,...,I; \; \ell = 1,2,3,4. \tag{6.86}$$

The system represented by Eq. (6.85) is called the *forward sensitivity system*, which can be solved, in principle, to compute the variations in the state functions for every variation in the model parameters. In turn, the solution of Eq. (6.85) can be used in Eq. (6.82) to compute the "indirect effect" term, $DR_{indirect}$. However, since there are many parameter variations to consider, solving Eq. (6.85) repeatedly to compute $DR_{indirect}$ becomes computationally impracticable. The need for solving Eq. (6.85) repeatedly to compute $DR_{indirect}$ can be circumvented by applying the Adjoint Sensitivity Analysis Procedure (ASAM) formulated by Cacuci (1981a, 1981b). The ASAM proceeds by forming the inner-product of Eq. (6.85) with a yet unspecified vector of the form $[\boldsymbol{\mu}_w, \boldsymbol{\tau}_w, \boldsymbol{\tau}_a, \mathbf{o}]^\dagger$, having the same structure as the vector $\mathbf{u} \triangleq (\mathbf{m}_w, \mathbf{T}_w, \mathbf{T}_a, \boldsymbol{\omega})^\dagger$, transposing the resulting scalar equation and using Eq. (6.82). Furthermore, by requiring that the vector $[\boldsymbol{\mu}_w, \boldsymbol{\tau}_w, \boldsymbol{\tau}_a, \mathbf{o}]^\dagger$ satisfy the following adjoint sensitivity system:

$$\begin{pmatrix} \mathbf{A}_1^\dagger & \mathbf{A}_2^\dagger & \mathbf{A}_3^\dagger & \mathbf{A}_4^\dagger \\ \mathbf{B}_1^\dagger & \mathbf{B}_2^\dagger & \mathbf{B}_3^\dagger & \mathbf{B}_4^\dagger \\ \mathbf{C}_1^\dagger & \mathbf{C}_2^\dagger & \mathbf{C}_3^\dagger & \mathbf{C}_4^\dagger \\ \mathbf{D}_1^\dagger & \mathbf{D}_2^\dagger & \mathbf{D}_3^\dagger & \mathbf{D}_4^\dagger \end{pmatrix} \begin{pmatrix} \boldsymbol{\mu}_w \\ \boldsymbol{\tau}_w \\ \boldsymbol{\tau}_a \\ \mathbf{o} \end{pmatrix} = \begin{pmatrix} \mathbf{R}_1 \\ \mathbf{R}_2 \\ \mathbf{R}_3 \\ \mathbf{R}_4 \end{pmatrix}, \tag{6.87}$$

it follows that the "indirect effect" term can be expressed in the form

$$DR_{indirect} = \boldsymbol{\mu}_w \cdot \mathbf{Q}_1 + \boldsymbol{\tau}_w \cdot \mathbf{Q}_2 + \boldsymbol{\tau}_a \cdot \mathbf{Q}_3 + \mathbf{o} \cdot \mathbf{Q}_4. \tag{6.88}$$

The system represented by Eq. (6.87) is called the *adjoint sensitivity system*, which –notably– is independent of parameter variations. Therefore, the adjoint sensitivity system needs to be solved only once, to compute the adjoint functions $[\boldsymbol{\mu}_w, \boldsymbol{\tau}_w, \boldsymbol{\tau}_a, \mathbf{o}]^\dagger$. In turn, the adjoint functions are used to compute $DR_{indirect}$, efficiently and exactly, using Eq. (6.88). As an illustrative example of computing response sensitivities using the adjoint sensitivity system, consider that the model response of interest is the air relative humidity, $RH^{(i)}$, in a generic control volume i, as given by Eq.(6.79). For this model response, the "direct effect" term, denoted as $D\left[RH^{(i)}\right]_{direct}$, is readily obtained in the form

$$D\left[RH^{(i)}\right]_{direct} = \frac{\partial\left(RH^{(i)}\right)}{\partial P_{atm}}\left(\delta P_{atm}\right) + \frac{\partial\left(RH^{(i)}\right)}{\partial a_0}\left(\delta a_0\right) + \frac{\partial\left(RH^{(i)}\right)}{\partial a_1}\left(\delta a_1\right), \quad i = 1,\ldots,I; \tag{6.89}$$

where

$$\frac{\partial\left(RH^{(i)}\right)}{\partial P_{atm}} = \frac{\partial}{\partial P_{atm}}\left[\frac{P_v\left(\omega^{(i)},\boldsymbol{\alpha}\right)}{P_{vs}\left(T_a^{(i)},\boldsymbol{\alpha}\right)}\times 100\right] = \frac{0.622}{\left(0.622+\omega^{(i)}\right)e^{a_0+\frac{a_1}{T_a^{(i)}}}}\times 100; \quad i = 1,\ldots,I; \tag{6.90}$$

$$\frac{\partial\left(RH^{(i)}\right)}{\partial a_0} = \frac{\partial}{\partial a_0}\left[\frac{P_v\left(\omega^{(i)},\boldsymbol{\alpha}\right)}{P_{vs}\left(T_a^{(i)},\boldsymbol{\alpha}\right)}\times 100\right] = -\frac{0.622P_{atm}}{\left(0.622+\omega^{(i)}\right)e^{a_0+\frac{a_1}{T_a^{(i)}}}}\times 100; \quad i = 1,\ldots,I; \tag{6.91}$$

$$\frac{\partial\left(RH^{(i)}\right)}{\partial a_1} = \frac{\partial}{\partial a_1}\left[\frac{P_v\left(\omega^{(i)},\boldsymbol{\alpha}\right)}{P_{vs}\left(T_a^{(i)},\boldsymbol{\alpha}\right)}\times 100\right] = -\frac{0.622P_{atm}}{\left(0.622+\omega^{(i)}\right)e^{a_0+\frac{a_1}{T_a^{(i)}}}}\frac{-1}{T_a^{(i)}}\times 100; \quad i = 1,\ldots,I. \tag{6.92}$$

On the other hand, the "indirect effect" term, denoted as $D\left[RH^{(i)}\right]_{indirect}$, is readily obtained in the form

$$D\left[RH^{(i)}\right]_{indirect} = \frac{\partial\left(RH^{(i)}\right)}{\partial \omega^{(i)}}\left(\delta\omega^{(i)}\right) + \frac{\partial\left(RH^{(i)}\right)}{\partial T_a^{(i)}}\left(\delta T_a^{(i)}\right); \quad i = 1,\ldots,I; \tag{6.93}$$

where

$$\begin{aligned}\frac{\partial\left(RH^{(i)}\right)}{\partial\omega^{(i)}}&=\frac{\partial}{\partial\omega^{(i)}}\left[\frac{P_v\left(\omega^{(i)},\boldsymbol{\alpha}\right)}{P_{vs}\left(T_a^{(i)},\boldsymbol{\alpha}\right)}\times 100\right]=\frac{100}{P_{vs}\left(T_a^{(i)},\alpha\right)}\frac{\partial P_v\left(\omega^{(i)},\boldsymbol{\alpha}\right)}{\partial\omega^{(i)}}\\&=\frac{0.622P_{atm}}{\left(0.622+\omega^{(i)}\right)^2 e^{a_0+\frac{a_1}{T_a^{(i)}}}}\times 100;\quad i=1,\dots,I;\end{aligned}\tag{6.94}$$

$$\begin{aligned}\frac{\partial\left(RH^{(i)}\right)}{\partial T_a^{(i)}}&=\frac{\partial}{\partial T_a^{(i)}}\left[\frac{P_v\left(\omega^{(i)},\boldsymbol{\alpha}\right)}{P_{vs}\left(T_a^{(i)},\boldsymbol{\alpha}\right)}\times 100\right]=100\times P_v\left(\omega^{(i)},\boldsymbol{\alpha}\right)\frac{\partial}{\partial T_a^{(i)}}\left[\frac{1}{P_{vs}\left(T_a^{(i)},\boldsymbol{\alpha}\right)}\right]\\&=\frac{0.622P_{atm}}{\left(0.622+\omega^{(i)}\right)e^{a_0+\frac{a_1}{T_a^{(i)}}}}\frac{a_1}{\left[T_a^{(i)}\right]^2}\times 100;\quad i=1,\dots,I.\end{aligned}\tag{6.95}$$

The units of the adjoint functions can be determined from Eq. (6.88) through dimensional analysis. Specifically, the units for the adjoint functions satisfy the following relation:

$$\left[\mu_w^{(i)}\right]=\frac{[R]}{[N_1]};\quad\left[\tau_w^{(i)}\right]=\frac{[R]}{[N_2]};\quad\left[\tau_a^{(i)}\right]=\frac{[R]}{[N_3]};\quad\left[o^{(i)}\right]=\frac{[R]}{[N_4]}\tag{6.96}$$

where "[R]" denotes the unit of the response R, while the units for the respective equations are as follows:

$$[N_1]=\frac{kg}{s};\quad[N_2]=\frac{J}{s};\quad[N_3]=[-];\quad[N_4]=\frac{J}{kg}\,.\tag{6.97}$$

Table 6.1 below lists the units of the adjoint functions for four responses: $R\triangleq T_a^{(1)}$, $R\triangleq T_w^{(50)}$, $R\triangleq RH^{(i)}$ and $R\triangleq m_w^{(50)}$, respectively, in which, $T_a^{(1)}$ denotes exit air temperature; $T_w^{(50)}$ denotes exit water temperature; $RH^{(50)}$ denotes exit air relative humidity; and $m_w^{(50)}$ denotes exit water mass flow rate.

Table 6.1. Units of the adjoint functions for different responses.

Responses	$\left[\mu_w^{(i)}\right]$	$\left[\tau_w^{(i)}\right]$	$\left[\tau_a^{(i)}\right]$	$\left[o^{(i)}\right]$
$R\triangleq T_a^{(1)}$	$K/(kg/s)$	$K/(J/s)$	K	$K/(J/kg)$
$R\triangleq T_w^{(50)}$	$K/(kg/s)$	$K/(J/s)$	K	$K/(J/kg)$
$R\triangleq RH^{(50)}$	$(kg/s)^{-1}$	$(J/s)^{-1}$	$-$	$(J/kg)^{-1}$
$R\triangleq m_w^{(50)}$	$-$	$(J/kg)^{-1}$	kg/s	$(kg/s)/(J/kg)$

Note that the adjoint sensitivity system represented by Eq. (6.87) is linear in the adjoint state functions, so it can be solved by using numerical methods appropriate for large-scale sparse linear systems. In particular for the results presented in this Chapter, Eq. (6.87) was solved using NSPCG (Oppe et al, 1986); 12 to 18 iterations sufficed for solving the adjoint system within convergence criterion of $\zeta = 10^{-12}$. The bar plots of the adjoint functions corresponding to the four measured responses of interest, namely: (i) the exit air temperature $R \triangleq T_a^{(1)}$; (ii) the outlet (exit) water temperature $R \triangleq T_w^{(50)}$; (iii) the exit air humidity ratio $R \triangleq RH^{(1)}$; and (iv) the outlet (exit) water mass flow rate $R \triangleq m_w^{(50)}$, are presented in Figures 6.8 through 6.11.

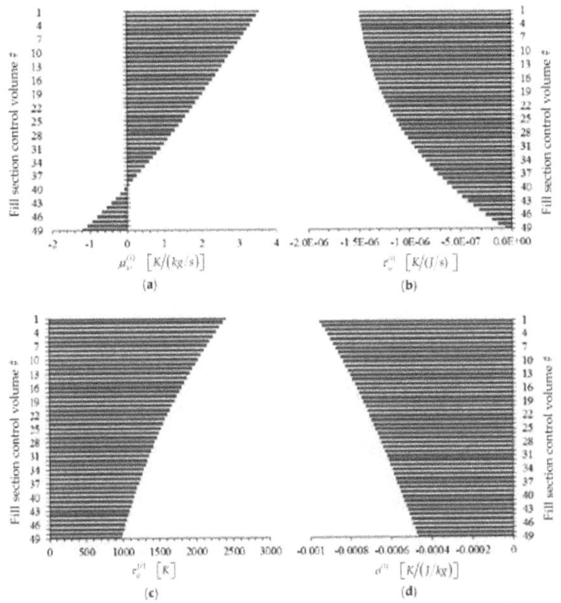

Figure 6.8: Bar plots of adjoint functions for the response $R \triangleq T_a^{(1)}$ ***as functions of the height of the cooling tower's fill section: (a)*** $\boldsymbol{\mu}_w \triangleq \left(\mu_w^{(1)},...,\mu_w^{(49)}\right)$, ***(b)*** $\boldsymbol{\tau}_w \triangleq \left(\tau_w^{(1)},...,\tau_w^{(49)}\right)$, ***(c)*** $\boldsymbol{\tau}_a \triangleq \left(\tau_a^{(1)},...,\tau_a^{(49)}\right)$, ***(d)*** $\mathbf{o} \triangleq \left(o^{(1)},...,o^{(49)}\right)$.

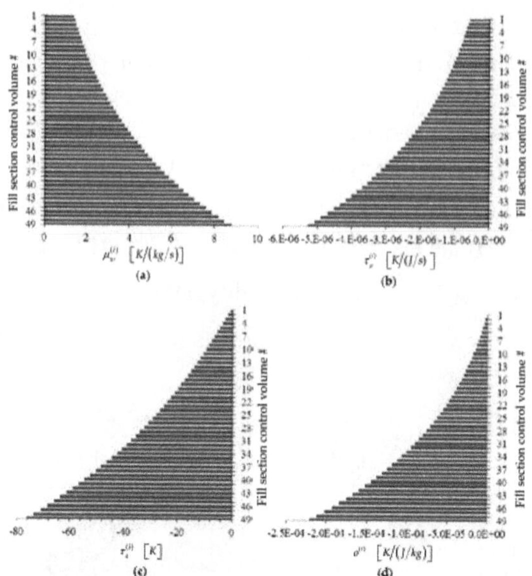

Figure 6.9: Bar plots of adjoint functions for the response $R \triangleq T_w^{(50)}$ ***as functions of the height of the cooling tower's fill section: (a)*** $\boldsymbol{\mu}_w \triangleq \left(\mu_w^{(1)},\ldots,\mu_w^{(49)}\right)$, ***(b)*** $\boldsymbol{\tau}_w \triangleq \left(\tau_w^{(1)},\ldots,\tau_w^{(49)}\right)$, ***(c)*** $\boldsymbol{\tau}_a \triangleq \left(\tau_a^{(1)},\ldots,\tau_a^{(49)}\right)$, ***(d)*** $\mathbf{o} \triangleq \left(o^{(1)},\ldots,o^{(49)}\right)$.

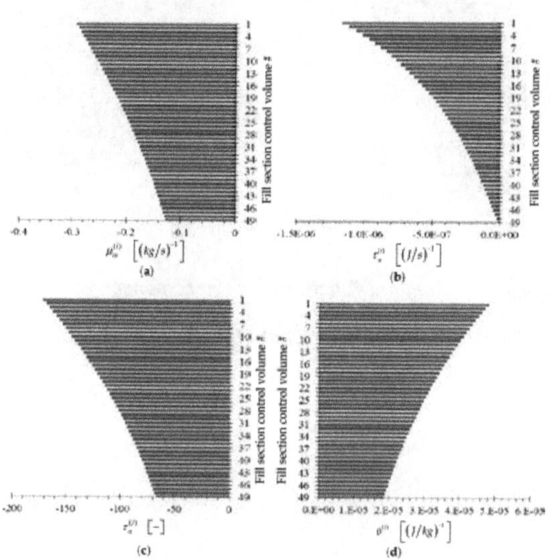

Figure 6.10: Bar plots of adjoint functions for the response $R \triangleq RH^{(1)}$ ***as functions of the height of the cooling tower's fill section: (a)*** $\boldsymbol{\mu}_w \triangleq \left(\mu_w^{(1)},\ldots,\mu_w^{(49)}\right)$, ***(b)*** $\boldsymbol{\tau}_w \triangleq \left(\tau_w^{(1)},\ldots,\tau_w^{(49)}\right)$, ***(c)*** $\boldsymbol{\tau}_a \triangleq \left(\tau_a^{(1)},\ldots,\tau_a^{(49)}\right)$, ***(d)*** $\mathbf{o} \triangleq \left(o^{(1)},\ldots,o^{(49)}\right)$.

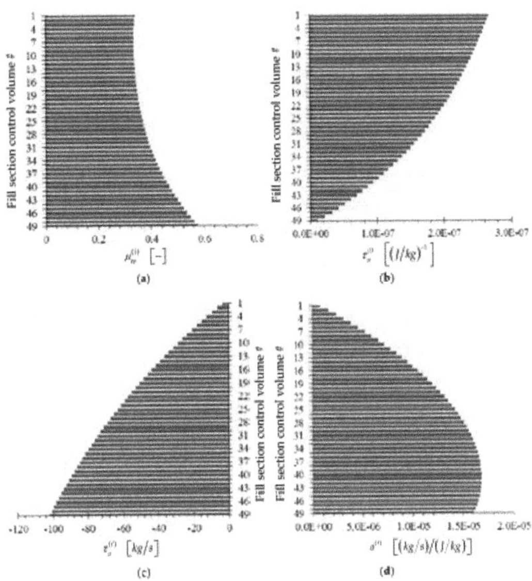

Figure 6.11: Bar plots of adjoint functions for the response $R \triangleq m_w^{(50)}$***, as functions of the height of the cooling tower's fill section: (a)*** $\boldsymbol{\mu}_w \triangleq \left(\mu_w^{(1)},...,\mu_w^{(49)}\right)$***, (b)*** $\boldsymbol{\tau}_w \triangleq \left(\tau_w^{(1)},...,\tau_w^{(49)}\right)$***, (c)*** $\boldsymbol{\tau}_a \triangleq \left(\tau_a^{(1)},...,\tau_a^{(49)}\right)$***, (d)*** $\mathbf{o} \triangleq \left(o^{(1)},...,o^{(49)}\right)$.

The numerical accuracy of the computed adjoint functions can be independently verified by first noting that Eqs. (6.80), (6.81) and (6.88) imply that

$$DR\left(\mathbf{m}_w^0,\mathbf{T}_w^0,\mathbf{T}_a^0,\boldsymbol{\omega}^0;\boldsymbol{\alpha}^0;\delta\mathbf{m}_w,\delta\mathbf{T}_w,\delta\mathbf{T}_a,\delta\boldsymbol{\omega};\delta\boldsymbol{\alpha}\right)=\sum_{j=1}^{N_\alpha} S_j\,\delta\alpha_j, \qquad (6.98)$$

where N_α denotes the total number of model parameters, and S_j denotes the "absolute sensitivity" of the response R with respect to the parameter α_j, and is defined as

$$S_j \triangleq \frac{\partial R}{\partial \alpha_j} - \sum_{i=1}^{I}\left(\mu_w^{(i)}\frac{\partial N_1^{(i)}}{\partial \alpha_j}+\tau_w^{(i)}\frac{\partial N_2^{(i)}}{\partial \alpha_j}+\tau_a^{(i)}\frac{\partial N_3^{(i)}}{\partial \alpha_j}+o^{(i)}\frac{\partial N_4^{(i)}}{\partial \alpha_j}\right). \qquad (6.99)$$

All of the derivatives with respect to the model parameter α_j on the right side of Eq.(6.99) are known quantities. On the other hand, the absolute response sensitivity S_j can be computed independently, as follows:

1. consider a small perturbation $\delta\alpha_j$ in the model parameter α_j;

2. re-compute the perturbed response $R\left(\alpha_j^0+\delta\alpha_j\right)$, where α_j^0 denotes the unperturbed parameter value;
3. use the finite difference formula

$$S_j^{FD}\cong\frac{R\left(\alpha_j^0+\delta\alpha_j\right)-R\left(\alpha_j^0\right)}{\delta\alpha_j}+O\left(\delta\alpha_j\right)^2. \tag{6.100}$$

4. use the approximate equality between Eqs. (6.100) and (6.99) to obtain independently the respective values of the adjoint function(s) being verified.

The verification procedure described in steps (1) through (4) above, will be illustrated in the remainder of this section, for verifying the adjoint functions depicted in Figures 6.8 through 6.11.

6.3.1 *Verification of the Adjoint Functions for the Outlet Air Temperature Response* $T_a^{(1)}$

When $R=T_a^{(1)}$, the quantities $r_\ell^{(i)}$ defined in Eq. (6.83) all vanish except for a single component, namely: $r_3^{(1)}\triangleq\partial R/\partial T_a^{(1)}=1$. Thus, the adjoint functions corresponding to the outlet air temperature response $T_a^{(1)}$ are computed by solving the adjoint sensitivity system given in Eq. (6.87) using $r_3^{(1)}\triangleq\partial R/\partial T_a^{(1)}=1$ as the only non-zero source term; for this case, the solution of Eq. (6.87) has been depicted in Figure 6.8.

(a) Verification of the adjoint function $o^{(49)}$

Note that the value of the adjoint function $o^{(49)}$ obtained by solving the adjoint sensitivity system given in Eq. (6.85) is $o^{(49)}=-4.687\times10^{-4}\left[K/(J/kg)\right]$, as indicated in Figure 6.8. Now select a variation $\delta T_{a,in}$ in the inlet air temperature $T_{a,in}$, and note that Eq. (6.99) yields the following expression for the sensitivity of the response $R=T_a^{(1)}$ to $T_{a,in}$:

$$S_{46}\triangleq\frac{\partial T_a^{(1)}}{\partial T_{a,in}}-\sum_{i=1}^{49}\left(\mu_w^{(i)}\frac{\partial N_1^{(i)}}{\partial T_{a,in}}+\tau_w^{(i)}\frac{\partial N_2^{(i)}}{\partial T_{a,in}}+\tau_a^{(i)}\frac{\partial N_3^{(i)}}{\partial T_{a,in}}+o^{(i)}\frac{\partial N_4^{(i)}}{\partial T_{a,in}}\right) \tag{6.101}$$
$$=0-o^{(49)}\frac{\partial N_4^{(49)}}{\partial T_{a,in}}=-o^{(49)}\left[C_p\left(\frac{T_a^{(49)}+tK}{2},\mathbf{\alpha}\right)+\omega_{in}\alpha_{1g}\right].$$

Re-writing Eq. (6.101) in the form

$$o^{(49)} = -\frac{S_{46}}{\partial N_4^{(49)}/\partial T_{a,in}} = -S_{46}\left[C_p\left(\frac{T_a^{(49)}+tK}{2},\boldsymbol{\alpha}\right)+\omega_{in}\alpha_{1g}\right]^{-1}, \qquad (6.102)$$

indicates that the value of the adjoint function $o^{(49)}$ could be computed independently if the sensitivity S_{46} were available, since the quantity $\partial N_4^{(49)}/\partial T_{a,in} = 1.0309\times10^3\left[J/(kg\cdot K)\right]$ is known. To first-order in the parameter perturbation, the finite-difference formula given in Eq. (6.100) can be used to compute the approximate sensitivity S_{46}^{FD}; subsequently, this value can be used in conjunction with Eq. (6.102) to compute a "finite-difference sensitivity" value, denoted as $\left[o^{(49)}\right]^{SFD}$, for the respective adjoint, which would be accurate up to second-order in the respective parameter perturbation:

$$\left[o^{(49)}\right]^{SFD} = -\frac{S_{46}^{FD}}{\partial N_4^{(49)}/\partial T_{a,in}} = -\left[\frac{T_{a,pert}^{(1)}-T_{a,nom}^{(1)}}{\delta T_{a,in}}\right]\left[\frac{\partial N_4^{(49)}}{\partial T_{a,in}}\right]^{-1} \qquad (6.103)$$

Numerically, the inlet air temperature $T_{a,in}\,(=T_{db})$ has the nominal ("base-case") value of $T_{a,in}^0 = 299.11[K]$. The corresponding nominal value $T_{a,nom}^{(1)}$ of the response $T_a^{(1)}$ is $T_{a,nom}^{(1)} = 297.4637\,[K]$. Consider next a perturbation $\delta T_{a,in} = (0.001)T_{a,in}^0$, for which the perturbed value of the inlet air temperature becomes $T_{a,in}^{pert} = T_{a,in}^0 + \delta T_{a,in} = 299.4091[K]$. Re-computing the perturbed response by solving Eqs. (6.2) –(6.13) with the value of $T_{a,in}^{pert}$ yields the "perturbed response" value $T_{a,pert}^{(1)} = 297.6073\,[K]$. Using now the nominal and perturbed response values together with the parameter perturbation in the finite-difference expression given in Eq. (6.100) yields the corresponding "finite-difference-computed sensitivity" $S_{46}^{FD} \triangleq \dfrac{T_{a,pert}^{(1)}-T_{a,nom}^{(1)}}{\delta T_{a,in}} = 0.4802$. Using this value together with the nominal values of the other quantities appearing in the expression on the right side of Eq. (6.103) yields $\left[o^{(49)}\right]^{SFD} = -4.658\times10^{-4}\left[K/(J/kg)\right]$. This result compares well with the value $o^{(49)} = -4.687\times10^{-4}\ \left[K/(J/kg)\right]$ obtained by solving the adjoint sensitivity system given in Eq. (6.85), cf., Figure 6.8. When solving this adjoint sensitivity system, the computation of $o^{(49)}$ depends on the previously computed adjoint functions $o^{(i)},\ i=1,...,I-1$; hence, the foregoing verification of the computational accuracy of $o^{(49)}$ also provides an indirect verification that the functions $o^{(i)},\ i=1,...,I-1,$ were also computed accurately.

(b) Verification of the adjoint function $\tau_a^{(49)}$

Note that the value of the adjoint function $\tau_a^{(49)}$ obtained by solving the adjoint sensitivity system given in Eq. (6.85) is $\tau_a^{(49)}=978.20\,[K]$, as indicated in Figure 6.8. Now select a variation $\delta\omega_{in}$ in the inlet air humidity ratio ω_{in}, and note that Eq. (6.99) yields the following expression for the sensitivity of the response $R=T_a^{(1)}$ to ω_{in}:

$$
\begin{aligned}
S_{48} &\triangleq \frac{\partial T_a^{(1)}}{\partial \omega_{in}} - \sum_{i=1}^{49}\left(\mu_w^{(i)}\frac{\partial N_1^{(i)}}{\partial \omega_{in}} + \tau_w^{(i)}\frac{\partial N_2^{(i)}}{\partial \omega_{in}} + \tau_a^{(i)}\frac{\partial N_3^{(i)}}{\partial \omega_{in}} + o^{(i)}\frac{\partial N_4^{(i)}}{\partial \omega_{in}}\right) \\
&= 0 - \left(\tau_a^{(49)}\frac{\partial N_3^{(49)}}{\partial \omega_{in}} + o^{(49)}\frac{\partial N_4^{(49)}}{\partial \omega_{in}}\right) = -\left[\tau_a^{(49)} + o^{(49)}h_{g,a}^{(50)}\left(T_{a,in},\boldsymbol{\alpha}\right)\right].
\end{aligned}
\tag{6.104}
$$

Re-writing Eq. (6.104) in the form

$$
\tau_a^{(49)} = -S_{48} - o^{(49)}h_{g,a}^{(50)}\left(T_{a,in},\boldsymbol{\alpha}\right), \tag{6.105}
$$

indicates that the value of the adjoint function $\tau_a^{(49)}$ could be computed independently if the sensitivity S_{48} were available, since the $o^{(49)}$ has been verified in (the previous) Section 6.2.1 (a) and the quantity $h_{g,a}^{(50)}\left(T_{a,in},\boldsymbol{\alpha}\right)$ is known. To first-order in the parameter perturbation, the finite-difference formula given in Eq. (6.100) can be used to compute the approximate sensitivity S_{48}^{FD}; subsequently, this value can be used in conjunction with Eq. (6.105) to compute a "finite-difference sensitivity" value, denoted as $\left[\tau_a^{(49)}\right]^{SFD}$, for the respective adjoint, which would be accurate up to second-order in the respective parameter perturbation:

$$
\left[\tau_a^{(49)}\right]^{SFD} = -S_{48}^{FD} - o^{(49)}h_{g,a}^{(50)}\left(T_{a,in},\boldsymbol{\alpha}\right). \tag{6.106}
$$

Numerically, the inlet air humidity ratio ω_{in} has the nominal ("base-case") value of $\omega_{in}^0 = 0.0137746$. The corresponding nominal value $T_{a,nom}^{(1)}$ of the response $T_a^{(1)}$ is

$T_{a,nom}^{(1)} = 297.4637\left[K\right]$. Consider next a perturbation $\delta\omega_{in} = \left(0.0063\right)\omega_{in}^{0}$, for which the perturbed value of the inlet air humidity ratio becomes $\omega_{in}^{pert} = \omega_{in}^{0} + \delta\omega_{in} = 0.0138612$. Re-computing the perturbed response by solving Eqs. (6.2) through (6.13) with the value of ω_{in}^{pert} yields the "perturbed response" value $T_{a,pert}^{(1)} = 297.4824\left[K\right]$. Using now the nominal and perturbed response values together with the parameter perturbation in the finite-difference expression given in Eq. (6.100) yields the corresponding "finite-difference-computed sensitivity" $S_{48}^{FD} \triangleq \dfrac{T_{a,pert}^{(1)} - T_{a,nom}^{(1)}}{\delta\omega_{in}} = 216.32\ \left[K\right]$. Using this value together with the nominal values of the other quantities appearing in the expression on the right side of Eq.(6.106) yields $\left[\tau_{a}^{(49)}\right]^{SFD} = 978.25\ \left[K\right]$. This result compares well with the value $\tau_{a}^{(49)} = 978.20\left[K\right]$ obtained by solving the adjoint sensitivity system given in Eq (6.85), cf. Figure 6.8. When solving this adjoint sensitivity system, the computation of $\tau_{a}^{(49)}$ depends on the previously computed adjoint functions $\tau_{a}^{(i)},\ i = 1,...,I-1$; hence, the foregoing verification of the computational accuracy of $\tau_{a}^{(49)}$ also provides an indirect verification that the functions $\tau_{a}^{(i)},\ i = 1,...,I-1$ were also computed accurately.

(c) Verification of the adjoint functions $\tau_{w}^{(1)}$ and $\mu_{w}^{(1)}$

Note that the values of the adjoint functions $\tau_{w}^{(1)}$ and $\mu_{w}^{(1)}$ obtained by solving the adjoint sensitivity system given in Eq. (6.85) are as follows: $\tau_{w}^{(1)} = -1.49067\times10^{-6}\left[K/\left(J/s\right)\right]$ and $\mu_{w}^{(1)} = 3.5735\left[K/\left(kg/s\right)\right]$, respectively, as indicated in Figure 6.8. Now select a variation $\delta T_{w,in}$ in the inlet water temperature $T_{w,in}$, and note that Eq. (6.99) yields the following expression for the sensitivity of the response $R = T_{a}^{(1)}$ to $T_{w,in}$:

$$\begin{aligned} S_{3} &\triangleq \frac{\partial T_{a}^{(1)}}{\partial T_{w,in}} - \sum_{i=1}^{49}\left(\mu_{w}^{(i)}\frac{\partial N_{1}^{(i)}}{\partial T_{w,in}} + \tau_{w}^{(i)}\frac{\partial N_{2}^{(i)}}{\partial T_{w,in}} + \tau_{a}^{(i)}\frac{\partial N_{3}^{(i)}}{\partial T_{w,in}} + o^{(i)}\frac{\partial N_{4}^{(i)}}{\partial T_{w,in}}\right) \\ &= 0 - \left(\mu_{w}^{(1)}\frac{\partial N_{1}^{(1)}}{\partial T_{w,in}} + \tau_{w}^{(1)}\frac{\partial N_{2}^{(1)}}{\partial T_{w,in}} + \tau_{a}^{(1)}\frac{\partial N_{3}^{(1)}}{\partial T_{w,in}} + o^{(1)}\frac{\partial N_{4}^{(1)}}{\partial T_{w,in}}\right). \end{aligned} \tag{6.107}$$

Since the adjoint functions $\tau_{a}^{(49)}$ and $o^{(49)}$ have been already verified as described in the foregoing, it follows that the computed values of adjoint functions

$\tau_a^{(1)} = 2410.83[K]$ $o^{(1)} = -9.5142 \times 10^{-4}\left[K/(J/kg)\right]$ can also be considered as being accurate, since they constitute the starting point for solving the adjoint sensitivity system in Eq. (6.85). Hence, the unknowns in Eq. (6.107) are the adjoint functions $\mu_w^{(1)}$ and $\tau_w^{(1)}$. A second equation involving solely these adjoint functions can be derived by selecting a perturbation, $\delta m_{w,in}$, in the inlet water mass flow rate, $m_{w,in}$, for which Eq. (6.99) yields the following expression for the sensitivity of the response $R = T_a^{(1)}$ to $m_{w,in}$:

$$\begin{aligned} S_{45} &\triangleq \frac{\partial T_a^{(1)}}{\partial m_{w,in}} - \sum_{i=1}^{49}\left(\mu_w^{(i)}\frac{\partial N_1^{(i)}}{\partial m_{w,in}} + \tau_w^{(i)}\frac{\partial N_2^{(i)}}{\partial m_{w,in}} + \tau_a^{(i)}\frac{\partial N_3^{(i)}}{\partial m_{w,in}} + o^{(i)}\frac{\partial N_4^{(i)}}{\partial m_{w,in}}\right) \\ &= 0 - \left(\mu_w^{(1)}\frac{\partial N_1^{(1)}}{\partial m_{w,in}} + \tau_w^{(1)}\frac{\partial N_2^{(1)}}{\partial m_{w,in}} + \tau_a^{(1)}\frac{\partial N_3^{(1)}}{\partial m_{w,in}} + o^{(1)}\frac{\partial N_4^{(1)}}{\partial m_{w,in}}\right). \end{aligned} \tag{6.108}$$

Numerically, the inlet water temperature, $T_{w,in}$, has the nominal ("base-case") value of $T_{w,in}^0 = 298.79[K]$, while the nominal ("base-case") value of the inlet water mass flow rate is $m_{w,in}^0 = 44.021229[kg/s]$. As before, the corresponding nominal value $T_{a,nom}^{(1)}$ of the response $T_a^{(1)}$ is $T_{a,nom}^{(1)} = 297.4637[K]$. Consider now a perturbation $\delta T_{w,in} = (0.00033)T_{w,in}^0$, for which the perturbed value of the inlet air temperature becomes $T_{w,in}^{pert} = T_{w,in}^0 + \delta T_{w,in} = 298.89[K]$. Re-computing the perturbed response by solving Eqs. (6.2) through (6.13) with the value of $T_{w,in}^{pert}$ yields the "perturbed response" value $T_{a,pert}^{(1)} = 297.4911[K]$. Using now the nominal and perturbed response values together with the parameter perturbation in the finite-difference expression given in Eq. (6.100) yields the corresponding "finite-difference-computed sensitivity" $S_3^{FD} \triangleq \dfrac{T_{a,pert}^{(1)} - T_{a,nom}^{(1)}}{\delta T_{w,in}} = 0.27428$.

Next, consider a perturbation $\delta m_{w,in} = (0.001)m_{w,in}^0$, for which the perturbed value of the inlet air temperature becomes $m_{w,in}^{pert} = m_{w,in}^0 + \delta m_{w,in} = 44.065[kg/s]$. Re-computing the perturbed response by solving Eqs. (6.2) through (6.13) with the value of $m_{w,in}^{pert}$ yields the "perturbed response" value $T_{a,pert}^{(1)} = 297.4646[K]$. Using now the nominal and perturbed response values together with the parameter perturbation in the finite-difference expression given in Eq. (6.100) yields the corresponding "finite-difference-computed sensitivity" $S_{45}^{FD} \triangleq \dfrac{T_{a,pert}^{(1)} - T_{a,nom}^{(1)}}{\delta m_{w,in}} = 0.02133\left[K/(kg/s)\right]$.

Inserting now all of the numerical values of the known quantities in Eqs. (6.107) and (6.108) yields the following system of coupled equations for obtaining $\mu_w^{(1)}$ and $\tau_w^{(1)}$:

$$\begin{aligned}0.27428 = -\Big[&0.0161\mu_w^{(1)} + 223597\tau_w^{(1)} + (2410.83)(-1.039\times10^{-4}) \\ &+(-9.5142\times10^{-4})(-264.65)\Big]\end{aligned} \tag{6.109}$$

$$\begin{aligned}0.02133 = -\Big[&-\mu_w^{(1)} + (-2.44038\times10^{6})\tau_w^{(1)} + (2410.83)(0.0064) \\ &+(-9.5142\times10^{-4})(16430.26)\Big]\end{aligned} \tag{6.110}$$

Solving Eqs. (6.109) and (6.110) yields $\mu_w^{(1)} = 3.5726\ [K/(kg/s)]$ and $\tau_w^{(1)} = 1.49017\times10^{-6}\ [K/(J/s)]$. These values compare well with the values $\mu_w^{(1)} = 3.5735[K/(kg/s)]$ and $\tau_w^{(1)} = -1.49067\times10^{-6}\ [K/(J/s)]$, respectively, which are obtained by solving the adjoint sensitivity system given in Eq. (6.85), cf. Figure 6.8.

6.3.2 *Verification of the Adjoint Functions for the Outlet Water Temperature Response* $T_w^{(50)}$

When $R = T_w^{(50)}$, the quantities $r_\ell^{(i)}$ defined in Eq. (6.83) all vanish except for a single component, namely: $r_2^{(49)} \triangleq \partial R/\partial T_w^{(50)} = 1$. Thus, the adjoint functions corresponding to the outlet water temperature response $T_w^{(50)}$ are computed by solving the adjoint sensitivity system given in Eq. (6.85) using $r_2^{(49)} \triangleq \partial R/\partial T_w^{(50)} = 1$, as the only non-zero source term; for this case, the solution of Eq. (6.87) has been depicted in Figure 6.9.

(a) Verification of the adjoint function $o^{(49)}$

Note that the value of the adjoint function $o^{(49)}$ obtained by solving the adjoint sensitivity system given in Eq. (6.85) is $o^{(49)} = -2.214\times10^{-4}\ [K/(J/kg)]$, as indicated in Figure 6.9. Now select a variation $\delta T_{a,in}$ in the inlet air temperature $T_{a,in}$, and note that Eq. (6.99) yields the following expression for the sensitivity of the response $R = T_w^{(50)}$ to $T_{a,in}$:

$$S_{46} \triangleq \frac{\partial T_w^{(50)}}{\partial T_{a,in}} - \sum_{i=1}^{49}\left(\mu_w^{(i)} \frac{\partial N_1^{(i)}}{\partial T_{a,in}} + \tau_w^{(i)} \frac{\partial N_2^{(i)}}{\partial T_{a,in}} + \tau_a^{(i)} \frac{\partial N_3^{(i)}}{\partial T_{a,in}} + o^{(i)} \frac{\partial N_4^{(i)}}{\partial T_{a,in}} \right)$$
$$= 0 - o^{(49)} \frac{\partial N_4^{(49)}}{\partial T_{a,in}} = -o^{(49)} \left[C_p \left(\frac{T_a^{(49)} + tK}{2}, \boldsymbol{\alpha} \right) + \omega_{in} \alpha_{1g} \right]. \tag{6.111}$$

Re-writing Eq. (6.111) in the form

$$o^{(49)} = -\frac{S_{46}}{\partial N_4^{(49)} / \partial T_{a,in}} = -S_{46} \left[C_p \left(\frac{T_a^{(49)} + tK}{2}, \boldsymbol{\alpha} \right) + \omega_{in} \alpha_{1g} \right]^{-1}, \tag{6.112}$$

indicates that the value of the adjoint function $o^{(49)}$ could be computed independently if the sensitivity S_{46} were available, since the quantity $\partial N_4^{(49)} / \partial T_{a,in} = 1.0309 \times 10^3 \left[J/(kg \cdot K) \right]$ is known. To first-order in the parameter perturbation, the finite-difference formula given in Eq. (6.100) can be used to compute the approximate sensitivity S_{46}^{FD}; subsequently, this value can be used in conjunction with Eq. (6.112) to compute a "finite-difference sensitivity" value, denoted as $\left[o^{(49)} \right]^{SFD}$, for the respective adjoint, which would be accurate up to second-order in the respective parameter perturbation:

$$\left[o^{(49)} \right]^{SFD} = -\frac{S_{46}^{FD}}{\partial N_4^{(49)} / \partial T_{a,in}} = -\left[\frac{T_{w,pert}^{(50)} - T_{w,nom}^{(50)}}{\delta T_{a,in}} \right] \left[\frac{\partial N_4^{(49)}}{\partial T_{a,in}} \right]^{-1}, \tag{6.113}$$

Numerically, the inlet air temperature $T_{a,in} \left(= T_{db} \right)$ has the nominal ("base-case") value of $T_{a,in}^0 = 299.11 [K]$. The corresponding nominal value $T_{w,nom}^{(50)}$ of the response $T_w^{(50)}$ is $T_{w,nom}^{(50)} = 294.579 [K]$. Consider next a perturbation $\delta T_{a,in} = (0.001) T_{a,in}^0$, for which the perturbed value of the inlet air temperature becomes $T_{a,in}^{pert} = T_{a,in}^0 + \delta T_{a,in} = 299.40911 [K]$. Re-computing the perturbed response by solving Eqs. (6.2) through (6.13) with the value of $T_{a,in}^{pert}$ yields the "perturbed response" value $T_{w,pert}^{(50)} = 294.645 [K]$. Using now the nominal and perturbed response values together with the parameter perturbation in the finite-difference expression given in Eq. (6.100) yields the corresponding "finite-difference-computed sensitivity" $S_{46}^{FD} \triangleq \dfrac{T_{w,pert}^{(50)} - T_{w,nom}^{(50)}}{\delta T_{a,in}} = 0.2207$. Using this value together with the nominal values of the other quantities appearing in the expression on the right side of Eq. (6.113) yields $\left[o^{(49)}\right]^{SFD} = -2.141\times10^{-4}\left[K/(J/kg)\right]$. This result compares well with the value $o^{(49)} = -2.214\times10^{-4}\left[K/(J/kg)\right]$ obtained by solving the adjoint sensitivity system given in Eq. (6.85), cf., Figure 6.9. When solving this adjoint sensitivity system, the computation of $o^{(49)}$ depends on the previously computed adjoint functions $o^{(i)},\ i=1,...,I-1$; hence, the foregoing verification of the computational accuracy of $o^{(49)}$ also provides an indirect verification that the functions $o^{(i)},\ i=1,...,I-1,$ were also computed accurately.

(b) Verification of the adjoint function $\tau_a^{(49)}$

Note that the value of the adjoint function $\tau_a^{(49)}$ obtained by solving the adjoint sensitivity system given in Eq. (6.85) is $\tau_a^{(49)} = -76.12\left[K\right]$, as indicated in Figure 6.9. Now select a variation $\delta\omega_{in}$ in the inlet air humidity ratio ω_{in}, and note that Eq. (6.99) yields the following expression for the sensitivity of the response $R = T_w^{(50)}$ to ω_{in}:

$$S_{48} \triangleq \frac{\partial T_w^{(50)}}{\partial\omega_{in}} - \sum_{i=1}^{49}\left(\mu_w^{(i)}\frac{\partial N_1^{(i)}}{\partial\omega_{in}} + \tau_w^{(i)}\frac{\partial N_2^{(i)}}{\partial\omega_{in}} + \tau_a^{(i)}\frac{\partial N_3^{(i)}}{\partial\omega_{in}} + o^{(i)}\frac{\partial N_4^{(i)}}{\partial\omega_{in}}\right)$$
$$= 0 - \left(\tau_a^{(49)}\frac{\partial N_3^{(49)}}{\partial\omega_{in}} + o^{(49)}\frac{\partial N_4^{(49)}}{\partial\omega_{in}}\right) = -\left[\tau_a^{(49)} + o^{(49)}\cdot h_{g,a}^{(50)}\left(T_{a,in},\boldsymbol{\alpha}\right)\right]. \tag{6.114}$$

Re-writing Eq. (6.114) in the form

$$\tau_a^{(49)} = -S_{48} - o^{(49)}\cdot h_{g,a}^{(50)}\left(T_{a,in},\boldsymbol{\alpha}\right), \tag{6.115}$$

indicates that the value of the adjoint function $\tau_a^{(49)}$ could be computed independently if the sensitivity S_{48} were available, since the $o^{(49)}$ has been verified in (the previous) Section 6.3.2 (a) and the quantity $h_{g,a}^{(50)}\left(T_{a,in},\alpha\right)$ is known. To first-order in the parameter perturbation, the finite-difference formula given in Eq. (6.100) can be used to compute the approximate sensitivity S_{48}^{FD}; subsequently, this value can be used in conjunction with Eq. (6.115) to compute a "finite-difference sensitivity" value, denoted as $\left[\tau_a^{(49)}\right]^{SFD}$, for the respective adjoint, which would be accurate up to second-order in the respective parameter perturbation:

$$\left[\tau_a^{(49)}\right]^{SFD} = -S_{48}^{FD} - o^{(49)} \cdot h_{g,a}^{(50)}\left(T_{a,in}, \boldsymbol{\alpha}\right), \tag{6.116}$$

Numerically, the inlet air humidity ratio ω_{in} has the nominal ("base-case") value of $\omega_{in}^{0} = 0.0137746$. The corresponding nominal value $T_{w,nom}^{(50)}$ of the response $T_{w}^{(50)}$ is $T_{w,nom}^{(50)} = 294.579029\left[K\right]$. Consider next a perturbation $\delta\omega_{in} = \left(0.0063\right)\omega_{in}^{0}$, for which the perturbed value of the inlet air humidity ratio becomes $\omega_{in}^{pert} = \omega_{in}^{0} + \delta\omega_{in} = 0.0138612$. Re-computing the perturbed response by solving Eqs. (6.2) through (6.13) with the value of ω_{in}^{pert} yields the "perturbed response" value $T_{w,pert}^{(50)} = 294.634438\left[K\right]$. Using now the nominal and perturbed response values together with the parameter perturbation in the finite-difference expression given in Eq. (6.100) yields the corresponding "finite-difference-computed sensitivity" $S_{48}^{FD} \triangleq \dfrac{T_{w,pert}^{(50)} - T_{w,nom}^{(50)}}{\delta\omega_{in}} = 639.98\left[K\right]$. Using this value together with the nominal values of the other quantities appearing in the expression on the right side of Eq. (6.116) yields $\left[\tau_a^{(49)}\right]^{SFD} = -75.64\left[K\right]$. This result compares well with the value $\tau_a^{(49)} = -76.12\left[K\right]$ obtained by solving the adjoint sensitivity system given in Eq. (6.85) cf. Figure 6.9. When solving this adjoint sensitivity system, the computation of $\tau_a^{(49)}$ depends on the previously computed adjoint functions $\tau_a^{(i)},\ i = 1,\ldots,I-1$; hence, the foregoing verification of the computational accuracy of $\tau_a^{(49)}$ also provides an indirect verification that the functions $\tau_a^{(i)},\ i = 1,\ldots,I-1$ were also computed accurately.

(c) *Verification of the adjoint functions* $\tau_w^{(1)}$ *and* $\mu_w^{(1)}$

Note that the values of the adjoint functions $\tau_w^{(1)}$ and $\mu_w^{(1)}$ obtained by solving the adjoint sensitivity system given in Eq. (6.85) are as follows: $\tau_w^{(1)} = -5.730\times 10^{-7}\left[K/(J/s)\right]$ and $\mu_w^{(1)} = 1.3996\left[K/(kg/s)\right]$, respectively, as indicated in Figure 6.9. Now select a variation $\delta T_{w,in}$ in the inlet water temperature $T_{w,in}$, and note that Eq. (6.99) yields the following expression for the sensitivity of the response $R = T_{w}^{(50)}$ to $T_{w,in}$:

$$
\begin{aligned}
S_3 &\triangleq \frac{\partial T_w^{(50)}}{\partial T_{w,in}} - \sum_{i=1}^{49}\left(\mu_w^{(i)} \frac{\partial N_1^{(i)}}{\partial T_{w,in}} + \tau_w^{(i)} \frac{\partial N_2^{(i)}}{\partial T_{w,in}} + \tau_a^{(i)} \frac{\partial N_3^{(i)}}{\partial T_{w,in}} + o^{(i)} \frac{\partial N_4^{(i)}}{\partial T_{w,in}} \right) \\
&= 0 - \left(\mu_w^{(1)} \frac{\partial N_1^{(1)}}{\partial T_{w,in}} + \tau_w^{(1)} \frac{\partial N_2^{(1)}}{\partial T_{w,in}} + \tau_a^{(1)} \frac{\partial N_3^{(1)}}{\partial T_{w,in}} + o^{(1)} \frac{\partial N_4^{(1)}}{\partial T_{w,in}} \right).
\end{aligned}
\tag{6.117}
$$

Since the adjoint functions $\tau_a^{(49)}$ and $o^{(49)}$ have been already verified, it follows that the computed values of adjoint functions $\tau_a^{(1)} = -0.88745[K]$ $o^{(1)} = -1.38335\times 10^{-6}\left[K/(J/kg)\right]$ can also be considered as being accurate, since they constitute the starting point for solving the adjoint sensitivity system in Eq. (6.85). Hence, the unknowns in Eq. (6.117) are the adjoint functions $\mu_w^{(1)}$ and $\tau_w^{(1)}$. A second equation involving solely these adjoint functions can be derived by selecting a perturbation, $\delta m_{w,in}$, in the inlet water mass flow rate, $m_{w,in}$, for which Eq. (6.99) yields the following expression for the sensitivity of the response $R = T_w^{(50)}$ to the parameter $m_{w,in}$:

$$
\begin{aligned}
S_{45} &\triangleq \frac{\partial T_w^{(50)}}{\partial m_{w,in}} - \sum_{i=1}^{49}\left(\mu_w^{(i)} \frac{\partial N_1^{(i)}}{\partial m_{w,in}} + \tau_w^{(i)} \frac{\partial N_2^{(i)}}{\partial m_{w,in}} + \tau_a^{(i)} \frac{\partial N_3^{(i)}}{\partial m_{w,in}} + o^{(i)} \frac{\partial N_4^{(i)}}{\partial m_{w,in}} \right) \\
&= 0 - \left(\mu_w^{(1)} \frac{\partial N_1^{(1)}}{\partial m_{w,in}} + \tau_w^{(1)} \frac{\partial N_2^{(1)}}{\partial m_{w,in}} + \tau_a^{(1)} \frac{\partial N_3^{(1)}}{\partial m_{w,in}} + o^{(1)} \frac{\partial N_4^{(1)}}{\partial m_{w,in}} \right).
\end{aligned}
\tag{6.118}
$$

Numerically, the inlet water temperature, $T_{w,in}$, has the nominal ("base-case") value of $T_{w,in}^0 = 298.79[K]$, while the nominal ("base-case") value of the inlet water mass flow rate is $m_{w,in}^0 = 44.021229[kg/s]$. As before, the corresponding nominal value $T_{w,nom}^{(50)}$ of the response $T_w^{(50)}$ is $T_{w,nom}^{(50)} = 294.579[K]$. Consider now a perturbation $\delta T_{w,in} = (0.00033)T_{w,in}^0$, for which the perturbed value of the inlet air temperature becomes $T_{w,in}^{pert} = T_{w,in}^0 + \delta T_{w,in} = 298.89[K]$. Re-computing the perturbed response by solving Eqs. (6.2) through (6.13) with the value of $T_{w,in}^{pert}$ yields the "perturbed response" value $T_{w,pert}^{(50)} = 294.5895[K]$. Using now the nominal and perturbed response values together with the parameter perturbation in the finite-difference expression given in Eq. (6.100) yields the corresponding "finite-difference-computed sensitivity" $S_3^{FD} \triangleq \dfrac{T_{w,pert}^{(50)} - T_{w,nom}^{(50)}}{\delta T_{w,in}} = 0.1049$.

Next, consider a perturbation $\delta m_{w,in} = (0.001) m_{w,in}^0$, for which the perturbed value of the inlet air temperature becomes $m_{w,in}^{pert} = m_{w,in}^0 + \delta m_{w,in} = 44.065252\ [kg/s]$. Re-computing the perturbed response by solving Eqs. (6.2) through (6.13) with the value of $m_{w,in}^{pert}$ yields the "perturbed response" value $T_{w,pert}^{(50)} = 294.5804[K]$. Using now the nominal and perturbed response values together with the parameter perturbation in the finite-difference expression given in Eq.(6.100) yields the corresponding "finite-difference-computed sensitivity" $S_{45}^{FD} \triangleq \dfrac{T_{w,pert}^{(50)} - T_{w,nom}^{(50)}}{\delta m_{w,in}} = 0.02986$. Inserting now all of the numerical values of the known quantities in Eqs. (6.117) and (6.118) yields the following system of coupled equations for obtaining the functions $\mu_w^{(1)}$ and $\tau_w^{(1)}$:

$$\begin{aligned} 0.1049 = -\Big[& 0.0161\mu_w^{(1)} + 223597\tau_w^{(1)} + (-0.88745)\times(-1.039\times10^{-4}) \\ & +(-1.38335\times10^{-6})\times(-264.65)\Big], \end{aligned} \tag{6.119}$$

$$\begin{aligned} 0.02986 = -\Big[& -\mu_w^{(1)} - 2.44038\times10^{6}\tau_w^{(1)} + (-0.88745)\times(0.0064) \\ & +(-1.38335\times10^{-6})\times(16430.26)\Big]. \end{aligned} \tag{6.120}$$

Solving Eqs. (6.119) and (6.120) yields $\mu_w^{(1)} = 1.3969\,[K/(kg/s)]$ and $\tau_w^{(1)} = -5.718\times10^{-7}\,[K/(J/s)]$. These values compare well with the values $\mu_w^{(1)} = 1.3996\,[K/(kg/s)]$ and $\tau_w^{(1)} = -5.730\times10^{-7}\,[K/(J/s)]$, respectively, which are obtained by solving the adjoint sensitivity system given in Eq. (6.85), cf. Figure 6.9.

6.3.3 Verification of the Adjoint Functions for the Outlet Air Relative Humidity Response $RH^{(1)}$

When $R = RH^{(1)}$, the quantities $r_\ell^{(i)}$ defined in Eq. (6.83) all vanish except for two components, namely:

$$
\begin{aligned}
r_3^{(1)} &\triangleq \frac{\partial\left(RH^{(1)}\right)}{\partial T_a^{(1)}} = \frac{\partial}{\partial T_a^{(1)}}\left[\frac{P_v\left(\omega^{(1)},\boldsymbol{\alpha}\right)}{P_{vs}\left(T_a^{(1)},\boldsymbol{\alpha}\right)}\times 100\right] \\
&= \frac{0.622 P_{atm}}{\left(0.622+\omega^{(1)}\right)e^{a_0+\frac{a_1}{T_a^{(1)}}}}\frac{a_1}{\left[T_a^{(1)}\right]^2}\times 100;
\end{aligned}
\tag{6.121}
$$

$$
\begin{aligned}
r_4^{(1)} &\triangleq \frac{\partial\left(RH^{(1)}\right)}{\partial \omega^{(1)}} = \frac{\partial}{\partial \omega^{(1)}}\left[\frac{P_v\left(\omega^{(1)},\boldsymbol{\alpha}\right)}{P_{vs}\left(T_a^{(1)},\boldsymbol{\alpha}\right)}\times 100\right] \\
&= \frac{0.622 P_{atm}}{\left(0.622+\omega^{(1)}\right)^2 e^{a_0+\frac{a_1}{T_a^{(1)}}}}\times 100.
\end{aligned}
\tag{6.122}
$$

Thus, the adjoint functions corresponding to the outlet air relative humidity response $RH^{(1)}$ are computed by solving the adjoint sensitivity system given in Eq. (6.85) using $r_3^{(1)}$ and $r_4^{(1)}$ as the only two non-zero source terms; for this case, the solution of Eq. (6.85) has been depicted in Figure 6.10.

(a) Verification of the adjoint function $o^{(49)}$

Note that the value of the adjoint function $o^{(49)}$ obtained by solving the adjoint sensitivity system given in Eq. (6.85) is $o^{(49)} = 1.860\times 10^{-5}\ \left[(J/kg)^{-1}\right]$, as indicated in Figure 6.10. Now select a variation $\delta T_{a,in}$ in the inlet air temperature $T_{a,in}$, and note that Eq. (6.99) yields the following expression for the sensitivity of the response $R = RH^{(1)}$ to $T_{a,in}$:

$$
\begin{aligned}
S_{46} &\triangleq \frac{\partial RH^{(1)}}{\partial T_{a,in}} - \sum_{i=1}^{49}\left(\mu_w^{(i)}\frac{\partial N_1^{(i)}}{\partial T_{a,in}} + \tau_w^{(i)}\frac{\partial N_2^{(i)}}{\partial T_{a,in}} + \tau_a^{(i)}\frac{\partial N_3^{(i)}}{\partial T_{a,in}} + o^{(i)}\frac{\partial N_4^{(i)}}{\partial T_{a,in}}\right) \\
&= 0 - o^{(49)}\frac{\partial N_4^{(49)}}{\partial T_{a,in}} = -o^{(49)}\left[C_p\left(\frac{T_a^{(49)}+tK}{2},\boldsymbol{\alpha}\right) + \omega_{in}\alpha_{1g}\right].
\end{aligned}
\tag{6.123}
$$

Re-writing Eq. (6.123) in the form

$$
o^{(49)} = -\frac{S_{46}}{\partial N_4^{(49)}/\partial T_{a,in}} = -S_{46}\left[C_p\left(\frac{T_a^{(49)}+tK}{2},\boldsymbol{\alpha}\right) + \omega_{in}\alpha_{1g}\right]^{-1}, \tag{6.124}
$$

indicates that the value of the adjoint function $o^{(49)}$ could be computed independently if the sensitivity S_{46} were available, since the quantity $\partial N_4^{(49)}/\partial T_{a,in} = 1.0309\times 10^3\left[J/(kg\cdot K)\right]$ is known. To first-order in the parameter perturbation, the finite-difference formula given in Eq. (6.100) can be used to compute the approximate sensitivity S_{46}^{FD}; subsequently, this value can be used in conjunction with Eq. (6.124) to compute a "finite-difference sensitivity" value, denoted as $\left[o^{(49)}\right]^{SFD}$, for the respective adjoint, which would be accurate up to second-order in the parameter perturbation:

$$\left[o^{(49)}\right]^{SFD} = -\frac{S_{46}^{FD}}{\partial N_4^{(49)}/\partial T_{a,in}} = -\left[\frac{RH_{pert}^{(1)} - RH_{nom}^{(1)}}{\delta T_{a,in}}\right]\left[\frac{\partial N_4^{(49)}}{\partial T_{a,in}}\right]^{-1}, \quad (6.125)$$

Numerically, the inlet air temperature $T_{a,in} = T_{db}$ has the nominal ("base-case") value of $T_{a,in}^0 = 299.11[K]$. The corresponding nominal value $RH_{nom}^{(1)}$ of the response $RH^{(1)}$ is $RH_{nom}^{(1)} = 86.11678\,\%$. Consider next a perturbation $\delta T_{a,in} = (0.001)T_{a,in}^0$, for which the perturbed value of the inlet air temperature becomes $T_{a,in}^{pert} = T_{a,in}^0 + \delta T_{a,in} = 299.40911[K]$. Re-computing the perturbed response by solving Eqs. (6.2) through (6.13) with the value of $T_{a,in}^{pert}$ yields the "perturbed response" value $RH_{pert}^{(1)} = 85.55717\%$. Using now the nominal and perturbed response values together with the parameter perturbation in the finite-difference expression given in Eq. (6.100) yields the corresponding "finite-difference-computed sensitivity" $S_{46}^{FD} \triangleq \dfrac{RH_{pert}^{(1)} - RH_{nom}^{(1)}}{\delta T_{a,in}} = -0.018709\left[\dfrac{1}{K}\right]$. Using this value together with the nominal values of the other quantities appearing in the expression on the right side of Eq.(6.125) yields $\left[o^{(49)}\right]^{SFD} = 1.815\times 10^{-5}\left[(J/kg)^{-1}\right]$. This result compares well with the value $o^{(49)} = 1.860\times 10^{-5}\left[(J/kg)^{-1}\right]$ obtained by solving the adjoint sensitivity system given in Eq (6.85), cf., Figure 6.10. When solving this adjoint sensitivity system, the computation of $o^{(49)}$ depends on the previously computed adjoint functions $o^{(i)},\ i = 1,\dots,I-1$; hence, the forgoing verification of the computational accuracy of $o^{(49)}$ also provides an indirect verification that the functions $o^{(i)},\ i = 1,\dots,I-1,$ were also computed accurately.

(b) Verification of the adjoint function $\tau_a^{(49)}$

Note that the value of the adjoint function $\tau_a^{(49)}$ obtained by solving the adjoint sensitivity system given in Eq. (6.85) is $\tau_a^{(49)} = -67.047$, as indicated in Figure 6.10. Now select a variation $\delta\omega_{in}$ in the inlet air humidity ratio ω_{in}, and note that Eq.(6.99) yields the following expression for the sensitivity of the response $R = RH^{(1)}$ to ω_{in}:

$$S_{48} \triangleq \frac{\partial RH^{(1)}}{\partial \omega_{in}} - \sum_{i=1}^{49}\left(\mu_w^{(i)} \frac{\partial N_1^{(i)}}{\partial \omega_{in}} + \tau_w^{(i)} \frac{\partial N_2^{(i)}}{\partial \omega_{in}} + \tau_a^{(i)} \frac{\partial N_3^{(i)}}{\partial \omega_{in}} + o^{(i)} \frac{\partial N_4^{(i)}}{\partial \omega_{in}} \right)$$
$$= 0 - \left(\tau_a^{(49)} \frac{\partial N_3^{(49)}}{\partial \omega_{in}} + o^{(49)} \frac{\partial N_4^{(49)}}{\partial \omega_{in}} \right) = -\left[\tau_a^{(49)} + o^{(49)} \cdot h_{g,a}^{(50)}\left(T_{a,in}, \boldsymbol{\alpha}\right) \right]. \tag{6.126}$$

Re-writing Eq. (6.126) in the form

$$\tau_a^{(49)} = -S_{48} - o^{(49)} \cdot h_{g,a}^{(50)}\left(T_{a,in}, \boldsymbol{\alpha}\right), \tag{6.127}$$

indicates that the value of the adjoint function $\tau_a^{(49)}$ could be computed independently if the sensitivity S_{48} were available, since the $o^{(49)}$ has been verified already and the quantity $h_{g,a}^{(50)}\left(T_{a,in}, \boldsymbol{\alpha}\right)$ is known. To first-order in the parameter perturbation, the finite-difference formula given in Eq. (6.100) can be used to compute the approximate sensitivity S_{48}^{FD}. Subsequently, this value can be used in conjunction with Eq. (6.127) to compute a "finite-difference sensitivity" value, denoted as $\left[\tau_a^{(49)}\right]^{SFD}$, for the respective adjoint, which would be accurate up to second-order in the parameter perturbation, as follows:

$$\left[\tau_a^{(49)}\right]^{SFD} = -S_{48}^{FD} - o^{(49)} \cdot h_{g,a}^{(50)}\left(T_{a,in}, \boldsymbol{\alpha}\right). \tag{6.128}$$

Numerically, the inlet air humidity ratio ω_{in} has the nominal ("base-case") value of $\omega_{in}^0 = 0.0137746$. The corresponding nominal value $RH_{nom}^{(1)}$ of the response $RH^{(1)}$ is $RH_{nom}^{(1)} = 86.11678\ \%$. Consider next a perturbation $\delta\omega_{in} = (0.0063)\omega_{in}^0$, for which the perturbed value of the inlet air humidity ratio becomes $\omega_{in}^{pert} = \omega_{in}^0 + \delta\omega_{in} = 0.0138612$. Re-computing the perturbed response by solving Eqs. (6.2) through (6.13) with the value of ω_{in}^{pert} yields the "perturbed response" value $RH_{pert}^{(1)} = 86.28967\%$. Using now the nominal and perturbed response values together with the parameter perturbation in the finite-difference expression given in

Eq. (6.100) yields the corresponding "finite-difference-computed sensitivity" $S_{48}^{FD} \triangleq \dfrac{RH_{pert}^{(1)} - RH_{nom}^{(1)}}{\delta\omega_{in}} = 19.6252$. Using this value together with the nominal values of the other quantities appearing in the expression on the right side of Eq. (6.128) yields $\left[\tau_a^{(49)}\right]^{SFD} = -67.034$. This result compares well with the value $\tau_a^{(49)} = -67.047$ obtained by solving the adjoint sensitivity system given in Eq.(6.85), cf. Figure 6.10. When solving this adjoint sensitivity system, the computation of $\tau_a^{(49)}$ depends on the previously computed adjoint functions $\tau_a^{(i)},\ i = 1,\ldots,I-1$; hence, the foregoing verification of the computational accuracy of $\tau_a^{(49)}$ also provides an indirect verification that the functions $\tau_a^{(i)},\ i = 1,\ldots,I-1$, were also computed accurately.

(c) Verification of the adjoint functions $\tau_w^{(1)}$ and $\mu_w^{(1)}$

Note that the values of the adjoint functions $\tau_w^{(1)}$ and $\mu_w^{(1)}$ obtained by solving the adjoint sensitivity system given in Eq. (6.85) are as follows: $\tau_w^{(1)} = -1.168 \times 10^{-8} \left[(J/s)^{-1}\right]$ and $\mu_w^{(1)} = -0.2926 \left[(kg/s)^{-1}\right]$, respectively, as indicated in Figure 6.10. Now select a variation $\delta T_{w,in}$ in the inlet water temperature $T_{w,in}$, and note that Eq. (6.99) yields the following expression for the sensitivity of the response $R = RH^{(1)}$ to $T_{w,in}$:

$$\begin{aligned} S_3 &\triangleq \frac{\partial RH^{(1)}}{\partial T_{w,in}} - \sum_{i=1}^{49} \left(\mu_w^{(i)} \frac{\partial N_1^{(i)}}{\partial T_{w,in}} + \tau_w^{(i)} \frac{\partial N_2^{(i)}}{\partial T_{w,in}} + \tau_a^{(i)} \frac{\partial N_3^{(i)}}{\partial T_{w,in}} + o^{(i)} \frac{\partial N_4^{(i)}}{\partial T_{w,in}} \right) \\ &= 0 - \left(\mu_w^{(1)} \frac{\partial N_1^{(1)}}{\partial T_{w,in}} + \tau_w^{(1)} \frac{\partial N_2^{(1)}}{\partial T_{w,in}} + \tau_a^{(1)} \frac{\partial N_3^{(1)}}{\partial T_{w,in}} + o^{(1)} \frac{\partial N_4^{(1)}}{\partial T_{w,in}} \right). \end{aligned} \tag{6.129}$$

Since the adjoint functions $\tau_a^{(49)}$ and $o^{(49)}$ have been already verified, it follows that the computed values of adjoint functions $\tau_a^{(1)} = -172.515$ and $o^{(1)} = 4.816 \times 10^{-5} \left[(J/kg)^{-1}\right]$ can also be considered as being accurate, since they constitute the starting point for solving the adjoint sensitivity system in Eq. (6.85). Hence, the unknowns in Eq. (6.129) are the adjoint functions $\mu_w^{(1)}$ and $\tau_w^{(1)}$. A second equation involving solely these adjoint functions can be derived by selecting a perturbation, $\delta m_{w,in}$, in the inlet water mass flow rate, $m_{w,in}$, for which Eq. (6.99) yields the following expression for the sensitivity of the response $R = RH^{(1)}$ to $m_{w,in}$:

$$S_{45} \triangleq \frac{\partial RH^{(1)}}{\partial m_{w,in}} - \sum_{i=1}^{49} \left(\mu_w^{(i)} \frac{\partial N_1^{(i)}}{\partial m_{w,in}} + \tau_w^{(i)} \frac{\partial N_2^{(i)}}{\partial m_{w,in}} + \tau_a^{(i)} \frac{\partial N_3^{(i)}}{\partial m_{w,in}} + o^{(i)} \frac{\partial N_4^{(i)}}{\partial m_{w,in}} \right)$$
$$= 0 - \left(\mu_w^{(1)} \frac{\partial N_1^{(1)}}{\partial m_{w,in}} + \tau_w^{(1)} \frac{\partial N_2^{(1)}}{\partial m_{w,in}} + \tau_a^{(1)} \frac{\partial N_3^{(1)}}{\partial m_{w,in}} + o^{(1)} \frac{\partial N_4^{(1)}}{\partial m_{w,in}} \right). \quad (6.130)$$

Numerically, the inlet water temperature, $T_{w,in}$, has the nominal ("base-case") value of $T_{w,in}^0 = 298.78[K]$, while the nominal ("base-case") value of the inlet water mass flow rate is $m_{w,in}^0 = 44.021229[kg/s]$. As before, the corresponding nominal value $RH_{nom}^{(1)}$ of the response $RH^{(1)}$ is $RH_{nom}^{(1)} = 86.11678\%$. Consider now a perturbation $\delta T_{w,in} = (0.00033)T_{w,in}^0$, for which the perturbed value of the inlet air temperature becomes $T_{w,in}^{pert} = T_{w,in}^0 + \delta T_{w,in} = 298.89[K]$. Re-computing the perturbed response by solving Eqs. (6.2) through (6.13) with the value of $T_{w,in}^{pert}$ yields the "perturbed response" value $RH_{pert}^{(1)} = 86.13838\%$. Using now the nominal and perturbed response values together with the parameter perturbation in the finite-difference expression given in Eq. (6.100) yields the corresponding "finite-difference-computed sensitivity" $S_3^{FD} \triangleq \dfrac{RH_{pert}^{(1)} - RH_{nom}^{(1)}}{\delta T_{w,in}} = 0.00216$.

Next, consider a perturbation $\delta m_{w,in} = (0.001)m_{w,in}^0$, for which the perturbed value of the inlet air temperature becomes $m_{w,in}^{pert} = m_{w,in}^0 + \delta m_{w,in} = 44.065252[kg/s]$. Re-computing the perturbed response by solving Eqs. (6.2) through (6.13) with the value of $m_{w,in}^{pert}$ yields the "perturbed response" value $RH_{pert}^{(1)} = 86.11725\%$. Using now the nominal and perturbed response values together with the parameter perturbation in the finite-difference expression given in Eq.(6.100) yields the corresponding "finite-difference-computed sensitivity" $S_{45}^{FD} \triangleq \dfrac{RH_{pert}^{(1)} - RH_{nom}^{(1)}}{\delta m_{w,in}} = 0.000107$. Inserting now all of the numerical values of the known quantities in Eqs. (6.129) and (6.130) yields the following system of coupled equations for obtaining the functions $\mu_w^{(1)}$ and $\tau_w^{(1)}$:

$$0.00216 = -\Big[0.0161\mu_w^{(1)} + 223597\tau_w^{(1)} + (-172.52)(-1.039 \times 10^{-4}) + (4.816 \times 10^{-5})(-264.65)\Big] \quad (6.131)$$

$$
\begin{aligned}
0.000107 = -\Big[& -\mu_w^{(1)} + (-2.44038\times 10^6)\tau_w^{(1)} + (-172.52)(0.0064) \\
& + (4.816\times 10^{-5})(16430.26)\Big]
\end{aligned} \tag{6.132}
$$

Solving Eqs. (6.131) and (6.132) yields $\mu_w^{(1)} = -0.2925\left[(kg/s)^{-1}\right]$ and $\tau_w^{(1)} = -1.173\times 10^{-8}\left[(J/s)^{-1}\right]$. These values compare well with the values $\mu_w^{(1)} = -0.2926\left[(kg/s)^{-1}\right]$ and $\tau_w^{(1)} = -1.168\times 10^{-8}\left[(J/s)^{-1}\right]$, respectively, which are obtained by solving the adjoint sensitivity system given in Eq. (6.85), cf. Figure 6.10.

6.3.4 Verification of the Adjoint Functions for the Outlet Water Mass Flow Rate Response $m_w^{(50)}$

When $R = m_w^{(50)}$, the quantities $r_\ell^{(i)}$ defined in Eq. (6.83) all vanish except for a single component, namely: $r_1^{(49)} \triangleq \partial R/\partial m_w^{(50)} = 1$. Thus, the adjoint functions corresponding to the outlet water mass flow rate response $m_w^{(50)}$ are computed by solving the adjoint sensitivity system given in Eq. (6.85) using $r_1^{(49)} \triangleq \partial R/\partial m_w^{(50)} = 1$. as the only non-zero source term; for this case, the solution of Eq. (6.85) has been depicted in Figure 6.11.

(a) Verification of the adjoint function $o^{(49)}$

Note that the value of the adjoint function $o^{(49)}$ obtained by solving the adjoint sensitivity system given in Eq. (6.85) is $o^{(49)} = 1.603\times 10^{-5}\ \left[(kg/s)/(J/kg)\right]$, as indicated in Figure 6.11. Now select a variation $\delta T_{a,in}$ in the inlet air temperature $T_{a,in}$, and note that Eq. (6.99) yields the following expression for the sensitivity of the response $R = m_w^{(50)}$ to $T_{a,in}$:

$$
\begin{aligned}
S_{46} & \triangleq \frac{\partial m_w^{(50)}}{\partial T_{a,in}} - \sum_{i=1}^{49}\left(\mu_w^{(i)}\frac{\partial N_1^{(i)}}{\partial T_{a,in}} + \tau_w^{(i)}\frac{\partial N_2^{(i)}}{\partial T_{a,in}} + \tau_a^{(i)}\frac{\partial N_3^{(i)}}{\partial T_{a,in}} + o^{(i)}\frac{\partial N_4^{(i)}}{\partial T_{a,in}}\right) \\
& = 0 - o^{(49)}\frac{\partial N_4^{(49)}}{\partial T_{a,in}} = -o^{(49)}\left[C_p\left(\frac{T_a^{(49)} + tK}{2}, \mathbf{\alpha}\right) + \omega_{in}\alpha_{1g}\right].
\end{aligned} \tag{6.133}
$$

Re-writing Eq.(6.133) in the form

$$o^{(49)} = -\frac{S_{46}}{\partial N_4^{(49)}/\partial T_{a,in}} = -S_{46}\left[C_p\left(\frac{T_a^{(49)}+tK}{2},\boldsymbol{\alpha}\right)+\omega_{in}\alpha_{1g}\right]^{-1} \tag{6.134}$$

indicates that the value of the adjoint function $o^{(49)}$ could be computed independently if the sensitivity S_{46} were available, since the quantity $\partial N_4^{(49)}/\partial T_{a,in} = 1.0309\times 10^3\left[J/(kg\cdot K)\right]$ is known. To first-order in the parameter perturbation, the finite-difference formula given in Eq. (6.100) can be used to compute the approximate sensitivity S_{46}^{FD}; subsequently, this value can be used in conjunction with Eq. (6.134) to compute a "finite-difference sensitivity" value, denoted as $\left[o^{(49)}\right]^{SFD}$, for the respective adjoint, which would be accurate up to second-order in the parameter perturbation:

$$\left[o^{(49)}\right]^{SFD} = -\frac{S_{46}^{FD}}{\partial N_4^{(49)}/\partial T_{a,in}} = -\left[\frac{m_{w,pert}^{(50)} - m_{w,nom}^{(50)}}{\delta T_{a,in}}\right]\left[\frac{\partial N_4^{(49)}}{\partial T_{a,in}}\right]^{-1}, \tag{6.135}$$

Numerically, the inlet air temperature $T_{a,in} = T_{db}$ has the nominal ("base-case") value of $T_{a,in}^0 = 299.11[K]$. The corresponding nominal value $T_{w,nom}^{(50)}$ of the response $m_w^{(50)}$ is $m_{w,nom}^{(50)} = 43.598097\,[kg/s]$. Consider next a perturbation $\delta T_{a,in} = (0.001)T_{a,in}^0$, for which the perturbed value of the inlet air temperature becomes $T_{a,in}^{pert} = T_{a,in}^0 + \delta T_{a,in} = 299.40911\,[K]$. Re-computing the perturbed response by solving Eqs. (6.2) through (6.13) with the value of $T_{a,in}^{pert}$ yields the "perturbed response" value $m_{w,pert}^{(50)} = 43.59293\,[kg/s]$. Using now the nominal and perturbed response values together with the parameter perturbation in the finite-difference expression given in Eq. (6.100) yields the corresponding "finite-difference-computed sensitivity" $S_{46}^{FD} \triangleq \dfrac{m_{w,pert}^{(50)} - m_{w,nom}^{(50)}}{\delta T_{a,in}} = -0.01728$. Using this value together with the nominal values of the other quantities appearing in the expression on the right side of Eq. (6.135) yields $\left[o^{(49)}\right]^{SFD} = 1.676\times 10^{-5}\left[(kg/s)/(J/kg)\right]$. This result compares well with the value $o^{(49)} = 1.603\times 10^{-5}$ $\left[(kg/s)/(J/kg)\right]$ obtained by solving the adjoint sensitivity system given in Eq. (6.85) cf., Figure 6.11. When solving this adjoint sensitivity system, the computation of $o^{(49)}$ depends on the previously computed adjoint functions $o^{(i)},\ i=1,\dots,I-1$; hence, the foregoing verification of the computational accuracy of $o^{(49)}$ also provides an indirect verification that the functions $o^{(i)},\ i=1,\dots,I-1$ were also computed accurately.

(b) Verification of the adjoint function $\tau_a^{(49)}$

Note that the value of the adjoint function $\tau_a^{(49)}$ obtained by solving the adjoint sensitivity system given in Eq. (6.85) is $\tau_a^{(49)} = -102.42\,[kg/s]$, as indicated in Figure 6.11. Now select a variation $\delta\omega_{in}$ in the inlet air humidity ratio ω_{in}, and note that Eq. (6.99) yields the following expression for the sensitivity of the response $R = m_w^{(50)}$ to ω_{in}:

$$\begin{aligned} S_{48} &\triangleq \frac{\partial m_w^{(50)}}{\partial \omega_{in}} - \sum_{i=1}^{49}\left(\mu_w^{(i)}\frac{\partial N_1^{(i)}}{\partial \omega_{in}} + \tau_w^{(i)}\frac{\partial N_2^{(i)}}{\partial \omega_{in}} + \tau_a^{(i)}\frac{\partial N_3^{(i)}}{\partial \omega_{in}} + o^{(i)}\frac{\partial N_4^{(i)}}{\partial \omega_{in}}\right) \\ &= 0 - \left(\tau_a^{(49)}\frac{\partial N_3^{(49)}}{\partial \omega_{in}} + o^{(49)}\frac{\partial N_4^{(49)}}{\partial \omega_{in}}\right) = -\left[\tau_a^{(49)} + o^{(49)}\cdot h_{g,a}^{(50)}\left(T_{a,in},\boldsymbol{\alpha}\right)\right]. \end{aligned} \tag{6.136}$$

Re-writing Eq. (6.136) in the form

$$\tau_a^{(49)} = -S_{48} - o^{(49)}\cdot h_{g,a}^{(50)}\left(T_{a,in},\boldsymbol{\alpha}\right), \tag{6.137}$$

indicates that the value of the adjoint function $\tau_a^{(49)}$ could be computed independently if the sensitivity S_{48} were available, since the $o^{(49)}$ has been verified previously and the quantity $h_{g,a}^{(50)}\left(T_{a,in},\alpha\right)$ is known. To first-order in the parameter perturbation, the finite-difference formula given in Eq. (6.100) can be used to compute the approximate sensitivity S_{48}^{FD}. Subsequently, this value can be used in conjunction with Eq. (6.137) to compute a "finite-difference sensitivity" value, denoted as $\left[\tau_a^{(49)}\right]^{SFD}$, for the respective adjoint, which would be accurate up to second-order in the parameter perturbation, so that:

$$\left[\tau_a^{(49)}\right]^{SFD} = -S_{48}^{FD} - o^{(49)}\cdot h_{g,a}^{(50)}\left(T_{a,in},\boldsymbol{\alpha}\right). \tag{6.138}$$

Numerically, the inlet air humidity ratio ω_{in} has the nominal ("base-case") value of $\omega_{in}^0 = 0.0137746$. The corresponding nominal value $m_{w,nom}^{(50)}$ of the response $m_w^{(50)}$ is $m_{w,nom}^{(50)} = 43.598097\,[kg/s]$. Consider next a perturbation $\delta\omega_{in} = (0.0063)\,\omega_{in}^0$, for which the perturbed value of the inlet air humidity ratio becomes $\omega_{in}^{pert} = \omega_{in}^0 + \delta\omega_{in} = 0.0138612$. Re-computing the perturbed response by solving Eqs. (6.2) through (6.13) with the value of ω_{in}^{pert} yields the "perturbed response" value $m_{w,pert}^{(50)} = 43.603424\,[kg/s]$. Using now the nominal and perturbed response values together with the parameter perturbation in the finite-difference expression given in Eq. (6.100) yields the corresponding "finite-difference-computed sensitivity" $S_{48}^{FD} \triangleq \frac{m_{w,pert}^{(50)} - m_{w,nom}^{(50)}}{\delta\omega_{in}} = 61.53\left[kg/s\right]$. Using this value together with the nominal values of the other quantities appearing in the expression on the right side of Eq(6.138) yields $\left[\tau_a^{(49)}\right]^{SFD} = -102.39\left[kg/s\right]$. This result compares well with the value $\tau_a^{(49)} = -102.42\left[kg/s\right]$ obtained by solving the adjoint sensitivity system given in Eq. (6.85), cf. Figure 6.11. When solving this adjoint sensitivity system, the computation of $\tau_a^{(49)}$ depends on the previously computed adjoint functions $\tau_a^{(i)},\ i = 1,...,I-1$; hence, the foregoing verification of the computational accuracy of $\tau_a^{(49)}$ also provides an indirect verification that the functions $\tau_a^{(i)},\ i = 1,...,I-1$ were also computed accurately.

(c) Verification of the adjoint functions $\tau_w^{(1)}$ and $\mu_w^{(1)}$

Note that the values of the adjoint functions $\tau_w^{(1)}$ and $\mu_w^{(1)}$ obtained by solving the adjoint sensitivity system given in Eq. (6.85) are as follows: $\tau_w^{(1)} = 2.6710\times 10^{-7}\left[(J/kg)^{-1}\right]$ and $\mu_w^{(1)} = 0.3377$, respectively, as indicated in Figure 6.11. Now select a variation $\delta T_{w,in}$ in the inlet water temperature $T_{w,in}$, and note that Eq. (6.99) yields the following expression for the sensitivity of the response $R = m_w^{(50)}$ to $T_{w,in}$:

$$
\begin{aligned}
S_3 &\triangleq \frac{\partial m_w^{(50)}}{\partial T_{w,in}} - \sum_{i=1}^{49}\left(\mu_w^{(i)}\frac{\partial N_1^{(i)}}{\partial T_{w,in}} + \tau_w^{(i)}\frac{\partial N_2^{(i)}}{\partial T_{w,in}} + \tau_a^{(i)}\frac{\partial N_3^{(i)}}{\partial T_{w,in}} + o^{(i)}\frac{\partial N_4^{(i)}}{\partial T_{w,in}}\right) \\
&= 0 - \left(\mu_w^{(1)}\frac{\partial N_1^{(1)}}{\partial T_{w,in}} + \tau_w^{(1)}\frac{\partial N_2^{(1)}}{\partial T_{w,in}} + \tau_a^{(1)}\frac{\partial N_3^{(1)}}{\partial T_{w,in}} + o^{(1)}\frac{\partial N_4^{(1)}}{\partial T_{w,in}}\right).
\end{aligned}
\tag{6.139}
$$

Since the adjoint functions $\tau_a^{(49)}$ and $o^{(49)}$ have been already verified, it follows that the computed values of adjoint functions $\tau_a^{(1)} = -3.1344\left[kg/s\right]$ $o^{(1)} = 8.1328\times 10^{-7}\left[(kg/s)/(J/kg)\right]$ can also be considered as being accurate, since they constitute the starting point for solving the adjoint sensitivity system in Eq. (6.85). Hence, the unknowns in Eq. (6.139) are the adjoint functions $\mu_w^{(1)}$ and $\tau_w^{(1)}$. A second equation involving solely these adjoint functions can be derived by selecting a perturbation, $\delta m_{w,in}$, in the inlet water mass flow rate, $m_{w,in}$, for which

Eq. (6.99) yields the following expression for the sensitivity of the response $R = m_w^{(50)}$ to $m_{w,in}$:

$$S_{45} \triangleq \frac{\partial m_w^{(50)}}{\partial m_{w,in}} - \sum_{i=1}^{49} \left(\mu_w^{(i)} \frac{\partial N_1^{(i)}}{\partial m_{w,in}} + \tau_w^{(i)} \frac{\partial N_2^{(i)}}{\partial m_{w,in}} + \tau_a^{(i)} \frac{\partial N_3^{(i)}}{\partial m_{w,in}} + o^{(i)} \frac{\partial N_4^{(i)}}{\partial m_{w,in}} \right)$$
$$= 0 - \left(\mu_w^{(1)} \frac{\partial N_1^{(1)}}{\partial m_{w,in}} + \tau_w^{(1)} \frac{\partial N_2^{(1)}}{\partial m_{w,in}} + \tau_a^{(1)} \frac{\partial N_3^{(1)}}{\partial m_{w,in}} + o^{(1)} \frac{\partial N_4^{(1)}}{\partial m_{w,in}} \right). \quad (6.140)$$

Numerically, the inlet water temperature, $T_{w,in}$, has the nominal ("base-case") value of $T_{w,in}^0 = 298.79[K]$, while the nominal ("base-case") value of the inlet water mass flow rate is $m_{w,in}^0 = 44.021229[kg/s]$. As before, the corresponding nominal value $m_{w,nom}^{(50)}$ of the response $m_w^{(50)}$ is $m_{w,nom}^{(50)} = 43.598097[kg/s]$. Consider now a perturbation $\delta T_{w,in} = (0.00033) T_{w,in}^0$, for which the perturbed value of the inlet air temperature becomes $T_{w,in}^{pert} = T_{w,in}^0 + \delta T_{w,in} = 298.89[K]$. Re-computing the perturbed response by solving Eqs. (6.2) through (6.13) with the value of $T_{w,in}^{pert}$ yields the "perturbed response" value $m_{w,pert}^{(50)} = 43.591565[kg/s]$. Using now the nominal and perturbed response values together with the parameter perturbation in the finite-difference expression given in Eq. (6.100) yields the corresponding "finite-difference-computed sensitivity" $S_3^{FD} \triangleq \frac{m_{w,pert}^{(50)} - m_{w,nom}^{(50)}}{\delta T_{w,in}} = -0.06531[(kg/s)/K]$.

Next, consider a perturbation $\delta m_{w,in} = (0.001) m_{w,in}^0$, for which the perturbed value of the inlet air temperature becomes $m_{w,in}^{pert} = m_{w,in}^0 + \delta m_{w,in} = 44.065252[kg/s]$. Re-computing the perturbed response by solving Eqs. (6.2) through (6.13) with the value of $m_{w,in}^{pert}$ yields the "perturbed response" value $m_{w,pert}^{(50)} = 43.641962[kg/s]$. Using now the nominal and perturbed response values together with the parameter perturbation in the finite-difference expression given in Eq. (6.100) yields the corresponding "finite-difference-computed sensitivity" $S_{45}^{FD} \triangleq \frac{m_{w,pert}^{(50)} - m_{w,nom}^{(50)}}{\delta m_{w,in}} = 0.99646$.

Inserting now all of the numerical values of the known quantities in Eqs. (6.139) and (6.140) yields the following system of coupled equations for obtaining $\mu_w^{(1)}$ and $\tau_w^{(1)}$:

$$-0.06531 = \left[0.0161\mu_w^{(1)} + 223597\tau_w^{(1)} + (-3.1344)(-1.039 \times 10^{-4}) + (8.1328 \times 10^{-7})(-264.65) \right], \quad (6.141)$$

$$0.99646 = -\Big[-\mu_w^{(1)} - 2.44038\times 10^6 \tau_w^{(1)} + (-3.1344)(0.0064) + (8.1328\times 10^{-7})(16430.26)\Big]. \tag{6.142}$$

Solving Eqs. (6.141) and (6.142) yields $\mu_w^{(1)} = 0.3373$ and $\tau_w^{(1)} = 2.6729\times 10^{-7}\ \left[(J/kg)^{-1}\right]$. These values compare well with the values $\mu_w^{(1)} = 0.3377$ and $\tau_w^{(1)} = 2.6710\times 10^{-7}\ \left[(J/kg)^{-1}\right]$, respectively, which are obtained by solving the adjoint sensitivity system given in Eq.(6.85), cf. Figure 6.11.

6.4 SENSITIVITY ANALYSIS RESULTS AND RANKINGS

Recall that the total sensitivity of a model response $R(\mathbf{m}_w,\mathbf{T}_w,\mathbf{T}_a,\boldsymbol{\omega};\boldsymbol{\alpha})$ to arbitrary variations in the model's parameters $\delta\boldsymbol{\alpha} \triangleq \left(\delta\alpha_1,\ldots,\delta\alpha_{N_\alpha}\right)$ and state functions $\delta\mathbf{m}_w,\delta\mathbf{T}_w,\delta\mathbf{T}_a,\delta\boldsymbol{\omega}$, around the nominal values $\left(\mathbf{m}_w^0,\mathbf{T}_w^0,\mathbf{T}_a^0,\boldsymbol{\omega}^0;\boldsymbol{\alpha}^0\right)$ of the parameters and state functions, is provided by the G-differential of the model's response to these variations. The expression of this G-differential was provided in Eqs. (6.80) and (6.88), which is reproduced below, for convenience:

$$\begin{aligned} &DR\left(\mathbf{m}_w^0,\mathbf{T}_w^0,\mathbf{T}_a^0,\boldsymbol{\omega}^0;\boldsymbol{\alpha}^0;\delta\mathbf{m}_w,\delta\mathbf{T}_w,\delta\mathbf{T}_a,\delta\boldsymbol{\omega};\delta\boldsymbol{\alpha}\right) \\ &= \sum_{i=1}^{N_\alpha}\left(\frac{\partial R}{\partial \alpha_i}\delta\alpha_i\right) + \boldsymbol{\mu}_w\cdot\mathbf{Q}_1 + \boldsymbol{\tau}_w\cdot\mathbf{Q}_2 + \boldsymbol{\tau}_a\cdot\mathbf{Q}_3 + \mathbf{o}\cdot\mathbf{Q}_4, \end{aligned} \tag{6.143}$$

where the vector $\left[\boldsymbol{\mu}_w,\boldsymbol{\tau}_w,\boldsymbol{\tau}_a,\mathbf{o}\right]^\dagger$ is the solution of the *adjoint sensitivity system* given in Eq. (6.87).

The explicit expressions of the vectors $\mathbf{Q}_\ell \triangleq \left(q_\ell^{(1)},\ldots,q_\ell^{(I)}\right)$, $\ell = 1,2,3,4$, are provided in Appendix 6. The model responses of interest in this work are the following quantities: (i) the outlet air temperature, $T_a^{(1)}$; (ii) the outlet water temperature, $T_w^{(50)}$; (iii) the outlet water flow rate, $m_w^{(50)}$; and (iv) the outlet air relative humidity, $RH^{(1)}$.

Note: the inlet air temperature $T_{a,in}$ and the dry-bulb temperature T_{db} appear in the model equations, for the sake of generality, as two distinct parameters. However, in the numerical computations to be presented in the remainder of this Chapter, the air inlet temperature $T_{a,in}$ (which is a boundary/inlet condition) is set to be identical to the dry-bulb temperature, T_{db}, so these two parameters become identical; therefore only the sensitivities of various responses to T_{db} will appear in the numerical results presented in the remaining Sections of this Chapter.

6.4.1 Relative sensitivities of the outlet air temperature, $T_a^{(1)}$

The sensitivities of the air outlet temperature with respect to all of the model's parameters have been computed using Eq. (6.143). The numerical results and ranking of the relative sensitivities, in descending order of their magnitudes, are provided in Table 6.14, below, along with their respective relative standard deviations.

Table 6.14: Ranked relative sensitivities of the outlet air temperature, $T_a^{(1)}$.

Rank #	Parameter (α_i)	Nominal Value	Rel. Sens. $RS(\alpha_i)$	Rel. std. dev. (%)
1	Air temperature (dry bulb), T_{db}	299.11 K	0.4829	1.39
2	Inlet water temperature, $T_{w,in}$	298.79 K	0.2756	0.57
3	Dew point temperature, T_{dp}	292.05 K	0.1834	0.81
4	$P_{vs}(T)$ parameter, a_0	25.5943	-0.0945	0.04
5	$P_{vs}(T)$ parameter, a_1	5229.89	0.0618	0.08
6	Inlet air humidity ratio, ω_{in}	0.0138	0.0100	14.93
7	Fan shroud inner diameter, D_{fan}	4.1 m	-0.0056	1.00
8	Water enthalpy $h_f(T)$ parameter, a_{1f}	4186.51	0.0050	0.04
9	Wetted fraction of fill surface area, w_{tsa}	1.0	-0.0049	0.00
10	Nusselt number, Nu	14.94	-0.0049	34.0
11	Fill section surface area, A_{surf}	14221 m^2	-0.0049	25.0

12	Dynamic viscosity of air at T=300K, μ	1.983E-5 kg/(m s)	0.0045	4.88
13	Nu parameter, $a_{1,Nu}$	0.0031498	-0.0045	31.75
14	Reynolds number, Re_d	4428	-0.0045	15.17
15	Fill section flow area, A_{fill}	67.29 m^2	0.0045	10.0
16	C_{pa}(T) parameter, $a_{0,cpa}$	1030.5	0.0032	0.03
17	Inlet water mass flow rate, $m_{w,in}$	44.02 kg/s	0.0031	5.0
18	h_g(T) parameter, a_{0g}	2005744	-0.0030	0.05
19	D_{av}(T) parameter, $a_{1,dav}$	2.65322	0.0028	0.11
20	Exit air speed at the shroud, V_{exit}	10.0 m/s	-0.0028	10.0
21	Inlet air mass flow rate, m_a	155.07 kg/s	-0.0028	10.26
22	Heat transfer coefficient multiplier, f_{ht}	1.0	-0.0026	50.0
23	Thermal conductivity of air at T=300K, k_{air}	0.02624 W/(m K)	-0.0026	6.04
24	Mass transfer coefficient multiplier, f_{mt}	1.0	-0.0022	50.0
25	Sherwood number, Sh	14.13	-0.0022	34.25
26	D_{av}(T) parameter, $a_{2,dav}$	-6.1681E-3	-0.0019	0.37
27	h_f(T) parameter, a_{0f}	1143423	-0.0017	0.05
28	D_{av}(T) parameter, $a_{0,dav}$	7.06085E-9	-0.0015	0
29	Atmospheric pressure, P_{atm}	100586 Pa	-0.0013	0.40
30	Kinematic viscosity of air at 300 K, ν	1.568E-5 m^2/s	-0.00074	12.09
31	Prandlt number of air at T=80 C, Pr	0.708	0.00074	0.71
32	Schmidt number, Sc	0.60	-0.00074	12.41
33	h_g(T) parameter, a_{1g}	1815.437	-0.00074	0.19
34	D_{av}(T) parameter, $a_{3,dav}$	6.55265E-6	0.00063	0.58
35	Nu parameter, $a_{2,Nu}$	0.9902987	-0.00032	33.02
36	Fill section equivalent diameter, D_h	0.0381 m	0.00032	1.0
37	C_{pa}(T) parameter, $a_{1,cpa}$	-0.19975	-0.00018	1.0

38	C_{pa} (T) parameter, $a_{2,cpa}$	3.9734E-4	0.00010	0.84
39	Sum of loss coefficients above fill, k_{sum}	10.0	0.000	50.0
40	Fill section frictional loss multiplier, f	4.0	0.000	50.0
41	Nu parameter, $a_{0,Nu}$	8.235	0.000	25.0
42	Nu parameter, $a_{3,Nu}$	0.023	0.000	38.26
43	Cooling tower deck width in x-dir, W_{dkx}	8.5 m	0.000	1.0
44	Cooling tower deck width in y-dir, W_{dky}	8.5 m	0.000	1.0
45	Cooling tower deck height above ground, Δz_{dk}	10.0 m	0.000	1.0
46	Fan shroud height, Δz_{fan}	3.0 m	0.000	1.0
47	Fill section height, Δz_{fill}	2.013 m	0.000	1.0
48	Rain section height, Δz_{rain}	1.633 m	0.000	1.0
49	Basin section height, Δz_{bs}	1.168 m	0.000	1.0
50	Drift eliminator thickness, Δz_{de}	0.1524 m	0.000	1.0
51	Wind speed, V_w	1.80 m/s	0.000	51.1

As the results in Table 6.14 indicate, the first 4 parameters (i.e., T_{db}, $T_{w,in}$, T_{dp}, a_0) have relative sensitivities between ca. 10% and 50%, and are therefore the most important for the air outlet temperature response, $T_a^{(1)}$. The two largest sensitivities have values of 48%, which means that a 1% change in T_{db} would induce a 0.48% change in $T_a^{(1)}$. The next two parameters (i.e., a_1 and ω_{in}) have relative sensitivities between 1% and 6%, and are therefore somewhat important. Parameters #7 through #15 (i.e.,. D_{fan}, a_{1f}, w_{tsa}, Nu, A_{surf}, μ, $a_{1,Nu}$, Re_d, A_{fill}) have relative sensitivities of the order of 0.5%. The remaining 36 parameters are relatively unimportant for this response, having relative sensitivities smaller than 1% of the largest relative sensitivity (with respect to T_{db}) for this response. Positive sensitivities imply that a positive change in the respective parameter would cause an increase in the response, while negative sensitivities imply that a positive change in the respective parameter would cause a decrease in the response.

6.4.2 Relative sensitivities of the outlet water temperature, $T_w^{(50)}$

The results and ranking of the relative sensitivities of the outlet water temperature with respect to the most important 11 parameters for this response are listed in Table 6.15. The largest sensitivity of $T_w^{(50)}$ is to the parameter T_{dp}, and has the value of 0.548; this means that a 1% increase in T_{db} would induce a 0.548% increase in $T_w^{(50)}$ The sensitivities to the remaining 40 model parameters have not been listed since they are smaller than 1% of the largest sensitivity (with respect to T_{dp}) for this response.

Table 6.15: Most important relative sensitivities of the outlet water temperature, $T_w^{(50)}$.

Rank #	Parameter (α_i)	Nominal value	Rel. Sens. $RS(\alpha_i)$	Rel. std. dev. (%)
1	Dew point temperature , T_{dp}	292.05 K	0.5482	0.81
2	Air temperature (dry bulb) , T_{db}	299.11 K	0.2244	1.39
3	P_{vs}(T) parameters, a_0	25.5943	-0.1949	0.04
4	P_{vs}(T) parameters, a_1	-5229.89	0.1282	0.08
5	Inlet water temperature, $T_{w,in}$	298.79 K	0.1066	0.57
6	Inlet air humidity ratio, ω_{in}	0.0138	0.0299	14.93
7	Fan shroud inner diameter, D_{fan}	4.1 m	-0.0085	1.00
8	Water enthalpy hf(T) parameter, a_{1f}	4186.51	0.0082	0.04
9	$D_{av}(T_{db})$ parameter, $a_{1,dav}$	2.653	0.0071	0.11
10	Enthalpy h_g(T) parameter, a_{0g}	2005744	-0.0062	0.05
11	Sherwood number, Sh	14.13	-0.0056	34.25

6.4.3 *Relative sensitivities of the outlet water mass flow rate, $m_w^{(50)}$*

The results and ranking of the relative sensitivities of the outlet water mass flow rate with respect to the most important 9 parameters for this response are listed in Table 6.16. This response is most sensitive to $m_{w,in}$ (a 1% increase in this parameter would cause a 1.01% increase in the response) and the second largest sensitivity is to the parameter $T_{w,in}$ (a 1% increase in this parameter would cause a 0.447% decrease in the response). The sensitivities to the remaining 42 model parameters have not been listed since they are smaller than 1% of the largest sensitivity (with respect to $m_{w,in}$) for this response.

Table 6.16: Most important relative sensitivities of the outlet water mass flow rate, $m_w^{(50)}$.

Rank #	Parameter (α_i)	Nominal value	Rel. Sens. $RS(\alpha_i)$	Rel. std. dev.(%)
1	Inlet water mass flow rate, $m_{w,in}$	44.02 kg/s	1.0060	5.00
2	Inlet water temperature, $T_{w,in}$	298.79 K	-0.4474	0.57
3	Dew point temperature , T_{dp}	292.05 K	0.3560	0.81
4	Pvs(T) parameters, a_0	25.5943	-0.1416	0.04
5	Air temperature (dry bulb) , T_{db}	299.11 K	-0.1184	1.39
6	Pvs(T) parameters, a_1	5229.89	0.0930	0.08
7	Inlet air humidity ratio, ω_{in}	0.0138	0.0195	14.93
8	Fan shroud inner diameter, D_{fan}	4.1 m	-0.0117	1.00
9	Inlet air mass flow rate, m_a	155.07 kg/s	-0.0058	10.26

6.4.4 Relative sensitivities of the outlet air relative humidity, $RH^{(1)}$

The results and ranking of the relative sensitivities of the outlet air relative humidity with respect to the most important 19 parameters for this response are listed in Table 6.17. The first two sensitivities of this response are quite large (relative sensitivities larger than unity are customarily considered to be very significant). In particular, an increase of 1% in T_{db} would cause a decrease in the response of 6.525%. On the other hand, an increase of 1% in T_{dp} would cause an increase of 5.75% in the response. The sensitivities to the remaining 32 model parameters have not been listed since they are smaller than 1% of the largest sensitivity (with respect to T_{db}) for this response.

Table 6.17: Most important relative sensitivities of the outlet air relative humidity, $RH^{(1)}$.

Rank #	**Parameter (α_i)**	**Nominal value**	**Rel. Sens. $RS(\alpha_i)$**	**Rel. std. dev. (%)**
1	Air temperature (dry bulb), T_{db}	299.11 K	-6.525	1.39
2	Dew point temperature, T_{dp}	292.05 K	5.750	0.81
3	Inlet water temperature, $T_{w,in}$	298.79 K	0.747	0.57
4	Inlet air humidity ratio, ω_{in}	0.0138	0.3141	14.93
5	$P_{vs}(T)$ parameters, a_0	25.5943	-0.3123	0.04
6	Wetted fraction of fill surface area, w_{tsa}	1.0	0.1487	0.00
7	Fill section surface area, A_{surf}	14221 m^2	0.1487	25.0
8	Nusselt number, Nu	14.94	0.1487	34.0
9	Dynamic viscosity of air at T=300 K, μ	1.983E-5 kg/(m s)	-0.1388	4.88
10	Nu parameters, $a_{1,Nu}$	0.0031498	0.1388	31.75
11	Fill section flow area, A_{fill}	67.29 m^2	-0.1388	10.0
12	Reynold's number, Re	4428	0.1388	15.17
13	$D_{av}(T_{db})$ parameter, $a_{1,dav}$	2.65322	-0.1297	0.11
14	Mass transfer coefficient multiplier, f_{mt}	1.0	0.1023	50.0

15	Sherwood number, Sh	14.13	0.1023	34.25
16	Atmosphere pressure, P_{atm}	100586 Pa	0.0992	0.40
17	$D_{av}(T_{db})$ parameter, $a_{2,dav}$	-6.1681E-3	0.0902	0.37
18	$D_{av}(T_{db})$ parameter, $a_{0,dav}$	7.06085E-9	0.0682	0.00
19	$P_{vs}(T)$ parameters, a_1	-5229.89	0.0681	0.08

Overall, the outlet air relative humidity, $RH^{(1)}$, displays the largest sensitivities, so this response is the most sensitive to parameter variations. The other responses, namely the outlet air temperature, the outlet water temperature, and the outlet water mass flow rate display sensitivities of comparable magnitudes.

6.5 A PRIORI STATISTICAL ANALYSIS OF EXPERIMENTALLY MEASURED MODEL RESPONSES AND PARAMETERS FOR SRNL F-AREA COOLING TOWERS

During the period from April, 2004 through August, 2004, a total of 8079 measured benchmark data sets for F-area cooling towers (fan-on case) were recorded every 15 fifteen minutes at SRNL. Among the 8079 measured data sets, 7688 of them are considered to represent the unsaturated conditions. Each of the data sets contained measurements for the following three responses: (i) outlet air temperature, denoted as $T_{a,out}^{meas}$; (ii) outlet water temperature, denoted as $T_{w,out}^{meas}$; and (iii) outlet air relative humidity, denoted as RH_{out}^{meas}. In addition, each of the data sets also contained measurements of the following four measured parameters: (i) the dry bulb air temperature; (ii) dew point temperature; (iii) inlet water temperature; and (iv) atmospheric pressure (which are denoted as T_{db}, T_{dp}, $T_{w,in}$ and P_{atm}, respectively). The measured outlet (exit) air relative humidity, RH^{meas}, was obtained using Hobo humidity sensors. The accuracy of these sensors is depicted in Figure 6.12, which indicates the following tolerances (standard deviations): ±2.5% for relative humidity from 10 to 90%; between ±2.5% and ±3.5% for relative humidity from 90% to 95%; and ±3.5% ~ ±4.0% from 95 to 100%. However, when exposed to relative humidity above 95%, the maximum sensor error may temporally increase by an additional 1%, so that the error can reach values between ±4.5% to ±5.0% for relative humidity from 95 to 100%.

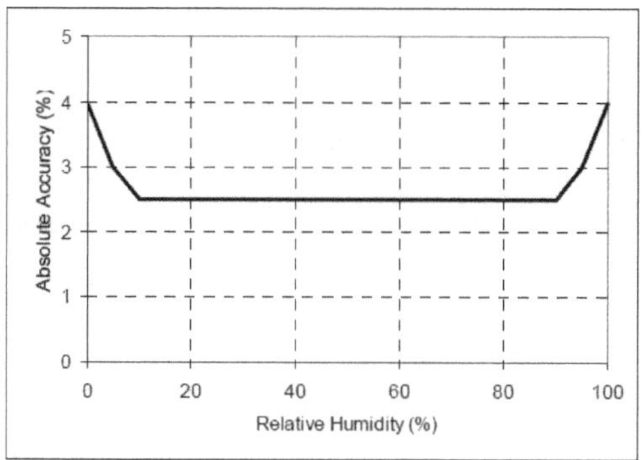

Figure 6.12: Humidity sensor accuracy plot (adopted from the specification of HOBO Pro v2).

The raw measured data was analyzed using the CTTool code (Aleman and Garrett, 2015) and 7668 measured values of the outlet (exit) air relative humidity, RH^{meas}, were considered to be "unsaturated," and are presented in the histogram plot shown in Figure 6.13. This plot, as well as all of the other histogram plots in this Subsection, have their total respective areas normalized to unity. The measured relative humidity RH^{meas} plotted in Figure 6.13 actually spans the range from 33.0% to 104.1%. In this range, 6975 data sets have their respective RH^{meas} less than 100% while the other 693 data sets have their respective RH^{meas} over 100%. This situation is nevertheless consistent with the range of the sensors when their tolerances (standard deviations) are taken into account, which would make it possible for a measurement with $RH^{meas}=105\%$ to be nevertheless "unsaturated". Consequently, all the 7668 benchmark data sets plotted in Figure 6.13 were considered as "unsaturated", since their respective RH^{meas} was less than 105%.

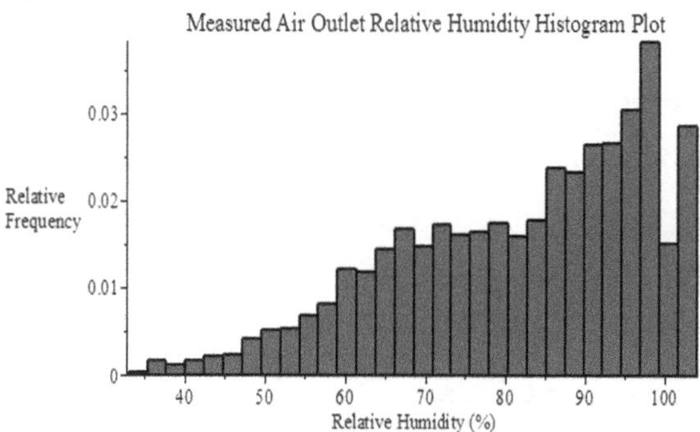

Figure 6.13: Histogram plot of the measured air outlet relative humidity, within the 7688 data sets collected by SRNL from F-Area cooling towers (unsaturated conditions).

The statistical properties of the (measured air outlet relative humidity) distribution shown in Figures 6.13 have been computed using standard packages, and are presented in Table 6.2. These statistical properties will be needed for the uncertainty quantification and predictive modeling computations.

Table 6.2: Statistics of the air outlet relative humidity distribution [%].

Min.	Max.	Range	Mean	Std. Dev.	Variance	Skewness	Kurtosis
33.0	104.1	71.1	81.98	15.63	244.44	-0.60	2.55

The histogram plots and their corresponding statistical characteristics of the 7668 data sets for the other measurements, namely for: the outlet air temperature, $T_{a,out(Tidbit)}$, measured using the "Tidbit" sensors; the outlet air temperature, $T_{a,out(Hobo)}$, measured using the "Hobo" sensors; and the outlet water temperature, $T_{w,out}^{meas}$, are reported below in Figures 6.14 through 6.17, and Tables 6.3 through 6.5, respectively.

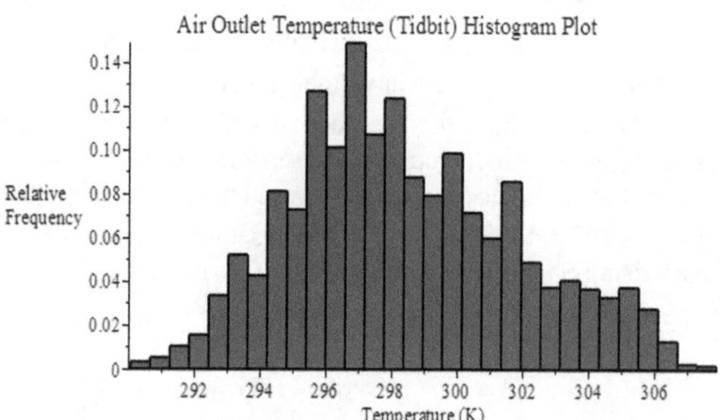

Figure 6.14: Histogram plot of the air outlet temperature measured using "Tidbit" sensors, within the 7688 data sets collected by SRNL from F-Area cooling towers (unsaturated conditions).

Table 6.3. Statistics of the air outlet temperature distribution [K], measured using "Tidbit" sensors.

Min.	Max.	Range	Mean	Std. Dev.	Variance	Skewness	Kurtosis
290.06	307.89	17.83	298.42	3.42	11.71	0.34	2.52

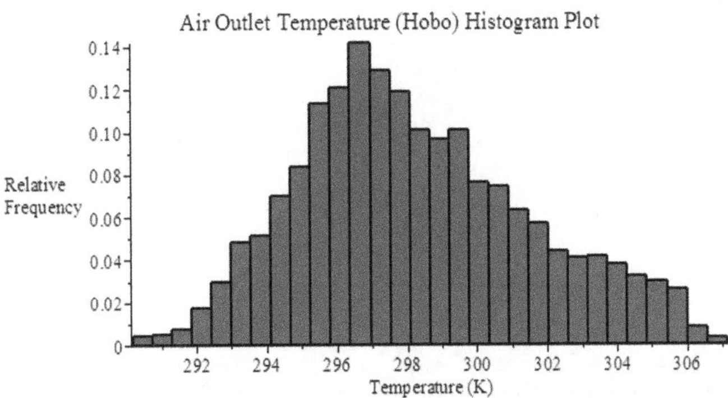

Figure 6.15: ***Histogram plot of the air outlet temperature measured using "Hobo" sensors, within the 7688 data sets collected by SRNL from F-Area cooling towers (unsaturated conditions).***

Table 6.4: ***Air outlet temperature distribution statistics [K], measured using "Hobo" sensors.***

Min.	Max.	Range	Mean	Std. Dev.	Variance	Skewness	Kurtosis
290.17	307.13	16.96	298.27	3.30	10.88	0.36	2.56

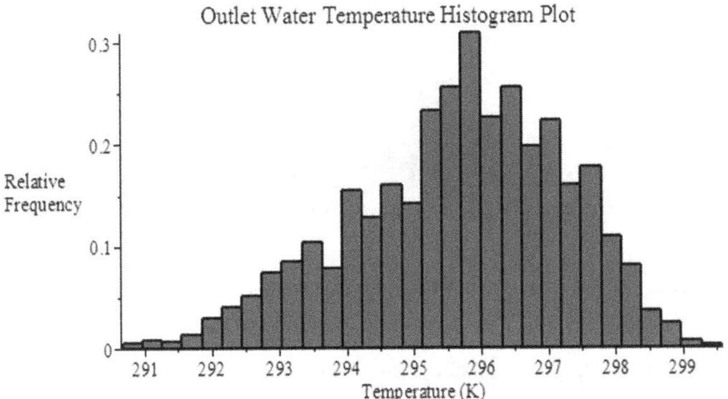

Figure 6.16: ***Histogram plot of water outlet temperature measurements, within the 7688 data sets collected by SRNL from F-Area cooling towers (unsaturated conditions).***

Table 6.5: ***Water outlet temperature distribution statistics [K].***

Min.	Max.	Range	Mean	Std. Dev.	Varian ce	Skewness	Kurtosis
290.67	299.57	8.90	295.68	1.58	2.48	-0.41	2.72

Ordering the above-mentioned four measured responses as follows: (i) outlet air temperature $T_{a,out(Tidbit)}$; (ii) outlet air temperature $T_{a,out(Hobo)}$; (iii) outlet water temperature $T_{w,out}^{meas}$; and (iv) outlet air relative humidity RH_{out}^{meas}, yields the following "measured response covariance matrix", denoted as $Cov\left(T_{a,out(Tidbit)}, T_{a,out(Hobo)}, T_{w,out}^{meas}, RH_{out}^{meas}\right)$:

$$Cov\left(T_{a,out(Tidbit)}, T_{a,out(Hobo)}, T_{w,out}^{meas}, RH_{out}^{meas}\right) = \begin{pmatrix} 11.71 & 11.23 & 3.57 & -44.76 \\ 11.23 & 10.88 & 3.52 & -42.94 \\ 3.57 & 3.52 & 2.48 & -5.31 \\ -44.76 & -42.94 & -5.31 & 244.44 \end{pmatrix}. \tag{6.144}$$

For the purposes of uncertainty quantification, data assimilation, model calibration and predictive modeling, the temperatures measurements provided by the "Tidbit" and "Hobo" sensors can be combined into an "averaged" data set of measured air outlet temperatures, which will be denoted as $T_{a,out}^{meas}$. The histogram plot and corresponding statistical characteristics of this averaged air outlet temperature are presented in Figure 6.17 and Table 6.6, respectively.

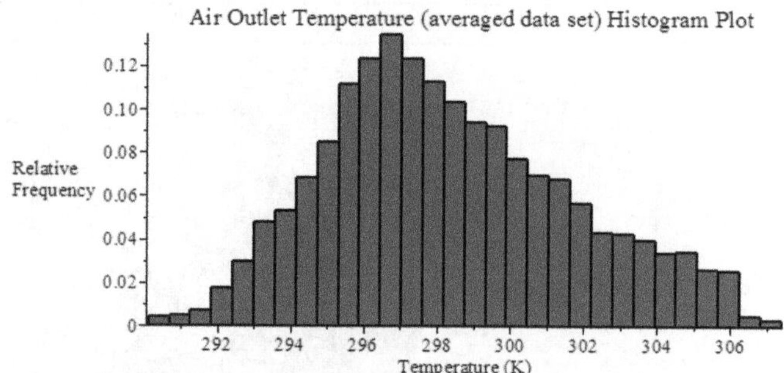

Figure 6.17: Histogram plot of air outlet temperatures averaged from Figure 6.14 and Figure 6.15.

Table 6.6: Statistics of the averaged air outlet temperature distribution [K].

Min.	Max.	Range	Mean	Std. Dev.	Variance	Skewness	Kurtosis
290.12	307.41	17.30	298.34	3.36	11.27	0.35	2.54

Computing the covariance matrix, denoted as $\left[Cov\left(T_{a,out}^{meas}, T_{w,out}^{meas}, RH_{out}^{meas}\right)\right]_{data}$, for all of the relevant experimental data for the averaged outlet air temperature $\left[T_{a,out}^{meas}\right]$, the outlet water temperature $\left[T_{w,out}^{meas}\right]$, and the outlet air relative humidity $\left[RH_{out}^{meas}\right]$, yields the following result:

$$\left[Cov\left(T_{a,out}^{meas}, T_{w,out}^{meas}, RH_{out}^{meas}\right)\right]_{data} = \begin{pmatrix} 11.27 & 3.55 & -43.85 \\ 3.55 & 2.48 & -5.31 \\ -43.85 & -5.31 & 244.44 \end{pmatrix}. \tag{6.145}$$

Comparing the results in Eqs. (6.144) and (6.145) shows that eliminating the second column and row in Eq. (6.144) yields a 3-by-3 matrix which has entries essentially equivalent to the covariance matrix in Eq. (6.145). In turn, this result indicates that the temperature distributions measured by the "Tidbit" and "Hobo" sensors, respectively, need not be treated as separate data sets for the purposes of uncertainty quantification and predictive modeling.

The sensors' standard deviations (namely: $\sigma_{sensor} = 0.2K$ for each of the responses $T_a^{(1)}$ and $T_w^{(50)}$, and $\sigma_{sensor} = 2.8\%$ for the response $RH^{(1)}$) have been taken into account for the data at the 100%-saturation point, by including the 693 data sets that have their respective measured relative humidity, RH^{meas}, between 100% and 104.1%. In addition, the respective sensors' uncertainties (standard deviations) must also be taken into account for the 6975 data sets that have their respective RH^{meas} less than 100%. Since the various measuring methods and devices are independent of each other, the standard deviation, $\sigma_{statistic}$, stemming from the statistical analysis of the 7668 benchmark data sets and the standard deviation, σ_{sensor}, stemming from the instrument's uncertainty are to be combined according to the well-known formula "addition of the variances of uncorrelated variates", namely:

$$\sigma = \sqrt{\sigma_{statistic}^{\ 2} + \sigma_{sensor}^{\ 2}}, \tag{6.146}$$

Using the relation in the above Eq. (6.146) in conjunction with the result presented in Eq.(6.145) will lead to an increase of the variances on the diagonal of the respective "measured covariance matrix", which will be denoted as $Cov\left(T_{a,out}^{meas}, T_{w,out}^{meas}, RH_{out}^{meas}\right)$. The final result obtained is

$$Cov\left(T_{a,out}^{meas}, T_{w,out}^{meas}, RH_{out}^{meas}\right) = \begin{pmatrix} 11.29 & 3.55 & -43.85 \\ 3.55 & 2.53 & -5.31 \\ -43.85 & -5.31 & 252.49 \end{pmatrix}. \tag{6.147}$$

The parameters $\alpha_1 \triangleq T_{db}, \alpha_2 \triangleq T_{dp}, \alpha_3 \triangleq T_{w,in}$, and $\alpha_4 \triangleq P_{atm}$ (i.e., the dry bulb air temperature, dew point temperature, inlet water temperature, and atmospheric pressure) were measured at the SRNL site at which the F-area cooling towers are located. Among the 8079 measured benchmark data sets, 7688 data sets are considered to represent "unsaturated conditions", which have been used to derive the statistical properties (means, variance and covariance, skewness and kurtosis) for these model parameters, as shown below in Figures 6.18 through 6.21 and Tables 6.10 through 6.13.

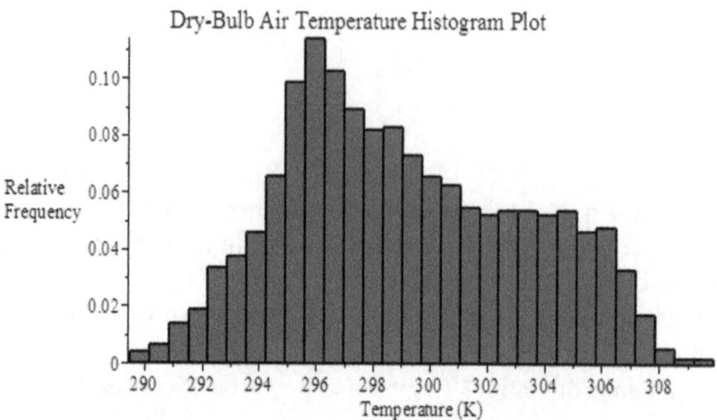

Figure 6.18: ***Histogram plot of dry-bulb air temperature data collected by SRNL from F-Area cooling towers (unsaturated conditions).***

Table 6.10: ***Statistics of the dry-bulb temperature (set to air inlet temperature) distribution [K].***

Min.	Max.	Range	Mean	Std. Dev.	Variance	Skewness	Kurtosis
289.50	309.91	20.41	299.11	4.17	17.37	0.25	2.18

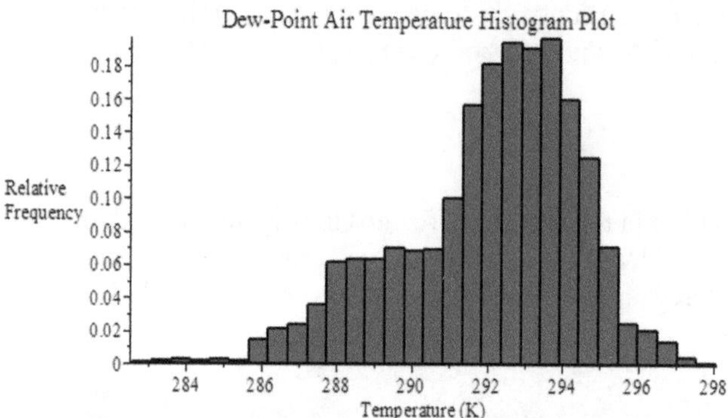

Figure 6.19: ***Histogram plot of dew-point air temperature data collected by SRNL from F-Area cooling towers (unsaturated conditions).***

Table 6.11: ***Statistics of the dew-point temperature distribution [K].***

Min.	Max.	Range	Mean	Std. Dev.	Variance	Skewness	Kurtosis
282.58	298.06	15.48	292.05	2.36	5.57	-0.66	3.10

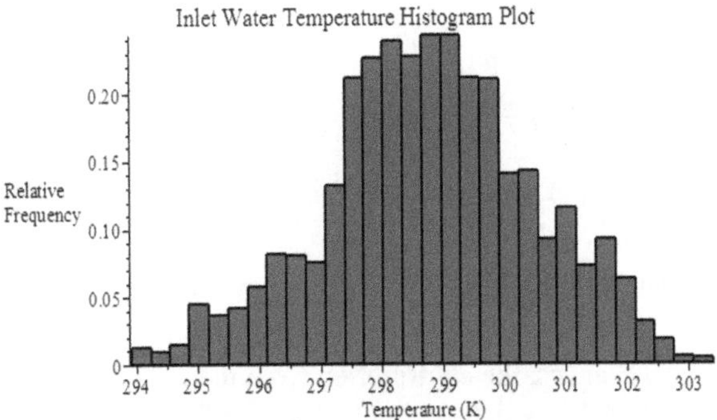

Figure 6.20: Histogram plot of inlet water temperature data collected by SRNL from F-Area cooling towers (unsaturated conditions).

Table 6.12: Statistics of the inlet water temperature distribution [K].

Min.	Max.	Range	Mean	Std. Dev.	Variance	Skewness	Kurtosis
293.93	303.39	9.46	298.79	1.70	2.90	-0.12	2.84

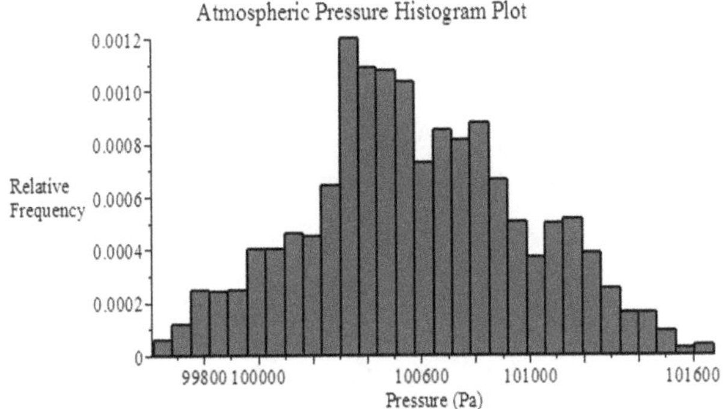

Figure 6.21: Histogram plot of atmospheric pressure data collected by SRNL from F-Area cooling towers (unsaturated conditions).

Table 6.13: Statistics of the atmospheric pressure distribution [Pa].

Min.	Max.	Range	Mean	Std. Dev.	Variance	Skewness	Kurtosis
99617	101677	2060	100586	401	160597	0.10	2.58

The 4-by-4 covariance matrix $Cov\left(T_{db};T_{dp};T_{w,in};P_{atm}\right)$ for the four correlated measured model parameters (dry-bulb air temperature T_{db}, dew-point air temperature T_{dp}, inlet water temperature $T_{w,in}$, and atmospheric air pressure P_{atm}) has also been computed from the measured data and is presented below:

$$Cov\left(T_{db};T_{dp};T_{w,in};P_{atm}\right)=\begin{pmatrix} 17.37 & 2.83 & 1.81 & -529.26 \\ 2.83 & 5.56 & 2.31 & -87.16 \\ 1.81 & 2.31 & 2.90 & -47.22 \\ -529.26 & -87.16 & -47.22 & 160597.01 \end{pmatrix}. \quad (6.148)$$

The covariance matrix $Cov\left(T_{db};T_{dp};T_{w,in};P_{atm}\right)$ neglects the uncertainty associated with sensor readings throughout the data collection period. When combining uncertainties by adding variances, the contribution from the sensors is 0.04 K for each of the first three parameters, which accounts for a maximum of ca. 1% of the total variance (for the inlet water temperature, specifically). The uncertainty in the atmospheric pressure sensor is at this time unknown. For these reasons, their contribution to overall uncertainty is considered insignificant at this time.

The *a priori* information needed for applying the BERRU predictive modeling methodology for the cooling tower can now be summarized as follows:

1 The mean values of the model parameters α_n, are provided in Table 6.14.

2 The parameter covariance matrix, $\mathbf{C}_{\alpha\alpha}$, is constructed using the covariance matrix $Cov\left(T_{db};T_{dp};T_{w,in};P_{atm}\right)$ provided in Eq. (6.148) and the variances of the remaining uncorrelated parameters, which have been provided in in Table 6.14, to obtain:

$$\mathbf{C}_{\alpha\alpha} \triangleq \begin{pmatrix} Var(\alpha_1) & Cov(\alpha_1,\alpha_2) & \bullet & Cov(\alpha_1,\alpha_{51}) \\ Cov(\alpha_2,\alpha_1) & Var(\alpha_2) & \bullet & Cov(\alpha_2,\alpha_{51}) \\ \bullet & \bullet & \bullet & \bullet \\ Cov(\alpha_{51},\alpha_1) & \bullet & \bullet & Var(\alpha_{51}) \end{pmatrix}$$
$$= \begin{pmatrix} 17.37 & 2.83 & 1.81 & -529.26 & 0 & \bullet & 0 \\ 2.83 & 5.56 & 2.31 & -87.16 & 0 & \bullet & 0 \\ 1.81 & 2.31 & 2.90 & -47.22 & 0 & \bullet & 0 \\ -529.26 & -87.16 & -47.22 & 160597.01 & 0 & \bullet & 0 \\ 0 & 0 & 0 & 0 & \bullet & \bullet & 0 \\ \bullet & \bullet & \bullet & \bullet & \bullet & \bullet & \bullet \\ 0 & 0 & 0 & 0 & 0 & \bullet & 25.81 \end{pmatrix}. \quad (6.149)$$

3 The nominal (average) values of the measured responses, which were presented in Tables 6.2, 6.5, and 6.6;

4 The covariances of the measured responses, $\mathbf{C}_{rr}$, which has been computed in Eq. (6.147);

5 The a priori response-parameter correlation matrix, denoted as $\mathbf{C}_{r\alpha}$, which is presented below:

$$\mathbf{C}_{r\alpha} \triangleq Cov\left(T_{a,out}^{meas}, T_{w,out}^{meas}, RH^{meas}, \alpha_1, \dots, \alpha_{51}\right)$$
$$= \begin{pmatrix} 12.96 & 3.51 & 2.33 & -447.09 & 0 & \cdots & 0 \\ 3.35 & 3.05 & 1.89 & -93.58 & 0 & \cdots & 0 \\ -54.16 & 1.73 & -2.27 & 1831.03 & 0 & \cdots & 0 \end{pmatrix}. \tag{6.150}$$

6 The matrix $\mathbf{S}_{r\alpha}^{N_r \times N_\alpha}$ of response sensitivities to parameters, the components of which are listed in Tables 6.14 through 6.17;

7 The covariance matrix of the computed responses, $\mathbf{C}_{rr}^{comp}$, which is computed as follows:

$$\mathbf{C}_{rr}^{comp} \equiv Cov\left(T_a^{(1)}, T_w^{(50)}, RH^{(1)}\right) = \mathbf{S}_{r\alpha}^{3\times 51}\mathbf{C}_{\alpha\alpha}\left(\mathbf{S}_{r\alpha}^{3\times 51}\right)^\dagger$$
$$= \begin{pmatrix} \frac{\partial T_a^{(1)}}{\partial \alpha_1}, \dots, \frac{\partial T_a^{(1)}}{\partial \alpha_{N\alpha}} \\ \frac{\partial T_w^{(50)}}{\partial \alpha_1}, \dots, \frac{\partial T_w^{(50)}}{\partial \alpha_{N\alpha}} \\ \frac{\partial RH^{(1)}}{\partial \alpha_1}, \dots, \frac{\partial RH^{(1)}}{\partial \alpha_{N\alpha}} \end{pmatrix} \mathbf{C}_{\alpha\alpha} \begin{pmatrix} \frac{\partial T_a^{(1)}}{\partial \alpha_1}, \dots, \frac{\partial T_a^{(1)}}{\partial \alpha_{N\alpha}} \\ \frac{\partial T_w^{(50)}}{\partial \alpha_1}, \dots, \frac{\partial T_w^{(50)}}{\partial \alpha_{N\alpha}} \\ \frac{\partial RH^{(1)}}{\partial \alpha_1}, \dots, \frac{\partial RH^{(1)}}{\partial \alpha_{N\alpha}} \end{pmatrix}^\dagger . \tag{6.151}$$
$$= \begin{pmatrix} 10.87 & 7.19 & -34.81 \\ 7.19 & 7.72 & -13.97 \\ -34.81 & -13.97 & 221.88 \end{pmatrix}.$$

6.6 BERRU PREDICTIVE MODELING OF COUNTER-FLOW COOLING TOWER

This Section presents the results of applying the BERRU-CMS methodology, in the simplified "One-Model Case" in Section 3.2.1 of Chapter 3, to perform predictive modeling of the counter-flow cooling tower model. For convenience, these equations are reproduced below:

$$\boldsymbol{\alpha}^{pred} = \boldsymbol{\alpha}^0 - \left(\mathbf{C}_{\alpha\alpha}\mathbf{S}_{r\alpha}^\dagger - \mathbf{C}_{\alpha r}\right)\left[\mathbf{D}_{rr}\right]^{-1}\left(\mathbf{r}^c - \mathbf{r}^m\right), \tag{6.152}$$

$$\mathbf{C}_{\alpha\alpha}^{pred} = \mathbf{C}_{\alpha\alpha} - \left(\mathbf{C}_{\alpha\alpha}\mathbf{S}_{r\alpha}^\dagger - \mathbf{C}_{\alpha r}\right)\left[\mathbf{D}_{rr}\right]^{-1}\left(\mathbf{C}_{\alpha\alpha}\mathbf{S}_{r\alpha}^\dagger - \mathbf{C}_{\alpha r}\right)^\dagger, \tag{6.153}$$

$$\mathbf{r}^{pred} = \mathbf{r}^m - \left(\mathbf{C}_{\alpha r}^\dagger\mathbf{S}_{r\alpha}^\dagger - \mathbf{C}_{rr}\right)\left[\mathbf{D}_{rr}\right]^{-1}\left(\mathbf{r}^c - \mathbf{r}^m\right), \tag{6.154}$$

$$\mathbf{C}_{rr}^{pred} = \mathbf{C}_{rr} - \left(\mathbf{C}_{\alpha r}^\dagger\mathbf{S}_{r\alpha}^\dagger - \mathbf{C}_{rr}\right)\left[\mathbf{D}_{rr}\right]^{-1}\left(\mathbf{C}_{\alpha r}^\dagger\mathbf{S}_{r\alpha}^\dagger - \mathbf{C}_{rr}\right)^\dagger, \tag{6.155}$$

$$\mathbf{C}_{\alpha r}^{pred} = \mathbf{C}_{\alpha r} - \left(\mathbf{C}_{\alpha\alpha}\mathbf{S}_{r\alpha}^{\dagger} - \mathbf{C}_{\alpha r}\right)\left[\mathbf{D}_{rr}\right]^{-1}\left(\mathbf{C}_{\alpha r}^{\dagger}\mathbf{S}_{r\alpha}^{\dagger} - \mathbf{C}_{rr}\right)^{\dagger}. \tag{6.156}$$

The a priori information described in the previous Subsection is has been used in conjunctions with Eqs. (6.152) and (6.153) to compute the best-estimate nominal parameter values and the corresponding best-estimate absolute standard deviations for these parameters, respectively. The resulting best-estimate nominal values are listed in Table 6.18, below. As the results in Table 6.18 indicate, the predicted best-estimate standard deviations are all smaller or at most equal to (i.e., left unaffected) the original standard deviations. The parameters are affected proportionally to the magnitudes of their corresponding sensitivities: the parameters experiencing the largest reductions in their predicted standard deviations are those having the largest sensitivities.

Table 6.18: Best-estimated nominal parameter values and their standard deviations.

i	**Independent Scalar Parameters (α_i)**	**Math Notation**	**Original Nominal Value**	**Original Absolute Std. Dev.**	**Best-estimated Nominal Value**	**Best-estimated Absolute Std. Dev.**
1	Air temperature (dry bulb), (K)	T_{db}	299.11	4.17	299.37	3.44
2	Dew point temperature (K)	T_{dp}	292.05	2.36	292.23	2.28
3	Inlet water temperature (K)	$T_{w,in}$	298.79	1.70	298.77	1.70
4	Atmospheric pressure (Pa)	P_{atm}	100586	401	100576	389
5	Wetted fraction of fill surface area	w_{tsa}	1	0	1	0
6	Sum of loss coefficients above fill	k_{sum}	10	5	10	5
7	Dynamic viscosity of air at T=300 K (kg/m s)	μ	$1.983\text{x}10^{-5}$	9.676E-7	$1.984\ \text{x}10^{-5}$	9.668E-7

8	Kinematic viscosity of air at T=300 K (m^2/s)	ν	1.568x10^{-5}	1.895 x10^{-6}	1.564 x10^{-5}	1.893 x10^{-6}
9	Thermal conductivity of air at T=300 K (W/m K)	k_{air}	0.02624	1.584 x10^{-3}	0.02625	1.583 x10^{-3}
10	Heat transfer coefficient multiplier	f_{ht}	1	0.5	1.0316	0.47
11	Mass transfer coefficient multiplier	f_{mt}	1	0.5	0.882	0.41
12	Fill section frictional loss multiplier	f	4	2	4	2.00
13	P_{vs}(T) parameters	a_0	25.5943	0.01	25.5943	0.01
14		a_1	-5229.89	4.4	-5229.92	4.40
15	C_{pa}(T) parameters	$a_{0,cpa}$	1030.5	0.2940	1030.5	0.294
16		$a_{1,cpa}$	-0.19975	0.0020	-0.19975	0.0020
17		$a_{2,cpa}$	3.9734x10^{-4}	3.345x10^{-6}	3.9734 x10^{-4}	3.345 x10^{-6}
18	D_{av}(T) parameters	$a_{0,dav}$	7.0608x10^{-9}	0	7.06085 x10^{-9}	0
19		$a_{1,dav}$	2.65322	0.003	2.65322	0.003
20		$a_{2,dav}$	-6.1681 x10^{-3}	2.3 x10^{-5}	-6.16806 x10^{-3}	2.3 x10^{-5}
21		$a_{3,dav}$	6.552659 x10^{-6}	3.8 x10^{-8}	6.552688 x10^{-6}	3.8 x 10^{-8}
22	h_f(T) parameters	a_{0f}	-1143423.8	543	-1143423.7	543
23		a_{1f}	4186.50768	1.8	4186.50818	1.8
24	h_g(T) parameters	a_{0g}	2005743.99	1046	2005743.80	1046
25		a_{1g}	1815.437	3.5	1815.436	3.5
26	Nu parameters	$a_{0,Nu}$	8.235	2.059	8.235	2.059

27		$a_{1,Nu}$	0.00314987	0.001	0.0030475	0.001
28		$a_{2,Nu}$	0.9902987	0.327	0.987827	0.327
29		$a_{3,Nu}$	0.023	0.0088	0.023	0.088
30	Cooling tower deck width in x-dir (m)	W_{dkx}	8.5	0.085	8.5	0.085
31	Cooling tower deck width in y-dir (m)	W_{dky}	8.5	0.085	8.5	0.085
32	Cooling tower deck height above ground (m)	Δz_{dk}	10	0.1	10	0.1
33	Fan shroud height (m)	Δz_{fan}	3.0	0.03	3.0	0.03
34	Fan shroud inner diameter (m)	D_{fan}	4.1	0.041	4.1	0.041
35	Fill section height (m)	Δz_{fill}	2.013	0.02013	2.013	0.02013
36	Rain section height (m)	Δz_{rain}	1.633	0.01633	1.633	0.01633
37	Basin section height (m)	Δz_{bs}	1.168	0.01168	1.168	0.01168
38	Drift eliminator thickness (m)	Δz_{de}	0.1524	0.001524	0.1524	0.001524
39	Fill section equivalent diameter (m)	D_h	0.0381	0.000381	0.0381	0.000381

40	Fill section flow area (m²)	A_{fill}	67.29	6.729	67.507	6.705
41	Fill section surface area (m²)	A_{surf}	14221	3555.3	13914	3463
42	Prandlt number of air at T=80 C	P_r	0.708	0.005	0.708	0.005
43	Wind speed (m/s)	V_w	1.80	0.92	1.80	0.92
44	Exit air speed at the shroud (m/s)	V_{exit}	10.0	1.0	9.978	1.0
i	**Boundary Parameters**	**Math. Notation**	**Original Nominal Value**	**Absolute Std. Dev.**	**Best-estimated Nominal Value**	**Best-estimated Absolute Std. Dev.**
45	Inlet water mass flow rate (kg/s)	$m_{w,in}$	44.02	2.201	44.05	2.199
46	Placeholder for $T_{a,in}$; NOT used for this computation (since $T_{a,in} = T_{db}$)					
47	Inlet air mass flow rate (kg/s)	m_a	155.07	15.91	154.70	15.87
48	Inlet air humidity ratio	ω_{in}	0.0138	0.00206	0.0142	0.00137
i	**Special Dependent Parameters**	**Math. Notation**	**Original Nominal Value**	**Absolute Std. Dev.**	**Best-estimated Nominal Value**	**Best-estimated Absolute Std. Dev.**
49	Reynold's number	Re_d	4428	671.6	4395	666.1
50	Schmidt number	Sc	0.60	0.074	0.5986	0.0739
51	Sherwood number	Sh	14.13	4.84	13.35	4.44
52	Nusselt number	Nu	14.94	5.08	14.34	4.83

The a priori information described in the previous Subsection has been used in conjunctions with Eqs. (6.156) to compute the best-estimate response-parameter correlation matrix, $\mathbf{C}_{\alpha r}^{pred}$. The non-zero elements with the largest magnitudes of $\mathbf{C}_{\alpha r}^{pred}$ are as follows:

$$\begin{aligned}
&rel.cor.(R_1,\alpha_4) = -0.278;\; rel.cor.(R_1,\alpha_{41}) = -0.070;\\
&rel.cor.(R_1,\alpha_{49}) = -0.039;\\
&rel.cor.(R_2,\alpha_4) = -0.108;\; rel.cor.(R_2,\alpha_{41}) = -0.019;\\
&rel.cor.(R_3,\alpha_4) = 0.232;\quad rel.cor.(R_3,\alpha_{41}) = 0.127;\\
&rel.cor.(R_3,\alpha_{49}) = 0.072.
\end{aligned} \tag{6.157}$$

The notation used in Eq. (6.157) is as follows: $R_1 \triangleq T_a^{(1)}, R_2 \triangleq T_w^{(50)}$, $R_3 \triangleq RH^{(1)}$; $\alpha_4 \triangleq P_{atm}$, $\alpha_{41} \triangleq A_{surf}$, and $\alpha_{49} \triangleq Re_d$.

The a priori information described in the previous Subsection has been used in conjunction with Eq. (6.154) to obtain the best-estimate nominal values of the (model responses) outlet air temperature, $T_a^{(1)}$, outlet water temperature $T_w^{(50)}$, and outlet air relative humidity, $RH^{(1)}$, which are summarized in Table 6.19. To facilitate comparison, the corresponding measured and computed nominal values are also presented in this table. Note that there are no direct measurements for the outlet water flow rate, $m_w^{(50)}$. For this response, therefore, the predicted best-estimate nominal value has been obtained by a forward re-computation using the best-estimate nominal parameter values listed in Table 6.18, while the predicted best estimate standard deviation for this response has been obtained by using "best-estimate" values in the following expression:

$$\left[\mathbf{C}_{rr}^{comp}\right]^{be} = \left[\mathbf{S}_{r\alpha}\right]^{be}\left[\mathbf{C}_{\alpha\alpha}\right]^{be}\left[\mathbf{S}_{r\alpha}^{\dagger}\right]^{be}. \tag{6.158}$$

Table 6.19: Computed, measured, and optimal best-estimate nominal values and standard deviations for the outlet air temperature, outlet water temperature, outlet air relative humidity, and outlet water flow rate responses.

Nominal Values and Standard Deviations	$T_a^{(1)}$ **[K]**	$T_w^{(50)}$ **[K]**	$RH^{(1)}$ **[%]**	$m_w^{(50)}$ **[kg/s]**
Measured				
nominal value	298.34	295.68	81.98	---
standard deviation	±3.36	±1.59	±15.89	---
Computed				
nominal value	297.46	294.58	86.12	43.60
standard deviation	±3.30	±2.78	±14.90	±2.21
Best-estimate				
nominal value	298.45	295.67	82.12	43.67
standard deviation	±2.59	±1.54	±12.05	±2.20

The results presented in Table 6.19 indicate that the predicted standard deviations are smaller than either the computed or the experimentally measured ones. This is indeed the consequence of using the BERRU-CMS methodology in conjunction with consistent (as opposed to discrepant) computational and experimental information. As mentioned earlier, measurements are available only for the three outlet responses: $T_a^{(1)}, T_w^{(50)}$ and $RH^{(1)}$. Otherwise, there are no direct measurements for the internal responses along the height of the fill section, namely: (i) the air temperature, $T_a^{(i)}, i = 2,...,I$, at the exit of each control volume; (ii) the water temperature, $T_w^{(i+1)}, i = 1,...,I-1$, at the exit of each control volume; and (iii) the air relative humidity, $RH^{(i)}, i = 2,...,I$, at the exit of each control volume. For these responses, therefore, the predicted best-estimate nominal value has been obtained by a forward re-computation using the best-estimate nominal parameter values, $\boldsymbol{\alpha}^{pred}$, as listed in Table 6.18, while the predicted best estimate standard deviation for these responses have been obtained by using "best-estimate" values in Eq. (6.158). The matrix of sensitivities $[\mathbf{S}_{r\alpha}]^{pred}$ which is used in Eq. (6.158) has been obtained for each of the responses $T_a^{(i)}, i = 2,...,I$, $T_w^{(i+1)}, i = 1,...,I-1$, and $RH^{(i)}, i = 2,...,I$ by performing computations using the adjoint sensitivity model detailed in Section 6.4, but at the best-estimate parameter values, rather than at the nominal parameter values (denoted as "re-computed" nominal parameter values and standard deviations). Figures 6.25 through 6.27 show the plots for the computed (depicted in black), best-estimate (depicted in red), and re-computed (depicted in green) nominal values and standard deviations for the air temperature $T_a^{(i)}, (i = 1,...,49)$, water temperature $T_w^{(i)}, (i = 2,...,50)$, and air humidity $RH^{(i)}, (i = 1,...,49)$, respectively, along the height of the fill section of the cooling tower. Note that the best-estimate nominal values and standard deviations are only available for the three outlet responses, i.e., $T_a^{(1)}, T_w^{(50)}$ and $RH^{(1)}$ (depicted in red).

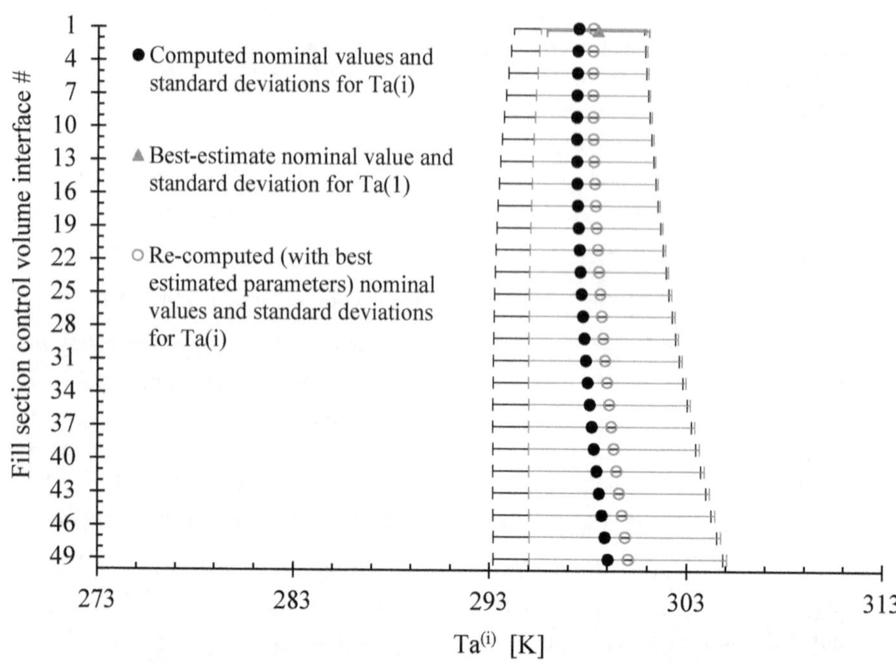

Figure 6.25: ***Computed (black), best-estimate (red), and re-computed (green; using best-estimate parameter values) nominal values and standard deviations for the air temperature,*** $T_a^{(i)}$, $(i = 1,...,49)$, ***at the exit of each control volume along the height of the fill section of the cooling tower.***

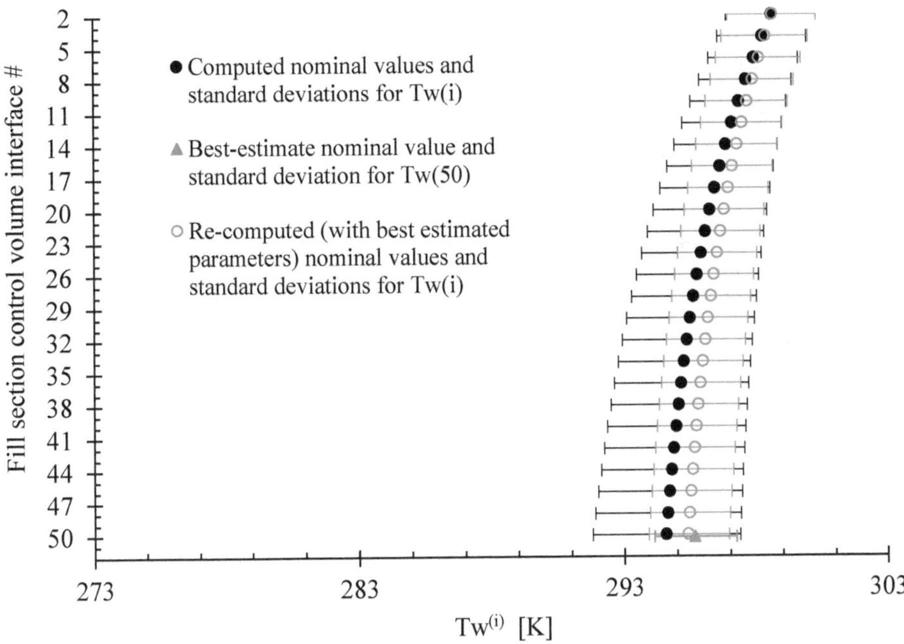

Figure 6.26: ***Computed (black), best-estimate (red), and re-computed (green; using best-estimate parameter values) nominal values and standard deviations for the water temperature,*** $T_w^{(i)}$, $(i = 2,...,50)$, ***at the exit of each control volume along the height of the fill section of the cooling tower.***

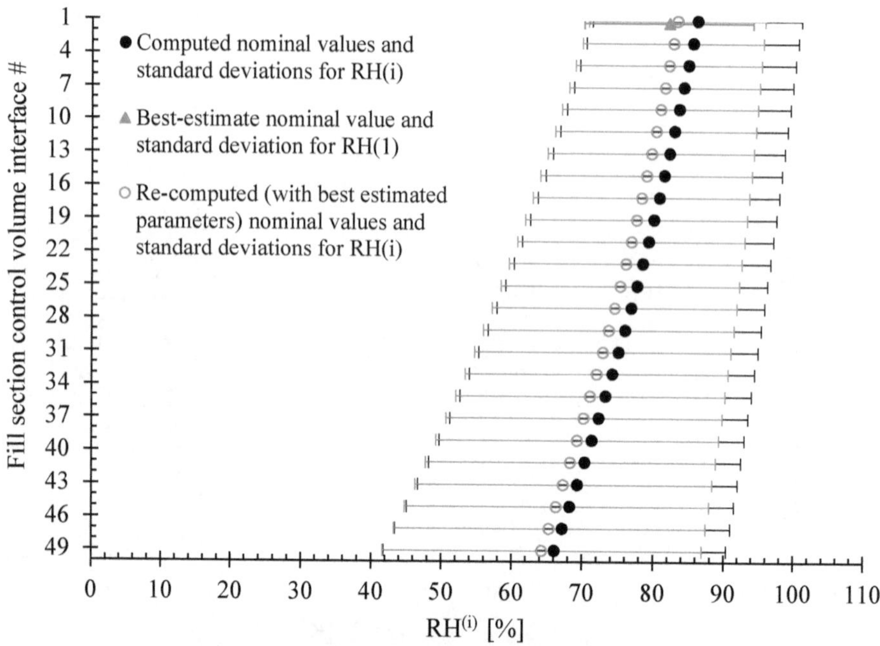

***Figure 6.27:** Computed (black), best-estimate (red), and re-computed (green; using best-estimate parameter values) nominal values and standard deviations for the air relative humidity, $RH^{(i)}$, $(i = 1,...,49)$, at the exit of each control volume along the height of the fill section of the cooling tower.*

The following major conclusions can be drawn from the results presented in this Chapter:

(i) The standard deviations predicted by the BERRU-CMS are smaller than either the computed or the experimentally measured ones at the locations where measurements are available.

(ii) Application of the BERRU-CMS methodology has improved the predicted standard deviations for the responses inside and along the height of the fill section, for which no measurements were available. Thus, the assimilation of measurements has reduced the uncertainties of the predicted internal responses well below the uncertainties in the computed responses (which stem from uncertainties in the model parameters).

(iii) The maximum reductions of uncertainties are always at the boundaries where direct measurements are available, and the amount of reductions decreases toward the inlets along the height of the fill section. For instance, as shown in Figure 6.25, a maximum of 19% reduction of the uncertainty is achieved for the response $T_a^{(1)}$ at the air exit of the fill section, and this reduction gradually decreases to 14% for the response $T_a^{(49)}$ near the air inlet of the fill section. Similarly, Figure 6.26 indicates that the maximum reduction of the uncertainty is around 45%, for the response $T_w^{(50)}$ at the water exit of the fill section, and this "uncertainty reduction" gradually diminishes to nearly 1% for the response $T_w^{(2)}$ near the water inlet of the fill section. Lastly, for the humidity responses shown in Figure 6.27, a maximum of 16% reduction is achieved for the response $RH^{(1)}$ at the air exit of the fill section, and this reduction gradually decreases to around 7% for the response $RH^{(49)}$ near the inlet of the fill section.

APPENDIX 6

The inlet air humidity ratio is defined as follows:

$$\omega_{in} = \frac{0.622 P_{vs}\left(T_{dp}, \boldsymbol{\alpha}\right)}{P_{atm} - P_{vs}\left(T_{dp}, \boldsymbol{\alpha}\right)} = \frac{0.622 e^{a_0 + \frac{a_1}{T_{dp}}}}{P_{atm} - e^{a_0 + \frac{a_1}{T_{dp}}}}. \tag{6.159}$$

The Reynold's number is defined (equivalently) as follows:

$$Re_d = \frac{D_h \left| m_a \right|}{\mu A_{fill}}. \tag{6.160}$$

The Schmidt number is defined as follows:

$$Sc = \frac{\nu}{D_{av}(T)}. \tag{6.161}$$

The Sherwood number is defined as follows:

$$Sh = Nu \left(\frac{Sc}{P_r} \right)^{\frac{1}{3}}. \tag{6.162}$$

The Nusselt number is defined as follows:

$$Nu = \begin{cases} a_{0,Nu} & Re_d < 2300 \\ a_{1,Nu} \cdot Re_d + a_{2,Nu} & 2300 \le Re_d \le 10000 \\ a_{3,Nu} \cdot Re_d^{\,0.8} \cdot Pr^{\frac{1}{3}} & Re_d > 10000 \end{cases}. \tag{6.163}$$

The inlet water mass flow rate is calculated using the following expression:

$$m_{w,in} = \rho\left(T_{w,in}\right) \cdot \frac{700.0}{15850.32}, \tag{6.164}$$

where

$$\begin{aligned} \rho\left(T_{w,in}\right) &= a_{1,\rho} + a_{2,\rho}\left(T_{w,in} - 273.15\right) \\ &\quad + a_{3,\rho}\left(T_{w,in} - 273.15\right)^2 + a_{4,\rho}\left(T_{w,in} - 273.15\right)^3; \\ a_{1,\rho} &= 1.0048897 \times 10^3;\ \ a_{2,\rho} = -2.6847207 \times 10^{-1}; \\ a_{3,\rho} &= -1.81136391 \times 10^{-3};\ \ a_{4,\rho} = -1.7041217 \times 10^{-6}. \end{aligned} \tag{6.165}$$

The inlet air mass flow rate is calculated using the following expression:

$$m_a = \rho\left(T_{db}\right) \cdot V_{exit} \cdot \frac{\pi D_{fan}^{\ 2}}{4} = \frac{P_{atm}}{R_{air} T_{db}} \cdot V_{exit} \cdot \frac{\pi D_{fan}^{\ 2}}{4}. \tag{6.166}$$

Derivatives of the liquid continuity equations with respect to model parameters

The derivatives of the "liquid continuity equations" with respect to the parameter $\alpha^{(1)} \triangleq T_{db}$ are as follows:

$$\begin{aligned} &\frac{\partial N_1^{(i)}}{\partial \alpha^{(1)}} = \frac{\partial N_1^{(i)}}{\partial T_{db}} \triangleq a_1^{i,1} \\ &= \frac{1}{\overline{R}}\left[\frac{P_{vs}^{(i+1)}\left(T_w^{(i+1)}, \boldsymbol{\alpha}\right)}{T_w^{(i+1)}} - \frac{\omega^{(i)} P_{atm}}{\left(0.622 + \omega^{(i)}\right) T_a^{(i)}}\right] \frac{\partial M\left(m_a, \boldsymbol{\alpha}\right)}{\partial D_{av}\left(T_{db}, \boldsymbol{\alpha}\right)} \cdot \frac{\partial D_{av}\left(T_{db}, \boldsymbol{\alpha}\right)}{\partial T_{db}}; \\ &\ell = 1;\ i = 1, \ldots, I;\ \ j = 1, \end{aligned} \tag{6.167}$$

where

$$\frac{\partial M\left(m_a, \boldsymbol{\alpha}\right)}{\partial D_{av}\left(T_{db}, \boldsymbol{\alpha}\right)} = \frac{2}{3} \cdot \frac{M\left(m_a, \boldsymbol{\alpha}\right)}{D_{av}\left(T_{db}, \boldsymbol{\alpha}\right)}, \tag{6.168}$$

$$\frac{\partial D_{av}\left(T_{db}, \boldsymbol{\alpha}\right)}{\partial T_{db}} = \frac{1.5 \cdot a_{0dav} T_{db}^{\ 0.5} - \left(a_{2dav} + 2 \cdot a_{3dav} T_{db}\right) D_{av}\left(T_{db}, \boldsymbol{\alpha}\right)}{a_{1dav} + a_{2dav} T_{db} + a_{3dav} T_{db}^{\ 2}}. \tag{6.169}$$

The derivatives of the "liquid continuity equations" with respect to the parameter $\alpha^{(2)} \triangleq T_{dp}$ are as follows:

$$\frac{\partial N_1^{(i)}}{\partial \alpha^{(2)}} = \frac{\partial N_1^{(i)}}{\partial T_{dp}} \triangleq a_1^{i,2} = 0; \quad \ell = 1; \, i = 1,\ldots,I; \;\; j = 2. \tag{6.170}$$

The derivatives of the "liquid continuity equations" with respect to the parameter $\alpha^{(3)} \triangleq T_{w,in}$ are as follows:

$$\frac{\partial N_1^{(1)}}{\partial \alpha^{(3)}} = \frac{\partial N_1^{(1)}}{\partial T_{w,in}} \triangleq a_1^{1,3} = -\frac{\partial m_{w,in}}{\partial T_{w,in}}; \quad \ell = 1; \, i = 1; \;\; j = 3, \tag{6.171}$$

where

$$\begin{aligned} \frac{\partial m_{w,in}}{\partial T_{w,in}} &= \frac{\partial}{\partial T_{w,in}}\left[\rho\left(T_{w,in}\right)\frac{700.0}{15850.32}\right] \\ &= \left[a_{2,\rho} + 2a_{3,\rho}\left(T_{w,in} - 273.15\right) + 3a_{4,\rho}\left(T_{w,in} - 273.15\right)^2\right]\frac{700.0}{15850.32}; \end{aligned} \tag{6.172}$$

and where

$$a_{2,\rho} = -0.26847207; \, a_{3,\rho} = -1.81136391 \times 10^{-3}; \, a_{4,\rho} = -1.7041217 \times 10^{-6}.$$

$$\frac{\partial N_1^{(i)}}{\partial \alpha^{(3)}} = \frac{\partial N_1^{(i)}}{\partial T_{w,in}} \triangleq a_1^{i,3} = 0; \quad \ell = 1; \, i = 2,\ldots,I; \;\; j = 3. \tag{6.173}$$

The derivatives of the "liquid continuity equations" with respect to the parameter $\alpha^{(4)} \triangleq P_{atm}$ are as follows:

$$\begin{aligned} &\frac{\partial N_1^{(i)}}{\partial \alpha^{(4)}} = \frac{\partial N_1^{(i)}}{\partial P_{atm}} \triangleq a_1^{i,4} = -\frac{M\left(m_a, \boldsymbol{\alpha}\right)}{\bar{R}} \frac{\omega^{(i)}}{T_a^{(i)}\left(0.622 + \omega^{(i)}\right)} \\ &+ \frac{1}{\bar{R}}\left[\frac{P_{vs}^{(i+1)}\left(T_w^{(i+1)}, \boldsymbol{\alpha}\right)}{T_w^{(i+1)}} - \frac{\omega^{(i)} P_{atm}}{\left(0.622 + \omega^{(i)}\right) T_a^{(i)}}\right] \frac{\partial M\left(m_a, \boldsymbol{\alpha}\right)}{\partial Nu\left(Re, \boldsymbol{\alpha}\right)} \frac{\partial Nu\left(Re, \boldsymbol{\alpha}\right)}{\partial m_a} \frac{\partial m_a}{\partial P_{atm}}, \\ &\ell = 1; \, i = 1,\ldots,I; \;\; j = 4, \end{aligned} \tag{6.174}$$

where

$$\frac{\partial M\left(Re, \boldsymbol{\alpha}\right)}{\partial Nu\left(Re, \boldsymbol{\alpha}\right)} = \frac{M\left(m_a, \boldsymbol{\alpha}\right)}{Nu\left(Re, \boldsymbol{\alpha}\right)}, \tag{6.175}$$

$$\frac{\partial Nu(Re,\boldsymbol{\alpha})}{\partial m_a} = \begin{cases} 0, & Re_d < 2300, \\ a_{1,Nu}\, Re(m_a,\boldsymbol{\alpha})/m_a, & 2300 \le Re_d \le 10000, \\ 0.8Nu(Re,\boldsymbol{\alpha})/m_a, & Re_d > 10000, \end{cases} \tag{6.176}$$

$$\frac{\partial m_a}{\partial P_{atm}} = \frac{1}{R_{air}T_{a,in}} V_{exit} \frac{\pi D_{fan}^2}{4}; \tag{6.177}$$

Note: The term on the right hand side of Eq. (6.177) stems from the following relation:

$$m_a = \rho(T_a) \cdot V_{exit} \cdot \frac{\pi D_{fan}^2}{4} = \frac{P_{atm}}{R_{air}T_a} \cdot V_{exit} \cdot \frac{\pi D_{fan}^2}{4}. \tag{6.178}$$

The derivatives of the "liquid continuity equations" with respect to the parameter $\alpha^{(5)} \triangleq w_{tsa}$ are as follows:

$$\frac{\partial N_1^{(i)}}{\partial \alpha^{(5)}} = \frac{\partial N_1^{(i)}}{\partial w_{tsa}} \triangleq a_1^{i,5} = \frac{1}{R}\left[\frac{P_{vs}^{(i+1)}\left(T_w^{(i+1)},\boldsymbol{\alpha}\right)}{T_w^{(i+1)}} - \frac{\omega^{(i)}P_{atm}}{\left(0.622+\omega^{(i)}\right)T_a^{(i)}}\right]\frac{\partial M(m_a,\boldsymbol{\alpha})}{\partial w_{tsa}}; \tag{6.179}$$

$$\ell = 1;\ i = 1,\ldots,I;\ j = 5,$$

where

$$\frac{\partial M(m_a,\boldsymbol{\alpha})}{\partial w_{tsa}} = \frac{M_{H2O} f_{mt} Nu(Re,\boldsymbol{\alpha})\left(\frac{\nu}{Pr}\right)^{\frac{1}{3}}\left[D_{av}(T_{db},\boldsymbol{\alpha})\right]^{\frac{2}{3}} A_{surf}}{D_h I}. \tag{6.180}$$

The derivatives of the "liquid continuity equations" with respect to the parameter $\alpha^{(6)} \triangleq k_{sum}$ are as follows:

$$\frac{\partial N_1^{(i)}}{\partial \alpha^{(6)}} = \frac{\partial N_1^{(i)}}{\partial k_{sum}} \triangleq a_1^{i,6} = 0;\quad \ell = 1;\ i = 1,\ldots,I;\ j = 6. \tag{6.181}$$

The derivatives of the "liquid continuity equations" with respect to the parameter $\alpha^{(7)} \triangleq \mu$ are as follows:

$$\frac{\partial N_1^{(i)}}{\partial \alpha^{(7)}} = \frac{\partial N_1^{(i)}}{\partial \mu} \triangleq a_1^{i,7}$$
$$= \frac{1}{\overline{R}} \left[\frac{P_{vs}^{(i+1)}\left(T_w^{(i+1)},\boldsymbol{\alpha}\right)}{T_w^{(i+1)}} - \frac{\omega^{(i)} P_{atm}}{\left(0.622+\omega^{(i)}\right) T_a^{(i)}} \right] \frac{\partial M\left(m_a,\boldsymbol{\alpha}\right)}{\partial \mu}; \tag{6.182}$$
$$\ell = 1;\ i = 1,\dots,I;\ j = 7,$$

where

$$\frac{\partial M\left(m_a,\boldsymbol{\alpha}\right)}{\partial \mu} = \begin{cases} 0, & Re_d < 2300, \\ -\dfrac{a_{1,Nu}\cdot M\left(m_a,\boldsymbol{\alpha}\right) Re\left(m_a,\boldsymbol{\alpha}\right)}{\mu Nu\left(Re,\boldsymbol{\alpha}\right)}, & 2300 \le Re_d \le 10000, \\ -0.8\dfrac{M\left(m_a,\boldsymbol{\alpha}\right)}{\mu}, & Re_d > 10000, \end{cases} \tag{6.183}$$

The derivatives of the "liquid continuity equations" with respect to the parameter $\alpha^{(8)} \triangleq \upsilon$ are as follows:

$$\frac{\partial N_1^{(i)}}{\partial \alpha^{(8)}} = \frac{\partial N_1^{(i)}}{\partial \upsilon} \triangleq a_1^{i,8} = \frac{1}{\overline{R}} \left[\frac{P_{vs}^{(i+1)}\left(T_w^{(i+1)},\boldsymbol{\alpha}\right)}{T_w^{(i+1)}} - \frac{\omega^{(i)} P_{atm}}{\left(0.622+\omega^{(i)}\right) T_a^{(i)}} \right] \frac{\partial M\left(m_a,\boldsymbol{\alpha}\right)}{\partial \upsilon}; \tag{6.184}$$
$$\ell = 1;\ i = 1,\dots,I;\ j = 8,$$

where

$$\frac{\partial M\left(m_a,\boldsymbol{\alpha}\right)}{\partial \upsilon} = \frac{1}{3}\frac{M\left(m_a,\boldsymbol{\alpha}\right)}{\upsilon}. \tag{6.185}$$

The derivatives of the "liquid continuity equations" with respect to the parameter $\alpha^{(9)} \triangleq k_{air}$ are as follows:

$$\frac{\partial N_1^{(i)}}{\partial \alpha^{(9)}} = \frac{\partial N_1^{(i)}}{\partial k_{air}} \triangleq a_1^{i,9} = 0;\quad \ell = 1;\ i = 1,\dots,I;\ j = 9. \tag{6.186}$$

The derivatives of the "liquid continuity equations" with respect to the parameter $\alpha^{(10)} \triangleq f_{ht}$ are as follows:

$$\frac{\partial N_1^{(i)}}{\partial \alpha^{(10)}} = \frac{\partial N_1^{(i)}}{\partial f_{ht}} \triangleq a_1^{i,10} = 0;\quad \ell = 1;\ i = 1,\dots,I;\ j = 10. \tag{6.187}$$

The derivatives of the "liquid continuity equations" with respect to the parameter $\alpha^{(11)} \triangleq f_{mt}$ are as follows:

$$\frac{\partial N_1^{(i)}}{\partial \alpha^{(11)}} = \frac{\partial N_1^{(i)}}{\partial f_{mt}} \triangleq a_1^{i,11} = \frac{1}{\overline{R}} \left[\frac{P_{vs}^{(i+1)}\left(T_w^{(i+1)}, \boldsymbol{\alpha}\right)}{T_w^{(i+1)}} - \frac{\omega^{(i)} P_{atm}}{\left(0.622 + \omega^{(i)}\right) T_a^{(i)}} \right] \frac{\partial M\left(m_a, \boldsymbol{\alpha}\right)}{\partial f_{mt}}; \quad (6.188)$$
$$\ell = 1; i = 1,\ldots,I; \; j = 11,$$

where

$$\frac{\partial M\left(m_a, \boldsymbol{\alpha}\right)}{\partial f_{mt}} = \frac{M_{H2O} Nu\left(Re, \boldsymbol{\alpha}\right)\left(\dfrac{\nu}{Pr}\right)^{\frac{1}{3}} \left[D_{av}\left(T_{db}, \boldsymbol{\alpha}\right)\right]^{\frac{2}{3}} w_{tsa} A_{surf}}{D_h I}. \quad (6.189)$$

The derivatives of the "liquid continuity equations" with respect to the parameter $\alpha^{(12)} \triangleq f$ are as follows:

$$\frac{\partial N_1^{(i)}}{\partial \alpha^{(12)}} = \frac{\partial N_1^{(i)}}{\partial f} \triangleq a_1^{i,12} = 0; \quad \ell = 1; i = 1,\ldots,I; \; j = 12. \quad (6.190)$$

The derivatives of the "liquid continuity equations" with respect to the parameter $\alpha^{(13)} \triangleq a_0$ are as follows:

$$\frac{\partial N_1^{(i)}}{\partial \alpha^{(13)}} = \frac{\partial N_1^{(i)}}{\partial a_0} \triangleq a_1^{i,13}$$
$$= \frac{M(m_a, \boldsymbol{\alpha})}{\overline{R}} \frac{1}{T_w^{(i+1)}} \frac{\partial P_{vs}^{(i+1)}\left(T_w^{(i+1)}, \boldsymbol{\alpha}\right)}{\partial a_0}; \quad \ell = 1; i = 1,\ldots,I; \; j = 13, \quad (6.191)$$

where

$$\frac{\partial P_{vs}^{(i+1)}\left(T_w^{(i+1)}, \boldsymbol{\alpha}\right)}{\partial a_0} = P_{vs}^{(i+1)}\left(T_w^{(i+1)}, \boldsymbol{\alpha}\right). \quad (6.192)$$

The derivatives of the "liquid continuity equations" with respect to the parameter $\alpha^{(14)} \triangleq a_1$ are as follows:

$$\frac{\partial N_1^{(i)}}{\partial \alpha^{(14)}} = \frac{\partial N_1^{(i)}}{\partial a_1} \triangleq a_1^{i,14} = \frac{M\left(m_a, \boldsymbol{\alpha}\right)}{\overline{R}} \frac{1}{T_w^{(i+1)}} \frac{\partial P_{vs}^{(i+1)}\left(T_w^{(i+1)}, \boldsymbol{\alpha}\right)}{\partial a_1}; \quad (6.193)$$
$$\ell = 1; i = 1,\ldots,I; \; j = 14,$$

where

$$\frac{\partial P_{vs}^{(i+1)}\left(T_w^{(i+1)},\boldsymbol{\alpha}\right)}{\partial a_1}=\frac{P_{vs}^{(i+1)}\left(T_w^{(i+1)},\boldsymbol{\alpha}\right)}{T_w^{(i+1)}}. \tag{6.194}$$

The derivatives of the "liquid continuity equations" with respect to the parameter $\alpha^{(15)} \triangleq a_{0,cpa}$ are as follows:

$$\frac{\partial N_1^{(i)}}{\partial \alpha^{(15)}}=\frac{\partial N_1^{(i)}}{\partial a_{0,cpa}} \triangleq a_1^{i,15}=0;\quad \ell=1;\, i=1,\dots,I;\;\; j=15. \tag{6.195}$$

The derivatives of the "liquid continuity equations" with respect to the parameter $\alpha^{(16)} \triangleq a_{1,cpa}$ are as follows:

$$\frac{\partial N_1^{(i)}}{\partial \alpha^{(16)}}=\frac{\partial N_1^{(i)}}{\partial a_{1,cpa}} \triangleq a_1^{i,16}=0;\quad \ell=1;\, i=1,\dots,I;\;\; j=16. \tag{6.196}$$

The derivatives of the "liquid continuity equations" with respect to the parameter $\alpha^{(17)} \triangleq a_{2,cpa}$ are as follows:

$$\frac{\partial N_1^{(i)}}{\partial \alpha^{(17)}}=\frac{\partial N_1^{(i)}}{\partial a_{2,cpa}} \triangleq a_1^{i,17}=0;\quad \ell=1;\, i=1,\dots,I;\;\; j=17. \tag{6.197}$$

The derivatives of the "liquid continuity equations" with respect to the parameter $\alpha^{(18)} \triangleq a_{0,dav}$ are as follows:

$$\begin{aligned}\frac{\partial N_1^{(i)}}{\partial \alpha^{(18)}}&=\frac{\partial N_1^{(i)}}{\partial a_{0,dav}} \triangleq a_1^{i,18}\\ &=\frac{1}{\overline{R}}\left[\frac{P_{vs}^{(i+1)}\left(T_w^{(i+1)},\boldsymbol{\alpha}\right)}{T_w^{(i+1)}}-\frac{\omega^{(i)}P_{atm}}{\left(0.622+\omega^{(i)}\right)T_a^{(i)}}\right]\frac{\partial M\left(m_a,\boldsymbol{\alpha}\right)}{\partial D_{av}\left(T_{db},\boldsymbol{\alpha}\right)}\cdot\frac{\partial D_{av}\left(T_{db},\boldsymbol{\alpha}\right)}{\partial a_{0,dav}};\\ &\ell=1;\, i=1,\dots,I;\;\; j=18,\end{aligned} \tag{6.198}$$

where $\dfrac{\partial M\left(m_a,\boldsymbol{\alpha}\right)}{\partial D_{av}\left(T_{db},\boldsymbol{\alpha}\right)}$ was defined previously in Eq. (6.175), and

$$\frac{\partial D_{av}\left(T_{db},\boldsymbol{\alpha}\right)}{\partial a_{0,dav}}=\frac{T_{db}^{\ 1.5}}{a_{1dav}+a_{2dav}T_{db}+a_{3dav}T_{db}^{\ 2}}. \tag{6.199}$$

The derivatives of the "liquid continuity equations" with respect to the parameter $\alpha^{(19)} \triangleq a_{1,dav}$ are as follows:

$$\begin{aligned}
&\frac{\partial N_1^{(i)}}{\partial \alpha^{(19)}} = \frac{\partial N_1^{(i)}}{\partial a_{1,dav}} \triangleq a_1^{i,19} \\
&= \frac{1}{R}\left[\frac{P_{vs}^{(i+1)}\left(T_w^{(i+1)},\boldsymbol{\alpha}\right)}{T_w^{(i+1)}} - \frac{\omega^{(i)}P_{atm}}{\left(0.622+\omega^{(i)}\right)T_a^{(i)}}\right]\frac{\partial M\left(m_a,\boldsymbol{\alpha}\right)}{\partial D_{av}\left(T_{db},\boldsymbol{\alpha}\right)}\cdot\frac{\partial D_{av}\left(T_{db},\boldsymbol{\alpha}\right)}{\partial a_{1,dav}}, \\
&\ell = 1;\ i = 1,\ldots,I;\ j = 19,
\end{aligned} \tag{6.200}$$

where $\frac{\partial M\left(m_a,\boldsymbol{\alpha}\right)}{\partial D_{av}\left(T_{db},\boldsymbol{\alpha}\right)}$ was defined previously in Eq. (6.168), and

$$\frac{\partial D_{av}\left(T_{db},\boldsymbol{\alpha}\right)}{\partial a_{1,dav}} = -\frac{a_{0dav}T_{db}^{\ \ 1.5}}{\left(a_{1dav}+a_{2dav}T_{db}+a_{3dav}T_{db}^{\ \ 2}\right)^2}. \tag{6.201}$$

The derivatives of the "liquid continuity equations" with respect to the parameter $\alpha^{(20)} \triangleq a_{2,dav}$ are as follows:

$$\begin{aligned}
&\frac{\partial N_1^{(i)}}{\partial \alpha^{(20)}} = \frac{\partial N_1^{(i)}}{\partial a_{2,dav}} \triangleq a_1^{i,20} \\
&= \frac{1}{R}\left[\frac{P_{vs}^{(i+1)}\left(T_w^{(i+1)},\boldsymbol{\alpha}\right)}{T_w^{(i+1)}} - \frac{\omega^{(i)}P_{atm}}{\left(0.622+\omega^{(i)}\right)T_a^{(i)}}\right]\frac{\partial M\left(m_a,\boldsymbol{\alpha}\right)}{\partial D_{av}\left(T_{db},\boldsymbol{\alpha}\right)}\cdot\frac{\partial D_{av}\left(T_{db},\boldsymbol{\alpha}\right)}{\partial a_{2,dav}}; \\
&\ell = 1;\ i = 1,\ldots,I;\ j = 20,
\end{aligned} \tag{6.202}$$

where $\frac{\partial M\left(m_a,\boldsymbol{\alpha}\right)}{\partial D_{av}\left(T_{db},\boldsymbol{\alpha}\right)}$ was defined previously in Eq. (6.168), and

$$\frac{\partial D_{av}\left(T_{db},\boldsymbol{\alpha}\right)}{\partial a_{2,dav}} = -\frac{a_{0dav}T_{db}^{\ \ 2.5}}{\left(a_{1dav}+a_{2dav}T_{db}+a_{3dav}T_{db}^{\ \ 2}\right)^2}. \tag{6.203}$$

The derivatives of the "liquid continuity equations" with respect to the parameter $\alpha^{(21)} \triangleq a_{3,dav}$ are as follows:

$$\frac{\partial N_1^{(i)}}{\partial \alpha^{(21)}} = \frac{\partial N_1^{(i)}}{\partial a_{3,dav}} \triangleq a_1^{i,21}$$

$$= \frac{1}{R}\left[\frac{P_{vs}^{(i+1)}\left(T_w^{(i+1)},\boldsymbol{\alpha}\right)}{T_w^{(i+1)}} - \frac{\omega^{(i)} P_{atm}}{\left(0.622+\omega^{(i)}\right)T_a^{(i)}}\right]\frac{\partial M\left(m_a,\boldsymbol{\alpha}\right)}{\partial D_{av}\left(T_{db},\boldsymbol{\alpha}\right)} \cdot \frac{\partial D_{av}\left(T_{db},\boldsymbol{\alpha}\right)}{\partial a_{3,dav}}; \quad (6.204)$$

$\ell = 1; i = 1,...,I; \ j = 21,$

where $\dfrac{\partial M\left(m_a,\boldsymbol{\alpha}\right)}{\partial D_{av}\left(T_{db},\boldsymbol{\alpha}\right)}$ was defined previously in Eq. (6.168), and

$$\frac{\partial D_{av}\left(T_{db},\boldsymbol{\alpha}\right)}{\partial a_{3,dav}} = -\frac{a_{0dav} T_{db}^{\ 3.5}}{\left(a_{1dav} + a_{2dav} T_{db} + a_{3dav} T_{db}^{\ 2}\right)^2}. \quad (6.205)$$

The derivatives of the "liquid continuity equations" with respect to the parameter $\alpha^{(22)} \triangleq a_{0f}$ are as follows:

$$\frac{\partial N_1^{(i)}}{\partial \alpha^{(22)}} = \frac{\partial N_1^{(i)}}{\partial a_{0f}} \triangleq a_1^{i,22} = 0; \quad \ell = 1; i = 1,...,I; \ j = 22. \quad (6.206)$$

The derivatives of the "liquid continuity equations" with respect to the parameter $\alpha^{(23)} \triangleq a_{1f}$ are as follows:

$$\frac{\partial N_1^{(i)}}{\partial \alpha^{(23)}} = \frac{\partial N_1^{(i)}}{\partial a_{1f}} \triangleq a_1^{i,23} = 0; \quad \ell = 1; i = 1,...,I; \ j = 23. \quad (6.207)$$

The derivatives of the "liquid continuity equations" with respect to the parameter $\alpha^{(24)} \triangleq a_{0g}$ are as follows:

$$\frac{\partial N_1^{(i)}}{\partial \alpha^{(24)}} = \frac{\partial N_1^{(i)}}{\partial a_{0g}} \triangleq a_1^{i,24} = 0; \quad \ell = 1; i = 1,...,I; \ j = 24. \quad (6.208)$$

The derivatives of the "liquid continuity equations" with respect to the parameter $\alpha^{(25)} \triangleq a_{1g}$ are as follows:

$$\frac{\partial N_1^{(i)}}{\partial \alpha^{(25)}} = \frac{\partial N_1^{(i)}}{\partial a_{1g}} \triangleq a_1^{i,25} = 0; \quad \ell = 1; i = 1,...,I; \ j = 25. \quad (6.209)$$

The derivatives of the "liquid continuity equations" with respect to the parameter $\alpha^{(26)} \triangleq a_{0,Nu}$ are as follows:

$$\frac{\partial N_1^{(i)}}{\partial \alpha^{(26)}} = \frac{\partial N_1^{(i)}}{\partial a_{0,Nu}} \triangleq a_1^{i,26}$$
$$= \frac{1}{R}\left[\frac{P_{vs}^{(i+1)}\left(T_w^{(i+1)},\boldsymbol{\alpha}\right)}{T_w^{(i+1)}} - \frac{\omega^{(i)}P_{atm}}{\left(0.622+\omega^{(i)}\right)T_a^{(i)}}\right]\frac{\partial M\left(m_a,\boldsymbol{\alpha}\right)}{\partial Nu\left(Re,\boldsymbol{\alpha}\right)}\frac{\partial Nu\left(Re,\boldsymbol{\alpha}\right)}{\partial a_{0,Nu}}; \quad (6.210)$$
$$\ell = 1;\ i = 1,\ldots,I;\ \ j = 26,$$

where $\frac{\partial M\left(m_a,\boldsymbol{\alpha}\right)}{\partial D_{av}\left(T_{db},\boldsymbol{\alpha}\right)}$ was defined previously in Eq. (6.168), and

$$\frac{\partial Nu\left(Re,\boldsymbol{\alpha}\right)}{\partial a_{0,Nu}} = \begin{cases} 1, & Re_d < 2300, \\ 0, & 2300 \le Re_d \le 10000,. \\ 0, & Re_d > 10000, \end{cases} \quad (6.211)$$

The derivatives of the "liquid continuity equations" with respect to the parameter $\alpha^{(27)} \triangleq a_{1,Nu}$ are as follows:

$$\frac{\partial N_1^{(i)}}{\partial \alpha^{(27)}} = \frac{\partial N_1^{(i)}}{\partial a_{1,Nu}} \triangleq a_1^{i,27}$$
$$= \frac{1}{R}\left[\frac{P_{vs}^{(i+1)}\left(T_w^{(i+1)},\boldsymbol{\alpha}\right)}{T_w^{(i+1)}} - \frac{\omega^{(i)}P_{atm}}{\left(0.622+\omega^{(i)}\right)T_a^{(i)}}\right]\frac{\partial M\left(m_a,\boldsymbol{\alpha}\right)}{\partial Nu\left(Re,\boldsymbol{\alpha}\right)}\frac{\partial Nu\left(Re,\boldsymbol{\alpha}\right)}{\partial a_{1,Nu}}; \quad (6.212)$$
$$\ell = 1;\ i = 1,\ldots,I;\ \ j = 27,$$

where $\frac{\partial M\left(m_a,\boldsymbol{\alpha}\right)}{\partial D_{av}\left(T_{db},\boldsymbol{\alpha}\right)}$ was defined previously in Eq. (6.168), and

$$\frac{\partial Nu\left(Re,\boldsymbol{\alpha}\right)}{\partial a_{1,Nu}} = \begin{cases} 0, & Re_d < 2300, \\ Re\left(m_a,\boldsymbol{\alpha}\right), & 2300 \le Re_d \le 10000, \\ 0, & Re_d > 10000, \end{cases} \quad (6.213)$$

The derivatives of the "liquid continuity equations" with respect to the parameter $\alpha^{(28)} \triangleq a_{2,Nu}$ are as follows:

$$\frac{\partial N_1^{(i)}}{\partial \alpha^{(28)}} = \frac{\partial N_1^{(i)}}{\partial a_{2,Nu}} \triangleq a_1^{i,28}$$
$$= \frac{1}{\overline{R}}\left[\frac{P_{vs}^{(i+1)}\left(T_w^{(i+1)},\boldsymbol{\alpha}\right)}{T_w^{(i+1)}} - \frac{\omega^{(i)}P_{atm}}{\left(0.622+\omega^{(i)}\right)T_a^{(i)}}\right]\frac{\partial M\left(m_a,\boldsymbol{\alpha}\right)}{\partial Nu\left(Re,\boldsymbol{\alpha}\right)}\frac{\partial Nu\left(Re,\boldsymbol{\alpha}\right)}{\partial a_{2,Nu}}; \quad (6.214)$$
$$\ell = 1;\ i = 1,\dots,I;\ \ j = 28,$$

where $\frac{\partial M\left(m_a,\boldsymbol{\alpha}\right)}{\partial D_{av}\left(T_{db},\boldsymbol{\alpha}\right)}$ was defined previously in Eq. (6.168), and

$$\frac{\partial Nu\left(Re,\boldsymbol{\alpha}\right)}{\partial a_{2,Nu}} = \begin{cases} 0, & Re_d < 2300, \\ 1, & 2300 \le Re_d \le 10000, \\ 0, & Re_d > 10000, \end{cases} \quad (6.215)$$

The derivatives of the "liquid continuity equations" with respect to the parameter $\alpha^{(29)} \triangleq a_{3,Nu}$ are as follows:

$$\frac{\partial N_1^{(i)}}{\partial \alpha^{(29)}} = \frac{\partial N_1^{(i)}}{\partial a_{3,Nu}} \triangleq a_1^{i,29}$$
$$= \frac{1}{\overline{R}}\left[\frac{P_{vs}^{(i+1)}(T_w^{(i+1)},\boldsymbol{\alpha})}{T_w^{(i+1)}} - \frac{\omega^{(i)}P_{atm}}{\left(0.622+\omega^{(i)}\right)T_a^{(i)}}\right]\frac{\partial M\left(m_a,\boldsymbol{\alpha}\right)}{\partial Nu\left(Re,\boldsymbol{\alpha}\right)}\frac{\partial Nu\left(Re,\boldsymbol{\alpha}\right)}{\partial a_{3,Nu}}; \quad (6.216)$$
$$\ell = 1;\ i = 1,\dots,I;\ \ j = 29,$$

where $\frac{\partial M\left(m_a,\boldsymbol{\alpha}\right)}{\partial D_{av}\left(T_{db},\boldsymbol{\alpha}\right)}$ was defined previously in Eq. (6.168), and

$$\frac{\partial Nu\left(Re,\boldsymbol{\alpha}\right)}{\partial a_{3,Nu}} = \begin{cases} 0, & Re_d < 2300, \\ 0, & 2300 \le Re_d \le 10000, \\ \left[Re\left(m_a,\boldsymbol{\alpha}\right)\right]^{0.8}\cdot Pr^{\frac{1}{3}}, & Re_d > 10000, \end{cases} \quad (6.217)$$

The derivatives of the "liquid continuity equations" with respect to the parameter $\alpha^{(30)} \triangleq W_{dkx}$ are as follows:

$$\frac{\partial N_1^{(i)}}{\partial \alpha^{(30)}} = \frac{\partial N_1^{(i)}}{\partial W_{dkx}} \triangleq a_1^{i,30} = 0; \quad \ell = 1;\ i = 1,\dots,I;\ \ j = 30. \quad (6.218)$$

The derivatives of the "liquid continuity equations" with respect to the parameter $\alpha^{(31)} \triangleq W_{dky}$ are as follows:

$$\frac{\partial N_1^{(i)}}{\partial \alpha^{(31)}} = \frac{\partial N_1^{(i)}}{\partial W_{dky}} \triangleq a_1^{i,31} = 0; \quad \ell = 1; i = 1,\ldots,I; \; j = 31. \tag{6.219}$$

The derivatives of the "liquid continuity equations" with respect to the parameter $\alpha^{(32)} \triangleq \Delta z_{dk}$ are as follows:

$$\frac{\partial N_1^{(i)}}{\partial \alpha^{(32)}} = \frac{\partial N_1^{(i)}}{\partial \Delta z_{dk}} \triangleq a_1^{i,32} = 0; \quad \ell = 1; i = 1,\ldots,I; \; j = 32. \tag{6.220}$$

The derivatives of the "liquid continuity equations" with respect to the parameter $\alpha^{(33)} \triangleq \Delta z_{fan}$ are as follows:

$$\frac{\partial N_1^{(i)}}{\partial \alpha^{(33)}} = \frac{\partial N_1^{(i)}}{\partial \Delta z_{fan}} \triangleq a_1^{i,33} = 0; \quad \ell = 1; i = 1,\ldots,I; \; j = 33. \tag{6.221}$$

The derivatives of the "liquid continuity equations" with respect to the parameter $\alpha^{(34)} \triangleq D_{fan}$ are as follows:

$$\begin{aligned} &\frac{\partial N_1^{(i)}}{\partial \alpha^{(34)}} = \frac{\partial N_1^{(i)}}{\partial D_{fan}} \triangleq a_1^{i,34} \\ &= \frac{1}{R}\left[\frac{P_{vs}^{(i+1)}\left(T_w^{(i+1)},\boldsymbol{\alpha}\right)}{T_w^{(i+1)}} - \frac{\omega^{(i)} P_{atm}}{\left(0.622+\omega^{(i)}\right)T_a^{(i)}}\right]\frac{\partial M\left(m_a,\boldsymbol{\alpha}\right)}{\partial Nu\left(Re,\boldsymbol{\alpha}\right)}\frac{\partial Nu\left(Re,\boldsymbol{\alpha}\right)}{\partial m_a}\frac{\partial m_a}{\partial D_{fan}}; \\ &\ell = 1; i = 1,\ldots,I; \; j = 34, \end{aligned} \tag{6.222}$$

where $\frac{\partial M\left(m_a,\boldsymbol{\alpha}\right)}{\partial Nu\left(Re,\boldsymbol{\alpha}\right)}$ and $\frac{\partial Nu\left(Re,\boldsymbol{\alpha}\right)}{\partial m_a}$ were defined previously in Eqs. (6.175) and (6.169), respectively, and where

$$\frac{\partial m_a}{\partial D_{fan}} = \frac{2m_a}{D_{fan}}. \tag{6.223}$$

The derivatives of the "liquid continuity equations" with respect to the parameter $\alpha^{(35)} \triangleq \Delta z_{fill}$ are as follows:

$$\frac{\partial N_1^{(i)}}{\partial \alpha^{(35)}} = \frac{\partial N_1^{(i)}}{\partial \Delta z_{fill}} \triangleq a_1^{i,35} = 0; \quad \ell = 1; i = 1,...,I; \; j = 35. \tag{6.224}$$

The derivatives of the "liquid continuity equations" with respect to the parameter $\alpha^{(36)} \triangleq \Delta z_{rain}$ are as follows:

$$\frac{\partial N_1^{(i)}}{\partial \alpha^{(36)}} = \frac{\partial N_1^{(i)}}{\partial \Delta z_{rain}} \triangleq a_1^{i,36} = 0; \quad \ell = 1; i = 1,...,I; \; j = 36. \tag{6.225}$$

The derivatives of the "liquid continuity equations" with respect to the parameter $\alpha^{(37)} \triangleq \Delta z_{bs}$ are as follows:

$$\frac{\partial N_1^{(i)}}{\partial \alpha^{(37)}} = \frac{\partial N_1^{(i)}}{\partial \Delta z_{bs}} \triangleq a_1^{i,37} = 0; \quad \ell = 1; i = 1,...,I; \; j = 37. \tag{6.226}$$

The derivatives of the "liquid continuity equations" with respect to the parameter $\alpha^{(38)} \triangleq \Delta z_{de}$ are as follows:

$$\frac{\partial N_1^{(i)}}{\partial \alpha^{(38)}} = \frac{\partial N_1^{(i)}}{\partial \Delta z_{de}} \equiv a_1^{i,38} = 0; \quad \ell = 1; i = 1,...,I; \; j = 38. \tag{6.227}$$

The derivatives of the "liquid continuity equations" with respect to the parameter $\alpha^{(39)} \triangleq D_h$ are as follows:

$$\frac{\partial N_1^{(i)}}{\partial \alpha^{(39)}} = \frac{\partial N_1^{(i)}}{\partial D_h} \triangleq a_1^{i,39} = \frac{1}{R}\left[\frac{P_{vs}^{(i+1)}\left(T_w^{(i+1)},\boldsymbol{\alpha}\right)}{T_w^{(i+1)}} - \frac{\omega^{(i)} P_{atm}}{\left(0.622+\omega^{(i)}\right)T_a^{(i)}}\right]\frac{\partial M\left(m_a,\boldsymbol{\alpha}\right)}{\partial D_h}, \tag{6.228}$$
$$\ell = 1; i = 1,...,I; \; j = 39,$$

where

$$\frac{\partial M\left(m_a,\boldsymbol{\alpha}\right)}{\partial D_h} = \begin{cases} -M\left(m_a,\boldsymbol{\alpha}\right)/D_h, & Re_d < 2300, \\ -\dfrac{a_{2,Nu} M\left(m_a,\boldsymbol{\alpha}\right)}{D_h Nu\left(Re,\boldsymbol{\alpha}\right)}, & 2300 \le Re_d \le 10000, \\ -0.2 \cdot M\left(m_a,\boldsymbol{\alpha}\right)/D_h, & Re_d > 10000, \end{cases} \tag{6.229}$$

The derivatives of the "liquid continuity equations" with respect to the parameter $\alpha^{(40)} \triangleq A_{fill}$ are as follows:

$$\frac{\partial N_1^{(i)}}{\partial \alpha^{(40)}} = \frac{\partial N_1^{(i)}}{\partial A_{fill}} \triangleq a_1^{i,40} = \frac{1}{R}\left[\frac{P_{vs}^{(i+1)}\left(T_w^{(i+1)},\boldsymbol{\alpha}\right)}{T_w^{(i+1)}} - \frac{\omega^{(i)} P_{atm}}{\left(0.622+\omega^{(i)}\right)T_a^{(i)}}\right]\frac{\partial M\left(m_a,\boldsymbol{\alpha}\right)}{\partial A_{fill}};$$
$$\ell = 1;\, i = 1,\ldots,I;\;\; j = 40, \tag{6.230}$$

where

$$\frac{\partial M\left(m_a,\boldsymbol{\alpha}\right)}{\partial A_{fill}} = \begin{cases} 0, & Re_d < 2300, \\ -\dfrac{a_{1,Nu} M\left(m_a,\boldsymbol{\alpha}\right) Re\left(m_a,\boldsymbol{\alpha}\right)}{Nu\left(Re,\boldsymbol{\alpha}\right) A_{fill}}, & 2300 \le Re_d \le 10000, \\ -0.8\cdot M\left(m_a,\boldsymbol{\alpha}\right)/\,A_{fill}, & Re_d > 10000, \end{cases} \tag{6.231}$$

The derivatives of the "liquid continuity equations" with respect to the parameter $\alpha^{(41)} \triangleq A_{surf}$ are as follows:

$$\frac{\partial N_1^{(i)}}{\partial \alpha^{(41)}} = \frac{\partial N_1^{(i)}}{\partial A_{surf}} \triangleq a_1^{i,41} = \frac{1}{R}\left[\frac{P_{vs}^{(i+1)}\left(T_w^{(i+1)},\boldsymbol{\alpha}\right)}{T_w^{(i+1)}} - \frac{\omega^{(i)} P_{atm}}{\left(0.622+\omega^{(i)}\right)T_a^{(i)}}\right]\frac{\partial M\left(m_a,\boldsymbol{\alpha}\right)}{\partial A_{surf}};$$
$$\ell = 1;\, i = 1,\ldots,I;\;\; j = 41, \tag{6.232}$$

where

$$\frac{\partial M\left(m_a,\boldsymbol{\alpha}\right)}{\partial A_{surf}} = \frac{M\left(m_a,\boldsymbol{\alpha}\right)}{A_{surf}}. \tag{6.233}$$

The derivatives of the "liquid continuity equations" with respect to the parameter $\alpha^{(42)} \triangleq Pr$ are as follows:

$$\frac{\partial N_1^{(i)}}{\partial \alpha^{(42)}} = \frac{\partial N_1^{(i)}}{\partial Pr} \triangleq a_1^{i,42} = \frac{1}{R}\left[\frac{P_{vs}^{(i+1)}\left(T_w^{(i+1)},\boldsymbol{\alpha}\right)}{T_w^{(i+1)}} - \frac{\omega^{(i)} P_{atm}}{\left(0.622+\omega^{(i)}\right)T_a^{(i)}}\right]\frac{\partial M\left(m_a,\boldsymbol{\alpha}\right)}{\partial Pr}; \tag{6.234}$$
$$\ell = 1;\, i = 1,\ldots,I;\;\; j = 42,$$

where

$$\frac{\partial M\left(m_a,\boldsymbol{\alpha}\right)}{\partial Pr} = \begin{cases} -M\left(m_a,\boldsymbol{\alpha}\right)/\left(3\,Pr\right), & Re_d \le 10000, \\ 0, & Re_d > 10000, \end{cases} \tag{6.235}$$

The derivatives of the "liquid continuity equations" with respect to the parameter $\alpha^{(43)} \triangleq V_w$ are as follows:

$$\frac{\partial N_1^{(i)}}{\partial \alpha^{(43)}} = \frac{\partial N_1^{(i)}}{\partial V_w} \triangleq a_1^{i,43} = 0; \quad \ell = 1; i = 1,\ldots,I; \; j = 43. \qquad (6.236)$$

The derivatives of the "liquid continuity equations" with respect to the parameter $\alpha^{(44)} \triangleq V_{exit}$ are as follows:

$$\frac{\partial N_1^{(i)}}{\partial \alpha^{(44)}} = \frac{\partial N_1^{(i)}}{\partial V_{exit}} \triangleq a_1^{i,44}$$

$$= \frac{1}{R}\left[\frac{P_{vs}^{(i+1)}\left(T_w^{(i+1)},\boldsymbol{\alpha}\right)}{T_w^{(i+1)}} - \frac{\omega^{(i)} P_{atm}}{\left(0.622+\omega^{(i)}\right)T_a^{(i)}}\right]\frac{\partial M\left(m_a,\boldsymbol{\alpha}\right)}{\partial Nu\left(Re,\boldsymbol{\alpha}\right)}\frac{\partial Nu\left(Re,\boldsymbol{\alpha}\right)}{\partial m_a}\frac{\partial m_a}{\partial V_{exit}}; \qquad (6.237)$$

$$\ell = 1; i = 1,\ldots,I; \; j = 44,$$

where $\frac{\partial M\left(m_a,\boldsymbol{\alpha}\right)}{\partial Nu\left(Re,\boldsymbol{\alpha}\right)}$ and $\frac{\partial Nu\left(Re,\boldsymbol{\alpha}\right)}{\partial m_a}$ were defined previously in Eqs. (6.175) and (6.176), respectively, and

$$\frac{\partial m_a}{\partial V_{exit}} = \frac{P_{atm}\pi D_{fan}^{\;2}}{4R_{air}T_{a,in}}. \qquad (6.238)$$

The derivatives of the "liquid continuity equations" with respect to the parameter $\alpha^{(45)} \triangleq m_{w,in}$ are as follows:

$$\frac{\partial N_1^{(1)}}{\partial \alpha^{(45)}} = \frac{\partial N_1^{(1)}}{\partial m_{w,in}} \triangleq a_1^{1,45} = -1; \quad \ell = 1; i = 1; \; j = 45, \qquad (6.239)$$

$$\frac{\partial N_1^{(i)}}{\partial \alpha^{(45)}} = \frac{\partial N_1^{(i)}}{\partial m_{w,in}} \triangleq a_1^{i,45} = 0; \quad \ell = 1; i = 2,\ldots,I; \; j = 45. \qquad (6.240)$$

The derivatives of the "liquid continuity equations" with respect to the parameter $\alpha^{(46)} \triangleq T_{a,in}$ are as follows:

$$\frac{\partial N_1^{(i)}}{\partial \alpha^{(46)}} = \frac{\partial N_1^{(i)}}{\partial T_{a,in}} \triangleq a_1^{i,46} = 0; \quad \ell = 1; i = 1,\ldots,I; \; j = 46. \qquad (6.241)$$

NOTE:the above derivative was NOT used in the computations for the numerical results presented in Chapter 6.

The derivatives of the "liquid continuity equations" with respect to the parameter $\alpha^{(47)} \triangleq m_a$ are as follows:

$$\begin{aligned}
\frac{\partial N_1^{(i)}}{\partial \alpha^{(47)}} &= \frac{\partial N_1^{(i)}}{\partial m_a} \triangleq a_1^{i,47} \\
&= \frac{1}{R}\left[\frac{P_{vs}^{(i+1)}\left(T_w^{(i+1)},\boldsymbol{\alpha}\right)}{T_w^{(i+1)}} - \frac{\omega^{(i)}P_{atm}}{\left(0.622+\omega^{(i)}\right)T_a^{(i)}}\right]\frac{\partial M\left(m_a,\boldsymbol{\alpha}\right)}{\partial Nu\left(Re,\boldsymbol{\alpha}\right)}\frac{\partial Nu\left(Re,\boldsymbol{\alpha}\right)}{\partial m_a}; \\
&\ell = 1;\, i = 1,\ldots,I;\;\; j = 47,
\end{aligned} \tag{6.242}$$

where $\frac{\partial M\left(m_a,\boldsymbol{\alpha}\right)}{\partial Nu\left(Re,\boldsymbol{\alpha}\right)}$ and $\frac{\partial Nu\left(Re,\boldsymbol{\alpha}\right)}{\partial m_a}$ were defined previously in Eqs. (6.175) and (6.176), respectively.

The derivatives of the "liquid continuity equations" with respect to the parameter $\alpha^{(48)} \triangleq \omega_{in}$ are as follows:

$$\frac{\partial N_1^{(i)}}{\partial \alpha^{(48)}} = \frac{\partial N_1^{(i)}}{\partial \omega_{in}} \triangleq a_1^{i,48} = 0;\;\; \ell = 1;\, i = 1,\ldots,I;\;\; j = 48. \tag{6.243}$$

The derivatives of the "liquid continuity equations" with respect to the parameter $\alpha^{(49)} \triangleq Re_d$ are as follows:

$$\begin{aligned}
\frac{\partial N_1^{(i)}}{\partial \alpha^{(49)}} &= \frac{\partial N_1^{(i)}}{\partial Re_d} \triangleq a_1^{i,49} \\
&= \frac{1}{R}\left[\frac{P_{vs}^{(i+1)}\left(T_w^{(i+1)},\boldsymbol{\alpha}\right)}{T_w^{(i+1)}} - \frac{\omega^{(i)}P_{atm}}{\left(0.622+\omega^{(i)}\right)T_a^{(i)}}\right]\frac{\partial M\left(m_a,\boldsymbol{\alpha}\right)}{\partial Nu\left(Re_d,\boldsymbol{\alpha}\right)}\frac{\partial Nu\left(Re_d,\boldsymbol{\alpha}\right)}{\partial Re_d}; \\
&\ell = 1;\, i = 1,\ldots,I;\;\; j = 49,
\end{aligned} \tag{6.244}$$

where $\frac{\partial M\left(m_a,\boldsymbol{\alpha}\right)}{\partial Nu\left(Re_d,\boldsymbol{\alpha}\right)}$ was defined in Eq. (6.175), and

$$\frac{\partial Nu\left(Re_d,\boldsymbol{\alpha}\right)}{\partial Re_d} = \begin{cases} 0, & Re_d < 2300, \\ a_{1,Nu}, & 2300 \le Re_d \le 10000, \\ 0.8a_{3,Nu}\,Re_d^{-0.2}\,Pr^{1/3}, & Re_d > 10000, \end{cases} \tag{6.245}$$

The derivatives of the "liquid continuity equations" with respect to the parameter $\alpha^{(50)} \triangleq Sc$ are as follows:

$$\frac{\partial N_1^{(i)}}{\partial \alpha^{(50)}} = \frac{\partial N_1^{(i)}}{\partial Sc} \triangleq a_1^{i,50} = \frac{1}{R}\left[\frac{P_{vs}^{(i+1)}\left(T_w^{(i+1)},\boldsymbol{\alpha}\right)}{T_w^{(i+1)}} - \frac{\omega^{(i)} P_{atm}}{\left(0.622+\omega^{(i)}\right)T_a^{(i)}}\right]\frac{\partial M\left(m_a,\boldsymbol{\alpha}\right)}{\partial Sc};$$
$$\ell = 1;\; i = 1,\ldots,I;\;\; j = 50, \tag{6.246}$$

where

$$\frac{\partial M\left(m_a,\boldsymbol{\alpha}\right)}{\partial Sc} = \frac{M\left(m_a,\boldsymbol{\alpha}\right)}{3Sc}. \tag{6.247}$$

The derivatives of the "liquid continuity equations" with respect to the parameter $\alpha^{(51)} \triangleq Sh$ are as follows:

$$\frac{\partial N_1^{(i)}}{\partial \alpha^{(51)}} = \frac{\partial N_1^{(i)}}{\partial Sh} \triangleq a_1^{i,51} = \frac{1}{R}\left[\frac{P_{vs}^{(i+1)}\left(T_w^{(i+1)},\boldsymbol{\alpha}\right)}{T_w^{(i+1)}} - \frac{\omega^{(i)} P_{atm}}{\left(0.622+\omega^{(i)}\right)T_a^{(i)}}\right]\frac{\partial M\left(m_a,\boldsymbol{\alpha}\right)}{\partial Sh};$$
$$\ell = 1;\; i = 1,\ldots,I;\;\; j = 51, \tag{6.248}$$

where

$$\frac{\partial M\left(m_a,\boldsymbol{\alpha}\right)}{\partial Sh} = \frac{M\left(m_a,\boldsymbol{\alpha}\right)}{Sh}. \tag{6.249}$$

The derivatives of the "liquid continuity equations" with respect to the parameter $\alpha^{(52)} \triangleq Nu$ are as follows:

$$\frac{\partial N_1^{(i)}}{\partial \alpha^{(52)}} = \frac{\partial N_1^{(i)}}{\partial Nu} \triangleq a_1^{i,52} = \frac{1}{R}\left[\frac{P_{vs}^{(i+1)}\left(T_w^{(i+1)},\boldsymbol{\alpha}\right)}{T_w^{(i+1)}} - \frac{\omega^{(i)} P_{atm}}{\left(0.622+\omega^{(i)}\right)T_a^{(i)}}\right]\frac{\partial M\left(m_a,\boldsymbol{\alpha}\right)}{\partial Nu\left(Re,\boldsymbol{\alpha}\right)};$$
$$\ell = 1;\; i = 1,\ldots,I;\;\; j = 52, \tag{6.250}$$

where $\dfrac{\partial M\left(m_a,\boldsymbol{\alpha}\right)}{\partial Nu\left(Re,\boldsymbol{\alpha}\right)}$ was defined in Eq. (6.175).

Derivatives of the liquid energy balance equations with respect to model parameters

The derivatives of the liquid energy balance equations with respect to the parameter $\alpha^{(1)} \triangleq T_{db}$ are as follows:

$$\frac{\partial N_2^{(i)}}{\partial \alpha^{(1)}} = \frac{\partial N_2^{(i)}}{\partial T_{db}} \triangleq a_2^{i,1} = 0; \quad \ell = 2;\ i = 1,\ldots,I;\ j = 1. \tag{6.251}$$

The derivatives of the liquid energy balance equations with respect to the parameter $\alpha^{(2)} \triangleq T_{dp}$ are as follows:

$$\frac{\partial N_2^{(i)}}{\partial \alpha^{(2)}} = \frac{\partial N_2^{(i)}}{\partial T_{dp}} \triangleq a_2^{i,2} = 0; \quad \ell = 2;\ i = 1,\ldots,I;\ j = 2. \tag{6.252}$$

The derivatives of the liquid energy balance equations with respect to the parameter $\alpha^{(3)} \triangleq T_{w,in}$ are as follows:

$$\begin{aligned} &\frac{\partial N_2^{(1)}}{\partial \alpha^{(3)}} = \frac{\partial N_2^{(1)}}{\partial T_{w,in}} \triangleq a_2^{1,3} \\ &= m_{w,in} \frac{\partial h_f^{(1)}\left(T_{w,in}, \boldsymbol{\alpha}\right)}{\partial T_{w,in}} + h_f^{(1)}\left(T_{w,in}, \boldsymbol{\alpha}\right) \frac{\partial m_{w,in}}{\partial T_{w,in}} - h_{g,w}^{(2)}\left(T_w^{(2)}, \boldsymbol{\alpha}\right) \frac{\partial m_{w,in}}{\partial T_{w,in}}; \\ &\ell = 2; i=1; j = 3, \end{aligned} \tag{6.253}$$

where $\frac{\partial m_{w,in}}{\partial T_{w,in}}$ was defined in Eq. (6.172), and

$$\frac{\partial h_f^{(1)}\left(T_{w,in}, \boldsymbol{\alpha}\right)}{\partial T_{w,in}} = a_{1f}, \tag{6.254}$$

$$\frac{\partial N_2^{(i)}}{\partial \alpha^{(3)}} = \frac{\partial N_2^{(i)}}{\partial T_{w,in}} \triangleq a_2^{i,3} = 0; \quad \ell = 2; i=2,\ldots I; j = 3. \tag{6.255}$$

The derivatives of the liquid energy balance equations with respect to the parameter $\alpha^{(4)} \triangleq P_{atm}$ are as follows:

$$\frac{\partial N_2^{(i)}}{\partial \alpha^{(4)}} = \frac{\partial N_2^{(i)}}{\partial P_{atm}} \triangleq a_2^{i,4} = -\left(T_w^{(i+1)} - T_a^{(i)}\right) \frac{\partial H(m_a, \boldsymbol{\alpha})}{\partial Nu(Re, \boldsymbol{\alpha})} \frac{\partial Nu(Re, \boldsymbol{\alpha})}{\partial m_a} \frac{\partial m_a}{\partial P_{atm}}; \quad (6.256)$$
$$\ell = 2; i = 1,...,I; \; j = 4,$$

where $\frac{\partial Nu(Re, \boldsymbol{\alpha})}{\partial m_a}$ and $\frac{\partial m_a}{\partial P_{atm}}$ were defined in Eqs. (6.176) and (6.177), respectively, and

$$\frac{\partial H(m_a, \boldsymbol{\alpha})}{\partial Nu(Re, \boldsymbol{\alpha})} = \frac{H(m_a, \boldsymbol{\alpha})}{Nu(Re, \boldsymbol{\alpha})}. \quad (6.257)$$

The derivatives of the liquid energy balance equations with respect to the parameter $\alpha^{(5)} \triangleq w_{tsa}$ are as follows:

$$\frac{\partial N_2^{(i)}}{\partial \alpha^{(5)}} = \frac{\partial N_2^{(i)}}{\partial w_{tsa}} \triangleq a_2^{i,5} = -\left(T_w^{(i+1)} - T_a^{(i)}\right) \frac{\partial H(m_a, \boldsymbol{\alpha})}{\partial w_{tsa}}; \; \ell = 2; i = 1,...,I; \; j = 5, \quad (6.258)$$

where

$$\frac{\partial H(m_a, \boldsymbol{\alpha})}{\partial w_{tsa}} = \frac{f_{ht} k_{air} A_{surf} Nu(Re, \boldsymbol{\alpha})}{D_h I}. \quad (6.259)$$

The derivatives of the liquid energy balance equations with respect to the parameter $\alpha^{(6)} \triangleq k_{sum}$ are as follows:

$$\frac{\partial N_2^{(i)}}{\partial \alpha^{(6)}} = \frac{\partial N_2^{(i)}}{\partial k_{sum}} \triangleq a_2^{i,6} = 0; \quad \ell = 2; i = 1,...,I; \; j = 6. \quad (6.260)$$

The derivatives of the liquid energy balance equations with respect to the parameter $\alpha^{(7)} \triangleq \mu$ are as follows:

$$\frac{\partial N_2^{(i)}}{\partial \alpha^{(7)}} = \frac{\partial N_2^{(i)}}{\partial \mu} \triangleq a_2^{i,7} = -\left(T_w^{(i+1)} - T_a^{(i+1)}\right) \frac{\partial H(m_a, \boldsymbol{\alpha})}{\partial \mu}; \; \ell = 2; i = 1,...,I; \; j = 7, \quad (6.261)$$

where

$$\frac{\partial H(m_a,\boldsymbol{\alpha})}{\partial \mu} = \begin{cases} 0 & Re_d < 2300, \\ -\frac{a_{1,Nu} \cdot H(m_a,\boldsymbol{\alpha}) Re(m_a,\boldsymbol{\alpha})}{\mu Nu(Re,\boldsymbol{\alpha})}, & 2300 \le Re_d \le 10000, \\ -0.8\frac{H(m_a,\boldsymbol{\alpha})}{\mu}, & Re_d > 10000, \end{cases} \tag{6.262}$$

The derivatives of the liquid energy balance equations with respect to the parameter $\alpha^{(8)} \triangleq \upsilon$ are as follows:

$$\frac{\partial N_2^{(i)}}{\partial \alpha^{(8)}} = \frac{\partial N_2^{(i)}}{\partial \upsilon} \triangleq a_2^{i,8} = 0; \quad \ell = 2; i = 1,\ldots,I; \; j = 8. \tag{6.263}$$

The derivatives of the liquid energy balance equations with respect to the parameter $\alpha^{(9)} \triangleq k_{air}$ are as follows:

$$\frac{\partial N_2^{(i)}}{\partial \alpha^{(9)}} = \frac{\partial N_2^{(i)}}{\partial k_{air}} \triangleq a_2^{i,9} = -\left(T_w^{(i+1)} - T_a^{(i)}\right)\frac{\partial H(m_a,\boldsymbol{\alpha})}{\partial k_{air}}; \; \ell = 2; i = 1,\ldots,I; \; j = 9, \tag{6.264}$$

where

$$\frac{\partial H(m_a,\boldsymbol{\alpha})}{\partial k_{air}} = \frac{H(m_a,\boldsymbol{\alpha})}{k_{air}} = \frac{f_{ht} w_{tsa} A_{surf} Nu(Re,\boldsymbol{\alpha})}{D_h I}. \tag{6.265}$$

The derivatives of the liquid energy balance equations with respect to the parameter $\alpha^{(10)} \triangleq f_{ht}$ are as follows:

$$\frac{\partial N_2^{(i)}}{\partial \alpha^{(10)}} = \frac{\partial N_2^{(i)}}{\partial f_{ht}} \equiv a_2^{i,10} = -\left(T_w^{(i+1)} - T_a^{(i)}\right)\frac{\partial H(m_a,\boldsymbol{\alpha})}{\partial f_{ht}}; \quad \ell = 2; i = 1,\ldots,I; \; j = 10, \tag{6.266}$$

where

$$\frac{\partial H(m_a,\boldsymbol{\alpha})}{\partial f_{ht}} = \frac{H(m_a,\boldsymbol{\alpha})}{f_{ht}} = \frac{k_{air} w_{tsa} A_{surf} Nu(Re,\boldsymbol{\alpha})}{D_h I}. \tag{6.267}$$

The derivatives of the liquid energy balance equations with respect to the parameter $\alpha^{(11)} \triangleq f_{mt}$ are as follows:

$$\frac{\partial N_2^{(i)}}{\partial \alpha^{(11)}} = \frac{\partial N_2^{(i)}}{\partial f_{mt}} \triangleq a_2^{i,11} = 0; \quad \ell = 2; i = 1,\ldots,I; \; j = 11. \tag{6.268}$$

The derivatives of the liquid energy balance equations with respect to the parameter $\alpha^{(12)} \triangleq f$ are as follows:

$$\frac{\partial N_2^{(i)}}{\partial \alpha^{(12)}} = \frac{\partial N_2^{(i)}}{\partial f} \triangleq a_2^{i,12} = 0; \quad \ell = 2; i = 1,\ldots,I; \; j = 12. \tag{6.269}$$

The derivatives of the liquid energy balance equations with respect to the parameter $\alpha^{(13)} \triangleq a_0$ are as follows:

$$\frac{\partial N_2^{(i)}}{\partial \alpha^{(13)}} = \frac{\partial N_2^{(i)}}{\partial a_0} \triangleq a_2^{i,13} = 0; \quad \ell = 2; i = 1,\ldots,I; \; j = 13. \tag{6.270}$$

The derivatives of the liquid energy balance equations with respect to the parameter $\alpha^{(14)} \triangleq a_1$ are as follows:

$$\frac{\partial N_2^{(i)}}{\partial \alpha^{(14)}} = \frac{\partial N_2^{(i)}}{\partial a_1} \triangleq a_2^{i,14} = 0; \quad \ell = 2; i = 1,\ldots,I; \; j = 14. \tag{6.271}$$

The derivatives of the liquid energy balance equations with respect to the parameter $\alpha^{(15)} \triangleq a_{0,cpa}$ are as follows:

$$\frac{\partial N_2^{(i)}}{\partial \alpha^{(15)}} = \frac{\partial N_2^{(i)}}{\partial a_{0,cpa}} \triangleq a_2^{i,15} = 0; \quad \ell = 2; i = 1,\ldots,I; \; j = 15. \tag{6.272}$$

The derivatives of the liquid energy balance equations with respect to the parameter $\alpha^{(16)} \triangleq a_{1,cpa}$ are as follows:

$$\frac{\partial N_2^{(i)}}{\partial \alpha^{(16)}} = \frac{\partial N_2^{(i)}}{\partial a_{1,cpa}} \triangleq a_2^{i,16} = 0; \quad \ell = 2; i = 1,\ldots,I; \; j = 16. \tag{6.273}$$

The derivatives of the liquid energy balance equations with respect to the parameter $\alpha^{(17)} \triangleq a_{2,cpa}$ are as follows:

$$\frac{\partial N_2^{(i)}}{\partial \alpha^{(17)}} = \frac{\partial N_2^{(i)}}{\partial a_{2,cpa}} \triangleq a_2^{i,17} = 0; \quad \ell = 2; i = 1,\ldots,I; \; j = 17. \tag{6.274}$$

The derivatives of the liquid energy balance equations with respect to the parameter $\alpha^{(18)} \triangleq a_{0,dav}$ are as follows:

$$\frac{\partial N_2^{(i)}}{\partial \alpha^{(18)}} = \frac{\partial N_2^{(i)}}{\partial a_{0,dav}} \triangleq a_2^{i,18} = 0; \quad \ell = 2; i = 1,\ldots,I; \; j = 18. \tag{6.275}$$

The derivatives of the liquid energy balance equations with respect to the parameter $\alpha^{(19)} \triangleq a_{1,dav}$ are as follows:

$$\frac{\partial N_2^{(i)}}{\partial \alpha^{(19)}} = \frac{\partial N_2^{(i)}}{\partial a_{1,dav}} \triangleq a_2^{i,19} = 0; \ell = 2; i = 1,\ldots,I; \; j = 19. \tag{6.276}$$

The derivatives of the liquid energy balance equations with respect to the parameter $\alpha^{(20)} \triangleq a_{2,dav}$ are as follows:

$$\frac{\partial N_2^{(i)}}{\partial \alpha^{(20)}} = \frac{\partial N_2^{(i)}}{\partial a_{2,dav}} \triangleq a_2^{i,20} = 0; \quad \ell = 2; i = 1,\ldots,I; \; j = 20. \tag{6.277}$$

The derivatives of the liquid energy balance equations with respect to the parameter $\alpha^{(21)} \triangleq a_{3,dav}$ are as follows:

$$\frac{\partial N_2^{(i)}}{\partial \alpha^{(21)}} = \frac{\partial N_2^{(i)}}{\partial a_{3,dav}} \triangleq a_2^{i,21} = 0; \quad \ell = 2; i = 1,\ldots,I; \; j = 21. \tag{6.278}$$

The derivatives of the liquid energy balance equations with respect to the parameter $\alpha^{(22)} \triangleq a_{0f}$ are as follows:

$$\begin{aligned} \frac{\partial N_2^{(i)}}{\partial \alpha^{(22)}} &= \frac{\partial N_2^{(i)}}{\partial a_{0f}} \triangleq a_2^{i,22} \\ &= m_w^{(i)} \frac{\partial h_f^{(i)}\left(T_w^{(i)}, \boldsymbol{\alpha}\right)}{\partial a_{0f}} - m_w^{(i+1)} \frac{\partial h_f^{(i+1)}\left(T_w^{(i+1)}, \boldsymbol{\alpha}\right)}{\partial a_{0f}} = m_w^{(i)} - m_w^{(i+1)}; \\ &\ell = 2; i = 1,\ldots,I; j = 22. \end{aligned} \tag{6.279}$$

The derivatives of the liquid energy balance equations with respect to the parameter $\alpha^{(23)} \triangleq a_{1f}$ are as follows:

$$\frac{\partial N_2^{(i)}}{\partial \alpha^{(23)}} = \frac{\partial N_2^{(i)}}{\partial a_{1f}} \triangleq a_2^{i,23} = m_w^{(i)} \frac{\partial h_f^{(i)}\left(T_w^{(i)},\boldsymbol{\alpha}\right)}{\partial a_{1f}} - m_w^{(i+1)} \frac{\partial h_f^{(i+1)}\left(T_w^{(i+1)},\boldsymbol{\alpha}\right)}{\partial a_{1f}}$$
$$= T_w^{(i)} m_w^{(i)} - T_w^{(i+1)} m_w^{(i+1)}; \quad \ell = 2; i = 1,\dots,I; j = 23. \tag{6.280}$$

The derivatives of the liquid energy balance equations with respect to the parameter $\alpha^{(24)} \triangleq a_{0g}$ are as follows:

$$\frac{\partial N_2^{(i)}}{\partial \alpha^{(24)}} = \frac{\partial N_2^{(i)}}{\partial a_{0g}} \triangleq a_2^{i,24} = -\left(m_w^{(i)} - m_w^{(i+1)}\right) \frac{\partial h_{g,w}^{(i+1)}\left(T_w^{(i+1)},\boldsymbol{\alpha}\right)}{\partial a_{0g}} = m_w^{(i+1)} - m_w^{(i)}; \tag{6.281}$$
$$\ell = 2; i = 1,\dots,I; j = 24.$$

The derivatives of the liquid energy balance equations with respect to the parameter $\alpha^{(25)} \triangleq a_{1g}$ are as follows:

$$\frac{\partial N_2^{(i)}}{\partial \alpha^{(25)}} = \frac{\partial N_2^{(i)}}{\partial a_{1g}} \triangleq a_2^{i,25}$$
$$= -\left(m_w^{(i)} - m_w^{(i+1)}\right) \frac{\partial h_{g,w}^{(i+1)}\left(T_w^{(i+1)},\boldsymbol{\alpha}\right)}{\partial a_{1g}} = -\left(m_w^{(i)} - m_w^{(i+1)}\right) T_w^{(i+1)}; \tag{6.282}$$
$$\ell = 2; i = 1,\dots,I; j = 25.$$

The derivatives of the liquid energy balance equations with respect to the parameter $\alpha^{(26)} \triangleq a_{0,Nu}$ are as follows:

$$\frac{\partial N_2^{(i)}}{\partial \alpha^{(26)}} = \frac{\partial N_2^{(i)}}{\partial a_{0,Nu}} \triangleq a_2^{i,26}$$
$$= -\left(T_w^{(i+1)} - T_a^{(i)}\right) \frac{\partial H\left(m_a,\boldsymbol{\alpha}\right)}{\partial Nu\left(Re,\boldsymbol{\alpha}\right)} \frac{\partial Nu\left(Re,\boldsymbol{\alpha}\right)}{\partial a_{0,Nu}}; \; \ell = 2; i = 1,\dots,I; \; j = 26, \tag{6.283}$$

where $\frac{\partial H\left(m_a,\boldsymbol{\alpha}\right)}{\partial Nu\left(Re,\boldsymbol{\alpha}\right)}$ was defined in Eq. (6.257) and $\frac{\partial Nu\left(Re,\boldsymbol{\alpha}\right)}{\partial a_{0,Nu}}$ was defined in Eq. (6.215).

The derivatives of the liquid energy balance equations with respect to the parameter $\alpha^{(27)} \triangleq a_{1,Nu}$ are as follows:

$$\frac{\partial N_2^{(i)}}{\partial \alpha^{(27)}} = \frac{\partial N_2^{(i)}}{\partial a_{1,Nu}} \triangleq a_2^{i,27} = -\left(T_w^{(i+1)} - T_a^{(i)}\right)\frac{\partial H(m_a,\boldsymbol{\alpha})}{\partial Nu(Re,\boldsymbol{\alpha})}\frac{\partial Nu(Re,\boldsymbol{\alpha})}{\partial a_{1,Nu}}; \quad \ell = 2; i = 1,\ldots,I; j = 27, \tag{6.284}$$

where $\frac{\partial H(m_a,\boldsymbol{\alpha})}{\partial Nu(Re,\boldsymbol{\alpha})}$ was defined in Eq. (6.257) and $\frac{\partial Nu(Re,\boldsymbol{\alpha})}{\partial a_{1,Nu}}$ was defined in Eq. (6.213).

The derivatives of the liquid energy balance equations with respect to the parameter $\alpha^{(28)} \triangleq a_{2,Nu}$ are as follows:

$$\frac{\partial N_2^{(i)}}{\partial \alpha^{(28)}} = \frac{\partial N_2^{(i)}}{\partial a_{2,Nu}} \triangleq a_2^{i,28} = -\left(T_w^{(i+1)} - T_a^{(i)}\right)\frac{\partial H(m_a,\boldsymbol{\alpha})}{\partial Nu(Re,\boldsymbol{\alpha})}\frac{\partial Nu(Re,\boldsymbol{\alpha})}{\partial a_{2,Nu}}; \quad \ell = 2; i = 1,\ldots,I; \; j = 28, \tag{6.285}$$

where $\frac{\partial H(m_a,\boldsymbol{\alpha})}{\partial Nu(Re,\boldsymbol{\alpha})}$ was defined in Eq. (6.257) and $\frac{\partial Nu(Re,\boldsymbol{\alpha})}{\partial a_{2,Nu}}$ was defined in Eq. (6.215).

The derivatives of the liquid energy balance equations with respect to the parameter $\alpha^{(29)} \triangleq a_{3,Nu}$ are as follows:

$$\frac{\partial N_2^{(i)}}{\partial \alpha^{(29)}} = \frac{\partial N_2^{(i)}}{\partial a_{3,Nu}} \triangleq a_2^{i,29} = -\left(T_w^{(i+1)} - T_a^{(i)}\right)\frac{\partial H(m_a,\boldsymbol{\alpha})}{\partial Nu(Re,\boldsymbol{\alpha})}\frac{\partial Nu(Re,\boldsymbol{\alpha})}{\partial a_{3,Nu}}; \quad \ell = 2; i = 1,\ldots,I; \; j = 29, \tag{6.286}$$

where $\frac{\partial H(m_a,\boldsymbol{\alpha})}{\partial Nu(Re,\boldsymbol{\alpha})}$ was defined in Eq. (6.257) and $\frac{\partial Nu(Re,\boldsymbol{\alpha})}{\partial a_{3,Nu}}$ was defined in Eq. (6.217).

The derivatives of the liquid energy balance equations with respect to the parameter $\alpha^{(30)} \triangleq W_{dkx}$ are as follows:

$$\frac{\partial N_2^{(i)}}{\partial \alpha^{(30)}} = \frac{\partial N_2^{(i)}}{\partial W_{dkx}} \triangleq a_2^{i,30} = 0; \quad \ell = 2; i = 1,\ldots,I; \; j = 30. \tag{6.287}$$

The derivatives of the liquid energy balance equations with respect to the parameter $\alpha^{(31)} \triangleq W_{dky}$ are as follows:

$$\frac{\partial N_2^{(i)}}{\partial \alpha^{(31)}} = \frac{\partial N_2^{(i)}}{\partial W_{dky}} \triangleq a_2^{i,31} = 0; \quad \ell = 2; i = 1,\ldots,I; \; j = 31. \tag{6.288}$$

The derivatives of the liquid energy balance equations with respect to the parameter $\alpha^{(32)} \triangleq \Delta z_{dk}$ are as follows:

$$\frac{\partial N_2^{(i)}}{\partial \alpha^{(32)}} = \frac{\partial N_2^{(i)}}{\partial \Delta z_{dk}} \triangleq a_2^{i,32} = 0; \quad \ell = 2; i = 1,...,I; \; j = 32. \tag{6.289}$$

The derivatives of the liquid energy balance equations with respect to the parameter $\alpha^{(33)} \triangleq \Delta z_{fan}$ are as follows:

$$\frac{\partial N_2^{(i)}}{\partial \alpha^{(33)}} = \frac{\partial N_2^{(i)}}{\partial \Delta z_{fan}} \triangleq a_2^{i,33} = 0; \quad \ell = 2; i = 1,...,I; \; j = 33. \tag{6.290}$$

The derivatives of the liquid energy balance equations with respect to the parameter $\alpha^{(34)} \triangleq D_{fan}$ are as follows:

$$\frac{\partial N_2^{(i)}}{\partial \alpha^{(34)}} = \frac{\partial N_2^{(i)}}{\partial D_{fan}} \triangleq a_2^{i,34} = -\left(T_w^{(i+1)} - T_a^{(i)}\right) \frac{\partial H(m_a, \boldsymbol{\alpha})}{\partial Nu(Re, \boldsymbol{\alpha})} \frac{\partial Nu(Re, \boldsymbol{\alpha})}{\partial m_a} \frac{\partial m_a}{\partial D_{fan}}; \quad \ell = 2; i = 1,...,I; \; j = 34, \tag{6.291}$$

where $\frac{\partial H(m_a, \boldsymbol{\alpha})}{\partial Nu(Re, \boldsymbol{\alpha})}$ was defined in Eq. (6.257), while $\frac{\partial Nu(Re, \boldsymbol{\alpha})}{\partial m_a}$ and $\frac{\partial m_a}{\partial D_{fan}}$ were defined in Eqs. (6.176) and (6.223), respectively.

The derivatives of the liquid energy balance equations with respect to the parameter $\alpha^{(35)} \triangleq \Delta z_{fill}$ are as follows:

$$\frac{\partial N_2^{(i)}}{\partial \alpha^{(35)}} = \frac{\partial N_2^{(i)}}{\partial \Delta z_{fill}} \triangleq a_2^{i,35} = 0; \quad \ell = 2; i = 1,...,I; \; j = 35. \tag{6.292}$$

The derivatives of the liquid energy balance equations with respect to the parameter $\alpha^{(36)} \triangleq \Delta z_{rain}$ are as follows:

$$\frac{\partial N_2^{(i)}}{\partial \alpha^{(36)}} = \frac{\partial N_2^{(i)}}{\partial \Delta z_{rain}} \triangleq a_2^{i,36} = 0; \quad \ell = 2; i = 1,...,I; \; j = 36. \tag{6.293}$$

The derivatives of the liquid energy balance equations with respect to the parameter $\alpha^{(37)} \triangleq \Delta z_{bs}$ are as follows:

$$\frac{\partial N_2^{(i)}}{\partial \alpha^{(37)}} = \frac{\partial N_2^{(i)}}{\partial \Delta z_{bs}} \triangleq a_2^{i,37} = 0; \quad \ell = 2; i = 1,\ldots,I; \; j = 37. \tag{6.294}$$

The derivatives of the liquid energy balance equations with respect to the parameter $\alpha^{(38)} \triangleq \Delta z_{de}$ are as follows:

$$\frac{\partial N_2^{(i)}}{\partial \alpha^{(38)}} = \frac{\partial N_2^{(i)}}{\partial \Delta z_{de}} \triangleq a_2^{i,38} = 0; \quad \ell = 2; i = 1,\ldots,I; \; j = 38. \tag{6.295}$$

The derivatives of the liquid energy balance equations with respect to the parameter $\alpha^{(39)} \triangleq D_h$ are as follows:

$$\begin{aligned} \frac{\partial N_2^{(i)}}{\partial \alpha^{(39)}} &= \frac{\partial N_2^{(i)}}{\partial D_h} \triangleq a_2^{i,39} \\ &= -\left(T_w^{(i+1)} - T_a^{(i)}\right)\frac{\partial H(m_a,\boldsymbol{\alpha})}{\partial D_h}; \; \ell = 2; i = 1,\ldots,I; \; j = 39, \end{aligned} \tag{6.296}$$

where

$$\frac{\partial H(m_a,\boldsymbol{\alpha})}{\partial D_h} = \begin{cases} -H(m_a,\boldsymbol{\alpha})/D_h, & Re_d < 2300, \\ -\dfrac{a_{2,Nu}H(m_a,\boldsymbol{\alpha})}{D_h Nu(Re,\boldsymbol{\alpha})}, & 2300 \le Re_d \le 10000, \\ -0.2H(m_a,\boldsymbol{\alpha})/D_h, & Re_d > 10000, \end{cases} \tag{6.297}$$

The derivatives of the liquid energy balance equations with respect to the parameter $\alpha^{(40)} \triangleq A_{fill}$ are as follows:

$$\frac{\partial N_2^{(i)}}{\partial \alpha^{(40)}} = \frac{\partial N_2^{(i)}}{\partial A_{fill}} \triangleq a_2^{i,40} = -\left(T_w^{(i+1)} - T_a^{(i)}\right)\frac{\partial H(m_a,\boldsymbol{\alpha})}{\partial A_{fill}}; \ell = 2; i = 1,\ldots,I; \; j = 40, \tag{6.298}$$

where

$$\frac{\partial H(m_a,\boldsymbol{\alpha})}{\partial A_{fill}} = \begin{cases} 0, & Re_d < 2300, \\ -\dfrac{a_{1,Nu}H(m_a,\boldsymbol{\alpha})Re(m_a,\boldsymbol{\alpha})}{A_{fill}Nu(Re,\boldsymbol{\alpha})}, & 2300 \le Re_d \le 10000, \\ -0.8H(m_a,\boldsymbol{\alpha})/A_{fill}, & Re_d > 10000, \end{cases} \tag{6.299}$$

The derivatives of the liquid energy balance equations with respect to the parameter $\alpha^{(41)} \triangleq A_{surf}$ are as follows:

$$\frac{\partial N_2^{(i)}}{\partial \alpha^{(41)}} = \frac{\partial N_2^{(i)}}{\partial A_{surf}} \triangleq a_2^{i,41} = -\left(T_w^{(i+1)} - T_a^{(i)}\right)\frac{\partial H(m_a,\boldsymbol{\alpha})}{\partial A_{surf}};\ \ell = 2;\ i = 1,\ldots,I;\ j = 41, \quad (6.300)$$

where

$$\frac{\partial H(m_a,\boldsymbol{\alpha})}{\partial A_{surf}} = \frac{H(m_a,\boldsymbol{\alpha})}{A_{surf}} = \frac{f_{ht} k_{air} w_{tsa} Nu(Re,\boldsymbol{\alpha})}{D_h I}. \quad (6.301)$$

The derivatives of the liquid energy balance equations with respect to the parameter $\alpha^{(42)} \triangleq Pr$ are as follows:

$$\frac{\partial N_2^{(i)}}{\partial \alpha^{(42)}} = \frac{\partial N_2^{(i)}}{\partial Pr} \triangleq a_2^{i,42} = -\left(T_w^{(i+1)} - T_a^{(i)}\right)\frac{\partial H(m_a,\boldsymbol{\alpha})}{\partial Pr};\ \ell = 2;\ i = 1,\ldots,I;\ j = 42, (6.302)$$

where

$$\frac{\partial H(m_a,\boldsymbol{\alpha})}{\partial Pr} = \begin{cases} 0, & Re_d \le 10000, \\ H(m_a,\boldsymbol{\alpha})/(3\,Pr), & Re_d > 10000, \end{cases} \quad (6.303)$$

The derivatives of the liquid energy balance equations with respect to the parameter $\alpha^{(43)} \triangleq V_w$ are as follows:

$$\frac{\partial N_2^{(i)}}{\partial \alpha^{(43)}} = \frac{\partial N_2^{(i)}}{\partial V_w} \triangleq a_2^{i,43} = 0;\ \ \ell = 2;\ i = 1,\ldots,I;\ \ j = 43. \quad (6.304)$$

The derivatives of the liquid energy balance equations with respect to the parameter $\alpha^{(44)} \triangleq V_{exit}$ are as follows:

$$\frac{\partial N_2^{(i)}}{\partial \alpha^{(44)}} = \frac{\partial N_2^{(i)}}{\partial V_{exit}} \triangleq a_2^{i,44} = -\left(T_w^{(i+1)} - T_a^{(i)}\right)\frac{\partial H(m_a,\boldsymbol{\alpha})}{\partial Nu(Re,\boldsymbol{\alpha})}\frac{\partial Nu(Re,\boldsymbol{\alpha})}{\partial m_a}\frac{\partial m_a}{\partial V_{exit}};$$
$$\ell = 2;\ i = 1,\ldots,I;\ \ j = 44, \quad (6.305)$$

where $\frac{\partial H(m_a,\boldsymbol{\alpha})}{\partial Nu(Re,\boldsymbol{\alpha})}$ was defined in Eq. (6.257), while $\frac{\partial Nu(Re,\boldsymbol{\alpha})}{\partial m_a}$ and $\frac{\partial m_a}{\partial V_{exit}}$ were defined previously in Eqs. (6.176) and (6.238), respectively.

The derivatives of the liquid energy balance equations with respect to the parameter $\alpha^{(45)} \triangleq m_{w,in}$ are as follows:

$$\frac{\partial N_2^{(1)}}{\partial \alpha^{(45)}} = \frac{\partial N_2^{(1)}}{\partial m_{w,in}} \triangleq a_2^{1,45} = h_f^{(1)}\left(T_{w,in},\boldsymbol{\alpha}\right) - h_{g,w}^{(2)}\left(T_w^{(2)},\boldsymbol{\alpha}\right) = T_{w,in}a_{1f} - a_{1g}T_w^{(2)} + a_{0f} - a_{0g}, \quad \ell = 2;\ i=1;\ j=45, \tag{6.306}$$

$$\frac{\partial N_2^{(i)}}{\partial \alpha^{(45)}} = \frac{\partial N_2^{(i)}}{\partial m_{w,in}} \triangleq a_2^{i,45} = 0; \quad \ell = 2;\ i = 2,\ldots,I;\ j = 45. \tag{6.307}$$

The derivatives of the liquid energy balance equations with respect to the parameter $\alpha^{(46)} \triangleq T_{a,in}$ are as follows:

$$\frac{\partial N_2^{(i)}}{\partial \alpha^{(46)}} = \frac{\partial N_2^{(i)}}{\partial T_{a,in}} \triangleq a_2^{i,46} = 0; \quad \ell = 2;\ i = 1,\ldots,I;\ j = 46. \tag{6.308}$$

NOTE:the above derivative was NOT used in the computations for the numerical results presented in Chapter 6.

The derivatives of the liquid energy balance equations with respect to the parameter $\alpha^{(47)} \triangleq m_a$ are as follows:

$$\frac{\partial N_2^{(i)}}{\partial \alpha^{(47)}} = \frac{\partial N_2^{(i)}}{\partial m_a} \triangleq a_2^{i,47} = -\left(T_w^{(i+1)} - T_a^{(i)}\right)\frac{\partial H(m_a,\boldsymbol{\alpha})}{\partial Nu(Re,\boldsymbol{\alpha})}\frac{\partial Nu(Re,\boldsymbol{\alpha})}{\partial m_a}; \quad \ell = 2;\ i = 1,\ldots,I;\ j = 47, \tag{6.309}$$

where $\frac{\partial H(m_a,\boldsymbol{\alpha})}{\partial Nu(Re,\boldsymbol{\alpha})}$ and $\frac{\partial Nu(Re,\boldsymbol{\alpha})}{\partial m_a}$ were defined previously in Eqs. (6.257) and (6.176), respectively.

The derivatives of the liquid energy balance equations with respect to the parameter $\alpha^{(48)} \triangleq \omega_{in}$ are as follows:

$$\frac{\partial N_2^{(i)}}{\partial \alpha^{(48)}} = \frac{\partial N_2^{(i)}}{\partial \omega_{in}} \triangleq a_2^{i,48} = 0; \quad \ell = 2; i = 1,...,I; \ j = 48. \tag{6.310}$$

The derivatives of the liquid energy balance equations with respect to the parameter $\alpha^{(49)} \triangleq Re_d$ are as follows:

$$\frac{\partial N_2^{(i)}}{\partial \alpha^{(49)}} = \frac{\partial N_2^{(i)}}{\partial Re_d} \triangleq a_2^{i,49} = -\left(T_w^{(i+1)} - T_a^{(i)}\right) \frac{\partial H(m_a, \boldsymbol{\alpha})}{\partial Nu(Re_d, \boldsymbol{\alpha})} \frac{\partial Nu(Re_d, \boldsymbol{\alpha})}{\partial Re_d}; \quad \ell = 2; i = 1,...,I; \ j = 49, \tag{6.311}$$

where $\dfrac{\partial H(m_a, \boldsymbol{\alpha})}{\partial Nu(Re_d, \boldsymbol{\alpha})}$ was defined in Eq. (6.257) and $\dfrac{\partial Nu(Re_d, \boldsymbol{\alpha})}{\partial Re_d}$ was defined in Eq. (6.245).

The derivatives of the liquid energy balance equations with respect to the parameter $\alpha^{(50)} \triangleq Sc$ are as follows:

$$\frac{\partial N_2^{(i)}}{\partial \alpha^{(50)}} = \frac{\partial N_2^{(i)}}{\partial Sc} \triangleq a_2^{i,50} = 0; \quad \ell = 2; i = 1,...,I; \ j = 50. \tag{6.312}$$

The derivatives of the liquid energy balance equations with respect to the parameter $\alpha^{(51)} \triangleq Sh$ are as follows:

$$\frac{\partial N_2^{(i)}}{\partial \alpha^{(51)}} = \frac{\partial N_2^{(i)}}{\partial Sh} \triangleq a_2^{i,51} = 0; \quad \ell = 2; i = 1,...,I; \ j = 51. \tag{6.313}$$

The derivatives of the liquid energy balance equations with respect to the parameter $\alpha^{(52)} \triangleq Nu$ are as follows:

$$\frac{\partial N_2^{(i)}}{\partial \alpha^{(52)}} = \frac{\partial N_2^{(i)}}{\partial Nu} \triangleq a_2^{i,52} = -\left(T_w^{(i+1)} - T_a^{(i)}\right) \frac{\partial H(m_a, \boldsymbol{\alpha})}{\partial Nu}; \ell = 2; i = 1,...,I; j = 52, \tag{6.314}$$

where $\dfrac{\partial H(m_a, \boldsymbol{\alpha})}{\partial Nu}$ was defined in Eq. (6.257).

Derivatives of the water vapor continuity equations with respect to model parameters

The derivatives of the water vapor continuity equations with respect to the parameter $\alpha^{(1)} \triangleq T_{db}$ are as follows:

$$\frac{\partial N_3^{(i)}}{\partial \alpha^{(1)}} = \frac{\partial N_3^{(i)}}{\partial T_{db}} \triangleq a_3^{i,1} = 0; \quad \ell = 3; i = 1,...,I; \ j = 1. \tag{6.315}$$

The derivatives of the water vapor continuity equations with respect to the parameter $\alpha^{(2)} \triangleq T_{dp}$ are as follows:

$$\frac{\partial N_3^{(i)}}{\partial \alpha^{(2)}} = \frac{\partial N_3^{(i)}}{\partial T_{dp}} \triangleq a_3^{i,2} = 0; \quad \ell = 3; i = 1,...,I-1; \ j = 2, \tag{6.316}$$

$$\frac{\partial N_3^{(I)}}{\partial \alpha^{(2)}} = \frac{\partial N_3^{(I)}}{\partial T_{dp}} \triangleq a_3^{I,2} = \frac{\partial \omega_{in}}{\partial T_{dp}}; \quad \ell = 3; i = I; \ j = 2, \tag{6.317}$$

where

$$\frac{\partial \omega_{in}}{\partial T_{dp}} = -\frac{0.622 a_1 P_{atm} e^{a_0 + \frac{a_1}{T_{dp}}}}{T_{tdp}{}^2 \left(P_{atm} - e^{a_0 + \frac{a_1}{T_{dp}}} \right)^2}. \tag{6.318}$$

The derivatives of the water vapor continuity equations with respect to the parameter $\alpha^{(3)} \triangleq T_{w,in}$ are as follows:

$$\frac{\partial N_3^{(1)}}{\partial \alpha^{(3)}} = \frac{\partial N_3^{(1)}}{\partial T_{w,in}} \triangleq a_3^{1,3} = \frac{1}{m_a} \frac{\partial m_{w,in}}{\partial T_{w,in}}; \quad \ell = 3; i=1; j = 3, \tag{6.319}$$

where $\dfrac{\partial m_{w,in}}{\partial T_{w,in}}$ was defined in Eq. (6.172).

$$\frac{\partial N_3^{(i)}}{\partial \alpha^{(3)}} = \frac{\partial N_3^{(i)}}{\partial T_{w,in}} \triangleq a_3^{i,3} = 0; \quad \ell = 3; i=2,...I; j = 3. \tag{6.320}$$

The derivatives of the water vapor continuity equations with respect to the parameter $\alpha^{(4)} \triangleq P_{atm}$ are as follows:

$$\frac{\partial N_3^{(i)}}{\partial \alpha^{(4)}} = \frac{\partial N_3^{(i)}}{\partial P_{atm}} \triangleq a_3^{i,4} = -\frac{m_w^{(i)} - m_w^{(i+1)}}{m_a^{\ 2}} \frac{\partial m_a}{\partial P_{atm}}; \ \ell = 3; i = 1,...,I-1; j = 4, \quad (6.321)$$

$$\frac{\partial N_3^{(I)}}{\partial \alpha^{(4)}} = \frac{\partial N_3^{(I)}}{\partial P_{atm}} \triangleq a_3^{I,4} = \frac{\partial \omega_{in}}{\partial P_{atm}} - \frac{m_w^{(i)} - m_w^{(i+1)}}{m_a^{\ 2}} \frac{\partial m_a}{\partial P_{atm}}; \ \ell = 3; i = I; \ j = 4, \quad (6.322)$$

where $\frac{\partial m_a}{\partial P_{atm}}$ was defined in Eq. (6.177) and

$$\frac{\partial \omega_{in}}{\partial P_{atm}} = -\frac{0.622 e^{a_0 + \frac{a_1}{T_{dp}}}}{\left(P_{atm} - e^{a_0 + \frac{a_1}{T_{dp}}}\right)^2}. \quad (6.323)$$

The derivatives of the water vapor continuity equations with respect to the parameter $\alpha^{(5)} \triangleq w_{tsa}$ are as follows:

$$\frac{\partial N_3^{(i)}}{\partial \alpha^{(5)}} = \frac{\partial N_3^{(i)}}{\partial w_{tsa}} \triangleq a_3^{i,5} = 0; \quad \ell = 3; i = 1,...,I; \ j = 5. \quad (6.324)$$

The derivatives of the water vapor continuity equations with respect to the parameter $\alpha^{(6)} \triangleq k_{sum}$ are as follows:

$$\frac{\partial N_3^{(i)}}{\partial \alpha^{(6)}} = \frac{\partial N_3^{(i)}}{\partial k_{sum}} \triangleq a_3^{i,6} = 0; \quad \ell = 3; i = 1,...,I; \ j = 6. \quad (6.325)$$

The derivatives of the water vapor continuity equations with respect to the parameter $\alpha^{(7)} \triangleq \mu$ are as follows:

$$\frac{\partial N_3^{(i)}}{\partial \alpha^{(7)}} = \frac{\partial N_3^{(i)}}{\partial \mu} \triangleq a_3^{i,7} = 0; \quad \ell = 3; i = 1,...,I; \ j = 7. \quad (6.326)$$

The derivatives of the water vapor continuity equations with respect to the parameter $\alpha^{(8)} \triangleq \upsilon$ are as follows:

$$\frac{\partial N_3^{(i)}}{\partial \alpha^{(8)}} = \frac{\partial N_3^{(i)}}{\partial \upsilon} \triangleq a_3^{i,8} = 0; \quad \ell = 3; i = 1,...,I; \ j = 8. \quad (6.327)$$

The derivatives of the water vapor continuity equations with respect to the parameter $\alpha^{(9)} \triangleq k_{air}$ are as follows:

$$\frac{\partial N_3^{(i)}}{\partial \alpha^{(9)}} = \frac{\partial N_3^{(i)}}{\partial k_{air}} \triangleq a_3^{i,9} = 0; \quad \ell = 3; i = 1,\ldots,I; \; j = 9. \tag{6.328}$$

The derivatives of the water vapor continuity equations with respect to the parameter $\alpha^{(10)} \triangleq f_{ht}$ are as follows:

$$\frac{\partial N_3^{(i)}}{\partial \alpha^{(10)}} = \frac{\partial N_3^{(i)}}{\partial f_{ht}} \triangleq a_3^{i,10} = 0; \quad \ell = 3; i = 1,\ldots,I; \; j = 10. \tag{6.329}$$

The derivatives of the water vapor continuity equations with respect to the parameter $\alpha^{(11)} \triangleq f_{mt}$ are as follows:

$$\frac{\partial N_3^{(i)}}{\partial \alpha^{(11)}} = \frac{\partial N_3^{(i)}}{\partial f_{mt}} \triangleq a_3^{i,11} = 0; \ell = 3; i = 1,\ldots,I; \; j = 11. \tag{6.330}$$

The derivatives of the water vapor continuity equations with respect to the parameter $\alpha^{(12)} \triangleq f$ are as follows:

$$\frac{\partial N_3^{(i)}}{\partial \alpha^{(12)}} = \frac{\partial N_3^{(i)}}{\partial f} \triangleq a_3^{i,12} = 0; \quad \ell = 3; i = 1,\ldots,I; \; j = 12. \tag{6.331}$$

The derivatives of the water vapor continuity equations with respect to the parameter $\alpha^{(13)} \triangleq a_0$ are as follows:

$$\frac{\partial N_3^{(i)}}{\partial \alpha^{(13)}} = \frac{\partial N_3^{(i)}}{\partial a_0} \triangleq a_3^{i,13} = 0; \quad \ell = 3; i = 1,\ldots,I-1; \; j = 13, \tag{6.332}$$

$$\frac{\partial N_3^{(I)}}{\partial \alpha^{(13)}} = \frac{\partial N_3^{(I)}}{\partial a_0} \triangleq a_3^{I,13} = \frac{\partial \omega_{in}}{\partial a_0}; \quad \ell = 3; i = I; \; j = 13, \tag{6.333}$$

where

$$\frac{\partial \omega_{in}}{\partial a_0} = \frac{0.622 P_{atm} e^{a_0 + \frac{a_1}{T_{dp}}}}{\left(P_{atm} - e^{a_0 + \frac{a_1}{T_{dp}}}\right)^2}; \tag{6.334}$$

The derivatives of the water vapor continuity equations with respect to the parameter $\alpha^{(14)} \triangleq a_1$ are as follows:

$$\frac{\partial N_3^{(i)}}{\partial \alpha^{(14)}} = \frac{\partial N_3^{(i)}}{\partial a_1} \triangleq a_3^{i,14} = 0; \quad \ell = 3; i = 1,...,I-1; \; j = 14, \qquad (6.335)$$

$$\frac{\partial N_3^{(I)}}{\partial \alpha^{(14)}} = \frac{\partial N_3^{(I)}}{\partial a_1} \triangleq a_3^{I,14} = \frac{\partial \omega_{in}}{\partial a_1}; \quad \ell = 3; i = I; \; j = 14, \qquad (6.336)$$

where

$$\frac{\partial \omega_{in}}{\partial a_1} = \frac{0.622 P_{atm} e^{a_0 + \frac{a_1}{T_{dp}}}}{T_{dp}\left(P_{atm} - e^{a_0 + \frac{a_1}{T_{dp}}}\right)^2}. \qquad (6.337)$$

The derivatives of the water vapor continuity equations with respect to the parameter $\alpha^{(15)} \triangleq a_{0,cpa}$ are as follows:

$$\frac{\partial N_3^{(i)}}{\partial \alpha^{(15)}} = \frac{\partial N_3^{(i)}}{\partial a_{0,cpa}} \triangleq a_3^{i,15} = 0; \quad \ell = 3; i = 1,...,I; \; j = 15. \qquad (6.338)$$

The derivatives of the water vapor continuity equations with respect to the parameter $\alpha^{(16)} \triangleq a_{1,cpa}$ are as follows:

$$\frac{\partial N_3^{(i)}}{\partial \alpha^{(16)}} = \frac{\partial N_3^{(i)}}{\partial a_{1,cpa}} \triangleq a_3^{i,16} = 0; \quad \ell = 3; i = 1,...,I; \; j = 16. \qquad (6.339)$$

The derivatives of the water vapor continuity equations with respect to the parameter $\alpha^{(17)} \triangleq a_{2,cpa}$ are as follows:

$$\frac{\partial N_3^{(i)}}{\partial \alpha^{(17)}} = \frac{\partial N_3^{(i)}}{\partial a_{2,cpa}} \triangleq a_3^{i,17} = 0; \quad \ell = 3; i = 1,...,I; \; j = 17. \qquad (6.340)$$

The derivatives of the water vapor continuity equations with respect to the parameter $\alpha^{(18)} \triangleq a_{0,dav}$ are as follows:

$$\frac{\partial N_3^{(i)}}{\partial \alpha^{(18)}} = \frac{\partial N_3^{(i)}}{\partial a_{0,dav}} \triangleq a_3^{i,18} = 0; \quad \ell = 3; i = 1,...,I; \; j = 18. \qquad (6.341)$$

The derivatives of the water vapor continuity equations with respect to the parameter $\alpha^{(19)} \triangleq a_{1,dav}$ are as follows:

$$\frac{\partial N_3^{(i)}}{\partial \alpha^{(19)}} = \frac{\partial N_3^{(i)}}{\partial a_{1,dav}} \triangleq a_3^{i,19} = 0; \quad \ell = 3; i = 1,\ldots,I; \; j = 19. \tag{6.342}$$

The derivatives of the water vapor continuity equations with respect to the parameter $\alpha^{(20)} \triangleq a_{2,dav}$ are as follows:

$$\frac{\partial N_3^{(i)}}{\partial \alpha^{(20)}} = \frac{\partial N_3^{(i)}}{\partial a_{2,dav}} \triangleq a_3^{i,20} = 0; \quad \ell = 3; i = 1,\ldots,I; \; j = 20. \tag{6.343}$$

The derivatives of the water vapor continuity equations with respect to the parameter $\alpha^{(21)} \triangleq a_{3,dav}$ are as follows:

$$\frac{\partial N_3^{(i)}}{\partial \alpha^{(21)}} = \frac{\partial N_3^{(i)}}{\partial a_{3,dav}} \triangleq a_3^{i,21} = 0; \quad \ell = 3; i = 1,\ldots,I; \; j = 21. \tag{6.344}$$

The derivatives of the water vapor continuity equations with respect to the parameter $\alpha^{(22)} \triangleq a_{0f}$ are as follows:

$$\frac{\partial N_3^{(i)}}{\partial \alpha^{(22)}} = \frac{\partial N_3^{(i)}}{\partial a_{0f}} \triangleq a_3^{i,22} = 0; \quad \ell = 3; i = 1,\ldots,I; j = 22. \tag{6.345}$$

The derivatives of the water vapor continuity equations with respect to the parameter $\alpha^{(23)} \triangleq a_{1f}$ are as follows:

$$\frac{\partial N_3^{(i)}}{\partial \alpha^{(23)}} = \frac{\partial N_3^{(i)}}{\partial a_{1f}} \triangleq a_3^{i,23} = 0; \quad \ell = 3; i = 1,\ldots,I; j = 23. \tag{6.346}$$

The derivatives of the water vapor continuity equations with respect to the parameter $\alpha^{(24)} \triangleq a_{0g}$ are as follows:

$$\frac{\partial N_3^{(i)}}{\partial \alpha^{(24)}} = \frac{\partial N_3^{(i)}}{\partial a_{0g}} \triangleq a_3^{i,24} = 0; \quad \ell = 3; i = 1,\ldots,I; j = 24. \tag{6.347}$$

The derivatives of the water vapor continuity equations with respect to the parameter $\alpha^{(25)} \triangleq a_{1g}$ are as follows:

$$\frac{\partial N_3^{(i)}}{\partial \alpha^{(25)}} = \frac{\partial N_3^{(i)}}{\partial a_{1g}} \triangleq a_3^{i,25} = 0; \quad \ell = 3; i = 1,...,I; j = 25. \qquad (6.348)$$

The derivatives of the water vapor continuity equations with respect to the parameter $\alpha^{(26)} \triangleq a_{0,Nu}$ are as follows:

$$\frac{\partial N_3^{(i)}}{\partial \alpha^{(26)}} = \frac{\partial N_3^{(i)}}{\partial a_{0,Nu}} \triangleq a_3^{i,26} = 0; \quad \ell = 3; i = 1,...,I; \; j = 26. \qquad (6.349)$$

The derivatives of the water vapor continuity equations with respect to the parameter $\alpha^{(27)} \triangleq a_{1,Nu}$ are as follows:

$$\frac{\partial N_3^{(i)}}{\partial \alpha^{(27)}} = \frac{\partial N_3^{(i)}}{\partial a_{1,Nu}} \triangleq a_3^{i,27} = 0; \quad \ell = 3; i = 1,...,I; j = 27. \qquad (6.350)$$

The derivatives of the water vapor continuity equations with respect to the parameter $\alpha^{(28)} \triangleq a_{2,Nu}$ are as follows:

$$\frac{\partial N_3^{(i)}}{\partial \alpha^{(28)}} = \frac{\partial N_3^{(i)}}{\partial a_{2,Nu}} \triangleq a_3^{i,28} = 0; \quad \ell = 3; i = 1,...,I; \; j = 28. \qquad (6.351)$$

The derivatives of the water vapor continuity equations with respect to the parameter $\alpha^{(29)} \triangleq a_{3,Nu}$ are as follows:

$$\frac{\partial N_3^{(i)}}{\partial \alpha^{(29)}} = \frac{\partial N_3^{(i)}}{\partial a_{3,Nu}} \triangleq a_3^{i,29} = 0; \quad \ell = 3; i = 1,...,I; \; j = 29. \qquad (6.352)$$

The derivatives of the water vapor continuity equations with respect to the parameter $\alpha^{(30)} \triangleq W_{dkx}$ are as follows:

$$\frac{\partial N_3^{(i)}}{\partial \alpha^{(30)}} = \frac{\partial N_3^{(i)}}{\partial W_{dkx}} \triangleq a_3^{i,30} = 0; \quad \ell = 3; i = 1,...,I; \; j = 30. \qquad (6.353)$$

The derivatives of the water vapor continuity equations with respect to the parameter $\alpha^{(31)} \triangleq W_{dky}$ are as follows:

$$\frac{\partial N_3^{(i)}}{\partial \alpha^{(31)}} = \frac{\partial N_3^{(i)}}{\partial W_{dky}} \triangleq a_3^{i,31} = 0; \quad \ell = 3; i = 1,...,I; \; j = 31. \tag{6.354}$$

The derivatives of the water vapor continuity equations with respect to the parameter $\alpha^{(32)} \triangleq \Delta z_{dk}$ are as follows:

$$\frac{\partial N_3^{(i)}}{\partial \alpha^{(32)}} = \frac{\partial N_3^{(i)}}{\partial \Delta z_{dk}} \triangleq a_3^{i,32} = 0; \quad \ell = 3; i = 1,...,I; \; j = 32. \tag{6.355}$$

The derivatives of the water vapor continuity equations with respect to the parameter $\alpha^{(33)} \triangleq \Delta z_{fan}$ are as follows:

$$\frac{\partial N_3^{(i)}}{\partial \alpha^{(33)}} = \frac{\partial N_3^{(i)}}{\partial \Delta z_{fan}} \triangleq a_3^{i,33} = 0; \quad \ell = 3; i = 1,...,I; \; j = 33. \tag{6.356}$$

The derivatives of the water vapor continuity equations with respect to the parameter $\alpha^{(34)} \triangleq D_{fan}$ are as follows:

$$\frac{\partial N_3^{(i)}}{\partial \alpha^{(34)}} = \frac{\partial N_3^{(i)}}{\partial D_{fan}} \triangleq a_3^{i,34} = -\frac{m_w^{(i)} - m_w^{(i+1)}}{m_a^{\;2}} \frac{\partial m_a}{\partial D_{fan}}; \; \ell = 3; i = 1,...,I; \; j = 34, \tag{6.357}$$

where $\frac{\partial m_a}{\partial D_{fan}}$ was defined in Eq. (6.223).

The derivatives of the water vapor continuity equations with respect to the parameter $\alpha^{(35)} \triangleq \Delta z_{fill}$ are as follows:

$$\frac{\partial N_3^{(i)}}{\partial \alpha^{(35)}} = \frac{\partial N_3^{(i)}}{\partial \Delta z_{fill}} \triangleq a_3^{i,35} = 0; \quad \ell = 3; i = 1,...,I; \; j = 35. \tag{6.358}$$

The derivatives of the water vapor continuity equations with respect to the parameter $\alpha^{(36)} \triangleq \Delta z_{rain}$ are as follows:

$$\frac{\partial N_3^{(i)}}{\partial \alpha^{(36)}} = \frac{\partial N_3^{(i)}}{\partial \Delta z_{rain}} \triangleq a_3^{i,36} = 0; \quad \ell = 3; i = 1,...,I; \; j = 36. \tag{6.359}$$

The derivatives of the water vapor continuity equations with respect to the parameter $\alpha^{(37)} \triangleq \Delta z_{bs}$ are as follows:

$$\frac{\partial N_3^{(i)}}{\partial \alpha^{(37)}} = \frac{\partial N_3^{(i)}}{\partial \Delta z_{bs}} \triangleq a_3^{i,37} = 0; \quad \ell = 3; i = 1,\ldots,I; \; j = 37. \qquad (6.360)$$

The derivatives of the water vapor continuity equations with respect to the parameter $\alpha^{(38)} \triangleq \Delta z_{de}$ are as follows:

$$\frac{\partial N_3^{(i)}}{\partial \alpha^{(38)}} = \frac{\partial N_3^{(i)}}{\partial \Delta z_{de}} \triangleq a_3^{i,38} = 0; \quad \ell = 3; i = 1,\ldots,I; \; j = 38. \qquad (6.361)$$

The derivatives of the water vapor continuity equations with respect to the parameter $\alpha^{(39)} \triangleq D_h$ are as follows:

$$\frac{\partial N_3^{(i)}}{\partial \alpha^{(39)}} = \frac{\partial N_3^{(i)}}{\partial D_h} \triangleq a_3^{i,39} = 0; \quad \ell = 3; i = 1,\ldots,I; \; j = 39. \qquad (6.362)$$

The derivatives of the water vapor continuity equations with respect to the parameter $\alpha^{(40)} \triangleq A_{fill}$ are as follows:

$$\frac{\partial N_3^{(i)}}{\partial \alpha^{(40)}} = \frac{\partial N_3^{(i)}}{\partial A_{fill}} \triangleq a_3^{i,40} = 0; \quad \ell = 3; i = 1,\ldots,I; \; j = 40. \qquad (6.363)$$

The derivatives of the water vapor continuity equations with respect to the parameter $\alpha^{(41)} \triangleq A_{surf}$ are as follows:

$$\frac{\partial N_3^{(i)}}{\partial \alpha^{(41)}} = \frac{\partial N_3^{(i)}}{\partial A_{surf}} \triangleq a_3^{i,41} = 0; \quad \ell = 3; i = 1,\ldots,I; \; j = 41. \qquad (6.364)$$

The derivatives of the water vapor continuity equations with respect to the parameter $\alpha^{(42)} \triangleq Pr$ are as follows:

$$\frac{\partial N_3^{(i)}}{\partial \alpha^{(42)}} = \frac{\partial N_3^{(i)}}{\partial Pr} \triangleq a_3^{i,42} = 0; \quad \ell = 3; i = 1,\ldots,I; \; j = 42. \qquad (6.365)$$

The derivatives of the water vapor continuity equations with respect to the parameter $\alpha^{(43)} \triangleq V_w$ are as follows:

$$\frac{\partial N_3^{(i)}}{\partial \alpha^{(43)}} = \frac{\partial N_3^{(i)}}{\partial V_w} \triangleq a_3^{i,43} = 0; \quad \ell = 3; i = 1,\ldots,I; \; j = 43. \qquad (6.366)$$

The derivatives of the water vapor continuity equations with respect to the parameter $\alpha^{(44)} \triangleq V_{exit}$ are as follows:

$$\frac{\partial N_3^{(i)}}{\partial \alpha^{(44)}} = \frac{\partial N_3^{(i)}}{\partial V_{exit}} \triangleq a_3^{i,44} = -\frac{m_w^{(i)} - m_w^{(i+1)}}{m_a^{\;2}} \frac{\partial m_a}{\partial V_{exit}}; \quad \ell = 3; i = 1,\dots,I; \; j = 44, \quad (6.367)$$

where $\frac{\partial m_a}{\partial V_{exit}}$ was defined in Eq. (6.238).

The derivatives of the water vapor continuity equations with respect to the parameter $\alpha^{(45)} \triangleq m_{w,in}$ are as follows:

$$\frac{\partial N_3^{(1)}}{\partial \alpha^{(45)}} = \frac{\partial N_3^{(1)}}{\partial m_{w,in}} \triangleq a_3^{1,45} = \frac{1}{m_a}; \quad \ell = 3; i=1; j = 45, \qquad (6.368)$$

$$\frac{\partial N_3^{(i)}}{\partial \alpha^{(45)}} = \frac{\partial N_3^{(i)}}{\partial m_{w,in}} \triangleq a_3^{i,45} = 0; \quad \ell = 3; i = 2,\dots,I; \; j = 45. \qquad (6.369)$$

The derivatives of the water vapor continuity equations with respect to the parameter $\alpha^{(46)} \triangleq T_{a,in}$ are as follows:

$$\frac{\partial N_3^{(i)}}{\partial \alpha^{(46)}} = \frac{\partial N_3^{(i)}}{\partial T_{a,in}} \triangleq a_3^{i,46} = 0; \quad \ell = 3; i = 1,\dots,I; \; j = 46. \qquad (6.370)$$

NOTE:the above derivative was NOT used in the computations for the numerical results presented in Chapter 6.

The derivatives of the water vapor continuity equations with respect to the parameter $\alpha^{(47)} \triangleq m_a$ are as follows:

$$\frac{\partial N_3^{(i)}}{\partial \alpha^{(47)}} = \frac{\partial N_3^{(i)}}{\partial m_a} \triangleq a_3^{i,47} = -\frac{m_w^{(i)} - m_w^{(i+1)}}{m_a^{\;2}}; \quad \ell = 3; i = 1,\dots,I; \; j = 47. \quad (6.371)$$

The derivatives of the water vapor continuity equations with respect to the parameter $\alpha^{(48)} \triangleq \omega_{in}$ are as follows:

$$\frac{\partial N_3^{(i)}}{\partial \alpha^{(48)}} = \frac{\partial N_3^{(i)}}{\partial \omega_{in}} \triangleq a_3^{i,48} = 0; \quad \ell = 3; i = 1,\dots,I-1; \; j = 48, \quad (6.372)$$

$$\frac{\partial N_3^{(I)}}{\partial \alpha^{(48)}} = \frac{\partial N_3^{(I)}}{\partial \omega_{in}} \triangleq a_3^{I,48} = 1; \quad \ell = 3; i = I; \; j = 48. \tag{6.373}$$

The derivatives of the water vapor continuity equations with respect to the parameter $\alpha^{(49)} \triangleq Re_d$ are as follows:

$$\frac{\partial N_3^{(i)}}{\partial \alpha^{(49)}} = \frac{\partial N_3^{(i)}}{\partial Re_d} \triangleq a_3^{i,49} = 0; \quad \ell = 3; i = 1,...,I; \; j = 49. \tag{6.374}$$

The derivatives of the water vapor continuity equations with respect to the parameter $\alpha^{(50)} \triangleq Sc$ are as follows:

$$\frac{\partial N_3^{(i)}}{\partial \alpha^{(50)}} = \frac{\partial N_3^{(i)}}{\partial Sc} \triangleq a_3^{i,50} = 0; \quad \ell = 3; i = 1,...,I; \; j = 50. \tag{6.375}$$

The derivatives of the water vapor continuity equations with respect to the parameter $\alpha^{(51)} \triangleq Sh$ are as follows:

$$\frac{\partial N_3^{(i)}}{\partial \alpha^{(51)}} = \frac{\partial N_3^{(i)}}{\partial Sh} \triangleq a_3^{i,51} = 0; \ell = 3; i = 1,...,I; \; j = 51. \tag{6.376}$$

The derivatives of the water vapor continuity equations with respect to the parameter $\alpha^{(52)} \triangleq Nu$ are as follows:

$$\frac{\partial N_3^{(i)}}{\partial \alpha^{(52)}} = \frac{\partial N_3^{(i)}}{\partial Nu} \triangleq a_3^{i,52} = 0; \quad \ell = 3; i = 1,...,I; \; j = 52. \tag{6.377}$$

Derivatives of the air and water vapor energy balance equations with respect to model parameters

Since $T_{db} \equiv T_{a,in}$ for the numerical results presented in Chapter 6, the derivatives of the air/water vapor energy balance equations with respect to the parameter $\alpha^{(1)} \triangleq T_{db}$ are as follows:

$$\frac{\partial N_4^{(i)}}{\partial \alpha^{(1)}} = \frac{\partial N_4^{(i)}}{\partial T_{db}} \triangleq a_4^{i,1} = 0; \quad \ell = 4; i = 1,...,I-1; \; j = 1, \tag{6.378}$$

$$\begin{aligned}\frac{\partial N_4^{(I)}}{\partial \alpha^{(1)}} &= \frac{\partial N_4^{(I)}}{\partial T_{db}} \triangleq a_4^{I,1} = C_p^{(I)}\left(\frac{T_a^{(I)}+273.15}{2},\boldsymbol{\alpha}\right) + \omega_{in}\frac{\partial h_{g,a}^{(I+1)}\left(T_{a,in},\boldsymbol{\alpha}\right)}{\partial T_{a,in}} \\ &= C_p^{(I)}\left(\frac{T_a^{(I)}+273.15}{2},\boldsymbol{\alpha}\right) + \omega_{in}a_{1g}; \quad \ell=4;\ i=I;\ j=1.\end{aligned} \tag{6.379}$$

The derivatives of the air/water vapor energy balance equations with respect to the parameter $\alpha^{(2)} \triangleq T_{dp}$ are as follows:

$$\frac{\partial N_4^{(i)}}{\partial \alpha^{(2)}} = \frac{\partial N_4^{(i)}}{\partial T_{dp}} \triangleq a_4^{i,2} = 0; \quad \ell=4;\ i=1,\ldots,I-1;\ j=2, \tag{6.380}$$

$$\frac{\partial N_4^{(I)}}{\partial \alpha^{(2)}} = \frac{\partial N_4^{(I)}}{\partial T_{dp}} \triangleq a_4^{I,2} = \frac{\partial \omega_{in}}{\partial T_{dp}}\left(a_{1\,g}T_{a,in} + a_{0\,g}\right); \quad \ell=4;\ i=I;\ j=2, \tag{6.381}$$

where $\dfrac{\partial \omega_{in}}{\partial T_{dp}}$ was defined in Eq. (6.318).

The derivatives of the air/water vapor energy balance equations with respect to the parameter $\alpha^{(3)} \triangleq T_{w,in}$ are as follows:

$$\begin{aligned}\frac{\partial N_4^{(1)}}{\partial \alpha^{(3)}} = \frac{\partial N_4^{(1)}}{\partial T_{w,in}} \triangleq a_4^{1,3} &= \frac{h_{g,w}^{(2)}\left(T_w^{(2)},\boldsymbol{\alpha}\right)}{m_a}\frac{\partial m_{w,in}}{\partial T_{w,in}} = \frac{a_{1g}T_w^{(2)} + a_{0g}}{m_a}\frac{\partial m_{w,in}}{\partial T_{w,in}}; \\ \ell &= 4;\ i=1;\ j=3,\end{aligned} \tag{6.382}$$

where $\dfrac{\partial m_{w,in}}{\partial T_{w,in}}$ was defined in Eq. (6.172), and

$$\frac{\partial N_4^{(i)}}{\partial \alpha^{(3)}} = \frac{\partial N_4^{(i)}}{\partial T_{w,in}} \triangleq a_4^{i,3} = 0; \quad \ell=4;\ i=2,\ldots,I;\ j=3. \tag{6.383}$$

The derivatives of the air/water vapor energy balance equations with respect to the parameter $\alpha^{(4)} \triangleq P_{atm}$ are as follows:

$$\frac{\partial N_4^{(i)}}{\partial \alpha^{(4)}} = \frac{\partial N_4^{(i)}}{\partial P_{atm}} \triangleq a_4^{i,4}$$
$$= \left(T_w^{(i+1)} - T_a^{(i)}\right)\left[\frac{1}{m_a}\frac{\partial H(m_a,\mathbf{\alpha})}{\partial Nu(Re,\mathbf{\alpha})}\frac{\partial Nu(Re,\mathbf{\alpha})}{\partial m_a}\frac{\partial m_a}{\partial P_{atm}} - \frac{H(m_a,\mathbf{\alpha})}{m_a^{\ 2}}\frac{\partial m_a}{\partial P_{atm}}\right]$$
$$-\frac{\left(m_w^{(i)} - m_w^{(i+1)}\right)h_{g,w}^{(i+1)}\left(T_w^{(i+1)},\mathbf{\alpha}\right)}{m_a^{\ 2}}\frac{\partial m_a}{\partial P_{atm}};\quad \ell = 4;\ i = 1,\ldots,I-1;\ j = 4,$$
(6.384)

$$\frac{\partial N_4^{(I)}}{\partial \alpha^{(4)}} = \frac{\partial N_4^{(I)}}{\partial P_{atm}} \triangleq a_4^{I,4}$$
$$= \left(T_w^{(i+1)} - T_a^{(i)}\right)\left[\frac{1}{m_a}\frac{\partial H(m_a,\mathbf{\alpha})}{\partial Nu(Re,\mathbf{\alpha})}\frac{\partial Nu(Re,\mathbf{\alpha})}{\partial m_a}\frac{\partial m_a}{\partial P_{atm}} - \frac{H(m_a,\mathbf{\alpha})}{m_a^{\ 2}}\frac{\partial m_a}{\partial P_{atm}}\right]$$
$$-\frac{\left(m_w^{(i)} - m_w^{(i+1)}\right)h_{g,w}^{(i+1)}\left(T_w^{(i+1)},\mathbf{\alpha}\right)}{m_a^{\ 2}}\frac{\partial m_a}{\partial P_{atm}} + \frac{\partial \omega_{in}}{\partial P_{atm}}\left(a_{1\,g}T_{a,in} + a_{0\,g}\right);\quad \ell = 4;\ i = I;\ j = 4,$$
(6.385)

where $\frac{\partial H(m_a,\mathbf{\alpha})}{\partial Nu(Re,\mathbf{\alpha})}$ and $\frac{\partial Nu(Re,\mathbf{\alpha})}{\partial m_a}$ were defined in Eqs. (6.257) and (6.176), respectively, while $\frac{\partial m_a}{\partial P_{atm}}$ and $\frac{\partial \omega_{in}}{\partial P_{atm}}$ were defined in Eqs. (6.177) and (6.323), respectively.

The derivatives of the air/water vapor energy balance equations with respect to the parameter $\alpha^{(5)} \triangleq w_{tsa}$ are as follows:

$$\frac{\partial N_4^{(i)}}{\partial \alpha^{(5)}} = \frac{\partial N_4^{(i)}}{\partial w_{tsa}} \triangleq a_4^{i,5} = \frac{\left(T_w^{(i+1)} - T_a^{(i)}\right)}{m_a}\frac{\partial H(m_a,\mathbf{\alpha})}{\partial w_{tsa}};\ \ell = 4;\ i = 1,\ldots,I;\ j = 5, \quad (6.386)$$

where $\frac{\partial H(m_a,\mathbf{\alpha})}{\partial w_{tsa}}$ was defined in Eq. (6.259).

The derivatives of the air/water vapor energy balance equations with respect to the parameter $\alpha^{(6)} \triangleq k_{sum}$ are as follows:

$$\frac{\partial N_4^{(i)}}{\partial \alpha^{(6)}} = \frac{\partial N_4^{(i)}}{\partial k_{sum}} \triangleq a_4^{i,6} = 0;\quad \ell = 4;\ i = 1,\ldots,I;\ j = 6. \quad (6.387)$$

The derivatives of the air/water vapor energy balance equations with respect to the parameter $\alpha^{(7)} \triangleq \mu$ are as follows:

$$\frac{\partial N_4^{(i)}}{\partial \alpha^{(7)}} = \frac{\partial N_4^{(i)}}{\partial \mu} \triangleq a_4^{i,7} = \frac{\left(T_w^{(i+1)} - T_a^{(i)}\right)}{m_a} \frac{\partial H\left(m_a, \mathbf{\alpha}\right)}{\partial \mu}; \; \ell = 4; i = 1,...,I; \; j = 7, \quad (6.388)$$

where $\frac{\partial H\left(m_a, \mathbf{\alpha}\right)}{\partial \mu}$ was defined in Eq. (6.262).

The derivatives of the air/water vapor energy balance equations with respect to the parameter $\alpha^{(8)} \triangleq \upsilon$ are as follows:

$$\frac{\partial N_4^{(i)}}{\partial \alpha^{(8)}} = \frac{\partial N_4^{(i)}}{\partial \upsilon} \triangleq a_4^{i,8} = 0; \;\; \ell = 4; i = 1,...,I; \; j = 8. \quad (6.389)$$

The derivatives of the air/water vapor energy balance equations with respect to the parameter $\alpha^{(9)} \triangleq k_{air}$ are as follows:

$$\frac{\partial N_4^{(i)}}{\partial \alpha^{(9)}} = \frac{\partial N_4^{(i)}}{\partial k_{air}} \triangleq a_4^{i,9} = \frac{\left(T_w^{(i+1)} - T_a^{(i)}\right)}{m_a} \frac{\partial H\left(m_a, \mathbf{\alpha}\right)}{\partial k_{air}}; \; \ell = 4; i = 1,...,I; \; j = 9, \quad (6.390)$$

where $\frac{\partial H\left(m_a, \mathbf{\alpha}\right)}{\partial k_{air}}$ was defined in Eq. (6.265).

The derivatives of the air/water vapor energy balance equations with respect to the parameter $\alpha^{(10)} \triangleq f_{ht}$ are as follows:

$$\frac{\partial N_4^{(i)}}{\partial \alpha^{(10)}} = \frac{\partial N_4^{(i)}}{\partial f_{ht}} \triangleq a_4^{i,10} = \frac{\left(T_w^{(i+1)} - T_a^{(i)}\right)}{m_a} \frac{\partial H\left(m_a, \mathbf{\alpha}\right)}{\partial f_{ht}}; \; \ell = 4; i = 1,...,I; \; j = 10, \quad (6.391)$$

where $\frac{\partial H\left(m_a, \mathbf{\alpha}\right)}{\partial f_{ht}}$ was defined in Eq. (6.267).

The derivatives of the air/water vapor energy balance equations with respect to the parameter $\alpha^{(11)} \triangleq f_{mt}$ are as follows:

$$\frac{\partial N_4^{(i)}}{\partial \alpha^{(11)}} = \frac{\partial N_4^{(i)}}{\partial f_{mt}} \triangleq a_4^{i,11} = 0; \;\; \ell = 4; i = 1,...,I; \; j = 11. \quad (6.392)$$

The derivatives of the air/water vapor energy balance equations with respect to the parameter $\alpha^{(12)} \triangleq f$ are as follows:

$$\frac{\partial N_4^{(i)}}{\partial \alpha^{(12)}} = \frac{\partial N_4^{(i)}}{\partial f} \triangleq a_4^{i,12} = 0;\ \ell = 4;\ i = 1,...,I;\ j = 12. \tag{6.393}$$

The derivatives of the air/water vapor energy balance equations with respect to the parameter $\alpha^{(13)} \triangleq a_0$ are as follows:

$$\frac{\partial N_4^{(i)}}{\partial \alpha^{(13)}} = \frac{\partial N_4^{(i)}}{\partial a_0} \triangleq a_4^{i,13} = 0;\quad \ell = 4;\ i = 1,...,I-1;\ j = 13, \tag{6.394}$$

$$\frac{\partial N_4^{(I)}}{\partial \alpha^{(13)}} = \frac{\partial N_4^{(I)}}{\partial a_0} \triangleq a_4^{I,13} = \frac{\partial \omega_{in}}{\partial a_0}\left(a_{1\,g} T_{a,in} + a_{0\,g}\right);\ \ell = 4;\ i = I;\ j = 13, \tag{6.395}$$

where $\frac{\partial \omega_{in}}{\partial a_0}$ was defined in Eq. (6.334).

The derivatives of the air/water vapor energy balance equations with respect to the parameter $\alpha^{(14)} \triangleq a_1$ are as follows:

$$\frac{\partial N_4^{(i)}}{\partial \alpha^{(14)}} = \frac{\partial N_4^{(i)}}{\partial a_1} \triangleq a_4^{i,14} = 0;\quad \ell = 4;\ i = 1,...,I-1;\ j = 14, \tag{6.396}$$

$$\frac{\partial N_4^{(I)}}{\partial \alpha^{(14)}} = \frac{\partial N_4^{(I)}}{\partial a_1} \triangleq a_4^{I,14} = \frac{\partial \omega_{in}}{\partial a_1}\left(a_{1\,g} T_{a,in} + a_{0\,g}\right);\ \ell = 4;\ i = I;\ j = 14, \tag{6.397}$$

where $\frac{\partial \omega_{in}}{\partial a_1}$ was defined in Eq. (6.337).

The derivatives of the air/water vapor energy balance equations with respect to the parameter $\alpha^{(15)} \triangleq a_{0,cpa}$ are as follows:

$$\begin{aligned}\frac{\partial N_4^{(i)}}{\partial \alpha^{(15)}} = \frac{\partial N_4^{(i)}}{\partial a_{0,cpa}} \triangleq a_4^{i,15} &= \left(T_a^{(i+1)} - T_a^{(i)}\right) \frac{\partial C_p^{(i)}\left(\frac{T_a^{(i)} + 273.15}{2}, \boldsymbol{\alpha}\right)}{\partial a_{0,cpa}} \\ &= T_a^{(i+1)} - T_a^{(i)};\ \ \ell = 4;\ i = 1,...,I;\ j = 15.\end{aligned} \tag{6.398}$$

The derivatives of the air/water vapor energy balance equations with respect to the parameter $\alpha^{(16)} \triangleq a_{1,cpa}$ are as follows:

$$\frac{\partial N_4^{(i)}}{\partial \alpha^{(16)}} = \frac{\partial N_4^{(i)}}{\partial a_{1,cpa}} \triangleq a_4^{i,16} = \left(T_a^{(i+1)} - T_a^{(i)}\right) \frac{\partial C_p^{(i)}\left(\frac{T_a^{(i)} + 273.15}{2}, \boldsymbol{\alpha}\right)}{\partial a_{1,cpa}}$$
$$= 0.5\left(T_a^{(i+1)} - T_a^{(i)}\right)\left(T_a^{(i)} + 273.15\right); \ \ell = 4; i = 1, \dots, I; \ j = 16. \quad (6.399)$$

The derivatives of the air/water vapor energy balance equations with respect to the parameter $\alpha^{(17)} \triangleq a_{2,cpa}$ are as follows:

$$\frac{\partial N_4^{(i)}}{\partial \alpha^{(17)}} = \frac{\partial N_4^{(i)}}{\partial a_{2,cpa}} \equiv a_4^{i,17} = \left(T_a^{(i+1)} - T_a^{(i)}\right) \frac{\partial C_p^{(i)}\left(\frac{T_a^{(i)} + 273.15}{2}, \boldsymbol{\alpha}\right)}{\partial a_{2,cpa}}$$
$$= 0.25\left(T_a^{(i+1)} - T_a^{(i)}\right)\left(T_a^{(i)} + 273.15\right)^2; \ell = 4; i = 1, \dots, I; \ j = 17. \quad (6.400)$$

The derivatives of the air/water vapor energy balance equations with respect to the parameter $\alpha^{(18)} \triangleq a_{0,dav}$ are as follows:

$$\frac{\partial N_4^{(i)}}{\partial \alpha^{(18)}} = \frac{\partial N_4^{(i)}}{\partial a_{0,dav}} \triangleq a_4^{i,18} = 0; \ \ell = 4; i = 1, \dots, I; \ j = 18. \quad (6.401)$$

The derivatives of the air/water vapor energy balance equations with respect to the parameter $\alpha^{(19)} \triangleq a_{1,dav}$ are as follows:

$$\frac{\partial N_4^{(i)}}{\partial \alpha^{(19)}} = \frac{\partial N_4^{(i)}}{\partial a_{1,dav}} \triangleq a_4^{i,19} = 0; \ \ell = 4; i = 1, \dots, I; \ j = 19. \quad (6.402)$$

The derivatives of the air/water vapor energy balance equations with respect to the parameter $\alpha^{(20)} \triangleq a_{2,dav}$ are as follows:

$$\frac{\partial N_4^{(i)}}{\partial \alpha^{(20)}} = \frac{\partial N_4^{(i)}}{\partial a_{2,dav}} \triangleq a_4^{i,20} = 0; \ \ell = 4; i = 1, \dots, I; \ j = 20. \quad (6.403)$$

The derivatives of the air/water vapor energy balance equations with respect to the parameter $\alpha^{(21)} \triangleq a_{3,dav}$ are as follows:

$$\frac{\partial N_4^{(i)}}{\partial \alpha^{(21)}} = \frac{\partial N_4^{(i)}}{\partial a_{3,dav}} \triangleq a_4^{i,21} = 0; \ \ell = 4; i = 1, \dots, I; \ j = 21. \quad (6.404)$$

The derivatives of the air/water vapor energy balance equations with respect to the parameter $\alpha^{(22)} \triangleq a_{0f}$ are as follows:

$$\frac{\partial N_4^{(i)}}{\partial \alpha^{(22)}} = \frac{\partial N_4^{(i)}}{\partial a_{0f}} \triangleq a_4^{i,22} = 0; \quad \ell = 4; \, i = 1,\dots,I; \; j = 22. \tag{6.405}$$

The derivatives of the air/water vapor energy balance equations with respect to the parameter $\alpha^{(23)} \triangleq a_{1f}$ are as follows:

$$\frac{\partial N_4^{(i)}}{\partial \alpha^{(23)}} = \frac{\partial N_4^{(i)}}{\partial a_{1f}} \triangleq a_4^{i,23} = 0; \quad \ell = 4; \, i = 1,\dots,I; \; j = 23. \tag{6.406}$$

The derivatives of the air/water vapor energy balance equations with respect to the parameter $\alpha^{(24)} \triangleq a_{0g}$ are as follows:

$$\frac{\partial N_4^{(i)}}{\partial \alpha^{(24)}} = \frac{\partial N_4^{(i)}}{\partial a_{0g}} \triangleq a_4^{i,24} = \omega^{(i+1)} - \omega^{(i)} + \frac{m_w^{(i)} - m_w^{(i+1)}}{m_a}; \, \ell = 4; \, i = 1,\dots,I; \, j = 24. \tag{6.407}$$

The derivatives of the air/water vapor energy balance equations with respect to the parameter $\alpha^{(25)} \triangleq a_{1g}$ are as follows:

$$\begin{aligned} &\frac{\partial N_4^{(i)}}{\partial \alpha^{(25)}} = \frac{\partial N_4^{(i)}}{\partial a_{1g}} \triangleq a_4^{i,25} \\ &= \omega^{(i+1)} T_a^{(i+1)} - \omega^{(i)} T_a^{(i)} + \frac{\left(m_w^{(i)} - m_w^{(i+1)}\right) T_w^{(i+1)}}{m_a}; \\ &\ell = 4; \, i = 1,\dots,I; \; j = 25. \end{aligned} \tag{6.408}$$

The derivatives of the air/water vapor energy balance equations with respect to the parameter $\alpha^{(26)} \triangleq a_{0,Nu}$ are as follows:

$$\begin{gathered} \frac{\partial N_4^{(i)}}{\partial \alpha^{(26)}} = \frac{\partial N_4^{(i)}}{\partial a_{0,Nu}} \triangleq a_4^{i,26} = \frac{\left(T_w^{(i+1)} - T_a^{(i)}\right)}{m_a} \frac{\partial H\left(m_a, \boldsymbol{\alpha}\right)}{\partial Nu\left(Re, \boldsymbol{\alpha}\right)} \frac{\partial Nu\left(Re, \boldsymbol{\alpha}\right)}{\partial a_{0,Nu}}; \\ \ell = 4; \, i = 1,\dots,I; \; j = 26, \end{gathered} \tag{6.409}$$

where $\frac{\partial H(m_a,\boldsymbol{\alpha})}{\partial Nu(Re,\boldsymbol{\alpha})}$ and $\frac{\partial Nu(Re,\boldsymbol{\alpha})}{\partial a_{0,Nu}}$ were defined previously in Eqs. (6.257) and (6.211) respectively.

The derivatives of the air/water vapor energy balance equations with respect to the parameter $\alpha^{(27)} \triangleq a_{1,Nu}$ are as follows:

$$\frac{\partial N_4^{(i)}}{\partial \alpha^{(27)}} = \frac{\partial N_4^{(i)}}{\partial a_{1,Nu}} \triangleq a_4^{i,27} = \frac{\left(T_w^{(i+1)} - T_a^{(i)}\right)}{m_a} \frac{\partial H(m_a,\boldsymbol{\alpha})}{\partial Nu(Re,\boldsymbol{\alpha})} \frac{\partial Nu(Re,\boldsymbol{\alpha})}{\partial a_{1,Nu}}; \quad \ell = 4;\ i = 1,\ldots,I;\ j = 27, \tag{6.410}$$

where $\frac{\partial H(m_a,\boldsymbol{\alpha})}{\partial Nu(Re,\boldsymbol{\alpha})}$ and $\frac{\partial Nu(Re,\boldsymbol{\alpha})}{\partial a_{1,Nu}}$ were defined previously in Eqs. (6.257) and (6.213), respectively.

The derivatives of the air/water vapor energy balance equations with respect to the parameter $\alpha^{(28)} \triangleq a_{2,Nu}$ are as follows:

$$\frac{\partial N_4^{(i)}}{\partial \alpha^{(28)}} = \frac{\partial N_4^{(i)}}{\partial a_{2,Nu}} \triangleq a_4^{i,28} = \frac{\left(T_w^{(i+1)} - T_a^{(i)}\right)}{m_a} \frac{\partial H(m_a,\boldsymbol{\alpha})}{\partial Nu(Re,\boldsymbol{\alpha})} \frac{\partial Nu(Re,\boldsymbol{\alpha})}{\partial a_{2,Nu}}; \quad \ell = 4;\ i = 1,\ldots,I;\ j = 28, \tag{6.411}$$

where $\frac{\partial H(m_a,\boldsymbol{\alpha})}{\partial Nu(Re,\boldsymbol{\alpha})}$ and $\frac{\partial Nu(Re,\boldsymbol{\alpha})}{\partial a_{2,Nu}}$ were defined previously in Eqs. (6.257) and (6.215), respectively.
The derivatives of the air/water vapor energy balance equations with respect to the parameter $\alpha^{(29)} \triangleq a_{3,Nu}$ are as follows:

$$\frac{\partial N_4^{(i)}}{\partial \alpha^{(29)}} = \frac{\partial N_4^{(i)}}{\partial a_{3,Nu}} \triangleq a_4^{i,29} = \frac{\left(T_w^{(i+1)} - T_a^{(i)}\right)}{m_a} \frac{\partial H(m_a,\boldsymbol{\alpha})}{\partial Nu(Re,\boldsymbol{\alpha})} \frac{\partial Nu(Re,\boldsymbol{\alpha})}{\partial a_{3,Nu}}; \quad \ell = 4;\ i = 1,\ldots,I;\ j = 29, \tag{6.412}$$

where $\frac{\partial H(m_a,\boldsymbol{\alpha})}{\partial Nu(Re,\boldsymbol{\alpha})}$ and $\frac{\partial Nu(Re,\boldsymbol{\alpha})}{\partial a_{3,Nu}}$ were defined previously in Eqs. (6.257) and (6.217), respectively.
The derivatives of the air/water vapor energy balance equations with respect to the parameter $\alpha^{(30)} \triangleq W_{dkx}$ are as follows:

$$\frac{\partial N_4^{(i)}}{\partial \alpha^{(30)}} = \frac{\partial N_4^{(i)}}{\partial W_{dkx}} \triangleq a_4^{i,30} = 0; \quad \ell = 4; i = 1,...,I; \; j = 30. \tag{6.413}$$

The derivatives of the air/water vapor energy balance equations with respect to the parameter $\alpha^{(31)} \triangleq W_{dky}$ are as follows:

$$\frac{\partial N_4^{(i)}}{\partial \alpha^{(31)}} = \frac{\partial N_4^{(i)}}{\partial W_{dky}} \triangleq a_4^{i,31} = 0; \quad \ell = 4; i = 1,...,I; \; j = 31. \tag{6.414}$$

The derivatives of the air/water vapor energy balance equations with respect to the parameter $\alpha^{(32)} \triangleq \Delta z_{dk}$ are as follows:

$$\frac{\partial N_4^{(i)}}{\partial \alpha^{(32)}} = \frac{\partial N_4^{(i)}}{\partial \Delta z_{dk}} \triangleq a_4^{i,32} = 0; \quad \ell = 4; i = 1,...,I; \; j = 32. \tag{6.415}$$

The derivatives of the air/water vapor energy balance equations with respect to the parameter $\alpha^{(33)} \triangleq \Delta z_{fan}$ are as follows:

$$\frac{\partial N_4^{(i)}}{\partial \alpha^{(33)}} = \frac{\partial N_4^{(i)}}{\partial \Delta z_{fan}} \triangleq a_4^{i,33} = 0; \quad \ell = 4; i = 1,...,I; \; j = 33. \tag{6.416}$$

The derivatives of the air/water vapor energy balance equations with respect to the parameter $\alpha^{(34)} \triangleq D_{fan}$ are as follows:

$$\begin{aligned} \frac{\partial N_4^{(i)}}{\partial \alpha^{(34)}} &= \frac{\partial N_4^{(i)}}{\partial D_{fan}} \triangleq a_4^{i,34} \\ &= \left(T_w^{(i+1)} - T_a^{(i)}\right)\left[\frac{1}{m_a}\frac{\partial H(m_a,\boldsymbol{\alpha})}{\partial Nu(Re,\boldsymbol{\alpha})}\frac{\partial Nu(Re,\boldsymbol{\alpha})}{\partial m_a}\frac{\partial m_a}{\partial D_{fan}} - \frac{H(m_a,\boldsymbol{\alpha})}{m_a^2}\frac{\partial m_a}{\partial D_{fan}}\right] \\ &\quad - \frac{\left(m_w^{(i)} - m_w^{(i+1)}\right)h_{g,w}^{(i+1)}\left(T_w^{(i+1)},\boldsymbol{\alpha}\right)}{m_a^2}\frac{\partial m_a}{\partial D_{fan}}; \quad \ell = 4; i = 1,...,I; \; j = 34, \end{aligned} \tag{6.417}$$

where $\frac{\partial H(m_a,\boldsymbol{\alpha})}{\partial Nu(Re,\boldsymbol{\alpha})}$ and $\frac{\partial Nu(Re,\boldsymbol{\alpha})}{\partial m_a}$ were defined previously in Eqs. (6.257) and (6.176), respectively, while $\frac{\partial m_a}{\partial D_{fan}}$ was defined in Eq. (6.223).

The derivatives of the air/water vapor energy balance equations with respect to the parameter $\alpha^{(35)} \triangleq \Delta z_{fill}$ are as follows:

$$\frac{\partial N_4^{(i)}}{\partial \alpha^{(35)}} = \frac{\partial N_4^{(i)}}{\partial \Delta z_{fill}} \triangleq a_4^{i,35} = 0; \quad \ell = 4; i = 1,...,I; \; j = 35. \tag{6.418}$$

The derivatives of the air/water vapor energy balance equations with respect to the parameter $\alpha^{(36)} \triangleq \Delta z_{rain}$ are as follows:

$$\frac{\partial N_4^{(i)}}{\partial \alpha^{(36)}} = \frac{\partial N_4^{(i)}}{\partial \Delta z_{rain}} \triangleq a_4^{i,36} = 0; \quad \ell = 4; i = 1,...,I; \; j = 36. \tag{6.419}$$

The derivatives of the air/water vapor energy balance equations with respect to the parameter $\alpha^{(37)} \triangleq \Delta z_{bs}$ are as follows:

$$\frac{\partial N_4^{(i)}}{\partial \alpha^{(37)}} = \frac{\partial N_4^{(i)}}{\partial \Delta z_{bs}} \triangleq a_4^{i,37} = 0; \quad \ell = 4; i = 1,...,I; \; j = 37. \tag{6.420}$$

The derivatives of the air/water vapor energy balance equations with respect to the parameter $\alpha^{(38)} \triangleq \Delta z_{de}$ are as follows:

$$\frac{\partial N_4^{(i)}}{\partial \alpha^{(38)}} = \frac{\partial N_4^{(i)}}{\partial \Delta z_{de}} \triangleq a_4^{i,38} = 0; \quad \ell = 4; i = 1,...,I; \; j = 38. \tag{6.421}$$

The derivatives of the air/water vapor energy balance equations with respect to the parameter $\alpha^{(39)} \triangleq D_h$ are as follows:

$$\frac{\partial N_4^{(i)}}{\partial \alpha^{(39)}} = \frac{\partial N_4^{(i)}}{\partial D_h} \triangleq a_4^{i,39} = \frac{\left(T_w^{(i+1)} - T_a^{(i)}\right)}{m_a} \frac{\partial H(m_a, \boldsymbol{\alpha})}{\partial D_h}; \ell = 4; i = 1,...,I; j = 39, \tag{6.422}$$

where $\frac{\partial H(m_a, \boldsymbol{\alpha})}{\partial D_h}$ was defined in Eq. (6.297).

The derivatives of the air/water vapor energy balance equations with respect to the parameter $\alpha^{(40)} \triangleq A_{fill}$ are as follows:

$$\frac{\partial N_4^{(i)}}{\partial \alpha^{(40)}} = \frac{\partial N_4^{(i)}}{\partial A_{fill}} \triangleq a_4^{i,40} = \frac{\left(T_w^{(i+1)} - T_a^{(i)}\right)}{m_a} \frac{\partial H(m_a, \boldsymbol{\alpha})}{\partial A_{fill}}; \ell = 4; i = 1,...,I; j = 40, \tag{6.423}$$

where $\frac{\partial H(m_a,\boldsymbol{\alpha})}{\partial A_{fill}}$ was defined in Eq. (6.299).

The derivatives of the air/water vapor energy balance equations with respect to the parameter $\alpha^{(41)} \triangleq A_{surf}$ are as follows:

$$\frac{\partial N_4^{(i)}}{\partial \alpha^{(41)}} = \frac{\partial N_4^{(i)}}{\partial A_{surf}} \triangleq a_4^{i,41} = \frac{\left(T_w^{(i+1)} - T_a^{(i)}\right)}{m_a}\frac{\partial H(m_a,\boldsymbol{\alpha})}{\partial A_{surf}}; \; \ell = 4; \; i = 1,\dots,I; \; j = 41, \quad (6.424)$$

where $\frac{\partial H(m_a,\boldsymbol{\alpha})}{\partial A_{surf}}$ was defined in Eq. (6.301).

The derivatives of the air/water vapor energy balance equations with respect to the parameter $\alpha^{(42)} \triangleq Pr$ are as follows:

$$\frac{\partial N_4^{(i)}}{\partial \alpha^{(42)}} = \frac{\partial N_4^{(i)}}{\partial Pr} \triangleq a_4^{i,42} = \frac{\left(T_w^{(i+1)} - T_a^{(i)}\right)}{m_a}\frac{\partial H(m_a,\boldsymbol{\alpha})}{\partial Pr}; \; \ell = 4; i = 1,\dots,I; j = 42, \quad (6.425)$$

where $\frac{\partial H(m_a,\boldsymbol{\alpha})}{\partial Pr}$ was defined in Eq. (6.303).

The derivatives of the air/water vapor energy balance equations with respect to the parameter $\alpha^{(43)} \triangleq V_w$ are as follows:

$$\frac{\partial N_4^{(i)}}{\partial \alpha^{(43)}} = \frac{\partial N_4^{(i)}}{\partial V_w} \triangleq a_4^{i,43} = 0; \; \ell = 4; \; i = 1,\dots,I; \; j = 43. \quad (6.426)$$

The derivatives of the air/water vapor energy balance equations with respect to the parameter $\alpha^{(44)} \triangleq V_{exit}$ are as follows:

$$\begin{aligned} \frac{\partial N_4^{(i)}}{\partial \alpha^{(44)}} &= \frac{\partial N_4^{(i)}}{\partial V_{exit}} \triangleq a_4^{i,44} \\ &= \left(T_w^{(i+1)} - T_a^{(i)}\right)\left[\frac{1}{m_a}\frac{\partial H(m_a,\boldsymbol{\alpha})}{\partial Nu(Re,\boldsymbol{\alpha})}\frac{\partial Nu(Re,\boldsymbol{\alpha})}{\partial m_a}\frac{\partial m_a}{\partial V_{exit}} - \frac{H(m_a,\boldsymbol{\alpha})}{m_a^{\,2}}\frac{\partial m_a}{\partial V_{exit}}\right] \quad (6.427) \\ &- \frac{\left(m_w^{(i)} - m_w^{(i+1)}\right)h_{g,w}^{(i+1)}\left(T_w^{(i+1)},\boldsymbol{\alpha}\right)}{m_a^{\,2}}\frac{\partial m_a}{\partial V_{exit}}; \quad \ell = 4; \; i = 1,\dots,I; \; j = 44, \end{aligned}$$

where $\frac{\partial H(m_a,\boldsymbol{\alpha})}{\partial Nu(Re,\boldsymbol{\alpha})}$ and $\frac{\partial Nu(Re,\boldsymbol{\alpha})}{\partial m_a}$ were defined previously in Eqs. (6.257) and (6.176), respectively, while $\frac{\partial m_a}{\partial V_{exit}}$ was defined in Eq. (6.238).

The derivatives of the air/water vapor energy balance equations with respect to the parameter $\alpha^{(45)} \triangleq m_{w,in}$ are as follows:

$$\frac{\partial N_4^{(1)}}{\partial \alpha^{(45)}} = \frac{\partial N_4^{(1)}}{\partial m_{w,in}} \triangleq a_4^{1,45} = \frac{h_{g,w}^{(2)}(T_w^{(2)},\boldsymbol{\alpha})}{m_a} = \frac{a_{1g}T_w^{(2)} + a_{0g}}{m_a}; \ \ell = 4;\ i = 1;\ j = 45, \quad (6.428)$$

$$\frac{\partial N_4^{(i)}}{\partial \alpha^{(45)}} = \frac{\partial N_4^{(i)}}{\partial m_{w,in}} \triangleq a_4^{i,45} = 0; \quad \ell = 4;\ i = 2,\ldots,I;\ j = 45. \quad (6.429)$$

The derivatives of the air/water vapor energy balance equations with respect to the parameter $\alpha^{(46)} \triangleq T_{a,in}$ are as follows:

$$\frac{\partial N_4^{(i)}}{\partial \alpha^{(46)}} = \frac{\partial N_4^{(i)}}{\partial T_{a,in}} \triangleq a_4^{i,46} = 0; \quad \ell = 4;\ i = 1,\ldots,I-1;\ j = 46, \quad (6.430)$$

$$\begin{aligned} \frac{\partial N_4^{(I)}}{\partial \alpha^{(46)}} &= \frac{\partial N_4^{(I)}}{\partial T_{a,in}} \triangleq a_4^{I,46} = C_p^{(I)}\left(\frac{T_a^{(I)} + 273.15}{2},\boldsymbol{\alpha}\right) + \omega_{in}\frac{\partial h_{g,a}^{(I+1)}(T_{a,in},\boldsymbol{\alpha})}{\partial T_{a,in}} \\ &= C_p^{(I)}\left(\frac{T_a^{(I)} + 273.15}{2},\boldsymbol{\alpha}\right) + \omega_{in}a_{1g}; \quad \ell = 4;\ i = I;\ j = 46. \end{aligned} \quad (6.431)$$

NOTE:the above derivative was NOT used in the computations for the numerical results presented in Chapter 6.

The derivatives of the air/water vapor energy balance equations with respect to the parameter $\alpha^{(47)} \triangleq m_a$ are as follows:

$$\begin{aligned} \frac{\partial N_4^{(i)}}{\partial \alpha^{(47)}} &= \frac{\partial N_4^{(i)}}{\partial m_a} \triangleq a_4^{i,47} \\ &= -\left(m_w^{(i)} - m_w^{(i+1)}\right)\frac{h_{g,w}^{(i+1)}\left(T_w^{(i+1)},\boldsymbol{\alpha}\right)}{m_a^2} \\ &+\left(T_w^{(i+1)} - T_a^{(i)}\right)\left[\frac{1}{m_a}\frac{\partial H(m_a,\boldsymbol{\alpha})}{\partial Nu(Re,\boldsymbol{\alpha})}\frac{\partial Nu(Re,\boldsymbol{\alpha})}{\partial m_a} - \frac{H(m_a,\boldsymbol{\alpha})}{m_a^2}\right]; \\ &\ell = 4;\ i = 1,\ldots,I;\ j = 47, \end{aligned} \quad (6.432)$$

where $\dfrac{\partial H(m_a,\boldsymbol{\alpha})}{\partial Nu(Re,\boldsymbol{\alpha})}$ and $\dfrac{\partial Nu(Re,\boldsymbol{\alpha})}{\partial m_a}$ were defined previously in Eqs. (6.257) and (6.176), respectively.

The derivatives of the air/water vapor energy balance equations with respect to the parameter $\alpha^{(48)} \triangleq \omega_{in}$ are as follows:

$$\frac{\partial N_4^{(i)}}{\partial \alpha^{(48)}} = \frac{\partial N_4^{(i)}}{\partial \omega_{in}} \triangleq a_4^{i,48} = 0; \quad \ell = 4;\, i = 1,\ldots,I-1;\; j = 48, \tag{6.433}$$

$$\frac{\partial N_4^{(I)}}{\partial \alpha^{(48)}} = \frac{\partial N_4^{(I)}}{\partial \omega_{in}} \triangleq a_4^{I,48} = h_{g,a}^{(I+1)}\left(T_{a,in},\boldsymbol{\alpha}\right); \quad \ell = 4;\, i = I;\; j = 48. \tag{6.434}$$

The derivatives of the air/water vapor energy balance equations with respect to the parameter $\alpha^{(49)} \triangleq Re_d$ are as follows:

$$\frac{\partial N_4^{(i)}}{\partial \alpha^{(49)}} = \frac{\partial N_4^{(i)}}{\partial Re_d} \triangleq a_4^{i,49} = \frac{\left(T_w^{(i+1)} - T_a^{(i)}\right)}{m_a} \frac{\partial H(m_a,\boldsymbol{\alpha})}{\partial Nu(Re_d,\boldsymbol{\alpha})} \frac{\partial Nu(Re_d,\boldsymbol{\alpha})}{\partial Re_d}; \quad \ell = 4;\, i = 1,\ldots,I;\; j = 49, \tag{6.435}$$

where $\dfrac{\partial H(m_a,\boldsymbol{\alpha})}{\partial Nu(Re_d,\boldsymbol{\alpha})}$ was defined in Eq. (6.257) and $\dfrac{\partial Nu(Re_d,\boldsymbol{\alpha})}{\partial Re_d}$ was defined in Eq. (6.245).

The derivatives of the air/water vapor energy balance equations with respect to the parameter $\alpha^{(50)} \triangleq Sc$ are as follows:

$$\frac{\partial N_4^{(i)}}{\partial \alpha^{(50)}} = \frac{\partial N_4^{(i)}}{\partial Sc} \triangleq a_4^{i,50} = 0; \quad \ell = 4;\, i = 1,\ldots,I;\; j = 50. \tag{6.436}$$

The derivatives of the air/water vapor energy balance equations with respect to the parameter $\alpha^{(51)} \triangleq Sh$ are as follows:

$$\frac{\partial N_4^{(i)}}{\partial \alpha^{(51)}} = \frac{\partial N_4^{(i)}}{\partial Sh} \triangleq a_4^{i,51} = 0; \quad \ell = 4;\, i = 1,\ldots,I;\; j = 51. \tag{6.437}$$

The derivatives of the air/water vapor energy balance equations with respect to the parameter $\alpha^{(52)} \triangleq Nu$ are as follows:

$$\frac{\partial N_4^{(i)}}{\partial \alpha^{(52)}} = \frac{\partial N_4^{(i)}}{\partial Nu} \triangleq a_4^{i,52} = \frac{\left(T_w^{(i+1)} - T_a^{(i)}\right)}{m_a} \frac{\partial H\left(m_a, \boldsymbol{\alpha}\right)}{\partial Nu\left(Re, \boldsymbol{\alpha}\right)}; \ell = 4; i = 1, \ldots, I; j = 52, \quad (6.438)$$

where $\dfrac{\partial H\left(m_a, \boldsymbol{\alpha}\right)}{\partial Nu\left(Re, \boldsymbol{\alpha}\right)}$ was defined in Eq. (6.257).

REFERENCES

Akdeniz, B., Ivanov, K.N. et al., 2005. Boiling Water Reactor Turbine Trip (TT) Benchmark, Volume II: Summary Results of Exercise 1, NEA/OECD.

Akdeniz, B., Ivanov, K.N. et al., 2006. Boiling Water Reactor Turbine Trip (TT) Benchmark, Volume III: Summary Results of Exercise 2, NEA/OECD.

Akdeniz, B., Ivanov, K.N. et al., 2010. Boiling Water Reactor Turbine Trip (TT) Benchmark, Volume IV: Summary Results of Exercise 3, NEA/OECD.

Aleman, S.E., Garrett, A.J., 2015. Operational Cooling Tower Model (CTTool v1.0), SRNL-STI-2015-00039, Revision 0, Savannah River National Laboratory, Savannah River, SC, USA, January 2015.

ANSYS, "CFX user's guide manual of modules and procedures," http://www1.ansys.com/customer/content/documentation.

Arslan, E., Cacuci, D. G., 2014. "Predictive Modeling of Liquid-Sodium Thermal-Hydraulics Experiments and Computations," Annals of Nuclear Energy, 63, 355–370 (2014)

Badea, A.F., Cacuci, D.G., 2017. "Predictive Uncertainty Reduction in Coupled Neutron-Kinetics/Thermal Hydraulics Modeling of the BWR-TT2 Benchmark," Nuclear Engineering and Design, 313 330–344 (2017).

Badea, M. C., Cacuci, D.G., Badea, A.F. 2012. "Best-Estimate Predictions and Model Calibration for Reactor Thermal-Hydraulics," Nucl. Sci. Eng., 172, 1-19 (2012).

Barhen, J., Cacuci, D.G., Wagschal, J. J., Bjerke, M. A., and Mullins, C.B., 1982. "Uncertainty Analysis of Time-Dependent Nonlinear Systems: Theory and Application to Transient Thermal Hydraulics," Nucl. Sci.Eng., 81, 23-44, (1982).

Bledsoe, K.C., Favorite, J. A., and Aldemir, T., 2011a. "Application of the Differential Evolution Method for Solving Inverse Transport Problems", Nucl. Sci. Eng., 169, 208 (2011).

Bledsoe, K.C., Favorite, J. A., and Aldemir, T., 2011b. "Using the Levenberg - Marquardt Method for Solutions of Inverse Transport Problems in One- and Two-Dimensional Geometries," Nucl. Techn., 176, 106-126 (2011).

D. G. Cacuci, *BERRU Predictive Modeling*,
https://doi.org/10.1007/978-3-662-58395-1

Cacuci, D.G., 1981a. “Sensitivity theory for nonlinear systems: I. Nonlinear functional analysis approach”. J., Math. Phys. 22, 2794–2802.

Cacuci, D.G., 1981b. “Sensitivity theory for nonlinear systems: II. Extensions to additional classes of responses”., J. Math. Phys. 22, 2803–2812.

Cacuci, D.G., 2003. Sensitivity and Uncertainty Analysis: Theory, Vol. 1, Chapman & Hall/CRC, Boca Raton (2003).

Cacuci, D.G., 2014a. “On the Ill-Posed Nature of Inverse Problems: An Exactly Solvable Paradigm Inverse Neutron Diffusion Problem Illustrating the Solution’s Non-Computability,” ANS RPSD 2014 - 18th Topical Meeting of the Radiation Protection & Shielding Division of ANS Knoxville, TN, September 14 – 18, 2014, on CD-ROM, American Nuclear Society, LaGrange Park, IL (2014).

Cacuci, D. G., 2014b. “Predictive modeling of coupled multi-physics systems: I. Theory.” Annals of Nuclear Energy 70, 266–278 (2014).

Cacuci, D. G., 2015a. “Second-order adjoint sensitivity analysis methodology (2nd-ASAM) for computing exactly and efficiently first- and second-order sensitivities in large-scale linear systems: I. Computational methodology,” J. Comp. Phys. 284, 687–699 (2015).

Cacuci D. G., 2015b. A Heat Transport Benchmark Problem for Predicting the Impact of Measurements on Experimental Facility Design, Nucl. Eng. Design, 300, 12-27 (2015).

Cacuci, D. G., 2016. “Second-order adjoint sensitivity analysis methodology (2nd-ASAM) for large-scale nonlinear systems: I. Theory,” Nucl. Sci. Eng. 184, 16–30 (2016).

Cacuci, D. G., 2017. “Inverse predictive modeling of radiation transport through optically thick media in the presence of counting uncertainties,” Nucl. Sci. Eng, 186, 199–223, 2017. http://dx.doi.org/10.1080/00295639.2017.1305244, 20 May 2017.

Cacuci, D. G., Arslan, E., 2014. “Reducing Uncertainties via Predictive Modeling: FLICA4 Calibration Using BFBT Benchmarks,” Nucl. Sci. Eng., 176, 339-349 (2014).

Cacuci, D. G., Badea, M. C., 2014. “Predictive Modeling of Coupled Multi-Physics Systems: II. Illustrative Application to Reactor Physics,” Annals of Nuclear Technology, Annals of Nuclear Energy, 70, 279-291 (2014).

Cacuci, D.G., and Di Rocco, F., 2017. "Predictive Modeling of a Buoyancy-Operated Cooling Tower Under Saturated Conditions: I. Adjoint Sensitivity Model," Nucl. Sci. Eng, 185, 484–548 (2017).

Cacuci, D.G. and Fang, R., 2016. "Predictive Modeling of a Paradigm Mechanical Cooling Tower: I. Adjoint Sensitivity Model", Energies, 9, 718-763, 2016; doi:10.3390/en9090718 (2016).

Cacuci, D. G., and Fang, R., 2017. "Sensitivity and Uncertainty Analysis of Counter-Flow Mechanical Draft Cooling Towers: I. Adjoint Sensitivity Analysis," Nuclear Technology, 198, 85-131 (2017).

Cacuci, D.G., Ionescu-Bujor, M., 2010a. "Model Calibration and Best-Estimate Prediction Through Experimental Data Assimilation: I. Mathematical Framework," Nucl. Sci. Eng., 165, 18-44 (2010). See also: Cacuci, D.G., and Ionescu-Bujor, M., 2010., "Sensitivity and Uncertainty Analysis, Data Assimilation and Predictive Best-Estimate Model Calibration", Chapter 17 in Vol.3, pp 1913 – 2051, Handbook of Nuclear Engineering, Cacuci, D. G., Editor, ISBN: 978-0-387-98150-5, Springer New York / Berlin (2010).

Cacuci, D.G., Ionescu-Bujor, M., 2010b. "On the evaluation of discrepant scientific data with unrecognized errors," Nucl. Sci. Eng. 165, 1–17 (2010).

Cacuci, D.G., Navon, M.I. and Ionescu-Bujor, M., 2013. Computational Methods for Data Evaluation and Assimilation. Chapman & Hall/CRC, Boca Raton, 2010.

Cacuci, D.G., Badea, A. F, Badea, M. C., and Peltz. J. J, 2016. "Efficient Computation of Operator-Type Response Sensitivities for Uncertainty Quantification and Predictive Modeling: Illustrative Application to a Spent Nuclear Fuel Dissolver Model," International Journal for Numerical Methods in Fluids, Published online in Wiley Online Library (wileyonlinelibrary.com). DOI: 10.1002/fld.4258, June 2016.

Cacuci, D. G., Fang, R., Badea, M. C., 2018. "MULTI-PRED: A Software Module for Predictive Modeling of Coupled Multi-Physics Systems," *Nucl. Sci. Eng*, **191**, 187-202 (2018).

Cecchini, I., Farinelli, U., Gandini, A., and Salvatores, M., 1964. A/Conf. 28/, Proc. 3rd Int. Conf. on Peaceful Uses of Atomic Energy, Vol. 2, p. 627, Geneva, 1964.

Di Rocco, F., and Cacuci, D. G., 2016. "Predictive Modeling of a Buoyancy-Operated Cooling Tower Under Unsaturated Conditions: Adjoint Sensitivity Model and Optimal Best-Estimate Results With Reduced Predicted Uncertainties," Energies, 9, 1028; doi:10.3390/en9121028, 1-52 (2016).

Di Rocco, F., Cacuci, D.G., and Badea, M. C., 2017. "Predictive Modeling of a Buoyancy-Operated Cooling Tower under Saturated Conditions: II. Optimal

Best-Estimate Results with Reduced Predicted Uncertainties," Nucl. Sci. Eng, 185, 549–603 (2017).

Dragt, J. B., et al, 1977. "Methods of Adjustment and Error Evaluation of Neutron Capture Cross Sections", Nucl. Sci. Eng., 62, 11 (1977).

Evans, R.T., and Cacuci, D.G., 2012. "A parallel Krylov-based adjoint sensitivity analysis procedure", Nuc. Sci. Eng., 172, pp. 216-222 (2012).

Evans, T.M., Stafford, A.S., Slaybaugh, R.N., and Clarno, K.T., 2010. "Denovo: A new three-dimensional parallel discrete ordinates code in SCALE", *Nuclear Technology*, **171**, pp. 171-200 (2010).

Fang, R., Cacuci, D. G., and Badea, M. C., 2016. "Predictive Modeling of a Paradigm Mechanical Cooling Tower: II. Optimal Best-Estimate Results with Reduced Predicted Uncertainties", Energies, 9, 764-813 (2016).

Fang, R., Cacuci, D. G., and Badea, M. C., 2017. "Sensitivity and Uncertainty Analysis of Counter-Flow Mechanical Draft Cooling Towers: II. Predictive Modeling", Nuclear Technology, 198, 132-192 (2017).

Fillion, Ph. et al., 2007. "FLICA4: Reference manual of modules and procedures," User Guide, Version V1.10.1, CEA, (2007).

Forster, E. and Greif, R., 1958. "Heat transfer to a boiling liquid; mechanism and correlations", Progress Report 7, US (1958).

Friedel, L., 1979. "Improved friction pressure drop correlations for horizontal and vertical two phase pipe flow", European Two Phase Flow Group Meeting, Ispra, 5-8 June, 1979.

Gandini, A. and Petilli, M., 1973. "AMARA: A Code Using the Lagrange Multipliers Method for Nuclear Data Adjustment", CNEN-RI/FI(73)39, Comitato Nazionale Energia Nucleare, Casaccia/Rome, Italy, 1973.

Garrett, A.J.; Parker M. J.; Villa-Aleman E. 2005. Savannah River site Cooling Tower Collection, SRNL-DOD-2005-07. Atmospheric Technologies Group, Savannah River National Laboratory, Aiken, SC, USA, May 2005.

Groeneveld, D.C. et al., 1996. "The 1995 look-up table for critical heat flux in tubes", Nuclear Engineering and Design, 163, 1–23, 1996.

Grundmann, U., Kliem, S., Rohde, U., 2004. Analysis of the Boiling Water Reactor Turbine Trip Benchmark with the Codes DYN3D and ATHLET/DYN3D, Nucl. Sci. Eng., 148, 226–234 (2004).

Humi, I, Wagschal, J.J., and Yeivin, Y., 1964. "Multi-group Constants From Integral Data", Proc. 3rd Int. Conf. on Peaceful Uses of Atomic Energy, Vol. 2, p. 398, Geneva, 1964;

ICSBEP, 2010. International Handbook of Evaluated Criticality Safety Benchmark Experiments, 2010. (ICSBEP), NEA/NSC/DOC(95)03, September 2010 Edition.

Inoue, A., Kurosu, T., Aoki, T., and Yagy, M., 1995. "Void Fraction Distribution in BWR Fuel Assembly and Evaluation of Sub-channel Code", Journal of Nuclear Science and Technology, July, 1995.

Ishii, M, 1977. "One dimensional drift-flux model and constitutive equations for relative motion between phases in various two-phase flow", Technical Report ANL-77-47, Argonne Nat. Lab., October 1977.

Knebel, J. U., 1993. "Experimentelle Untersuchungen in turbulenten Auftriebsstrahlen in Natrium," Report KFK-5175, (1993).

Kuroi, H. and Mitani, H., 1975. "Adjustment to Cross-Section Data to Fit Integral Experiments by Least Squares Method", J. Nuc. Sci. Technology, 12, 663, 1975.

Lahoz, W., Khattatov, B., Ménard, R. (Editors), 2010. Data Assimilation: Making Sense of Observations, Springer Verlag (2010).

Latten, C., Cacuci, D. G., 2014. Predictive Modeling of Coupled Systems: Uncertainty Reduction Using Multiple Reactor Physics Benchmarks, Nucl. Sci. Eng., 178, 156–171 (2014).

Levenberg–Marquardt methods are due to Levenberg (1944) and Marquardt (1963). See: K. Levenberg, "A method for the solution of certain nonlinear problems in least squares," Quart. Appl. Math., 2, 164–168 (1944); and D.W. Marquardt, "An algorithm for least-squares estimation of nonlinear parameters," J. Soc. Indust. Appl. Math., 11, 431–441 (1963).

Lewis, B. E. and Weber, F. E., 1980. "A Mathematical Model for Liquid Flow Transients in a Rotary Dissolver", ORNL/TM-7490, Oak Ridge National Laboratory (1980).

Lewis, J. M., Lakshmivarahan, S., and Dhall, S.K., 2006. "Dynamic Data Assimilation: A Least Square Approach," Cambridge University Press, Cambridge (2006).

Mattingly, J. K., 2015, private communication, North Carolina State University.

Moeller, R., 1989. "TEGENA Detaillierte experimentelle Untersuchungen der Temperatur- und Geschwindigkeitsuntersuchungen in Stabbündel- Geometrien mit turbulenter Natriumströmung," Wissenschaftliche Berichte KFK (1989).

McCormick, N.J., 1992. "Inverse Radiative Transfer Problems: A Review", Nucl. Sci. Eng., 112, 185 (1992).

Molochnikov, Yu. S., 1982. Generalization of experimental data on volumetric void fraction for sub-cooled boiling of water (in Russian). Teplohenergetika No.7, p. 47.

Neykov, B. et al., 2006. "NUPEC BWR Full-size Fine-mesh Bundle Test (BFBT) Bench mark, Nuclear Science", NEA/NSC/DOC(2005)5, ISBN 92-64-01088-2, NEA No. 6212, OECD 2006.

Oppe, T. C., W. D. Joubert, and D. R. Kincaid. 1988. "A Package for Solving Large Sparse Linear Systems by Various Iterative Methods", NSPCG User's Guide, Version 1.0. Center for Numerical Analysis, the University of Texas at Austin, April 1988.

Peltz, J.J. and Cacuci, D. G, 2016a. "Predictive Modeling Applied to a Paradigm Spent Fuel Dissolver Model. I: Adjoint Sensitivity and Uncertainty Analysis," Nucl. Sci. Eng, 183, 305-331. dx.doi.org/10.13182/NSE15-98 (2016).

Peltz, J.J. and Cacuci, D. G, 2016b. "Inverse Predictive Modeling of a Spent Fuel Dissolver Model," Nucl. Sci. Eng, 184, 1–15, 2016.

Peltz, J. J., Cacuci, D. G., Badea, A. F. and Badea, M. C., 2016. "Predictive Modeling Applied to a Spent Fuel Dissolver Model: II. Uncertainty Quantification and Reduction," Nucl. Sci. Eng, 183, 332-346. dx.doi.org/10.13182/NSE15-99 (2016).

Práger, T. and Kelemen, F. D., 2013. "Adjoint Methods and their Application in Earth Sciences", Chapter 4, Part A, pp 203 - 275, in Advanced Numerical Methods for Complex Environmental Models: Needs and Availability, I. Faragó, Á. Havasi, and Z. Zlatev, Eds., Bentham Science Publishers, Bussum, The Netherlands, 2013.

Rowlands, J., et al, 1973. "The Production and Performance of the Adjusted Cross-Section Set FGL5", Proc. Int. Symposium "Physics of Fast Reactors", Tokio, 1973. ROOT: CERN http://www.cern.ch/root.

Sanchez, R. and McCormick, N.J., 2008. "On the Uniqueness of the Inverse Source Problem for Linear Particle Transport Theory", TTSP, 37, 236 (2008).

SCALE: A comprehensive modeling and simulation suite for nuclear safety analysis and design, 2011. ORNL/TM-2005/39, Version 6.1, June 2011. Available from

Radiation Safety Information Computational Center at Oak Ridge National Laboratory as CCC-785.

Saad, Y. and Schultz, M.H. 1986. GMRES: A Generalized Minimal Residual Algorithm for Solving Nonsymmetric Linear Systems, SIAM J. Sci. Stat. Comp 7, No. 3, 856-869, 1986.

Shannon, C. E., 1948. Bell Syst. Tech. J., Vol. 27, pp. 379 and 623.

Solis, J., Ivanov, K.N., et al., 2001. Boiling Water Reactor Turbine Trip (TT) Benchmark, Volume I: Final Specifications, NEA/OECD.

Sudo, Y. and Kaminaga, M., 1993. "A New CHF Correlation Scheme Proposed for Vertical Rectangular Channels Heated From Both Sides in Nuclear Research Reactors", Journal of Heat Transfer, 115, May, 1993.

Tarantola, A., 2005. Inverse Problem Theory and Methods for Model Parameter Estimation, Society for Industrial and Applied Mathematics (2005).

Tichonov, A. N., 1963. "Regularization of Non-Linear Ill-Posed Problems", Doklady Akademii Nauk, 49(4), 1963. See also: Tichonov, "Solution of Incorrectly Formulated Problems and the Regularization Method", Soviet Math. Doklady 4, 1035 (1963).

TRAC-PF1/MOD2 Theory Manual, Appendix B, Material Properties, NUREG/CR-5673, 1993.

Usachhev, L.N., 1964. "Perturbation Theory for the Breeding Ratio and for Other Number Ratios Pertaining to Various Reactor Processes", J. Nuc. Energy Part A/B, 18, 571, 1964.

Weisbin, C.R., et al., 1978. "Application of Sensitivity and Uncertainty Methodology to Fast Reactor Integral Experiment Analysis", Nucl. Sci. Eng., 66, 307, 1978.

Zuber, N. and Findlay, J.A., 1995. "Average volumetric concentration in two-phase flow systems", Journal of Heat Transfer, 87, 453–468, 1995.

CPSIA information can be obtained
at www.ICGtesting.com
Printed in the USA
LVHW060014030119
602570LV00003B/72/P

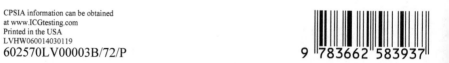

9 783662 583937